Taking on Climate Change Through Green Taxation

Ana Arromba Dinis
Research Centre on Accounting and Taxation, Polytechnic Institute of Cávado and Ave, Portugal

Fátima David
Research Centre on Accounting and Taxation, Polytechnic Institute of Cávado and Ave, Portugal

Liliana Pereira
Research Centre on Accounting and Taxation, Polytechnic Institute of Cávado and Ave, Portugal

Sara Luís Dias
Polytechnic Institute of Cávado and Ave, Portugal

A volume in the Advances in Public Policy and Administration (APPA) Book Series

IGI Global
PUBLISHER of TIMELY KNOWLEDGE

Published in the United States of America by
IGI Global
Information Science Reference (an imprint of IGI Global)
701 E. Chocolate Avenue
Hershey PA, USA 17033
Tel: 717-533-8845
Fax: 717-533-8661
E-mail: cust@igi-global.com
Web site: http://www.igi-global.com

Library of Congress Cataloging-in-Publication Data

Names: Dinis, Ana Cristina dos Santos Arromba, editor. | David, Fátima, 1968- editor. | Pereira, Liliana, 1978- editor. | Dias, Sara Luís, 1985- editor.
Title: Taking on climate change through green taxation / edited by Ana Arromba Dinis, Fátima David, Liliana Pereira, and Sara Luís Dias.
Description: Hershey, PA : Information Science Reference, [2023] | Includes bibliographical references and index. | Summary: "Taking on Climate Change Through Green Taxation provides applied research on increasing green tax literacy to build the capacity of companies to adopt sustainable practices in favor of environmental protection, to raise companies' awareness of sustainable reporting, and to increase international discussion on the issue of environmental taxation and its impact on more sustainable business decisions. Led by business experts with over 20 years of experience, this book will cover topics such as corporate social responsibility, environmental tax management, and sustainable tax policy. This resource is ideal for policymakers, corporate governance and social responsibility professionals, and researchers interested in taxation, accounting, auditing, finance, corporate governance, and corporate social responsibility"-- Provided by publisher.
Identifiers: LCCN 2023022607 (print) | LCCN 2023022608 (ebook) | ISBN 9781668485927 (hardcover) | ISBN 9781668485934 (paperback) | ISBN 9781668485941 (ebook)
Subjects: LCSH: Environmental impact charges. | Sustainability--Economic aspects. | Taxation--Environmental aspects. | Environmental policy.
Classification: LCC HJ5316 .T35 2023 (print) | LCC HJ5316 (ebook) | DDC 336.2/783337--dc23/eng/20230615
LC record available at https://lccn.loc.gov/2023022607
LC ebook record available at https://lccn.loc.gov/2023022608

This book is published in the IGI Global book series Advances in Public Policy and Administration (APPA) (ISSN: 2475-6644; eISSN: 2475-6652)

British Cataloguing in Publication Data
A Cataloguing in Publication record for this book is available from the British Library.

For electronic access to this publication, please contact: eresources@igi-global.com.

Advances in Public Policy and Administration (APPA) Book Series

ISSN:2475-6644
EISSN:2475-6652

Mission

Proper management of the public sphere is necessary in order to maintain order in modern society. Research developments in the field of public policy and administration can assist in uncovering the latest tools, practices, and methodologies for governing societies around the world.

The **Advances in Public Policy and Administration (APPA) Book Series** aims to publish scholarly publications focused on topics pertaining to the governance of the public domain. APPA's focus on timely topics relating to government, public funding, politics, public safety, policy, and law enforcement is particularly relevant to academicians, government officials, and upper-level students seeking the most up-to-date research in their field.

Coverage

- Government
- Law Enforcement
- Political Economy
- Politics
- Public Administration
- Public Funding
- Public Policy
- Resource Allocation
- Urban Planning

IGI Global is currently accepting manuscripts for publication within this series. To submit a proposal for a volume in this series, please contact our Acquisition Editors at Acquisitions@igi-global.com or visit: http://www.igi-global.com/publish/.

The Advances in Public Policy and Administration (APPA) Book Series (ISSN 2475-6644) is published by IGI Global, 701 E. Chocolate Avenue, Hershey, PA 17033-1240, USA, www.igi-global.com. This series is composed of titles available for purchase individually; each title is edited to be contextually exclusive from any other title within the series. For pricing and ordering information please visit http://www.igi-global.com/book-series/advances-public-policy-administration/97862. Postmaster: Send all address changes to above address.

Titles in this Series

For a list of additional titles in this series, please visit: www.igi-global.com/book-series

Global Perspectives on the Emerging Trends in Public Diplomacy
Samet Kavoğlu (Marmara University, Turkey) and Ergün Köksoy (Marmara University, Trkey)
Information Science Reference • © 2023 • 282pp • H/C (ISBN: 9781668491614) • US $240.00

Insights and Explorations in Democracy, Political Unrest, and Propaganda in Elections
Opeyemi Idowu Aluko (Ajayi Crowther University, Nigeria) and Lalekan Bolu Oluwadele (Walden University USA)
Information Science Reference • © 2023 • 377pp • H/C (ISBN: 9781668486290) • US $215.00

Historical and Political Analysis on Power Balances and Deglobalization
Emily Brynn Stacey (Rose State College, USA)
Information Science Reference • © 2024 • 300pp • H/C (ISBN: 9781668456330) • US $205.00

Integrating Principles of Success in Planning Denial and Deception Missions
Eugenie de Silva (Booz Allen Hamilton, USA)
Information Science Reference • © 2024 • 330pp • H/C (ISBN: 9781668468401) • US $215.00

Governmental Power Market-Ing in the VU-CHAOS World
Kittisak Jermsittiparsert (University of City Island, Cyprus) Srirath Gohwong (Universitas Muhammadiyah Sidenreng Rappang, Indonesia) Shayut Pavapanunkul (Universitas Muhammadiyah Sidenreng Rappang, Indonesia) and Nasaran Mahittichatkul (Universitas Muhammadiyah Sidenreng Rappang, Indonesia)
Information Science Reference • © 2024 • 360pp • H/C (ISBN: 9781668472804) • US $205.00

Reimagining Systems Thinking in a Post-Pandemic World
M. Elizabeth Azukas (East Stroudsburg University of Pennsylvania, USA) and Minkyoung Kim (University of West Florida, USA)
Information Science Reference • © 2023 • 325pp • H/C (ISBN: 9781668472859) • US $215.00

Public Policy's Role in Achieving Sustainable Development Goals
Mahani Hamdan (Universiti Brunei Darussalam, Brunei) Mohammad Anshari (Universiti Brunei Darussalam, Brunei) Norainie Ahmad (Universiti Brunei Darussalam. Brunei) and Emil Ali (Universiti Brunei Darussalam. Brunei)
Information Science Reference • © 2023 • 330pp • H/C (ISBN: 9781668489031) • US $235.00

Using Crises and Disasters as Opportunities for Innovation and Improvement
Saeed Siyal (Beijing University of Chemical Technology, China)
Information Science Reference • © 2023 • 300pp • H/C (ISBN: 9781668495223) • US $240.00

IGI Global
PUBLISHER of TIMELY KNOWLEDGE

701 East Chocolate Avenue, Hershey, PA 17033, USA
Tel: 717-533-8845 x100 • Fax: 717-533-8661
E-Mail: cust@igi-global.com • www.igi-global.com

Editorial Advisory Board

Table of Contents

Detailed Table of Contents

Green business in Colombia is a commercial, social, and environmental strategy that seeks to strengthen companies with an environmentally responsible approach, which is why the Colombian government establishes some legal and institutional instruments to benefit these organizations. This chapter describes an analysis of the financial and tax support received by these types of companies and the impact it has on the Colombian economy. Colombia has a tax statute and in it, everything related to green taxes, green tax benefits, and financial support for the energy transition, and other variables that are studied and described in this text.

One of the main climatic concerns is the increase of CO2 concentrations in the atmosphere. In the attempt and need to combat this problem, several regulatory policies have been applied in several countries by the governments, with the carbon tax being one of the most economical and effective policies in reducing CO2 emissions, thus leading to an increase in scientific research on this topic. Therefore, the main objective of this study is characterizing the existing research on the theme Carbon Tax. A sample of 121 scientific articles was collected, where several characteristics related to the content of published articles and respective authors were subsequently identified. The results obtained reveal that the number of publications has been increasing, and the themes most related to this investigation are: cap-and-trade, carbon dioxide emissions, climate change, computable general equilibrium (cge) model and emissions. Regarding the profile of researchers, the type of collective authorship is the most common and the country with the most affiliated authors is China.

The Portuguese Decree-Law No. 102-D/2020, of December 10 (PCM, 2020b), dictates the implementation of biowaste selective collection networks or the separation and recycling of biowaste at the source until December 31, 2023. This chapter aims to analyze the financial support provided to municipalities (local authorities) for implementing biowaste collection solutions, such as on public roads or through door-to-door collection, and to citizens for adopting biowaste separation and recycling practices through domestic or community composting. The authors also explore the application of the polluter-pays principle, which requires citizens to pay environmental taxes and fees to support sustainable development. Regarding this focus, the chapter highlights the legal, financial, and fiscal measures used to support biowaste collection and recycling efforts in Portugal, providing insight into the steps needed to achieve a sustainable environment while addressing the economic and social needs of the population.

It is a well-known fact that the automotive sector plays a key role in a country's socio-economic development, due to the large contribution of vehicles to greenhouse gas emissions (GHG). Alternatives such as car-sharing, public transport, cycling, the transition to electric vehicles, or walking can help reduce emissions. Therefore, it becomes a sector where environmental regulation should be demanded. This research aims to contribute to the tax literature by extending previous studies on corporate taxation by evaluating green tax incentives as an effective environmental tax measure. Considering taxis as an important part of urban public transport systems, the authors set out to investigate Portuguese taxi companies that have taken advantage of tax incentives for less environmentally polluting vehicles. The authors conclude that the income tax level, the economic profit, and the number of employees, are determining factors in these firms 'access to the tax incentives under study.

The "green taxation," which usually includes taxes on energy, transport, pollution, and resources, is being used to achieve environmental and sustainability purposes. The European Union is promoting this strategy, creating different provisions related to the usage of taxes to change environmental behaviors. Portugal, as a member-estate of European Union, follows this path establishing a policy where taxes are being used for these purposes. However, "green taxation" is also a significant form of revenue and, currently, it is difficult to define the criteria for the creation of these taxes, as they are not considering the ability to pay (one of the main principles in the creation of taxes) affecting the equitable distribution of the tax burden. The authors can conclude that, with the masking of "taxes required for environmental protection," these taxes are being used as an additional source of revenue, distorting the principles underlying the creation and collection of taxes.

This chapter aims to analyze the environmental tax's contribution to reducing carbon dioxide emissions in Portugal and Spain. Moreover, this study analyses the impact of the use of renewable energies and the gross domestic product (GDP) on carbon dioxide emissions. The data are collected from the Organization for Economic Cooperation and Development (OECD) and Word Bank indicators from 1990 to 2019. The methodology is the time series (two- stage least squares; fully modified OLS; DOLS- dynamic OLS canonical cointegration regression, and granger causality). The unit-roots test demonstrates that the variables used in this investigation are integrated with the first differences. The results demonstrate the importance of the environmental tax and the use of renewable energies in reducing carbon dioxide emissions, both in Portugal and Spain. The results also show the positive effect of economic growth on Spain's increase in carbon dioxide emissions.

Environmental taxation is an important tool in the hands of member states to combat against climate change and promote environmental sustainability. This chapter describes in detail the taxes that are part of the Portuguese tax system and that have the social function of protecting the environment. For this purpose, a background of the Portuguese Green Tax Reform is provided and each of the environmental taxes considered to be the main burdens of the Portuguese tax system are explained. This study aims to contribute to providing academics and practitioners with more information on environmental taxation and to motivate regulators to develop more articulated green taxation strategies in the future.

Tax transparency is often associated with standardized reporting. Among the goals of the 2030 agenda, goal 12.6 encourages organizations to adopt sustainable practices and integrate sustainability information into their reporting cycle. The Global Reporting Initiative (GRI) published the GRI 207- Tax (2019), aiming to increase transparency and accountability in corporate tax practices, by providing disclosure guidance for companies to disclose information about their tax strategy, governance, and risk management. As it is a recent standard, research on GRI-207 is still scarce. Thus, this chapter aims to contribute to the debate on the link between corporate reporting and fiscal responsibility by providing an overview on the background and structure of the GRI 207, which could serve as a basis for future research in the field of tax disclosure, as well as contributing to the policy debate. This chapter could have political and practical implications, as transparency in the disclosure of tax-related issues in corporate reporting could mitigate the reputational tax risk of companies.

As the global tourism industry continues to grow, concerns about its impact on the environment and local communities have increased. In response to these concerns, many destinations have implemented tourist taxes as a policy tool to manage the negative impacts of tourism on destinations and promote sustainable tourism. This chapter presents a systematic literature review on the relationship between tourist taxes and sustainability, using the Bibliometrix R software and a content analysis. The aim of this review is to provide a comprehensive analysis of the existing research, identifying gaps in the literature and suggesting areas for future research. The results demonstrate two main categories, concerning demand and supply dimensions. The issue of sustainability is transversal to the articles analyzed, highlighting its evolution in a holistic way and encompassing the economic, social, and environmental pillars.

As a result of major climate problems, this chapter aims to analyze the ways of using taxation to achieve its objectives, especially to combat externalities. The methodology used is bibliographical. The first chapter explained the social function of the tribute in the socio-environmental state, which would have as its function the fulfillment of its socio-environmental function. The second chapter deals with the objectives of environmental taxation, which is with the generation of dividends, benefits, for society with the correction of externalities. And the last chapter shows the ways to use environmental taxation to reduce externalities, such as use of direct taxation on emissions, use taxes, indirect taxation, taxation at different rates, funds. Finally, an analysis is made of the best way to use environmental taxation.

Preface

As editors of this reference book, it is with great enthusiasm that we present to you the culmination of collaborative efforts aimed at exploring the crucial intersection of environmental issues, taxation, and sustainability. In a world increasingly concerned about climate change, the imperative to address ecological challenges has become more pressing than ever before. Governments and businesses worldwide are recognizing the need for swift action to transition towards a sustainable future, and green taxation has emerged as a promising tool in this endeavor.

Sustainability, often associated solely with environmental responsibility, encompasses a broader scope, including social and economic aspects. Businesses striving for sustainability must embrace not only environmentally conscious practices but also foster positive relationships with stakeholders and uphold transparent and equitable economic practices. Consequently, the demand for corporate transparency has led to the emergence of sustainability and Corporate Social Responsibility (CSR) reports, enabling companies to showcase their efforts towards social and environmental dimensions, such as human rights protection, climate change adaptation, and community engagement.

Embracing this vision of sustainability, governments have proactively implemented tax measures to promote greener consumption and production patterns. By aligning tax policies with environmental objectives, nations aspire to create an energy-efficient economy that fosters responsible resource usage. Green tax incentives play a pivotal role in this transformation, encouraging businesses and investors to adopt sustainable practices. These incentives have the potential to drive significant changes in behavior, reduce carbon emissions, and contribute to a more sustainable future for our planet.

Through this edited reference book, we embark on a journey of applied research, drawing upon the collective expertise and insights of researchers and professionals who share a keen interest in environmental taxation and sustainable business practices. Our primary goal is to enhance green tax literacy and empower companies to make informed decisions that prioritize environmental protection. Additionally, we aim to deepen the understanding of sustainable reporting and foster international dialogue on the critical role of environmental taxation in shaping sustainable business strategies.

The audience for this book encompasses a wide range of professionals and scholars from various disciplines. Researchers in taxation, accounting, auditing, finance, corporate governance, and corporate social responsibility will find valuable insights and empirical evidence within these pages. Moreover, professionals engaged in corporate governance and social responsibility, such as accountants, auditors, tax officers, counselors, directors, and managers, will discover practical guidance on integrating green taxation principles into their organizations' frameworks. Lastly, policymakers in the realm of taxation will benefit from the comprehensive analysis of sustainable tax policies and their potential impact on the global pursuit of environmental preservation.

The core topics covered in this book include Green Taxation, Environmental Tax Incentives, Corporate Social Responsibility and Taxation, Environmental Tax Compliance Strategies, Sustainable Tax Policy, Environmental Tax Management, and the application of GRI Standards for enhanced tax transparency.

Chapter 1 delves into the concept of green business in Colombia, which focuses on environmentally responsible approaches for companies. It explores the legal and institutional instruments set by the Colombian government to support these organizations. The chapter provides an analysis of the financial and tax support received by green businesses and its impact on the overall Colombian economy. It also examines the potential of environmental taxes in generating alternative sources of public revenues to achieve sustainable public finances while guiding consumer behavior.

Chapter 2 addresses the pressing concern of increasing CO2 concentrations in the atmosphere and the regulatory policies, particularly carbon taxes, implemented by governments worldwide to combat climate change. The chapter presents a comprehensive characterization of existing scientific research on the topic of carbon taxes. It includes an analysis of 121 scientific articles, identifying themes related to carbon tax, cap-and-trade, climate change, and emissions. Additionally, the chapter sheds light on the profile of researchers involved in this research and their affiliations.

Chapter 3 provides an overview of corporate climate change-related accounting and taxation practices in Portugal and Spain. It delves into the respective accounting treatment for Greenhouse (GHG) emissions and highlights the differences in the recognition of emission allowances as inventories or intangible assets in the two countries. Furthermore, the chapter explores the tax treatment of emission allowances and introduces the reverse-charge mechanism on CO2 emissions trading within Europe.

Chapter 4 focuses on the tax regime applicable to Environmental Non-Governmental Organizations (NGOs) in Portugal. It examines the framework specifically designed for these organizations under Portuguese tax laws and the tax benefits that may be recognized to support their environmental protection initiatives. The chapter highlights how the tax regime can act as an instrument for the state's intervention towards environmental preservation and sustainability.

Chapter 5 provides a comprehensive description of the environmental tax reform implemented in Western Balkan nations. It examines the effectiveness of environmental taxes in promoting green initiatives and addressing environmental protection challenges. The chapter explores how environmental taxation can influence production and consumption choices to move towards more eco-friendly practices.

Chapter 6 offers a bibliometric overview of available articles in the SCOPUS database related to green tax and double dividend research. It identifies influential authors, prominent journals, and trending topics within the field. The chapter presents key focus areas related to green tax and double dividends, providing valuable insights into the most influential aspects of the subject.

Chapter 7 emphasizes the significance of digitalization for environmental sustainability, particularly for Portuguese micro and small to medium-sized enterprises (SMEs). It focuses on tax incentives aimed at promoting the digital transformation of accounting practices and explores their adoption among Portuguese businesses. The chapter highlights the role of governments in fostering digitalization for greener and more sustainable practices.

Chapter 8 delves into the concept of green taxes and their role as financial instruments in combating environmental pollution. The chapter discusses how green taxes aim to tax environmentally harmful activities while incentivizing eco-friendly practices through economic incentives. It also explores how these taxes address market failures in considering ecological impacts resulting from production and consumption activities.

Chapter 9 focuses on the financial support provided to municipalities and citizens in Portugal for implementing biowaste collection solutions and adopting biowaste separation and recycling practices. The chapter examines the application of the polluter-pays principle in supporting sustainable biowaste management efforts, shedding light on the legal, financial, and fiscal measures employed to achieve environmental goals.

Chapter 10 investigates the effectiveness of green tax incentives in promoting environmentally friendly vehicles among Portuguese taxi companies. It explores how tax incentives are adopted based on factors such as income tax level, economic profit, and the number of employees. The chapter highlights the role of tax measures in encouraging less environmentally polluting practices in the taxi sector.

Chapter 11 delves into the concept of “green taxation” and its usage to achieve environmental and sustainability objectives. It discusses how some environmental taxes, although intended for environmental protection, are used as an additional source of revenue, potentially distorting the principles underlying tax creation and collection. The chapter raises questions about the equitable distribution of the tax burden and its impact on tax justice.

Chapter 12 analyzes the contribution of environmental taxes to reducing carbon dioxide emissions in Portugal and Spain. It also explores the impact of renewable energies and Gross Domestic Product (GDP) on carbon dioxide emissions. The chapter employs various time series methodologies to assess the role of environmental taxes and renewable energies in curbing carbon emissions.

Chapter 13 provides a detailed overview of environmental taxes in Portugal, examining the role of taxation in promoting environmental protection and sustainability. It explains the background of the Portuguese Green Tax Reform and elucidates the environmental taxes that constitute significant burdens in the country’s tax system. The chapter aims to provide valuable information for academics, practitioners, and policymakers interested in environmental taxation.

Chapter 14 explores the link between corporate reporting and fiscal responsibility through the lens of GRI 207 - Tax, a standard developed by the Global Reporting Initiative (GRI). This chapter outlines the structure and purpose of GRI 207, focusing on its guidance for companies to disclose tax-related information. The chapter highlights the potential implications of tax transparency in corporate reporting on reputational tax risk.

Chapter 15 conducts a systematic literature review on the relationship between tourist taxes and sustainability. It investigates the impact of tourist taxes as policy tools to manage the negative impacts of tourism on destinations and promote sustainable tourism. The chapter identifies key themes concerning the demand and supply dimensions related to sustainability and highlights the evolving significance of sustainability in the tourism industry.

Chapter 16 explores the utilization of taxation as a means to achieve climate change objectives, particularly in combating externalities. It discusses the social function of taxation in the Socio-Environmental State, the objectives of environmental taxation, and the ways in which it can be used to reduce externalities. The chapter highlights how environmental taxation is essential for sustainable development and environmental protection.

We extend our sincere gratitude to all the contributors who have lent their expertise and time to make this volume a reality. Their commitment to advancing the understanding of green taxation has been instrumental in shaping the diverse perspectives presented herein.

In conclusion, we hope this edited reference book serves as a valuable resource for academics, professionals, and policymakers alike, inspiring meaningful action and dialogue to address climate change through effective and innovative taxation strategies. Let us join hands in building a sustainable future for generations to come.

Ana Arromba Dinis
Research Centre on Accounting and Taxation, Polytechnic Institute of Cávado and Ave, Portugal

Fátima David
Research Centre on Accounting and Taxation, Polytechnic Institute of Cávado and Ave, Portugal

Liliana Pereira
Research Centre on Accounting and Taxation, Polytechnic Institute of Cávado and Ave, Portugal

Sara Luís Dias
Research Centre on Accounting and Taxation, Polytechnic Institute of Cávado and Ave, Portugal

Chapter 1
Analysis of Financial and Tax Support for Green Businesses in Colombia

Franz Esteban Rodríguez Franky
https://orcid.org/0000-0002-5253-3015
Fundación Universitaria Compensar, Colombia

ABSTRACT

Green business in Colombia is a commercial, social, and environmental strategy that seeks to strengthen companies with an environmentally responsible approach, which is why the Colombian government establishes some legal and institutional instruments to benefit these organizations. This chapter describes an analysis of the financial and tax support received by these types of companies and the impact it has on the Colombian economy. Colombia has a tax statute and in it, everything related to green taxes, green tax benefits, and financial support for the energy transition, and other variables that are studied and described in this text.

INTRODUCTION

The green economy is a vital tool for the changes that every government seeks in favor of international agreements and treaties for the mitigation of greenhouse gases. The energy transition, the model of human well-being and the protection by human resources is undoubtedly a necessity in conventional production models, in which exploitation and linear production must be oriented towards a much more circular and environmentally friendly environment. Currently, governments are looking for political, social and economic alternatives that allow markets to regulate a much more conservative and rational consumption of ecological resources. In Colombia there is Law 2099 of 2021, which mainly regulates investments in sustainable projects related to green and blue hydrogen, as well as various tax benefits that are applicable to companies that focus on activities aimed at caring for the environment, capture and management of CO2, use and storage of carbon, among others. For Antón Alvaro (2022) "the Green

DOI: 10.4018/978-1-6684-8592-7.ch001

Deal highlights the potential of environmental taxes to generate alternative sources of public revenue that contribute to achieving sustainable public finances, while sending appropriate price signals to consumers to guide their behavior" (p.2).

Corson et al (2013) state that "in the last two decades, the incorporation of market logics into environmental and conservation policy has led to a reconceptualization of 'nature'. Resulting constructions as ecosystem services and derived from biodiversity". (p. 1). This same author reports that the financial and tax mechanisms that should be used to strengthen green businesses should focus on actions to reduce emissions from deforestation, forest degradation, species banks, among others.

One of the contributions provided by the chapter is the analysis of sustainable development and environmental protection, in which it has become a global concern. Green businesses, also known as sustainable enterprises or eco-enterprises, play a critical role in the transition to a greener and more sustainable economy. To promote their growth and contribute to the protection of the environment, several countries have implemented financial and tax support policies. In this sense, this text reviews the existing literature and the background related to these initiatives.

Background

The background reveals that the need to financially support green businesses arises from the growing awareness of the negative effects of conventional business practices on the environment. In addition, international agreements, such as the Kyoto Protocol and the Paris Agreement, have prompted governments to adopt measures that foster the transition to a low-carbon economy. These factors have led to the creation of specific programs and policies to support green businesses (Smith, 2018; Johnson & Thompson, 2020).

The literature shows that there are different forms of financial support for green businesses. Governments and financial institutions offer a variety of financial instruments, such as preferential loans, grants, tax incentives and venture capital, to facilitate the creation and expansion of sustainable enterprises. These programs seek to reduce economic barriers and promote innovation in clean technologies and sustainable practices (Jones, 2019; Brown et al., 2021). In addition, an increase in private equity investment in green businesses has been observed, supported by growing demand from environmentally conscious consumers and the potential profitability of these companies. Investors are recognizing that green businesses can generate both environmental and economic benefits in the long term (García et al., 2022). Governments use tax incentives to encourage the adoption of sustainable business practices. These incentives can include tax exemptions or reductions for companies that implement clean technologies, use renewable energy sources, or reduce their carbon emissions. Emissions trading schemes have also been established to promote greenhouse gas reductions (Lee, 2017; Martinez & Lopez, 2020).

Tax support for green businesses not only aims to drive the adoption of sustainable practices, but also to foster green job creation and strengthen the competitiveness of enterprises in the globalized economy (Chen & Wang, 2018; Kim & Park, 2021). Programs and policies implemented by governments, together with private investment, have contributed to the growth of sustainable enterprises and the mitigation of negative impacts on the environment. However, the importance of continuously evaluating and improving these initiatives to maximize their effectiveness and achieve a successful transition to a green and sustainable economy is highlighted (Baker & Green, 2019; Liu et al., 2022). On the other hand, this chapter presents the strategies developed by the Colombian government to strengthen this type of companies; Therefore, the main objective is to analyze the support and fiscal and financial benefits that are generated to green, green or green SMEs, in such a way that it describes the impact on the market and business models.

Finally, the chapter presents a series of rules, statutes and other legal instruments that allow Colombian companies to identify what kind of financial and tax benefits they have depending on their business model. It is important to note that these regulations are updated periodically in Colombia, without modifying their essence in relation to the reduction in the generation of carbon dioxide, methane and other gases that generate global warming. Likewise, a series of concepts, opinions and texts related to the understanding of green business models, their impact on the Colombian economy and the future that awaits considering the energy transformation promoted by the National Government of Colombia are presented. The reflections and conclusions given are part of a research study conducted by Fundación Universidad Compensar, the Business Administration program and green organizations in the Colombian capital.

SUSTAINABLE BUSINESS METHODS

Green business is a type of business model that has been building social, economic, environmental and competitive advantage impact; This has allowed it to cover new markets and capture the attention of those present, in which the differential of goods and services focuses on the quality of the good, the positive impact on the environment and the rational use of natural resources. However, this chapter cites a number of authors who describe the advantages, characteristics, and challenges that this type of enterprise has in international, national, and regional economies.

Focusing the discussion around the critical thinking of some institutions and writers, academics and government entities that affirm that the central axis of global warming mitigation are eco-friendly economies or social inclusion, the relationship with the volume of units of goods and services that can be produced based on the constant and growing demand of high-consumption countries has a great debate. For this reason, the development of sustainable production must be considered in a comprehensive manner, along with the expansion of social protection mechanisms, natural resources, human and environmental heritage of the regions. Mechanisms that allow a more systemic and logical dimension in the mobilization of financial and tax projects that favor this type of venture.

Sustainable Markets

Sustainable markets are all those spaces in which the supply of goods and services is developed to contemplate the care of ecosystems and the mitigation of the overexploitation of natural resources. According to Segrelles (2009) "The environmental services business generates millions of dollars in profits around the planet, which means that there is an unusual proliferation of economic, strategic and geopolitical interests linked to it, whose beneficiaries are the large international banks" (p.4). This makes it possible to identify the importance of both who produces the goods and services and the type of market to which they are offered. Antonio Ubieto (2022), states that sustainable processes began in the mid-1990s, when the appearance of the "4R" (Reuse-Repair-Recycle-Reduce), began to transform production and management processes in organizations. With the passage of time, the author reports that today there is the use of the "7R" (Reuse-Repair-Recycle-Recycle-Reduce-Renew-Recover-Redesign).

These types of characteristics undoubtedly allow entrepreneurs to start from much more objective business ideas and oriented towards a socially responsible impact. It is also necessary to mention that sustainable markets must be evaluated in such a way as to determine their influence on the different business spheres "social-financial-competitive environment and innovation". For Castellani & Sala (2012) "the evaluation of the sustainability of human activities requires integrating as far as possible tools and methodologies to achieve a comprehensive assessment of impacts, both environmental and socioeconomic" (p.135). MSMEs (small and medium-sized enterprises) must generate business strategies that not only generate competitive advantage but must also do so from a productive approach or even from the transformation of new production models that allow a better and adequate application of sustainability in the face of scenarios of health, financial and / or geopolitical restrictions (Velásquez Chacón, 2022, p. 2).

Salas-Arbeláez et al. (2017). Quoted by Velásquez Chacón (2022), he reports that "several studies have shown that companies with an organizational culture and focused on sustainability experience positive effects on performance, in this sense this could favor institutional objectives, and depending on management, encourage companies to be innovative, with high returns" (p.3). In this sense, ECLAC (2020) suggests that companies supported by national governments and focused on a circular economy, bioeconomy or sustainable production must have evidence on the dimension of social, environmental and financial good. This evidence allows the generation of traceability processes in social responsibility actions and various actions.

Undoubtedly, sustainable markets have a large space for participation in a high-growth world population, but more aware of the production method, protection of flora and fauna, overexploitation of natural resources, equity, social equality and competitive prices. Dvarskas (2017) indicates that "in national, regional and local decisions of growth and development, there is often a trade-off between the direct monetary benefit derived from economic development activity" (p.163), so sustainable markets must have that balance between financial and social benefit.

Green Market

Green markets are a 21st century concept, indicating spaces where supply and demand are in favor of acquiring and selling green goods and services, made of recycling or other ecological features. It is worth remembering that the need to protect the environment is not a current need, this derives from a neologism of the 60s in which concern about the excess generation of greenhouse gases, sudden changes in climate and other anthropic processes generate changes in ecosystems. For Sadeleer (2022):

Although environmental protection is not a recent concern, it has taken on a new urgency in recent years, marked by the need to find universal solutions to global warming, biodiversity loss and natural resource depletion. Not a day goes by without being informed of new threats, unsuspected risks, oil spills, disappearance of wildlife, devastating fires, melting glaciers, tropospheric smog, endocrine disruptors. (page 151)

The emergence of green markets then becomes an alternative for companies to initiate a transformation in the commercialization of ecological goods and services, not as a trend or market fad but as a competitive strategy to strengthen small and medium entrepreneurs. In Latin America, specifically in Colombia, green markets have been gaining importance in the regional, sectorized and community economy, which seeks to generate trade and fair prices between the producer and the buyer. Jaramillo & Patino (2020), state that green markets bring with them the so-called "green consumers", people, companies and other agents that consume biodegradable, ecological and organic goods and services.

Products that result in companies working in a network of agricultural, livestock, agro-industrial and industrial practices friendly to the environment, in the reduction of the use of pesticides or chemical inputs, processes of polluting synthetic molecules in rivers and soils and / or any other factor that threatens or affects the flora and fauna. According to the World Bank (2012), cited by Gutman and López (2017) they cite that "inclusive green growth is not a new development, avoiding irreversible environmental costs" (p. 14). However, there is a growing demand in the market for inputs available for the production of goods and services, green businesses are a sustainable alternative for the exploitation of these natural resources, where life cycle approaches, types of production processes and administrative actions are part of the essence of the business model. The United Nations is one of the international organizations most concerned with the development of linear productive changes, more sustainable production and the mitigation of greenhouse gas production.

However, all these efforts are limited if governments do not develop policies to improve production processes based on the conservation of natural resources. Green business is an opportunity for sustainable development, and international consumer trends favor the creation of goods and services that meet these needs.

But it is not only about green production, the research process of academia, companies, and the country itself, must be aimed at strengthening the entrepreneur and his growth opportunities as a green model. For Potluri & Phani (2020) cited by Pertuz et al. (2021) "Green businesses can be understood as those that aim to achieve economic growth through the rational use of natural resources and avoiding pollution" (p. 746). These factors give green models a different dynamic than those developed by traditional companies, especially those based on segmented and specialized markets.

For Mora & Cañon (2017) "The policy guidelines and the intention of the Regional Autonomy Corporation of Cundinamarca – CAR, contemplates in its action plan, environmental agendas as agreements established between the productive sectors, through their unions and companies". (p.5). In Colombia, focusing on the development of environmentally friendly goods and services is no longer considered a utopia in the productive sector, but a strategy to gain competitive advantage. These behaviors are visible to consumers looking for products and services that cause minimal damage to the environment and its natural resources; Public policies play a critical role in the success of these business models to better model economic performance and maximize resources.

They have different characteristics, combined with the development of environmentally friendly and profitable productive enterprises. However, national green business plans are included in the Organization for Economic Cooperation and Development's (OECD) Green Growth Declaration, an instrument that countries insist seeks to focus their efforts on developing green growth strategies.

Bioeconomy

The National Sustainable BioTrade Program is another strategy in which the Colombian government seeks to promote support not only for the producer of sustainable goods and services, but also to encourage buyers to buy products focused on responsible consumption. Bio-trade's processes highlight its development in marketing strategies and ongoing revenue on behalf of organic products. For Primo Tapia (2017), biotrade and green market processes must be linked as a priority in the agenda of governmental, political and financial developments in the regions, so that they generate a level of national and international competitiveness; These business opportunities should be seen as a potential in the natural capital, human capital and biodiverse ecosystems of the producing areas. (p. 108).

Therefore, organic production must be an effort, which must be combined with a culture of responsible, social and environmental consumption in which small and medium-sized enterprises providing goods and services lead to green markets, bioeconomy and circular economic success. For the United Nations Conference on Trade and Development (2006), the biotrade value chain approach begins with administrative, commercial and legal mechanisms that can be used to strengthen production chains, the implementation of good practices in the use of natural resources, the conservation of biodiversity and the payment of ecological debt owed to ecosystems (p.3). The responsible production, marketing and consumption of sustainable goods and services can be better elucidated by considering the environmental value, productive potential and added value of Colombian biotrade companies. In summary, biotrade, responsible production, marketing and consumption of sustainable goods and services can be better elucidated if the environmental value, productive potential and added value of Colombian biotrade companies are taken into account, this translates into green business opportunities, from biodiversity resources added value is generated in companies.

FINANCIAL AND ECONOMIC SUPPORT FOR THE GREEN BUSINESS MODEL IN COLOMBIA

Understand the functioning of government structures in Colombia that financially and economically favor companies that offer ecological goods and services in the Colombian market. This is one of the analytical variables in this chapter. It is important to mention that Colombia, being a South American country and belonging to that block of emerging countries with high environmental and human wealth, is seen as a potential for life and growth in capital investment. Colombia has a green financing program that allows entrepreneurs to obtain an economic reinforcement focused on sustainable life cycles of products and services.

For the Colombian Association of Banking and Financial Institutions, Asobancaria (2021) "In addition to adopting sustainable production processes already established, the green financing line aims to encourage the creation and development of project companies that promote environmental protection" (para. 3). The Colombian organization is also linking the types of projects that are eligible for green financing lines through bank credit, which could mitigate negative environmental impacts and adapt to climate change from the country's productive actions, as shown in Table 1.

Table 1. Green credits for types of green businesses in Colombia

	Sustainable infrastructure	
Financing for	Ecotourism	
companies with a	Implementation of Technologies	for
friendly Life cycle	Environmental protection	
approach.	Energy efficiency	
	Renewable energy generation	

Source: Own elaboration based on Asobancaria (2021).

It is important to mention that these types of benefits mentioned above are linked to a political strategy in which the strengthening of entrepreneurship is its focus. That is why in Colombia to identify this type of financial and economic support we start from the Colombian association of banks "ASOBANCARIA", which is an organization that favors credits and loans for those SMEs that are focused on the protection and conservation of natural resources. According to Asobancaria (2021), the main tax benefits in addition to the financial support generated for green organizations are:

Environmental credit benefits (Tax benefits are defined as deductions, exemptions, and special treatments that result in a reduction of taxpayers' tax liabilities.)

- Incentives such as the reimbursement of a part of the investment, depending on the reduction of the environmental impact.
- Some financial institutions have preferential rates.
- Tax incentives: Articles 424 paragraph 7 and 428 subparagraphs f) and (i) of the Tax Statute establish that articles, equipment and / or machinery used for environmental improvement processes are excluded from sales tax (VAT) under the terms and conditions of the regulations.
- Technological update
- Reduction of environmental impacts
- Increase in the productivity of the company and improvement of the work environment.

In Colombia, companies that invest in the control and improvement of the environment have the right to deduct from their annual income the value of their investments for the corresponding fiscal year, but not to exceed 20% of their net income. Some regulations allow to deepen the exploration of the tax benefits derived from the implementation of green business models, Decree No. 3172 of 2003, Resolution No. 779 of 2012 and Resolution No. 186 of 2012 grant authority in terms of tax benefits for this type of business. According to Rúa-Usme et al (2019) in Colombia, "initiatives are emerging where the Bank of Colombia will be the protagonist, mainly through proposals to create new green credit lines" (p. 2). From this, the country has public entities that are guarantors in compliance with legal requirements as an option for tax and financial benefits. The main ones are:

- Regional Environmental Authorities (RCA)
- Urban Environmental Authorities (AUU)
- Ministry of Environment and Sustainable Development (through ANLA)

The Colombian tax statute is the legal form that allows green SMEs to benefit from this type of process. When looking for other alternatives to financial support and citing the work of Bancóldex (2018) "First Green Bond Report", they affirm that the support generated by this entity through green bonds aims to support companies that generate positive environmental benefits, contributing to change. Achieve a low-carbon economy through more sustainable productivity, circular production, more efficient use of energy, incorporating sustainable practices, including the substitution of polluting and inefficient technologies. Also, actions related to pollution control and ultimately the implementation of clean technologies. For Bancóldex (2018), "with the issuance of the green bond, the entity consolidates its strategic focus on promoting business growth, increasing the productivity and competitiveness of Colombian companies, and promoting the rational use of resources and social welfare". (p. 5).

What is described in this chapter reflects an exhaustive study of the evaluation of the sustainability of green business models in Bogotá - Colombia. This study allowed not only to analyze the fiscal and tax benefits of green organizations, but also to evaluate the activities that guarantee sustainability, such as organizational products, from different management perspectives. The reuse, valorization, recycling and final disposal of services through the implementation of eco-efficiency measures in organizational processes, understanding eco-efficiency as a link between the added and created value and the additional environmental impact required to generate it, is another of the characteristics studied in the project. These types of results show that Eco-SMEs understand their business models and are focused on strengthening these activities, but it is worth noting that this evaluation is for institutions that have been in the market for 1 to 3 years and more than 3 years. Now, giving continuity to the above in relation to the tax incentives and economic benefits of this type of companies; It is important to mention that the different ways in which industrialists and their production processes contribute to the improvement of the environment are important to mitigate the effects of Callentamiento, for this, nothing better than to encourage them fiscally to exercise actions and strategies for a circular production system.

Colombian law allows for environmental tax incentives, such as tax incentives, VAT exemptions, accelerated depreciation of assets, and tax exemptions. The environmental revenue benefit includes a 25% tax credit for investments made to control, protect and improve the environment. For the Non-Conventional Energy Consumption Fund (FNCE), the benefit is 50% of the total investment. As for the exemption from VAT, this incentive allows exemption from VAT applicable to the purchase of improvement equipment or machinery, environmental control and surveillance systems. For FNCE, revenues also include services related to the start-up of the project. However, FNCE's power generation projects are subject to accelerated asset depreciation incentives and tariff exemptions. First, beneficiaries can obtain a fixed annual interest rate of up to 20%, and the plants, equipment and engineering works necessary for the initial investment and operation of the project are used. Secondly, the benefit of exemption from customs duties on imported equipment or machinery. Table 2 shows in detail the fiscal and tax benefits of green companies and the legal and juridical supports that support them in Colombia, an exercise that was carried out due to research actions.

Table 2. Fiscal and tax benefits for green companies in Colombia

ARTICLE	REGULATIOS	BENEFIT	FEATURE
1	Article 1.2.1.18.70 of Decree 1625 of 201 Law 1715 of 2014.	Income tax deduction for investments made in energy efficiency management projects (E.E.)	Income tax deduction of 50% of the value of the investment certified by the Mining and Energy Planning Unit (UPME).
2	Art. 14 of Law 1715 of 2014 Article 1.2.1.1.18.70 of Decree 1625 of 2016.	Accelerated deduction for depreciation for investments in non-conventional energy sources - FNCE	It is an accounting incentive in which the taxpayer is allowed to take the deduction for accelerated depreciation of equipment and civil works necessary for the pre-investment, investment and operation of energy generation with FNCE.
3	Article 1.2.1.1.18.52 of Decree 2205 of 2017 Art. 258 of the Tax Statute.	Deduction of income tax for investments made in the control, conservation and improvement of the environment	Income tax discount equivalent to 25% of the value of the investment certified by the National Environmental Licensing Authority - ANLA or the Competent Environmental Authority.

continues on following page

Table 2. Continued

ARTICLE	REGULATIOS	BENEFIT	FEATURE
4	Article 1.2.1.1.18.52 of Decree 2205 of 2017 Art. Article 258 of the Tax Statute Article 1.2.1.1.18.52 of Decree 2205 of 2017	Income tax rebate for investments made in energy efficiency management projects (E.E.)	Income tax discount equivalent to 25% of the value of the investment certified by the National Environmental Licensing Authority - ANLA.
5	E.T. Art. 240, paragraph 5 letter F	Decrease in income tax rate for ecotourism projects	Returned taxed at a rate of 9% for new projects of theme parks, ecotourism and agrotourism and new nautical springs that are built in municipalities of up to two hundred thousand inhabitants.
6	Art. 477 E.T. number 1 Article 6 of Decree 939 of 2004	VAT exemption for the purchase of biofuels of plant or animal origin	VAT exemption on the purchase of biofuels of plant or animal origin for use in domestically produced diesel engines for blending with ACPM.
7	Article 1.2.1.1.18.70 of Decree 1625 of 2016.	VAT exemption on domestic or imported equipment and services for FNCE investments.	The VAT exclusion applies to national or imported equipment, elements, machinery and services that are intended for pre-investment and investment, for the production and use of energy of the FNCE (Art. 1.3.1.12.21 of Decree 1625 of 2020).
8	Article 424 of the Tax Statute	Exclusion of VAT for agricultural and other goods	Exclusion of VAT on the sale or import of goods for the agricultural sector and others. These goods are not subject to value added tax (VAT).
9	Article 3 of Resolution 2000 of 2017	VAT exclusion for investments in environmental control and monitoring systems to meet the commitments of the Montreal Protocol.	Exclusion of VAT on the importation of elements, machinery and equipment for the Environmental Control and Monitoring System for the protection of the earth's ozone layer.
10	Articles 3 of Resolution 2000 of 2017 Art. 428 numeral f of the T.E. and articles 1 and 2 of Decree 1564 of 2017.	VAT exclusion for investments in Environmental control and monitoring systems for the development of recycling and waste processing activities, among others.	Exclusion of VAT on the importation of elements, machinery or equipment, provided that they are not produced in the country, destined to the Environmental Control and Monitoring System for the development of recycling and processing of garbage or waste. activities, among others.
11	Articles 1 and 2 of Decree 1564 of 2017 No. 7 of Art. 424 T.E.	VAT exclusion for investments made in energy efficiency management projects (E.E.)	Exclusion of VAT for the acquisition of national or imported equipment and elements for the construction, installation, assembly and operation of Environmental Control and Monitoring Systems resulting from Efficient Energy Management projects.
12	Articles 3 of Resolution 2000 of 2017 No. 7 of Art. 424 E.T. and Articles 1 and 2 of Decree 1564 of 2017.	Exclusion of VAT for the acquisition of equipment and elements for the environmental control and surveillance system.	Exclusion of VAT for the acquisition of national or imported equipment and elements for the construction, installation, assembly and operation of Environmental Control and Monitoring Systems.
13	Article 1.2.1.1.18.79 of Decree 1625 Nov 2016 Numerals 16 and 17 of article 5 of Law 1715 of 2014.	Tariff exemption for imports of equipment intended exclusively for investments in NCEF	The tariff exemption applies to the importation of machinery, equipment, materials and inputs exclusively for pre-investment work and investment in FNCE projects, provided that they are NOT produced by the national industry and the only way to acquire them in Colombia is by import.
14	Article 14 Law 299 of 1996	Exemption from property tax for Conservation of biodiversity in flora	Exemption from property tax for owning or owning properties that develop flora conservation activities (botanical gardens or properties intended for these purposes).
15	Article 1 of Decree 2755 of 2003 article 2 of Decree 2755 of 2003	Incentives for investment in alternative energy - clean development mechanism (CDM)	Exempt income from the sale of electricity generated from wind resources, biomass or agricultural waste, made only by generating companies, for a period of fifteen (15) years.
16	Article 2 of Decree 2051 of 2019 Article 3 of Decree 2051 of 2019	Tariff reductions for electric or natural gas vehicles	0% tariffs on the import of electric vehicles. 5% tariffs on imports of vehicles with exclusive operation of natural gas engines.

Source: Authors.

To understand a little more about how these benefits work in Colombia, we must break certain differences in favor of participating and accepting the fiscal support generated by the Colombian government.

Investments in environmental control: These are those that face the implementation of an environmental control system designed to achieve measurable and verifiable results for the reduction of the demand for natural resources, renewable energy products or prevention and / or reduction and / or improvement of the quality of electricity production, liquid waste, atmospheric emissions or solid waste. It also works on the control of investments in the environment, it can be put into production, at the source and / or at the end of the manufacturing process, this is called environmental control.

Investments in environmental improvement: Develops processes aimed at renewing, complementing, reserving and protecting renewable natural resources and the environment, as well as direct environmental benefits. These investments are directly associated with environmental control, understood as a measurable and verifiable set of results obtained through the implementation of an environmental control system. The reduction of the demand for renewable natural resources to prevent and/or reduce the generation of liquid waste, air emissions or solid waste, and also includes collection, inspection, processing, monitoring, supervision or surveillance.

Certificate of investment in environmental control and improvement: This is an administrative act issued by the Department of Environmental Protection, which makes it clear that investment in environmental control and improvement is adequate in accordance with the standards and requirements established by Decree No. 2205.

Figure 1. Breakdown of net income
Source: Authors.

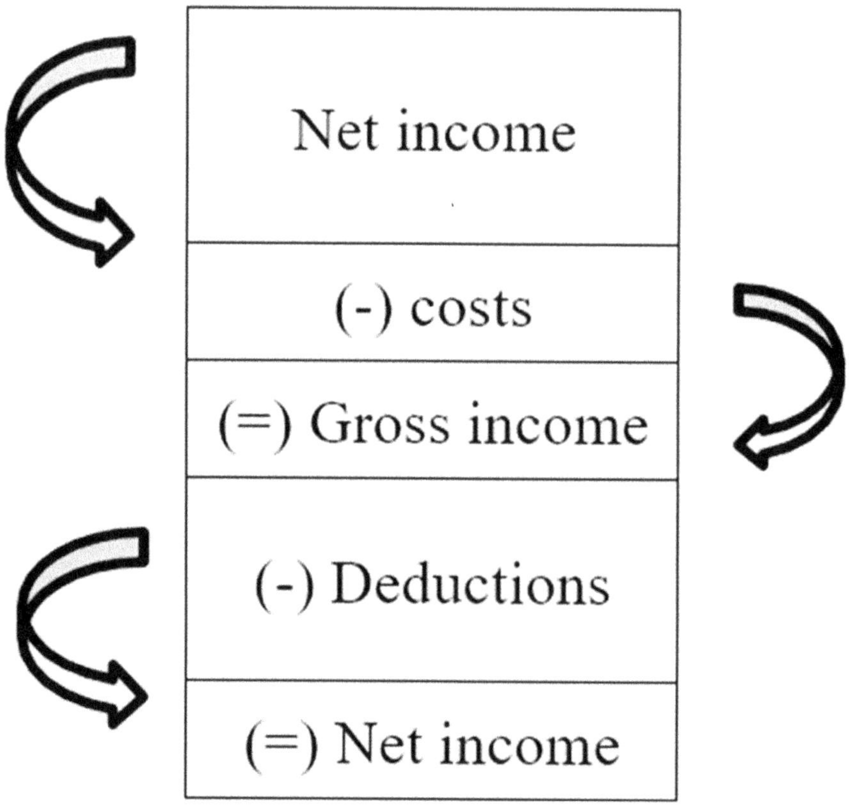

Net income: Net income is made up of gross income less deductions that have a causal relationship with income-generating activities. Net income is taxable income, and the respective rates are applied, except when there is exempt income, in which case it is subtracted to determine the taxable income, as shown in Figure 1. (PRESIDENCY OF THE REPUBLIC OF COLOMBIA. Ministry of Finance and Public Credit. Decree 624 of 1989. By which the Tax Statute of the taxes administered by the General Directorate of National Taxes is issued. Article 178)

Despite the impact that tax benefits have had on companies that develop environmentally friendly goods and services, it is not only exponential, since the records year after year in the last decade in Colombia show that no more than 7% of recognized and legalized SMEs in the country benefit from this type of actions. On the other hand, according to the National Survey of Perception Mission of Green Growth, carried out by the National Planning Department in 2019, 66.2% of companies in Colombia have been affected by changes in the climate which has led to implement this type of business strategies. Green businesses are established in 750 municipalities in the 32 departments of Colombia and have an impact in terms of employment generation, with 71,519 associates and 44,623 registered jobs. It is also noteworthy that these green businesses boost the economy and contribute to the economic development of the country, with annual sales exceeding 700,511 million pesos. With the aim of strengthening these sustainable ventures, initiatives have been implemented such as the BioExpo fairs, which takes place in Medellín and is considered the most important in the sector in Latin America. In its ninth edition in 2022, this fair managed to close business for more than 9,800 million pesos and sales close to 400 million were registered. Another outstanding event is BioSinú, a fair dedicated to environmental education and green business, which was attended by more than 3,600 students, teachers, professionals and entrepreneurs.

However, in Colombia, various green tax discount measures are being implemented as part of the government's strategy to encourage the adoption of sustainable business practices and promote the development of a green economy. These measures seek to incentivize companies to implement clean technologies, reduce their carbon emissions and promote energy efficiency. According to data from the Ministry of Environment and Sustainable Development of Colombia, it is estimated that until 2021, more than 1,000 companies in the country have accessed tax benefits related to green taxes (Ministry of Environment and Sustainable Development, 2021). These tax benefits include exemptions or reductions from income tax, VAT and other local taxes for companies that meet the criteria set in terms of environmental sustainability.

The economic impact of these measures is significant. According to a report by the Inter-American Development Bank (IDB), it is estimated that the total amount of green tax discounts in Colombia exceeds 300 billion Colombian pesos (Inter-American Development Bank, 2019). These tax credits allow companies to invest in sustainability projects, improve their competitiveness and contribute to the protection of the environment. In addition to tax discounts, emissions trading schemes have been implemented in Colombia. These schemes set limits on greenhouse gas emissions and allow companies to buy and sell emission permits. This creates an additional incentive for companies to reduce their emissions and adopt more sustainable practices. The application of these green tax discount measures has been a key factor in boosting business sustainability in Colombia. It has encouraged the adoption of clean technologies, the reduction of polluting emissions and the development of more environmentally responsible business practices. Table 3 shows a repository of companies that have benefited from this type of financial discounts when they work their production and business operation with a sustainable life cycle approach.

Table 3. Companies benefiting from green taxes in Colombia

Name of the business organization	Economic sector to which it belongs	Commercial document - NIT	Green business model		Process by which you have a discount on rent	Environmental certification to access the tax benefit -	Regulatory basis	Value to which a discount of 25% is made in rent according to E.T (millions of pesos $ cop)	Value in discount in rent - millions of pesos $ cop.	Discount value on rent - € euros
			Yes	No						
Reciclados industriales de colombia s.a.s.	Construction sector and materials and aggregates subsector	9003609541	X		Machinery for the use of construction and demolition waste -rcd.	2278 of 2018	Article 255 et discount on rent	$ 725,000,000.00	$ 181,250,000.00	$ 37,386.55
Reciclados industriales de colombia s.a.s.	Construction sector and materials and aggregates subsector	9003609541	X		Machinery, equipment and infrastructure required directly and exclusively for the operation or execution of environmental control systems.	2391 of 2018	Article 255 et discount on rent	$ 805.000.000,00	$ 201.250.000,00	$ 41.511,96
Ciudad limpia bogota s.a e.s.p.	Services sector and cleaning and cleaning subsector	830048122		X	Equipment and elements for the collection, transport and compaction of solid waste and containers for the collection of usable solid waste within the framework of the integral management of solid waste	1892 of 2019	Article 255 et discount on rent	$ 40.355.000.000,00	$ 10.088.750.000,00	$ 2.081.012,79
Reciclados industriales de colombia s.a.s	Construction sector and materials and aggregates subsector	9003609541	X		Investments in environmental control which will be part of the plant for the use of waste from construction and demolition – RCD of the company Reciclados Industriales de Colombia S.A.S.	1320 of 2020	Article 255 et discount on rent	$ 1.807.000.000,00	$ 451.750.000,00	$ 93.182,76
American lighting s.a.s.	Industrial sector and electrical subsector	806010696		X	High efficiency LED luminaires and photocontrols, intended for the development of the "facatativá public lighting modernization project".	2020053993 2020	Article 255 et discount on rent	$ 9.968.000.000,00	$ 2.492.000.000,00	$ 514.026,40
Aseo internacional s.a. e.s.p.					Elements and equipment for the collection, transport and compaction of solid waste with final disposal in the new mondoñedo landfill and plastic containers for the correct segregation and separation of solid waste in the municipality of Soacha - Cundinamarca.	2020110169 2020	Article 255 et discount on rent		$ -	$ -
Servigenerales s.a. e.s.p.	Service sector and cleaning and cleaning subsector	830024104			Equipment and elements for solid waste management in the municipality of Facatativá - Cundinamarca.	2020094181 2020	Article 255 et discount on rent	$ 20.384.000.000,00	$ 5.096.000.000,00	$ 1.051.155,12

continues on following page

Table 3. Continued

Name of the business organization	Economic sector to which it belongs	Commercial document - NIT	Green business model		Process by which you have a discount on rent	Environmental certification to access the tax benefit -	Regulatory basis	Value to which a discount of 25% is made in rent according to E.T (millions of pesos $ cop)	Value in discount in rent - millions of pesos $ cop.	Discount value on rent - € euros
			Yes	No						
Emgesa s.a e.s.p.	Services sector and electricity subsector	860063875		X	Elements and equipment for the implementation of the second stage of the BEPP (Best Environment Protection Practices) industrial reconversion project in the thermoelectric plant termozipa, located in the municipality of Tocancipá - Cundinamarca.	2020082525 2020	Article 255 et discount on rent	$ 705.080.310,00	$ 176.270.077,50	$ 36.359,34
Distriaromas de colombia s.a.s	Industrial sector and chemical and fragrance subsector	830018274		X	One (1) 100% electric vehicle BYD brand to attend the transport and sales operations of the company Distriaromas de Colombia S.A.S. in the areas of Bogotá, Soacha, Cota and Funza in the department of Cundinamarca.	2020110581 2020	Article 255 et discount on rent	$ 1.523.000.000,00	$ 380.750.000,00	$ 78.537,54
Constructora conconcreto s.a.	Construction sector and materials and aggregates subsector	890901110		X	Production plant of asphalt mixtures, destined to carry out the use of residual construction materials located in the municipality of Nilo - Cundinamarca.	1866 of 2020	Article 255 et discount on rent	$ 50.055.000,00	$ 12.513.750,00	$ 2.581,22
Reciclene s.a.s.	Industrial sector and eco-industrial subsector	800226901	X		One (1) functional unit of recycling and use of plastics to be installed in the plant of the company Reciclene S.A.S., in the municipality of Tocancipá - Cundinamarca.	2021033433 2021	Article 255 et discount on rent	$ 16.080.000,00	$ 4.020.000,00	$ 829,21
Pelpak s.a.	Industrial sector and plastics subsector	860070605		X	A functional unit for the process of using recycled plastic material (high density polyethylene) to be installed in the Pelpak S.A. plant in the municipality of Tocancipá - Cundinamarca.	2021143443 2021	Article 255 et discount on rent	$ 11.497.000,00	$ 2.874.250,00	$ 592,87
Coliseo live s.a.s.	Industrial sector and civil works subsector	901306573		X	Equipment for the development of the energy efficiency project: “Intramural thermal district - Coliseo Arena Bogotá”, located in the municipality of Cota - Cundinamarca.	2021229793 2021	Article 255 et discount on rent	-$ 2.963.000,00	-$ 740.750,00	-$ 152,79

continues on following page

Table 3. Continued

Name of the business organization	Economic sector to which it belongs	Commercial document - NIT	Green business model		Process by which you have a discount on rent	Environmental certification to access the tax benefit -	Regulatory basis	Value to which a discount of 25% is made in rent according to E.T (millions of pesos $ cop)	Value in discount in rent - millions of pesos $ cop.	Discount value on rent - € euros
			Yes	No						
Coliseo live s.a.s.	Industrial sector and civil works subsector	901306573		X	Equipment for the development of the energy efficiency project: "LED lighting system and control system - Coliseo Arena Bogotá" located in the municipality of Cota - Cundinamarca	2021251374 2021	Article 255 et discount on rent	-$ 2.963.000,00	-$ 740.750,00	-$ 152,79
Coliseo live s.a.s.	Industrial sector and civil works subsector	901306573		X	Equipment for the development of the energy efficiency project: "Electric motors and variable speed drives – Coliseo Arena Bogotá", located in the municipality of Cota - Cundinamarca	2021263381 2021	Article 255 et discount on rent	-$ 2.963.000,00	-$ 740.750,00	-$ 152,79
Coliseo live s.a.s.	Industrial sector and civil works subsector	901306573		X	Equipment for the development of the energy efficiency project: "coliseo arena bogotá" located in the municipality of cota - cundinamarca.	2021284753 2021	Article 255 et discount on rent	-$ 2.963.000,00	-$ 740.750,00	-$ 152,79
Aguas de la sabana de bogota s.a. e.s.p.	Services sector subsector	900021737		X	Equipment for the operation and expansion of the wastewater treatment plant of the company Aguas de la Sabana de Bogotá S.A. E.S.P., located in the municipality of Cota – Cundinamarca.	2021087692 2021	Article 255 et discount on rent	$ 10,271,000.00	$ 2,567,750.00	$ 529.65

Source: Authors.

Group of companies benefited in the income tax for accreditation of investments in control, conservation and improvement of the environment based on article 255 of the Tax Statute of Colombia and resolution 509 of April 3, 2018: the data represented here relate the years from 2018 to 2022.

That is why Colombia's natural treasures integrate the system of national protected areas; coastal erosion management; reducing deforestation and monitoring it; protecting water sources; Protection of strategic ecosystems, especially moors. To this are added the tax benefits associated with climate change measures and their respective monitoring, reporting and verification, and payment for environmental services, as well as the implementation of all agreements to end armed conflicts and establish a stable and lasting peace that meets the standards of environmental sustainability carried out by companies in Colombia. According to Rodríguez Cely (2015):

The deductions of income generated by environmental investments for the control and improvement of the environment are contemplated in article 158-2 of the ET Tax Statute; It specifies that legal entities that make direct investments in environmental control and improvement will have the right to deduct annually from their income the value of said investments made in the respective fiscal year, after accreditation by the respective environmental authority, which will take into account the direct environmental benefits associated with said investments. (page 32)

One of the greatest benefits of tax exemptions is the ease of growth of economic activity and the promotion of the development of specific regions, which seeks in the first place a starting point of environmental protection that makes its structure fairer, this is precisely what green and sustainable organizations categorized in Colombia do. It is important to mention that these companies, apart from generating positive impacts on the environment, contribute to the internalization of negative external environmental factors, being companies with high corporate responsibility. According to (Ortega, et al., 2000), cited by Rodríguez Cely (2015) Colombian tax incentives are appropriate for them, the objectives are established as long as the uncollected resources are collected efficiently, in relation to investments in the environmental field. However, these appropriate environmental incentives must be reflected through transparency and minimal transaction costs. The application of tax incentives for environmental protection contributes to the reduction of polluting missions and decisions by economic operators on strategies to reduce environmental impact. Polluters may decide to reduce their costs and incentives for investment and participation in new environmental technologies.

Consequently, this type of incentive stimulates the consumption and production of environmentally friendly goods, encourages investment in certain industries or activities, and facilitates international certifications when organizations seek international markets. Although not everything is one hundred percent pleasant, one of the main disadvantages of environmental incentives is that there is no high level of certainty about their effectiveness in reducing environmental impact, since it is the same organizations that guarantee the impact on the generation of gases.greenhouse effect; That is why the regional corporations in Colombia, which are the guarantor entities to verify this type of actions that are responsible for confirming environmental investment in favor of friendly practices with the country's natural resources, have a great responsibility for this. Figure 2 describes the characteristics of tax incentives and how they apply to Colombian consumption.

Figure 2. Characteristics of tax incentives for green companies in Colombia
Source: Authors.

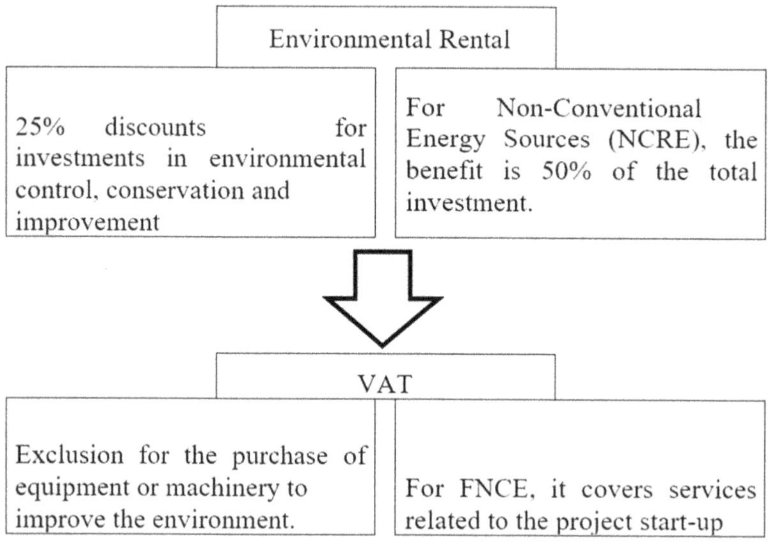

Disclosure of Tax Incentives for Green Businesses in Colombia

The green business model is expanding throughout the Colombian geography, this is happening for different reasons; one of them is that this way of doing business with sustainable practices, is becoming a good alternative for the growth and development of the country, so the Ministry of Environment and Sustainable Development created the Green Business Office, a commitment of the National Government that is part of a strategy "to promote the sustainable use of biodiversity in competitive and sustainable production processes linked to green business product categories and at the same time compete in a national and international market that demands sustainable products". (Ministry of Science, Technology and Innovation, 2017). However, regulations have not yet been established for the dissemination of information for the needs of its stakeholders, however, through the promulgation of the document "National Green Business Plan", issued by the Ministry of Environment and Sustainable Development in 2014, verification mechanisms have been indicated for the criteria that characterize green companies (see Table 4).

Compliance covering tax benefits is verified by the office in charge of validation, this unit is known as the regional green business window, which can take into account company statements, test certificates or analysis of materials or products; it must be environmental and/or social in nature and receive a national or international seal of approval or an environmental mark; the technical concept of the authority or body competent on the particular subject that is. Presenting; and any other supporting documentation to demonstrate compliance. (Ministry of Environment and Sustainable Development, 2014, p. 50). Currently, companies that are considered green in Colombia do not have to disclose information about their sustainability, except for the verification mentioned above, and there is no evidence that these institutions have adopted this practice to publish their reports, considering that this business model will be extended in the near future, considering that one of the main objectives of the new government of Colombia is to reduce environmental impact., support green enterprises, increase the tax incentive base for environmentally friendly companies and support the creation of favorable environmental management scenarios.

Table 4. Criteria for the identification of green companies

FEATURE	DESCRIPTION
Economy viabhim of the business	A company is economically viable when the resources generated by its sales are sufficient to meet the financial needs that arise and the remuneration of its employees and/or the distribution of profits among its partners or shareholders.
Positive environmental impact of the good or service	The good or service must demonstrate that it generates a positive impact on the environment and that it contributes to the conservation and preservation of ecosystem services (provisioning, regulation, support, cultural) in its Area of influence.
Life cycle approach of the good or service	A Green Business must ensure the identification of the main aspects and impacts associated with the life cycle of the good or service, ensuring the analysis of all stages: extraction and processing of raw materials, production, transport and distribution, use, maintenance and final disposal. Once the impacts are known, Green Business must work to reduce them, directly or indirectly, through proper management with suppliers and users.
Life	Those goods with a longer shelf life than conventional products represent a lower demand for resources to manufacture new products with the same characteristics and Reduce the generation of waste or residues when disposing of such goods.
No use of hazardous substances or materials	This criterion refers to goods or services that have traditionally used substances or materials that are hazardous to health and the environment and that through changes in their design or technology have achieved their replacement. There is currently a global trend in this direction and such substitutions imply a criterion of sustainability.
Recyclability of materials and/or use of recycled materials	Recyclability (of the product, packaging, packaging) is identified when, in the selection of materials, alternatives with properties superior to those of conventional inputs are considered in terms of their recovery and reincorporation into the production cycle at the end of their useful life. Similarly This criterion also considers those goods that use recycled materials for their manufacture.
Efficient and sustainable use of resources to produce the good or service	This criterion is related to the manufacturing stage of the product in which all measures are taken to use natural resources (water, energy, raw materials) in the least quantity and in the most sustainable way possible, achieving a rational and efficient production process. Related to this criterion are the use of cleaner technologies in industries or the sustainable management of ecosystems from which hydrobiological resources, flora or fauna originate.
Corporate Social Responsibility	It applies to companies that, for the production and provision of their goods or services, consider ethical treatment and fair working conditions for their direct employees and shareholders.
Social and environmental responsibility in the company's value chain	This criterion occurs in those companies that, for the production and provision of their goods or services, consider promoting the development of their value chain and share good practices of sustainability and corporate social responsibility with their suppliers, *outsourced services* and customers.
Social and environmental responsibility outside the company	This occurs in companies that, for the production and provision of their goods or services, manage their impacts in areas of external influence: communities, State, consumers and society in general.
Social or environmental communication attributes associated with the good or service.	The communication strategies of the social or environmental attributes of a Green Business good or service must be consistent with the management that the company develops to guarantee sustainability and must involve the different actors in the life cycle. It is therefore of the utmost importance that the attribute highlighting these aspects is accurate and not misleading and that truthful support is ensured. when they are verified by consumers or any actor present in the chain, in order to protect their rights.
Environmental or social schemes, programs or recognitions implemented or received.	Companies can demonstrate the implementation of sustainability-oriented models or programs (with or without certification), as well as other recognitions received from third parties, such as verifications, mentions, prizes or endorsements.

Source: Own elaboration based on the existing information in the National Plan for Green Enterprises, (Ministry of Environment and Sustainable Development, 2014).

RECOMMENDATIONS

Sustainability assessments and the analysis carried out within them are not only a discreet activity, but also very influential as a business model. Therefore, analyzing the impacts on the market, seen from the volumes of consumption and profitability generated, will allow in the future to adjust the benefits and tax incentives for companies that develop environmental conservation practices.

The emergence of a profitable economy is related to the environment, the payment of ecological debt, the rational use of natural resources and investment in the reduction of greenhouse gases that cause global warming, are alternatives that are necessary today and that will contribute to the sustainability of planet earth. These are the main characteristics of the development of green business models in Colombia. However, this investment has existed for many years in the development of policies that allow innovation in business sustainability. "Through the National Development Plan (PND) 2014-2018, Colombia uses the Green Growth (GC) approach as one of the pillars on which it bases its socioeconomic development" (Ministry of National Planning, 2016, p. 4). Based on the economic strategy established by Law 1753 approved by the Congress of the Republic of Colombia. The government has carried out productive, social and environmental actions that have contributed to the development of these business models. However, we know that in addition to regulated efforts, these efforts are more beneficial to entrepreneurs based on different and unique ideas who engage in ventures that gain greater exposure in international and consumer markets over time.

In Colombia, as in many Latin American countries, green growth and production must guarantee not only stable incomes and volumes, but also be guarantors of the long-term economic, social and environmental well-being of the population. "In countries that have undertaken reform processes aimed at green growth, the most common challenges have been found in measures such as pricing or the implementation of market-based instruments to reflect the true social cost of pollution" (OECD 2015) cited by (National Planning Department, 2016. p.9). As we have seen, the path of the process of sustainable development in Colombia is an arduous, harmonious and extensive work, in which broad proposals must be sought in favor of the economic and commercial growth of the national territory. The green model is no longer a utopia, but a model based on the internal wisdom of the Colombian regions, where the needs of the market exceed the world and in which it is not ashamed of its ecoproductive system. Therefore, the promotion of different sustainability-oriented approaches was a mandatory decision and meets the requirements of environmental protection. For Qin, Harrison & Chen (2019) cited by Martinez, et al. (2021) "In this sense, environmental responsibility is conceptualized as a business practice based on green management to obtain potential benefits, including product differentiation and reputation improvement." (page 534)

Regarding the specific forecasts of eco-enterprises and the results of the sustainability assessments presented in this document, it should be noted that economies based on such organizations tend to increase the level of productivity and competitiveness of regions and societies. This will also bring greater financial, economic and fiscal benefits for organizations wishing to enter the system of circular production and protection of natural resources. Table 5 shows some of the non-economic benefits that are also generated by the implementation of this type of business strategies.

Table 5. Non-economic benefits of green business models in Colombia

IMPACT	DESCRIPTION
Resilient infrastructure	This Be avoid blockade Countries in an energy dependency based on fossil fuels or Emissions intensive. Way.
Poverty reduction	Depending on the design of VC policies and accompanying measures, results will vary. This will affect how the The benefits and costs of development are distributed.
Natural assets in good condition	Based on this, human growth and well-being can be built to provide the inputs of tradable goods as well as the ecosystem services on which quality of life depends.
Safer livelihoods for those who depend on the sustainable management of natural resources such as agricultural land, fisheries and forests.	It will curb the extension of the agricultural frontier, improve production methods, improve access to drinking water, increase air quality, reduce rural poverty, among others.
Greater access to drinking water and sanitation services, diversity in energy sources and greater energy security.	It will generate the reduction of polluting and GHG emissions, improvements in terms of health and productivity.
Reducing vulnerability to climate Change and natural disasters	Social environmental and Economic Disasters are avoidable.

Fuente: Elaboración propia basada en OCDE - Putting Green Growth at the heart of development (2013).

FUTURE TRENDS IN RESEARCH

The projection of green businesses is one of the actions that all governments should draw, if what is sought is to mitigate the effects of global warming. This set of measures effectively promotes the development of business models in the protection of the environment. A change in the concept of production of goods and services, where the focus should be on green growth rather than economic growth; from capital- and labor-intensive production to innovation-based production, energy-dependent production, to energy self-sufficiency in renewable energy; This leads to national productivity being put into the equation "doing more with less" is a challenge faced by young entrepreneurs and startups.

The role of financial institutions, fiscal regulators and other economic agents associated with fiscal forecasts, economic uncertainty and the depletion of natural resources will undoubtedly be a challenge for any government focused on the development of a purely green economy. The processes of innovation, research and social development are important factors for the green model to grow exponentially, but as long as it continues to generate productive predictions in the short term. On the other hand, if this type of strategy is not supported in relation to the extractive model in a circular model, any effort made to mitigate greenhouse gases will be little. In Colombia, different strategies can be implemented to predict the profitability and sustainability of green SME models. The Green Growth Potential Assessment (GGPA) method can be used for this purpose. This approach allows the analysis and evaluation of the potential of green products in markets with economic growth trends. In addition, this methodology developed by the Global Green Growth Institute (GGGI) identifies key factors that lead to the development of optimal strategies to strengthen green models.

CONCLUSION

Beyond linear production and its traditional economic backwardness, sustainable and circular production processes are not only gaining momentum in business and academia, but also allow business strategies to focus their efforts on communicating with customers and/or consumers of goods and services. For this, processes of research, development, science and technology, sustainability measurement, market and consumer behavior, innovation, remanufacturing, operations management, big data and recycling are used. These are variables of economic and productive trend that contribute to the sustainable development of small and medium-sized enterprises. The circular economy and green companies have an economic power that comes from negotiating with customers, power over suppliers and the ability to identify alternative products, factors that generate fiscal and tax benefits provided by governments, essential axes for the business model to last over time.

However, the implementation of sustainable economic models tends to directly affect markets, consumption patterns and the impact of greenhouse gas production. On the other hand, the value activities that these companies carry out in their productive activities allow them to develop a clear strategy for their business and environment. Companies, as a differentiator not only in the production of goods and services, but also as an articulator in consumption, must use marketing as a differentiating factor that allows green companies to expand in sales and profitability. This is why green SMEs, if they want to have a greater financial impact not only through sales but also through fiscal and tax benefits, must develop campaigns that strengthen the attractiveness of the territory, and this plays an important role in the development of marketing plans and programs developed by each city, Municipality or department. The creation of differentiated value seeks that the services or products of the territory provide consumers, tourists or investors with recreational, entertainment, cultural, commercial spaces and a different and friendly image that represents the territory and its society. Likewise, the trade cooperation strategies developed in the region should be oriented towards the formulation of sustainable projects in the region to promote participation and openness to such projects.

Finally, it can be concluded that, when approaching and accessing the financial sector, the green financing system -green bonds-, the process of evaluating costs and benefits of the different interventions carried out by organizations, the distribution of costs and benefits among the different actors and the gross profit or investment; are one of the most important actions when establishing the viability of green models in Colombia. However, the evaluation and comparison of rates in the sale of goods and services is one of the least used green business activities, and the need to strengthen these organizations in these areas, even more so when looking at the financial and economic level, where sustainability is considered a strategic axis to reduce operating costs and increase profitability.

The author thanks in this chapter to the sustainable organizations in Colombia, their importance and the need to strengthen them, as well as to the 69% of the green SMEs that have been participating in the research processes to the more than 50 students of the Business Administration Program of the Fundación Universidad Compensar, colleagues among others for the support and participation in the academic impact that has been had in the business and financial analysis. that green companies in Colombia have had.

REFERENCES

Antón, Á. A. (2022). Review of the EU energy taxation framework in the context of the Green Deal. Towards a green, efficient, socially just and inclusive economic transition. *International Visual Culture Review*, *1*(2), 2–13. doi:10.37467/revvisual.v9.3670

Antonio-Paulo, U. A. (2022). *"4R" sustainability and quality teaching using Free Software*. University of Zaragoza. doi:10.37467/revtechno.v11.3794

Baker, R., & Green, J. (2019). The role of financial incentives in promoting green businesses. *Journal of Sustainable Business*, *15*(3), 187–202.

Bancoldex. (2018). *First Green Bonds Report*. Bancoldex - Banco De Comercio Exterior De Colombia, Bogotá, Colombia. https://www.bancoldex.com/sites/default/files/11097_green_bonds_report_-_bancoldex_2018_copy.pdf

Barragán Terán, D., & Muñoz Ávila, L. (2018). Forest governance in Colombia and Ecuador. *Journal of Environmental Law*, (09), 93–117. doi:10.5354/0719-4633.2018.49747

Brown, M., Smith, K., & Johnson, L. (2021). Government support for green businesses: A comparative analysis. *Environmental Policy Review*, *27*(2), 45–60.

Castellani, V., & Sala, S. (2012). Ecological Footprint and Life Cycle Analysis in the sustainability assessment of tourism activities. *Ecological Indicators*, *16*, 135–147. doi:10.1016/j.ecolind.2011.08.002

Chen, H., & Wang, Y. (2018). The impact of tax incentives on green businesses: Evidence from a panel data analysis. *Journal of Environmental Economics*, *42*(1), 78–92.

Corson, C., MacDonald, K. I., & Neimark, B. (2013). Grasping "green": Markets, environmental governance and the materialization of natural capital. *Human Geographies*, *6*(1), 1–15. doi:10.1177/194277861300600101

D'Atri, A. M. (2017). Keucheyan, Razmig. (2016). Nature is a battlefield. Finance, ecological crisis and new green wars. *Yearbook of the Faculty of Human Sciences*, *13*(13), 140. doi:10.19137/an1310

N. De Sadeleer. (2022). Harmonisation of law, internal market and environment: the challenge of implementing the Green Deal. https://doi.org/ doi:10.20318/cdt.2022.6681

Dvarskas, A. (2017). Dynamically linking economic models to ecological condition for coastal zone management: Application to sustainable tourism planning. *Journal of Environmental Management*, *188*, 163–172. doi:10.1016/j.jenvman.2016.12.014 PMID:27978443

ECLAC. (2020). *MSMEs and COVID-19*. ECLAC. https://www.cepal.org/es/euromipyme/mipymes-covid-19

Chacón, E. (2022). Mypes and sustainability in times of pandemic in Arequipa, Peru A structural equation model. *Visual Review*. doi:10.37467/revvisual.v9.3624

García, A. (2022). Private investment in green businesses: Trends and drivers. *Journal of Sustainable Finance*, *18*(4), 223–240.

Fede Organicos. (2008). *Green and sustainable business in Colombia.* FO. http://www.fedeorganicos.com/wp-content/uploads/2016/07/los-negocios-verdes-y-sostenibles-en-colombia-minambiente.pdf

Gutman, V., & López, A. (2017). Green production and eco-innovation. *Eco-innovation and green production: a review of policies in Latin America and the Caribbean.* Santiago: ECLAC. https://repositorio.cepal.org/bitstream/handle/11362/41748/S1700072_es.pdf?sequence=1&isAllowed=y

Inter-American Development Bank. (2019). *Nationally determined contributions in Colombia.* Inter-American Development Bank.

Jaramillo, C., & Patiño, A. (2020). Industrial engineering and the international green market. *Journal of Engineering Sciences of Athenea, 1*(2), 37–40. https://athenea.autanabooks.com/index.php/revista/article/view/11/22

Jones, P. (2019). Financial instruments for supporting green businesses: A systematic review. *Environmental Finance Review, 25*(1), 56–72.

Kim, S., & Park, J. (2021). The effects of tax incentives on green business performance: Evidence from a quasi-experimental design. *Journal of Sustainable Development, 37*(4), 123–137.

Lee, C. (2017). Tax policy for promoting green businesses: Lessons from international experience. *Green Economy Journal, 12*(2), 89–105.

Liu, Q. (2022). Evaluating the effectiveness of financial and tax incentives for green businesses: A meta-analysis. *Journal of Sustainable Business, 18*(1), 35–50.

Martinez, M., & López, R. (2020). Carbon pricing and emissions trading: Implications for green businesses. *Environmental Economics Review, 28*(3), 152–168.

Martínez-Rubio, K., Delgado-Cruz, A., & Vargas-Martínez, E. E. (2021). Adoption of green technologies and their influence on environmental responsibility practices. Perceptions of hotel workers. *Management Studies, 37*(161), 532–541. doi:10.18046/j.estger.2021.161.4071

Ministry of Environment and Sustainable Development. (2021). *Report on green taxes in Colombia.* MESD.

Ministry of Environment. (2017). *Portfolio of goods and services of Green Business.* MoE. http://www.minambiente.gov.co/index.php/negocios-verdes-y-sostenibles/negocios-verdes/portafolio-de-bienes-y-servicios-de-negocios-verdes#portafolio-de-negocios

Ministry of the Environment. (2010). *Colombian green businesses.* MoE. http://www.minambiente.gov.co/index.php/negocios-verdes-y-sostenibles/negocios-verdes

National Planning Department. (2020). *Green growth policy.* NPD. https://colaboracion.dnp.gov.co/CDT/Prensa/Presentaci%C3%B3n%20Pol%C3%ADtica%20de%20Crecimiento%20Verde.pdf

Organization for Economic Cooperation and Development. (2011). *Towards Green Growth A summary for policymakers.* Council at OECD ministerial level. https://www.oecd.org/greengrowth/49709364.pdf

Organization for Economic Cooperation and Development. (2013). *Putting green growth at the heart of development.* OECD Publications.

Pertuz, V., Francisco Miranda, L., & Sánchez Buitrago, J. O. (2021). Towards a conceptual understanding of green entrepreneurship. *Revista Venezolana de Gerencia*, *26*(94), 745–761. doi:10.52080/rvgv26n94.23

Ramos Díaz, J. (2016). Green Economy and Employment: The labour potential of the "Ecological Transition" in Spain. *Notebooks of Labour Relations*, *34*(2). Advance online publication. doi:10.5209/CRLA.53436

Rua-Usme, D., Ospina, A., Ramírez, C., Ortiz, P., & Patiño-Murillo, J. A. (2019). Scenario for the development of the line of green financial products in the Colombian economy. *CINTEX Magazine*, *24*(2), 64–90. doi:10.33131/24222208.352

Segrelles, J. A. (2009). A reflection on the unsustainability of tourism activities in rural and natural areas. The cases of ecotourism and deep ecology*. *Human Geographies*, *2*(1), 1–21. doi:10.1177/194277860900200109

Smith, D. (2018). Government policies for supporting green businesses: An international perspective. *Journal of Sustainable Development*, *35*(2), 87–102.

UNCTAD BioTrade Initiative. (2007). Biotrade principles and criteria. United Nations Conference on Trade and Development. *BioTrade Initiative* (pp. 6-14). Geneva, Switzerland: UNCTAD. http://www.asocam.org/sites/default/files/publicaciones/files/2807acffecd17b29a5d3ee02ca4395a5.pdf

Saber Más Ser Más. (2021). *What are green financing lines?* ASOBANCARIA. https://www.sabermassermas.com/que-son-lineas-de-financiamiento-verde/

World Bank. (2017) *Green bonds*. World Bank. https://www.bancomundial.org/es/results/2017/12/01/green-bonds

ADDITIONAL READING

Aparicio, A. (2022). *The road ahead: narratives and imaginaries of the value of biodiversity in shaping bioeconomy policy in Colombia, Tapuya*. Latin American Science, Technology and Society. doi:10.1080/25729861.2022.2059137

Casillas-González, A. C., Pérez-Camacho, M. N., García-López, M., Cátedra-Cerón, M. M., Capote, C., Martín-Jiménez, I., & Villanueva, A. J. (2023). Circular bioeconomy strategies. A comparative analysis at European level. *ITEA-Agricultural Economic Technical Information*, *118*(4), 613–630. doi:10.12706/itea.2022.016

Paşnicu, D., Ghenţa, M., & Matei, A. (2019). Transition to the bioeconomy: Perceptions and behaviours in Central and Eastern Europe. *Amfiteatru Economic*, *21*(50), 9–23. doi:10.24818/EA/2019/50/9

Pavone, V., & Goven, J. (Eds.). (2017). *Bioeconomies: life, technology and capital in the 21st century.* Springer International., doi:10.1007/978-3-319-55651-2

Petersen, A., & Krisjansen, I. (2015). Assembling 'the Bioeconomy': Harnessing the Power of Promising Life Sciences. *Journal of Sociology (Melbourne, Vic.)*, *51*(1), 28–46. doi:10.1177/1440783314562314

Radice, M., Scalvenzi, L., & Gutiérrez, D. (2020). Ethnopharmacology, bioactivity and phytochemistry of Maxillaria densa Lindl. Scientific journal and Biotrading in the Neotropics. *Forestry Colombia.*, *23*(2), 20–33. doi:10.14483/2256201X.15924

Reardon, S. (2018). FARC and the Forest: Peace Is Destroying Colombia's Jungle and Opening It to Science. *Nature*, *558*(7709), 169–170. doi:10.1038/d41586-018-05397-2 PMID:29895926

Staffas, L., Gustavsson, M., & McCormick, K. (2013). Strategies and policies for the bioeconomy and the bioeconomy: An analysis of official national approaches. Sustainability, [e-journal] 5(6), pp.2751-2769. doi:10.3390/su5062751

Staffas, L., Gustavsson, M., & McCormick, K. (2013). Strategies and policies for the bioeconomy and the bioeconomy: An analysis of official national approaches. *Sustainability, [e-journal]* *5*(6), 2751–2769. doi:10.3390/su5062751

Stern, T., Ploll, U., Spies, R., Schwarzbauer, P., Hesser, F., & Ranacher, L. (2018). Understanding perceptions of the bioeconomy in Austria: an exploratory case study. *Sustainability, [e-journal], 10(*11), pp.1-17. . doi:10.3390/su10114142

KEY TERMS AND DEFINITIONS

Agro-Industrial Food: It has to do with the transformation and processing of products from ecological, organic and biological production systems, or biodiversity products, intended for human consumption (Ministry of the Environment, 2016).

Direct Environmental Benefits: These investments are directly related to environmental control; direct environmental benefits are understood as the set of measurable and verifiable results that are achieved with the implementation of an environmental control system. (Regional Autonomy Corporation of Cundinamarca, 2021).

Ecolabeling: It is a mechanism that allows differentiation, through a distinctive (seal or label), those products that are characterized by being less polluting or have been manufactured making sustainable use of natural resources, from products with similar characteristics, but that do not contemplate environmental criteria. (Ministry of Environment, 2016).

Ecosystem Services: Ecosystem services are the direct and indirect benefits that humanity receives from biodiversity and are the result of the interaction between the different components, structures and functions that constitute biodiversity. Ecosystem services have been recognized as the bridge between biodiversity and humans. (ONVS, 2014).

Green Businesses: To economic activities in which goods or services are offered, which generate positive environmental impact and incorporate good environmental, social and economic practices applied in a systemic manner, with a life cycle approach, which contribute to ensuring the conservation of the environment as natural capital that supports the development of the territory. (Ministry of Environment, 2016).

Investment Certificate for Environmental Control and Improvement: It is the administrative act issued by the environmental authorities where it is established that investments in environmental control and improvement comply with the criteria and requirements established in Decree 2205 of 2017 that regulates said article. (Regional Autonomy Corporation of Cundinamarca, 2021)

Investments in Environmental Control: Wings oriented to the implementation of environmental control systems, aimed at achieving measurable and verifiable results in the reduction of the demand for renewable natural resources, or prevention and / or reduction in the generation and / or improvement of the quality of liquid waste, atmospheric emissions or solid waste (Regional Autonomy Corporation of Cundinamarca, 2021).

Investments in Environmental Improvement: They are necessary to develop processes aimed at the restoration, regeneration, repopulation, preservation and conservation of renewable natural resources and the environment. (Regional Autonomy Corporation of Cundinamarca, 2021).

Nature Tourism: That economic activity "whose offer of products and services is developed around a natural attraction and is governed by principles of sustainability". This modality includes the by-products of ecotourism, rural tourism and adventure tourism (Ministry of Environment, 2016).

Positive Environmental Impact: Conservation, Reduction of pressure on the resource, Repopulation and maintenance of the natural base, Change of non-renewable materials to renewables, Reduction of pollution, Improvement of natural resource conditions, Maintenance of native biodiversity, Maintenance of ecosystem services, Reduction of greenhouse gas emissions, Change from non-renewable to renewable energy sources, Environmental education and culture, Respect for traditional knowledge and friendly cultural practices (ONVS, 2014).

Products and By-Products Derived From Wildlife Populations: Products derived from the direct extraction of natural populations of wildlife, under management and extraction measures that allow the maintenance and conservation of the species and its habitat (Ministry of the Environment, 20 16).

Products and By-Products Derived From Wildlife Production Systems: These are products derived from wildlife production systems, in a closed cycle and without affecting natural populations (Ministry of the Environment, 20 to 16).

Value Chain: In the context of the BioTrade Initiative and the BioTrade Facilitation Program, the strengthening of value chains is used as a mechanism to facilitate the articulation between actors in a production chain; for the implementation of good practices related to the sustainable use and conservation of biodiversity; and for the equitable distribution of environmental means, social and economic benefits among participants (UNCTAD, 2007).

Chapter 2
Carbon Tax:
Characterization of Research

Paulo Jorge Dias
https://orcid.org/0000-0002-0351-9751
Instituto Universitário de Lisboa (ISCTE-IUL), Portugal

Andreia Rocha
Instituto Universitário de Lisboa (ISCTE-IUL), Portugal

ABSTRACT

One of the main climatic concerns is the increase of CO2 concentrations in the atmosphere. In the attempt and need to combat this problem, several regulatory policies have been applied in several countries by the governments, with the carbon tax being one of the most economical and effective policies in reducing CO2 emissions, thus leading to an increase in scientific research on this topic. Therefore, the main objective of this study is characterizing the existing research on the theme Carbon Tax. A sample of 121 scientific articles was collected, where several characteristics related to the content of published articles and respective authors were subsequently identified. The results obtained reveal that the number of publications has been increasing, and the themes most related to this investigation are: cap-and-trade, carbon dioxide emissions, climate change, computable general equilibrium (cge) model and emissions. Regarding the profile of researchers, the type of collective authorship is the most common and the country with the most affiliated authors is China.

INTRODUCTION

Over the past few years, extreme weather events and significant weather anomalies have often been observed, such as high temperatures, high precipitation, and strong winds.

The discussion about climate change is not a recent topic, but it is a problem that has damaged the economy worldwide, affecting the lives of communities. Associated with this theme is global warming caused by Greenhouse Gas (GHG) emissions from human activities, with carbon dioxide (CO2) being

DOI: 10.4018/978-1-6684-8592-7.ch002

the gas that presents the highest levels of emissions into the atmosphere. Among the determining factors of these emissions are the burning of fossil fuels, industrial processes, livestock, and waste treatments.

Thus, to counter the negative impacts that CO2 emissions have on ecosystems, the functioning of socio-economic systems and human health and well-being, the need has arisen to implement government policies to reduce such emissions into the atmosphere. Currently, the two main policies used are the carbon tax and emissions trading, also known as "cap-and-trade".

Since pre-industrial times, humanity has caused a great increase in GHG emissions, with CO2 being the main gas generated and the one that contributes the most to global warming.

The high concentrations of CO2 cause air pollution, acid rain, increased temperature on Earth, resulting in major environmental degradations at the level of ecosystems and landscapes. Thus, to reduce CO2 emissions it is necessary not only the privilege of renewable energy sources, but also the adoption of stricter government policies to achieve this goal as soon as possible. As the carbon tax is one of these policies that has been shown to be quite effective in reducing CO2 emissions, it has caused scientific research on this topic to increase over the years, thus contributing to a strong source of data.

Within the framework of the United Nations Convention, urgent action is one of the main sustainable development goals to combat this problem and its impacts, thus giving rise to the adoption of an instrument for global action, the Kyoto Protocol, which entered into force in February 2005 and the Paris Agreement, approved in November 2016 as an urgent need to end the rise in global temperature and solve all challenges related to climate change. This agreement stands out for the possibility of reducing CO2 emissions so that the global average temperature does not exceed 2ºC.

The main objective of this work is to characterize the existing research on the theme "Carbon Tax" through the review of articles published in the Scopus database between 1995 and mid-2020, to verify the trends of publications of scientific articles on the carbon tax in management and accounting over 25 years. It is also intended to analyze a set of bibliometric indicators referring to the characteristics of the content of the published articles and the profile and characteristics of the researchers.

The evaluation of scientific production is fundamental not only for academic students to be able to follow the relevant literature related to a specific theme, but also for the recognition of researchers and educational institutions, being based on bibliometric indicators.

Thus, the present study presents a scientific relevance since, when the trends of publications related to carbon taxation are identified, it serves as support to researchers in future research.

LITERATURE REVIEW

In December 1997, the signing of the Kyoto Protocol[1] under the United Nations Framework Convention on Climate Change had as its fundamental objective to encourage developed countries to limit their GHG emissions. The global plan to reduce carbon emissions began with the establishment of this Protocol, stipulating that more than 30 countries were required to reduce their major GHG emissions by at least 5.2% in the period between 2008 and 2012 compared to 1990 levels (Hu et al., 2020; Zhou et al., 2018; Yang et al., 2017).

Likewise, the implementation of the Paris Agreement[2] adopted in December 2015, represented an important advance at the international level seeking to give a global and effective response to the urgent need to prevent the global average temperature from continuously rising and to act on the challenges related to climate change (APA, 2020). As a result, several developed and developing countries have committed themselves to reducing their carbon emissions.

Thus, increased concerns over the past few years about environmental protection, due to severe global climate change, have put pressure on countries to move towards a low-carbon economy and sustainable development (Yu et al., 2020; Chen and Hu, 2018; Meng et al., 2018; Zhuo and Zhang, 2017).

A low-carbon economy advocates technological and institutional innovation to achieve the goal of reducing the consumption of fossil fuels such as coal and oil. According to Xu et al. (2020: 1), "the essence of developing a low carbon economy is to change the way energy is consumed", therefore, it becomes necessary for future economic development the existence of a low-carbon economy, with lower energy consumption and pollution (Xu et al., 2020; Wang et al., 2017).

According to Haites (2018), the main policies that have been applied in many countries, separately or together, to control carbon emissions and fight climate change are the Carbon Tax and the Emissions Trading Scheme (ETS), also called "cap-and-trade" " (Zhao et al., 2020; Zhang and Zhang, 2020; Yu et al., 2020; Datta et al., 2020; Saelim, 2019; Ndhaief et al., 2019; Lee, 2019; Zhou and Zhang, 2017; Yang et al., 2017; Xu et al., 2016; Pillay and Buys, 201 5, 2013; Hou et al., 2015; Zakeri et al., 2015; Dwyer et al., 201 3, 2012; Andrew et al., 2010). Both policies are among the most effective tools for reducing carbon emissions (Zhao et al., 2020; Mishra et al., 2020; Shi et al., 2019; Lin and Xu, 2019; Kalita and Barua, 2019; Meng et al., 2018; Xu et al. 2016; Hou et al., 2015; Williams et al., 2014).

Several empirical studies have focused on the economic and environmental impacts of these two emissions policy regimes through different methods and research cases, however, they are all in tune with their implementation, i.e. it allows to reduce CO2 emissions but will impede economic development (Zhang and Zhang, 2020; Haites, 2018).

The most important carbon market in operation is the European Union's Emissions Trading System – EU ETS, launched in 2005, which operates through the cap-and-trade system. It is a flexible mechanism developed by the EU, foreseen in the context of the Kyoto Protocol, and was the first intra-Community market instrument for regulating GHG emissions (Ndhaief et al., 2019; Dwyer et al., 2013), allowing participating companies the ability to buy and sell emission allowances. That is, a fixed quantity of allowances is allocated free of charge to each of the EU Member States, on which the installations covered must not exceed this limit, however, if there is a difference between the allocated quantity and the verified emissions, the installations will be able to acquire or sell allowances to other companies within the EU (Mishra et al., 2020; Shuang et al., 2019; Lee, 2019; Yu et al., 2018; Xu et al., 2016; Pillay and Buys, 2015; Jeffrey and Perkins, 2015; Pillay and Buys, 2013; Dwyer et al., 2012).

The "cap-and-trade" system generates pressures on companies, in that it imposes significant fines on them for excessive pollution, but it also creates incentives, for the financial reward derived from the sale of surplus licenses, with the aim of encouraging them to have appropriate environmental initiatives (Zakeri et al., 2015). Thus, companies end up reducing their carbon emissions not only as a social responsibility, but also for economic benefits (Zhou and Zhang, 2017).

This emissions trading scheme has been introduced by many countries, such as Norway, Finland, Switzerland, Denmark, Sweden, Italy, Slovenia, the UK, the USA, Canada, among others (Mishra et al., 2020; Hu et al., 2020; Andrew et al., 2010).

The carbon tax policy consists of a form of explicit carbon pricing directly related to the level of CO2 emissions, usually expressed per ton of CO2 equivalent (tCO2e), derived from the burning of fossil fuels, creating incentives for emitters to switch to less GHG-intensive production in order to protect the environment and mitigate global warming (Shi et al., 2019; Lin and Xu, 2019; Lee, 2019; Yu et al.,

2018; Zhou and Zhang, 2017; Kortum and Weisbach, 2017; Kuo et al., 2016; Liang et al., 2016; Xu et al., 2016; Liu et al., 2015; Pillay and Buys, 2015; Zakeri et al., 2015; Jeffrey and Perkins, 2015; Nordhaus, 2010; Zhang and Baranzini, 2004).

Nordhaus (1992) has shown that it is possible to slow global warming with an adequate level of carbon tax (Hu et al., 2020), as it could lead carbon-intensive companies to change their long-term business strategies by altering their production process to reduce their emissions. Otherwise, if the tax level is inadequate, there is a risk that companies will choose to pay this tax rather than try to reduce carbon emissions (Kuo et al., 2016).

Many researchers argue that tax policy is easier to implement compared to cap-and-trade, given that it can rely on administrative energy tax structures and does not require the establishment of an emissions trading market (Saelim, 2019; Park et al., 2015; Pillay and Buys, 2013).

This incentive-based policy instrument (Hou et al., 2015) is advocated as one of the most popular policies to reduce carbon emissions in the atmosphere and has often been recognized as economical, thus attracting more and more wide-ranging academic discussions related to this topic (Liang et al., 2016).

The first author to mention the importance of creating a tax per unit of measurement of pollution was Arthur Cecil Pigou (1920) to combat the negative effects of economic activities on the environment (Jeffrey and Perkins, 2015), in a model he called the "Pigouvian Tax".

Carbon taxation is thus a market-based policy mechanism (Saelim, 2019; Yu et al., 2018; Huo et al., 2015) used by many countries to meet various objectives, including environmental protection and revenue raising (Jeffrey and Perkins, 2015).

From an economic perspective, the carbon tax is an example of a "Pigouvian Tax", which has been considered as an important means to encourage consumers to change their behaviour towards a more efficient use of natural resources in order to reduce carbon emissions and consequently the effects of global warming (Apleni et al., 2017; Liang et al., 2016; van Heerden et al., 2016; Zakeri et al., 2015; Jeffrey and Perkins, 2015; Dwyer et al., 2013).

Therefore, to reduce the burden of this tax, companies can reduce their emissions by investing in "green" technology at the level of their supply chain, for example in the production process, storage, and transport (Datta et al., 2020; Yu et al., 2018; van Heerden et al., 2016; Metcalf and Weisbach, 2009).

According to Hou et al. (2015), the carbon tax can be positive, when it takes the form of punishment for companies that harm the environment, or negative, when it is presented as a subsidy for companies that care about environmental sustainability.

Kalita and Barua (2019) state that imposing a price on carbon would be crucial to decarbonizing the environment, with the Social Cost of Carbon (SCC) approach being the most popular method in determining the optimal price that should be applied. The SCC must be equal to or less than the revenues generated by the implementation of the carbon tax for this policy instrument to be viable (Pillay and Buys, 2013).

The carbon tax policy thus entails several benefits, namely in the increase of state revenue, in the disincentive to the use of fossil fuels that consequently leads to the fall of CO2 emissions, and in the promotion of a more environmental industry (Jin et al., 2018) contributing to a decrease in air pollution and leading in turn to improvements in social welfare.

On the other hand, it has differential implications for households and industries (Metcalf, 2014). There are many studies that explore the environmental and economic impacts of a carbon tax on important macroeconomic variables, such as GDP, production, and employment, using one of the main analytical tools considered powerful and highly credible, the[3] Computable General Equilibrium (CGE) model) (Zhang and Zhang, 2020b; Li and Masui, 2019; Liang et al., 2016; Koks et al., 2016; Dwyer et al., 2013).

This model is descended from another widely used by the scientific community, the Input-Output (I-O) model,[4] often praised for its simplicity and ability to analyze cost-price economic interdependencies between sectors and regions (Chang and Han, 2020; Koks et al., 2016).

Most of these studies find that such a policy has high economic costs, reductions in GDP, consumption, increases in energy costs, and additional investments that become necessary to produce cleaner energy (Zhu et al., 2020b; Peng et al., 2017; Dwyer et al., 2013, 2012).

For example, Li and Masui (2019) studied the socio-economic and environmental impacts of carbon tax policy on China's electricity sector concluding that the carbon tax caused a GDP loss of between 0.1% and 0.7%, with energy-intensive sectors being more impacted than the agricultural and service sectors which, on the contrary, they suffer a slight increase in production. In addition, the tax pushes the electricity sector to transition to "green" technologies, electricity from coal and natural gas will gradually transform into other clean technologies to produce electricity. As stated by Zuo et al. (2016: 1484) "CO 2 emission reduction and carbon taxes can both help to reduce CO2 emissions and increase diversification of the power generation structure to include a greater share of renewables".

The carbon tax policy is, however, widely criticized for its potential regressiveness (Williams et al., 2015), being more harmful to low-income populations, since it requires them to pay a high percentage, generating a disproportionate tax burden (Williams et al., 2014; Gale et al., 2013). In addition, taxpayers have the perception that the carbon tax functions simply as a tax scheme to raise revenue which distorts the desired effect of the policy as an incentive to non-pollution.

The impact of a carbon tax is also reflected in the decrease in companies' profits, since it influences their cost structure, thus being a factor that can compromise organizational competitiveness (Jin et al., 2018; Gray and Metcalf, 2017; Liang et al., 2016; Hovelaque and Bironneau, 2015; Baranzini et al., 2000).

However, a socially responsible organization is an essential aspect not only of the corporate environment and to ensure long-term sustainable development but also to attract more investors, because, according to Church et al. (2019), this type of companies attracts all types of investors, both those who care only about financial performance and those who give importance to corporate social responsibility, however, the latter tend to invest more. Therefore, an organization must be financially secure as well as reduce undesirable environmental impacts and act according to the expectations of all stakeholders, harmonizing economic development, environmental quality, and social well-being (Pillay and Buys, 2015)

Zhou and Zhang (2017), Hovelaque and Bironneau (2015) argue that it is possible to mitigate CO2 emissions through operational adjustments, such as changes in batch sizes or order quantities, and that it can be an alternative or complement to more expensive investments in carbon reduction technologies.

According to Costello et al. (2019), research and development (R&D) enables significant efficiency gains for a company, since it can encourage them to commercialize new economic technologies with the ability to reduce new carbon emissions, place carbon harmlessly underground or remove it from the atmosphere. An example of this is Carbon Capture and Storage (CCS)[5] which has gained great attention as a good and effective technology for decarbonizing the electricity sector and other industries (Moioli and Pellegrini, 2020).

Thus, although the implementation of a carbon tax influences corporate profits, according to Porter's hypothesis, it promotes the innovation capacity of a country or sector of activity, thus leading to an increase in competitiveness (Ma et al., 2020).

METHODOLOGY

To achieve the objectives of this study, the methodology that will be applied to the data obtained will be bibliometrics, which is one of the main tools to evaluate scientific production.

Thus, the present research proceeded to the collection of articles published in several scientific journals on the theme "Carbon Tax" until August 31, 2020, using the digital repository of contents of the Scopus database, having established the following research criteria: all documents had to present the keyword "Carbon Tax ", be of the scientific article type and belong to the area of study Business, Management and Accounting, having obtained a total of 137 scientific articles. We also withdrew 16 articles because they were unavailable, thus reaching a total sample of 121 scientific articles eligible for this study.

Table 1 shows all articles eligible for the study, as well as the author(s) and respective year of publication.

Table 1. Articles eligible for the study, divided by title, author(s) and year of publication

Article title	Author(s)	Year of publication
Carbon emissions reduction and financial effects of a cap and tax system on an operating supply chain in the cement sector	Cadavid-Giraldo, N., Velez-Gallego, M.C., Restrepo-Boland, A.	2020
Should firms provide online return service for remanufactured products?	Cao, K., Xu, Y., Wang, J.	2020
Rethinking the choice of carbon tax and carbon trading in China	Jia, Z., Lin, B.	2020
Research on a single policy or policy mix in carbon emissions reduction	Zhao, L., Yang, C., Su, B., Zeng, S.	2020
Examining the economic and environmental effects of emissions policies in China: A Bayesian DSGE model	Zhang, J., Zhang, Y.	2020
Optimal carbon reduction level and ordering quantity under financial constraints	Cao, K., Xu, B., He, Y., Xu, Q.	2020
Integrated optimization of railway freight operation, planning and pricing based on carbon emission reduction policies;	Li, L., Zhang, X.	2020
Environmental issue in an integrated production and maintenance control of unreliable manufacturing/remanufacturing systems	Ndhaief, N., Nidhal, R., Hajji, A., Bistorin, O.	2020
Carbon tax and sustainable facility location: The role of production technology	Gaigné, C., Hovelaque, V., Mechouar, Y.	2020
Collaborative optimal carbon tax rate under economic and energy price shocks: A dynamic stochastic general equilibrium model approach	Chan, Y.T.	2020
A sustainable production-inventory model for a controllable carbon emissions rate under shortages	Mishra, U., Wu, J.-Z., Sarkar, B.	2020
Fixed and Capture Level Reduction operating modes for carbon dioxide removal in a Natural Gas Combined Cycle power plant	Moioli, S., Pellegrini, L.A.	2020
Chinese tourism economic change under carbon tax scenarios	Zhang, J., Zhang, Y.	2020
The short-term roles of sectors during a carbon tax on Chinese economy based on complex network: An in-process analysis	Ma, N., Li, H., Wang, Y., Zhang, J., Li, Z., Arif, A.	2020
Impacts of horizontal integration on social welfare under the interaction of carbon tax and green subsidies	Xu, C., Wang, C., Huang, R.	2020
Collaboration in a low-carbon supply chain with reference emission and cost learning effects: Cost sharing versus revenue sharing strategies	Yu, B., Wang, J., Lu, X., Yang, H.	2020
Cost-push impact of taxing carbon in China: A price transmission perspective	Chang, N., Han, C.	2020
Challenges and Development Perspectives of the Tax System	Niemann, R., Schreiber, U.	2020
The Role of Universities in Private Environmental Governance Experimentalism	Light, S.E.	2020
A hybrid carbon policy inventory model with emission sourcebased green investments	Datta, T.K., Nath, P., Dutta Choudhury, K.	2020

continues on following page

Table 1. Continued

Article title	Author(s)	Year of publication
Green financial behavior and green development strategy of Chinese power companies in the context of carbon tax	Zhu, N., Bu, Y., Jin, M., Mbroh, N.	2020
Carbon tax or cap-and-trade: Which is more viable for Chinese remanufacturing industry?	Hu, X., Yang, Z., Sun, J., Zhang, Y.	2020
A simulation study of China's imposing carbon tax against American carbon tariffs	Zhu, N., Qian, L., Jiang, D., Mbroh, N.	2020
Optimal trade-in and warranty period strategies for new and remanufactured products under carbon tax policy	Cao, K., Xu, B., Wang, J.	2020
Evolutionary macroeconomic assessment of employment and innovation impacts of climate policy packages	Rengs, B., Scholz-Wäckerle, M., van den Bergh, J.	2020
Flexible versus simple trade-in strategy for remanufacturing	Yang, Z., Hu, X., Sun, J., Zhang, Y.	2020
How to set the proper level of carbon tax in the context of Chinese construction sector? A CGE analysis	Shi, Q., Ren, H., Cai, W., Gao, J.	2019
Essay on climate apocalypse and a carbon tax	Costello, K.W.	2019
The Role of Universities in Private Environmental Governance Experimentalism	Light, S.E.	2019
Green scheduling model of shuttle tanker fleet considering carbon tax and variable speed factor	Xin, X., Wang, X., Tian, X., Chen, Z., Chen, K.	2019
Carbon tax incidence on household demand: Effects on welfare, income inequality and poverty incidence in Thailand	Saelim, S.	2019
Exploring the green total factor productivity of China's metallurgical industry under carbon tax: A perspective on factor substitution	Lin, B., Xu, M.	2019
Determining a Carbon Tax Rate for India in the Context of Global Climate Change	Kalita, U., Barua, N.A.	2019
Techno-economic optimization and social costs assessment of microgrid-conventional grid integration using genetic algorithm and Artificial Neural Networks: A case study for two US cities	Nagapurkar, P., Smith, J.D.	2019
A stochastic reverse logistics production routing model with emissions control policy selection	Shuang, Y., Diabat, A., Liao, Y.	2019
Operational and environmental decisions for a two-stage supply chain under vendor managed consignment inventory partnership	Hariga, M., Babekian, S., Bahroun, Z.	2019
Service and pricing strategies in online retailing under carbon emission regulation	Ding, Y., Jin, M.	2019
Optimizing a sustainable logistics problem in a renewable energy network using a genetic algorithm	Sadeghi, J., Haapala, K.R.	2019
Economic and environmental models for cold products with varying time demand	Shamayleh, A., Hariga, M., As'ad, R., Diabat, A.	2019
Will carbon tax affect the strategy and performance of low-carbon technology sharing between enterprises?	Wang, M., Li, Y., Li, M., Shi, W., Quan, S.	2019
Optimal Climate Policy and the Future of World Economic Development	Budolfson, M., Dennig, F., Fleurbaey, M., Scovronick, N., Siebert, A., Spears, D., Wagner, F.	2019
Assessing the impacts of China's environmental tax using a dynamic computable general equilibrium model	Li, G., Masui, T.	2019
Production decision model with carbon tax for the knitted footwear industry under activity-based costing	Tsai, W.-H., Jhong, S.-Y.	2019
Production and pricing decisions in a dual-channel supply chain under remanufacturing subsidy policy and carbon tax policy	Cao, K., He, P., Liu, Z.	2019
Towards a sustainable and inclusive low-carbon economy: Why carbon taxes, and not schemes of emission trading, are a cost-effective economic instrument to curb greenhouse gas emissions	Ionescu, L.	2019
Investing in carbon emissions reduction in the EOQ model	Lee, J.-Y.	2019

continues on following page

Table 1. Continued

Article title	Author(s)	Year of publication
Using evolutionary game theory to study governments and manufacturers' behavioral strategies under various carbon taxes and subsidies	Chen, W., Hu, Z.-H.	2018
A carbon-constrained EOQ model with uncertain demand for remanufactured products	Liao, H., Deng, Q.	2018
A fuzzy goal programme with carbon tax policy for Brownfield Tyre remanufacturing strategic supply chain planning	Saxena, L.K., Jain, P.K., Sharma, A.K.	2018
Coordinated Pricing Analysis with the Carbon Tax Scheme in a Supply Chain	Ma, X., Ho, W., Ji, P., Talluri, S.	2018
Shelf space allocation problem under carbon tax and emission trading policies	Yu, V.F., Maglasang, R., Tsao, Y.-C.	2018
Pricing decisions and social welfare in a supply chain with multiple competing retailers and carbon tax policy	Zhou, Y., Hu, F., Zhou, Z.	2018
The unpaid social cost of carbon Introducing a framework to estimate "legal looting" in the fossil fuel industry	Linnenluecke, M., Smith, T., Whaley, R.E.	2018
Retailer-driven carbon emission abatement with consumer environmental awareness and carbon tax: Revenue-sharing versus Cost-sharing	Yang, H., Chen, W.	2018
Low-carbon product selection with carbon tax and competition: Effects of the power structure	Meng, X., Yao, Z., Nie, J., Zhao, Y., Li, Z.	2018
Service competition in an online duopoly market	Ding, Y., Gao, X., Huang, C., Shu, J., Yang, D.	2018
The effect of governmental policies of carbon taxes and energy-saving subsidies on enterprise decisions in a two-echelon supply chain	Yuyin, Y., Jinxi, L.	2018
Carbon tax, tourism CO2 emissions and economic welfare	Zhang, J., Zhang, Y.	2018
Determining the optimal carbon tax rate based on data envelopment analysis	Jin, M., Shi, X., Emrouznejad, A., Yang, F.	2018
Make or buy? It is the question: A study in the presence of carbon tax	Meng, X., Yao, Z., Nie, J., Zhao, Y.	2018
Assigning firm-level GHGE reductions based on national goals - Mathematical model & empirical evidence	Jackson, J., Belkhir, L.	2018
The cost of policy uncertainty in electric sector capacity planning: Implications for instrument choice	Leibowicz, B.D.	2018
Carbon pricing in the Northeast: Looking through a legal lens	Milne, J.E.	2017
Integration and consolidation in air freight shipment planning: An economic and environmental perspective	Zhou, G., Zhang, Y.	2017
The role of revenue sharing and first-mover advantage in emission abatement with carbon tax and consumer environmental awareness	Yang, H., Luo, J., Wang, H.	2017
The impacts of carbon pricing on coastal megacities: A CGE analysis of Singapore	Li, Y., Su, B.	2017
Performance evaluation of climate policies in China: A study based on an integrated assessment model	Peng, P., Zhu, L., Fan, Y.	2017
Environmental taxation and mergers in oligopoly markets with product differentiation	Fikru, M.G., Gautier, L.	2017
Regulatory impact on voluntary climate change–related reporting by Australian government-owned corporations	Liu, Z., Abhayawansa, S., Jubb, C., Perera, L.	2017
The design of border adjustments for carbon prices	Kortum, S., Weisbach, D.	2017
Carbon tax competitiveness concerns: Assessing a best practices carbon credit	Gray, W.B., Metcalf, G.E.	2017
WTO law constraints on border tax adjustment and tax credit mechanisms to reduce the competitive effects of carbon taxes	Trachtman, J.P.	2017
An assessment of decarbonization in the strategic Indian electricity generation sector	Vedachalam, N., Surendar, S., Srinivasalu, S.	2017
Supply chain enterprise operations and government carbon tax decisions considering carbon emissions	Wang, C., Wang, W., Huang, R.	2017
Cumulative emissions, unburnable fossil fuel, and the optimal carbon tax	van der Ploeg, F., Rezai, A.	2017
Cutting carbon emissions from electricity generation	Procter, R.J.	2017

continues on following page

Table 1. Continued

Article title	Author(s)	Year of publication
The implications of the proposed carbon tax in South Africa's tourism industry	Apleni, L., Mxunyelwa, S., Vallabh, D.	2017
The impact of the Australian carbon tax on the tourism industry	Meng, S., Pham, T.	2017
Unintended possible consequences of fuel input taxes for individual investments in greenhouse gas mitigation technologies and the resulting emissions	Klingelhöfer, H.E.	2017
Do carbon taxes work? Analysis of government policies and enterprise strategies in equilibrium	Kuo, T.C., Hong, I.-H., Lin, S.C.	2016
Ambiguous tipping points	Lemoine, D., Traeger, C.P.	2016
Multi-objective optimization of the operating conditions in a cutting process based on low carbon emission costs	Liu, Z.-J., Sun, D.-P., Lin, C.-X., Zhao, X.-Q., Yang, Y.	2016
Low-carbon city logistics distribution network design with resource deployment	Yang, J., Guo, J., Ma, S.	2016
Australian climate concern and the 'attitude–behaviour gap'	Higham, J., Kings, A., Cohen, S.A.	2016
China's future energy mix and emissions reduction potential: a scenario analysis incorporating technological learning curves	Zou, H., Du, H., Broadstock, D.C., Guo, J., Gong, Y., Mao, G.	2016
Addressing the competitiveness effects of taxing carbon in China: domestic tax cuts versus border tax adjustments	Liang, Q.-M., Wang, T., Xue, M.-M.	2016
Joint production and pricing decisions for multiple products with cap-and-trade and carbon tax regulations	Xu, X., Xu, X., He, P.	2016
The economic and environmental effects of a carbon tax in South Africa: A dynamic CGE modelling approach	van Heerden, J., Blignaut, J., Bohlmann, H., Cartwright, A., Diederichs, N., Mander, M.	2016
Investigating the impact of carbon tax on socially responsible corporate governance: The case of south African motor vehicle manufacturers	Pillay, S., Buys, P.	2015
Carbon taxes and U.S. fiscal reform	McKibbin, W.J., Morris, A.C., Wilcoxen, P.J., Cai, Y.	2015
The initial incidence of a carbon tax across income groups	Williams, R.C., Gordon, H., Burtraw, D., Carbone, J.C., Morgenstern, R.D.	2015
Carbon taxes and fiscal reform in the United States	Jorgenson, D.W., Goettle, R.J., Ho, M.S., Wilcoxen, P.J.	2015
Low-carbon product multi-objective optimization design for meeting requirements of enterprise, user and government	Xu, Z.-Z., Wang, Y.-S., Teng, Z.-R., Zhong, C.-Q., Teng, H.-F.	2015
CO2 mitigation potential of CCS in China and an evaluation based on an integrated assessment model	Zhu, L., Duan, H.-B., Fan, Y.	2015
An analysis of company choice preference to carbon tax policy in China	Liu, X., Wang, C., Niu, D., Suk, S., Bao, C.	2015
Will export rebate policy be effective for CO2 emissions reduction in China? CEEPA-based analysis	Fan, J.-L., Liang, Q.-M., Wang, Q., Zhang, X., Wei, Y.-M.	2015
Optimal Decisions of Countries with Carbon Tax and Carbon Tariff	Hou, Y., Jia, M., Tian, X., Wei, F., Wei, K.	2015
The carbon-constrained EOQ model with carbon emission dependent demand	Hovelaque, V., Bironneau, L.	2015
Carbon emissions in a multi-echelon production-inventory model with lead time constraints	Hammami, R., Nouira, I., Frein, Y.	2015
Tactical supply chain planning under a carbon tax policy scheme: A case study	Fahimnia, B., Sarkis, J., Choudhary, A., Eshragh, A.	2015
Carbon pricing versus emissions trading: A supply chain planning perspective	Zakeri, A., Dehghanian, F., Fahimnia, B., Sarkis, J.	2015
A carbon market sensitive optimization model for integrated forward–reverse logistics	Choudhary, A., Sarkar, S., Settur, S., Tiwari, M.K	2015
System Dynamics Approach for the Impacts of FINEX Technology and Carbon Taxes on Steel Demand: Case Study of the POSCO	Jeong, S.-J.	2015
The association between energy taxation, participation in an emissions trading system, and the intensity of carbon dioxide emissions in the European Union	Jeffrey, C., Perkins, J.D.	2015

continues on following page

Table 1. Continued

Article title	Author(s)	Year of publication
Supply Chain Design and Carbon Penalty: Monopoly vs. Monopolistic Competition	Park, S.J., Cachon, G.P., Lai, G., Seshadri, S.	2015
Using the tax system to address competition issues with a carbon tax	Metcalf, G.E.	2014
The initial incidence of a carbon tax across U.S. States	Williams, R.C., Gordon, H., Burtraw, D., Carbone, J.C., Morgenstern, R.D.	2014
Carbon tax, corporate carbon profile and financial return	Luo, L., Tang, Q.	2014
Retail store density and the cost of greenhouse gas emissions	Cachon, G.P.	2014
Downscaling long term socio-economic scenarios at city scale: A case study on Paris	Viguié, V., Hallegatte, S., Rozenberg, J.	2014
Carbon tax pricing and the social cost of carbon: The case in the South African motor vehicle manufacturing industry	Pillay, S., Buys, P.W.	2013
Economic Impacts of a Carbon Tax on the Australian Tourism Industry	Dwyer, L., Forsyth, P., Spurr, R., Hoque, S.	2013
Forging cleaner production: the importance of academic-practitioner links for successful sustainability embedded carbon accounting	Burritt, R.L., Tingey-Holyoak, J.	2012
Wither Australian Tourism? Implications of the Carbon Tax	Dwyer, L., Forsyth, P., Spurr, R.	2012
The financials of constructing a solar PV for net-zero energy operations on college campuses	Kwan, C.L., Kwan, T.J.	2011
Carbon tax: Challenging neoliberal solutions to climate change	Andrew, J., Kaidonis, M.A., Andrew, B.	2010
Economic impact of the integration of alternative vehicle technologies into the New Zealand vehicle fleet	Leaver, J., Gillingham, K.	2010
Carbon taxes to move toward fiscal sustainability	Nordhaus, W.D.	2010
Public policies against global warming: a supply side approach	Sinn, H.-W.	2008
Tourists' Perception of International Air Travel's Impact on the Global Climate and Potential Climate Change Policies	Becken, S.	2007
The Political Economy of Implementing Environmental Taxes	Pearson, M.	1995

The indicators of scientific activity used in the investigation were the number and distribution of published works, the productivity of the authors, the collaboration in the authorship of the works and the geographical collaboration of the works.

The indicators of scientific impact are subdivided into indicators of the impact of the works in the scientific community and indicators of the impact of the sources of publication. As indicators of impact of the works, the number of citations received from the articles was analyzed, and as indicators of impact of the sources the influence of the scientific journals was evaluated as well as their impact factor using the CiteScore metric.

Finally, about the indicators of thematic associations, an analysis of the keywords of the scientific articles was carried out.

FINDINGS

This study comprises a time horizon of 25 years, with the first scientific article being published in 1995 and the most recent one being analyzed in 2020. Scientific publications on the subject "Carbon Tax" have been increasing over the years, growing on average 14% per year, and most of the articles were published in the last 5 years (105 articles).

Figure 1. Number and evolution of articles published per year

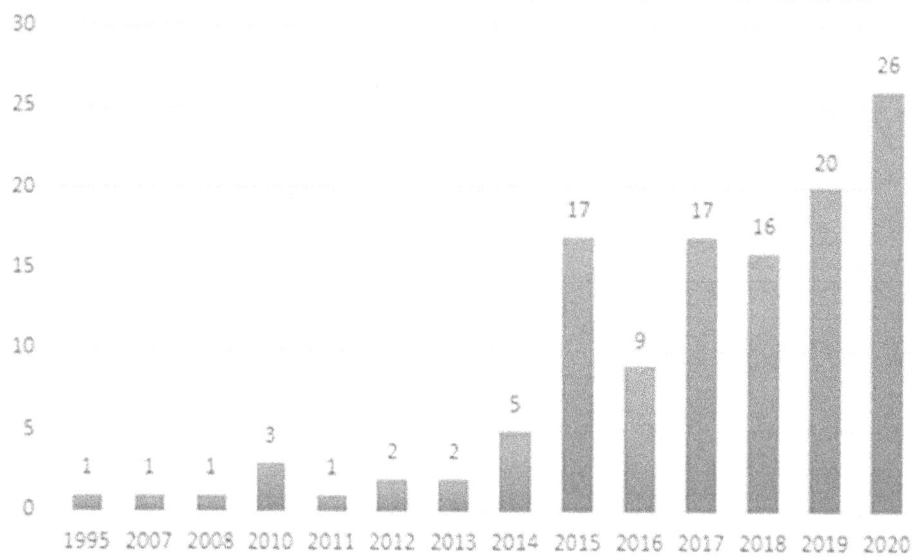

The largest increase in the number of scientific publications was in 2015 compared to 2014, in 12 articles, reaching 17 published articles. This can be explained by an important event that emerged in 2015, related to environmental concern, the approval of the Paris Agreement. However, the following year, in 2016, this number decreased significantly to a total of 9 scientific articles, thus breaking the positive trend in that same year. In addition to that year, in the years 2011 and 2018 there were also decreases in the number of articles published compared to previous years, but not significant, in 2 and 1 article, respectively.

The productivity of the authors is an indicator that allows us to understand how much more they contribute their knowledge to the scientific community on a given topic. In this study, it was found that 281 authors contributed to the total sample of 121 scientific publications, most of them (230 authors corresponding to 82% of the total) contributing only one scientific article and the remaining 51 (18%) with two or more articles (Figure 2).

It is possible to observe that Zhang, Y. is the author who contributes the most to the research on carbon taxation with 6 scientific articles, followed by Cao, K. and Zhang, J.with 4 articles and Jin, M., Wang, C., Wang, J. and Yang, H.with 3 articles. The remaining 44 and 230 authors contributed only 2 and 1 scientific articles, respectively, the latter accounting for more than half of the articles published per author (66.9%).

The most common type of authorship is collective, representing 84% of the sample. Within the collective authorship, most articles are written by 2 authors or by 4 or more authors (29.8%), followed by articles with 3 authors (24.8%). Regarding individual authorship, this has only a representation of 16% in the research in "Carbon Tax", with 19 scientific articles (Figure 3 and Table 3).

Figure 2. Productivity of the authors

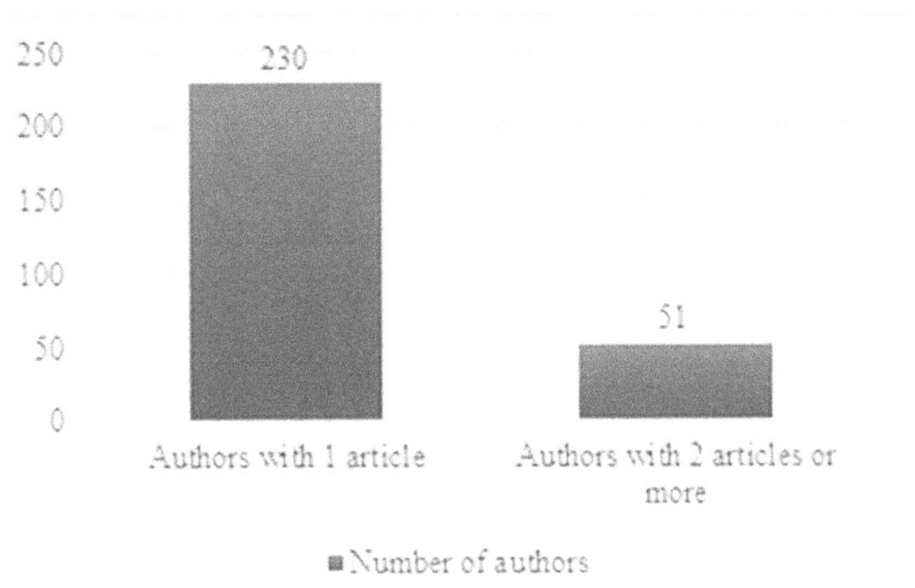

Figure 3. Type of authorship

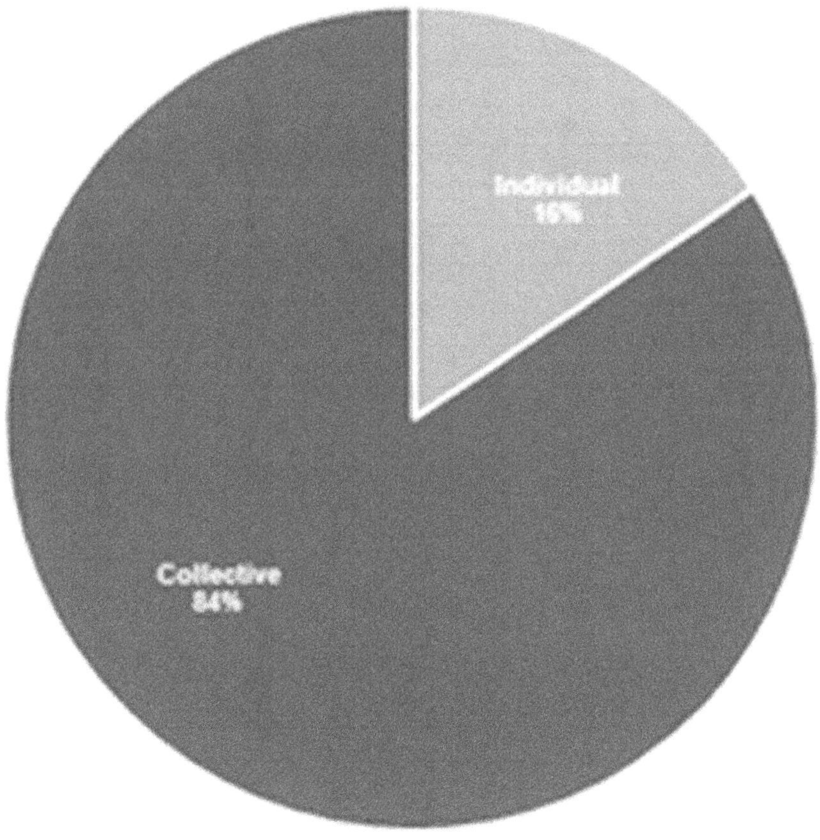

Table 2. Number of articles published per author

Author(s)	Number of articles	%		Author(s)	Number of articles	%
Zhang, Y.	6	1,74%		Li, Z.	2	0,58%
Dog, K.	4	1,16%		Liang, Q.-M.	2	0,58%
Zhang, J.	4	1,16%		Light, S.E.	2	0,58%
Jin, M.	3	0,87%		Lin, B.	2	0,58%
Wang, C.	3	0,87%		Liu, Z.	2	0,58%
Wang, J.	3	0,87%		Mbroh, N.	2	0,58%
Yang, H.	3	0,87%		Meng, X.	2	0,58%
Burtraw, D.	2	0,58%		Metcalf, G.E.	2	0,58%
Cachon, G.P.	2	0,58%		Morgenstern, R.D.	2	0,58%
Carbone, J.C.	2	0,58%		Nie, J.	2	0,58%
Chen, W.	2	0,58%		Pillay, S.	2	0,58%
Choudhary, A.	2	0,58%		Sarkis, J.	2	0,58%
Diabat, A.	2	0,58%		Spurr, R.	2	0,58%
Ding, Y.	2	0,58%		Su, B.	2	0,58%
Dwyer, L.	2	0,58%		Sun, J.	2	0,58%
Fahimnia, B.	2	0,58%		Tian, X.	2	0,58%
Fan, Y.	2	0,58%		Wilcoxen, P.J.	2	0,58%
Forsyth, P.	2	0,58%		Williams, R.C.	2	0,58%
Gordon, H.	2	0,58%		Xu, B.	2	0,58%
Guo, J.	2	0,58%		Yang, Z.	2	0,58%
Hariga, M.	2	0,58%		Yao, Z.	2	0,58%
He, P.	2	0,58%		Zhang, X.	2	0,58%
Hovelaque, V.	2	0,58%		Zhao, Y.	2	0,58%
Hu, X.	2	0,58%		Zhu, L.	2	0,58%
Huang, R.	2	0,58%		Zhu, N.	2	0,58%
Li, Y.	2	0,58%		Other authors	1	66,86%
TOTAL					344	100,00%

Table 3. Number of authors per article

Type of authorship	Number of articles	%
Articles with 1 author	19	15,70%
Articles with 2 authors	36	29,75%
Articles with 3 authors	30	24,79%
Articles with 4 authors or more	36	29,75%
TOTAL	121	100,00%

To understand the relationship between the authors, a bibliographic map or co-authorship network was constructed. Figure 4 shows the associations of the most productive authors, that is, who contribute to the research on carbon taxation with 3 or more scientific articles, through clusters. In this map, each circle represents a researcher, and its size is greater the more articles are published by that author, and the lines represent the link between the authors when they publish articles together.

Thus, it is verified that there are 4 clusters (4 different colors), the largest consisting of 3 authors (Cao, K., Wang, J. and Yang, H.), followed by the cluster with 2 authors (Zhang, Y. and Zhang, J.) and the remaining two consisting only of 1 author (Jin, M. and Wang, C.), meaning the latter who published articles individually or do not associate with any of the other authors covered in the bibliographic map.

As for Zhang, Y. and Zhang, J. it was possible to verify that they published 3 articles together, namely, the article "Carbon tax, tourism CO2 emissions and economic welfare" in 2018 and the articles "Chinese tourism economic change under carbon tax scenarios" and "Examining the economic and environmental effects of emissions policies in China: A Bayesian DSGE model" in 2020.

Figure 4. Co-Authorship associations of the most productive authors

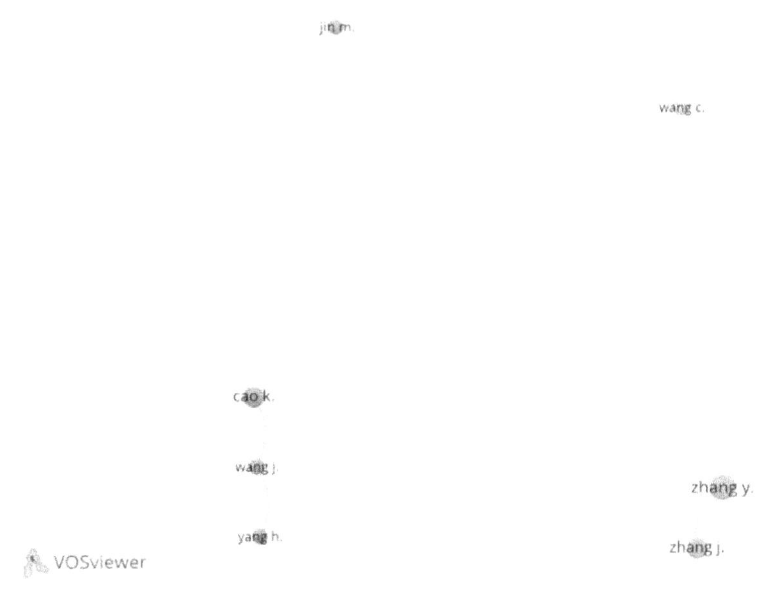

After performing the analysis of co-authorship of the most productive authors in the research in "Carbon Tax" to recognize their relationship in the execution and joint publication of scientific articles, it is also pertinent to understand that researchers are often cited by these same authors in their scientific publications, since they are equally relevant for investigations inserted in this theme.

Thus, the VOSviewer software was used again to construct a co-citation network map of the most cited authors in the "Carbon Tax" investigation, that is, with at least 24 citations (Figure 5). Co-citation occurs when two authors are cited simultaneously by a third author (Lima and Leocádio, 2017).

As evidenced by Figure 4, the 60 co-citation associations were identified among 13 authors grouped into 3 clusters. The largest cluster, represented by the red color, consists of 6 authors (Benjafaar, S., Chen, X., Xu, X., He, P., Li, Y. and Wang, S.) mostly focused on the following research topics: "Environmental Markets", "Pollution Tax" and "Trade Regulations". In this cluster it was possible to identify two co-authorship associations, one between Benjafaar, S. and Chen, X. in the publication of the article "The carbon-constrained EOQ" in 2013 and another between Xu, X. and He, P.with the article "Joint production and pricing decisions for multiple products with cap-and-trade and carbon tax regulations" in 2016.

Following is the green cluster composed of 5 authors (Wang, C., Lin, B., Metcalf, G.E., Zhang, X. and Dwyer, L.) whose topics of interest are "Pollution Tax", "Environmental Tax Reform" and "Double Dividend", and finally, the blue cluster, marked by 2 authors (Fahimnia, B. and Sarkis, J.) who jointly published in 2015 the articles "Carbon pricing versus emissions trading: A supply chain planning perspective" and "Tactical supply chain planning under a carbon tax policy scheme: A case study", focusing on "Green Supply Chain Management", "Environmentally Preferable Purchasing" and "Green Practices".

Figure 5. Co-Citation associations of the most cited authors

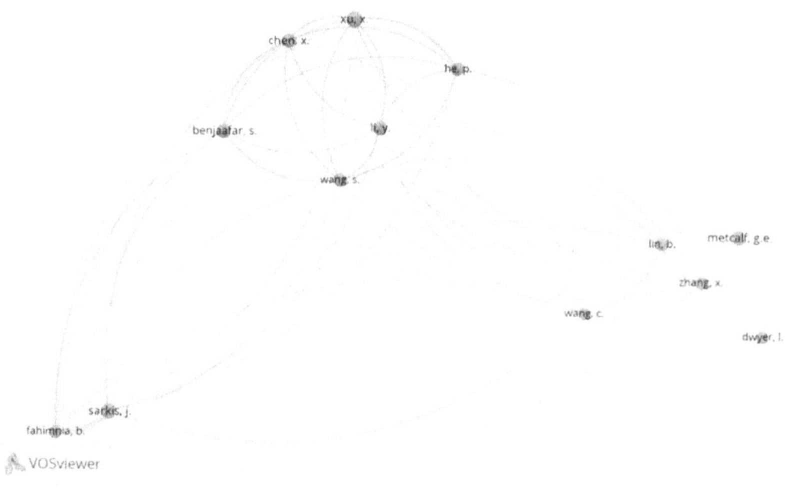

The affiliation of the authors was identified through the country of university ties and/or another institution to which the researcher is affiliated on the date the article was published. It should be noted that in this study there is 1 author who is affiliated with three countries (van der Ploeg, F.), 8 authors who are affiliated with two countries (Cai, W., Diabat, A., Hallegatte, S., Liu, Z., McKibbin, W.J. , Mishra, U. , Traeger, C.P. , van den Bergh, J.) and 3 authors who are not affiliated with any (Diederichs, N., Mander, M. and Pearson, M.). Thus, it is understood that the total number of affiliated authors (288 authors) does not correspond to the total number of authors who contribute to the research in "Carbon Tax" (281 authors).

Through Figure 6 it is observed that this study covers the five existing continents, with the Asian continent having a greater number of affiliated authors, with a representation of 55% (157 authors), more specifically in China with 112 affiliated authors (Table 4), followed by the American continent with 21% (60 authors). In contrast, it was found that Europe, Oceania and Africa have the lowest number of affiliated authors with a representation of only 11% (31 authors), 10% (29 authors) and 4% (11 authors), respectively. This predominance may be related to the fact that the Chinese government is considering the possible implementation of a carbon tax on CO2 emissions.

Figure 6. Authors' affiliation by continent

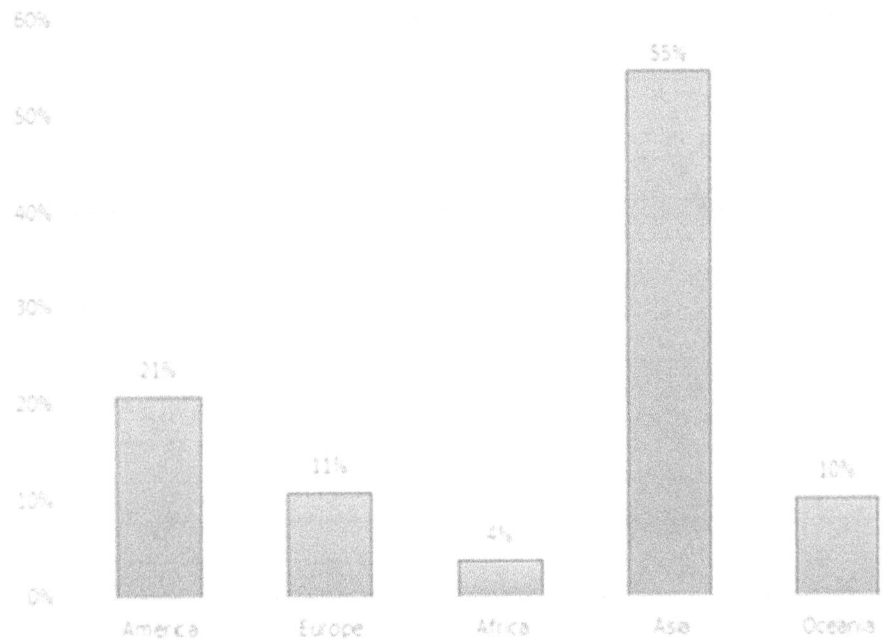

It is also verified that within America, the country that most published scientific articles were the USA, with a representation of 19% (53 affiliated authors). On the European continent, France stands out with 13 affiliated authors and in Oceania, Australia with 26 authors. On the other hand, in Africa there is no predominance of one country in relation to others, because the only country with affiliated authors on this continent was South Africa, with 11 authors. It should also be noted that Spain, Norway, Romania, Hong Kong, Russia, and Thailand are countries where there was only one affiliated author (0.4% of the total number of affiliated authors).

Just as the collaboration analysis was done in the authorship of the works, the same analysis can also be applied to the identified geographies to understand the level of communication between the countries (Liao et al., 2018) as well as the most influential countries in this research.

Thus, considering the continents and countries previously analyzed, a network map was constructed with the geographical collaborations of the countries with at least two published scientific articles (Figure 7). There was a total of 17 countries grouped into 7 clusters, with 32 geographical associations.

Table 4. Authors' affiliation by country

Continent/Country	Number of affiliated authors	%
America	60	20,83%
Canada	4	1,39%
United States of America	53	18,40%
Colombia	3	1,04%
Europe	31	10,76%
Germany	2	0,69%
Austria	5	1,74%
Spain	1	0,35%
France	13	4,51%
Holland	2	0,69%
Italy	2	0,69%
Norway	1	0,35%
United Kingdom	4	1,39%
Romania	1	0,35%
Africa	11	3,82%
South Africa	11	3,82%
Asia	157	54,51%
China	112	38,89%
South Korea	2	0,69%
United Arab Emirates	7	2,43%
Hong Kong	1	0,35%
India	15	5,21%
Iran	2	0,69%
Japan	4	1,39%
Russia	1	0,35%
Singapore	3	1,04%
Thailand	1	0,35%
Taiwan	9	3,13%
Oceania	29	10,07%
Australia	26	9,03%
New Zealand	3	1,04%
TOTAL	288	100,00%

It is noted that the USA is the country with the highest number of links with other countries regarding scientific publications related to the theme "Carbon Tax", with 9 associations, having the strongest relationship with China (link strength of 7, that is, 7 articles shared in a total of 21) and then with Australia (link strength of 5).

India is the second country with the most associations after the US, with 7 links (UAE, USA, Singapore, UK, Austria, South Korea, and Taiwan), followed by China with 6 links (Japan, Canada, USA, UAE, UK and Singapore) and Australia with 5 links (USA, New Zealand, Netherlands, Austria and UK). Most of the associations in India and half of the associations in China are with nearby countries belonging to the same continent (Asian continent) and Australia's links are mainly with European countries. However, they all have the strongest relationship with the US, which means that the advantage of geographical proximity between the countries is not an influencing factor in the collaborations for scientific production in "Carbon Tax".

Figure 7. Geographical associations

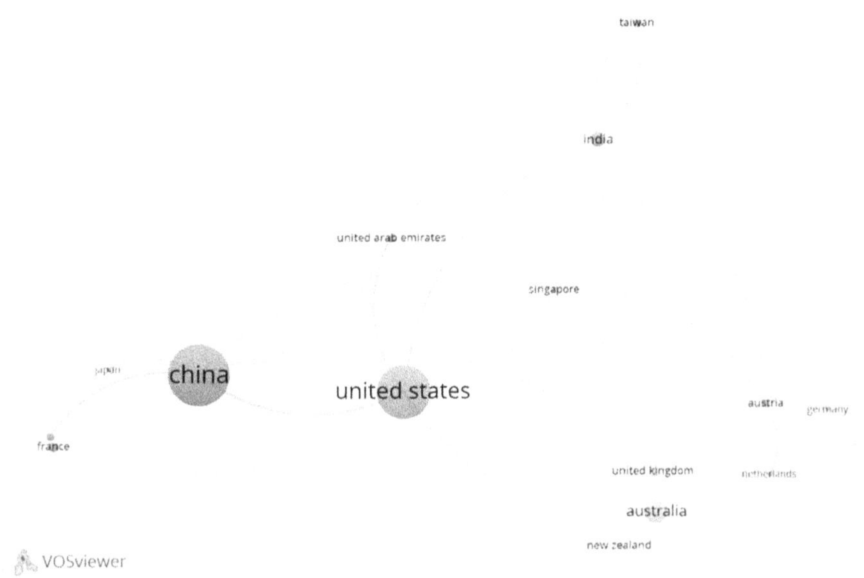

Scientific impact indicators are divided into indicators of the impact of work on the scientific community and the impact of publication sources (Costa et al., 2012). In this study, as an indicator of the impact of the works will be analyzed the number of citations received from the scientific articles referring to the H-Index of the study to complement this analysis, and as indicators of impact of the sources will be investigated the influence of scientific journals and their impact factor CiteScore.

Through Figure 8 it is possible to observe that the article that has the highest number of citations is the "Public policies against global warming: a supply side approach" published by Sinn, H. (2008) with 350 citations, followed by the article "Tourists' Perception of International Air Travel's Impact on the Global Climate and Potential Climate Change Policies" published by Becken, S. (2007), with 234 citations.

Thus, it is verified that the author with the highest number of citations (Sinn, H.) does not correspond to the most productive author (Zhang, Y) that was observed in point 4.1.2. regarding the productivity of the authors.

Figure 8. Map of citation densities per scientific article

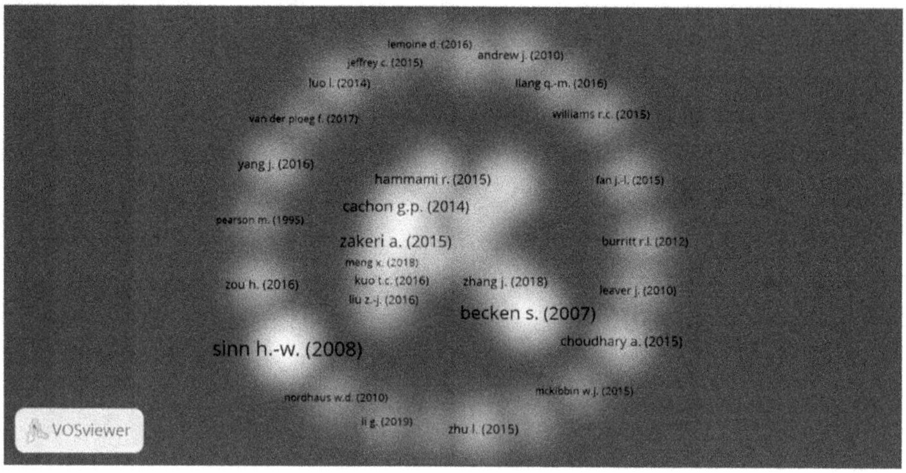

The H-Index is a metric first introduced by physicist Jorge Hirsch in 2005, aiming to measure the productivity and impact of a researcher, based on his most cited articles (Costa et al., 2012). This indicator is made available by the Scopus database and brings together the most cited articles of each researcher and the number of citations obtained. Applying to the set of scientific publications that constitute the sample of this study, it is concluded that it presents an H-Index of 27, that is, of the 121 scientific articles, 27 of these have at least 27 citations each (Figure 9).

Figure 9. H-Index of the scientific articles that constitute the study sample
Source: Scopus

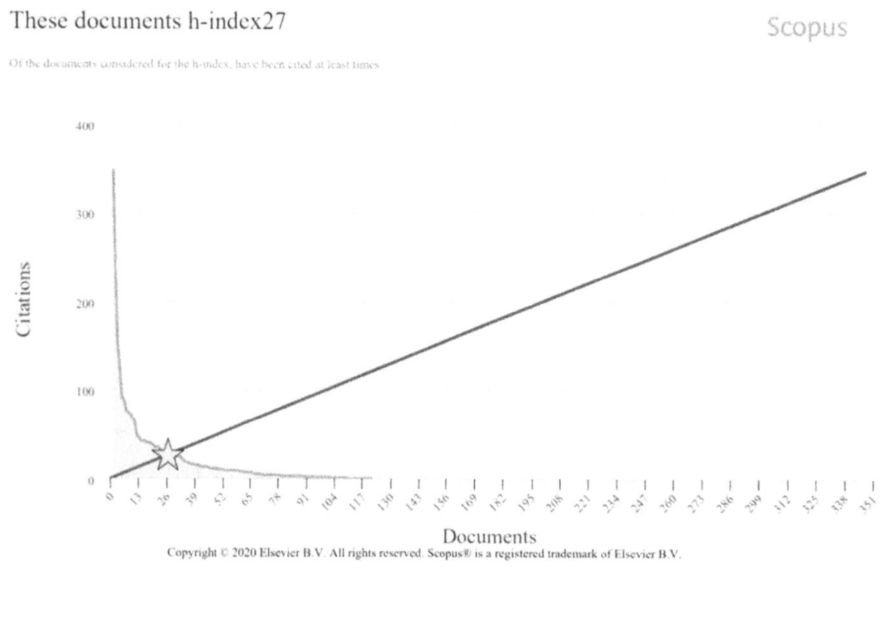

The analysis of Table 5 shows that the 121 articles that make up the sample of this study are published in 40 scientific journals. Within these, the journal that most published scientific articles on carbon taxation was the Journal of Cleaner Production, with 48 articles, representing about 40% of the total sample.

The second journal with the most publications was the International Journal of Production Economics with 11 articles (9.1%), followed by the National Tax Journal with 9 articles (7.4%). The remaining 44% of the articles are published in journals with 4 or fewer publications on this topic, and more than half (65%) of these journals only published 1 scientific article.

Table 5. Distribution of articles published by scientific journal

Journal	Number of articles published	%
Journal of Cleaner Production	48	39,67%
International Journal of Production Economics	11	9,09%
National Tax Journal	9	7,44%
Electricity Journal	4	3,31%
International Journal of Production Research	3	2,48%
Technological Forecasting and Social Change	3	2,48%
Journal of the Operational Research Society	3	2,48%
Current Issues in Tourism	2	1,65%
International Tax and Public Finance	2	1,65%
Journal of Economic Behavior and Organization	2	1,65%
Omega (United Kingdom)	2	1,65%
OPSEARCH	2	1,65%
Organization and Environment	2	1,65%
South African Journal of Economic and Management Sciences	2	1,65%
Accounting Research Journal	1	0,83%
African Journal of Hospitality, Tourism and Leisure	1	0,83%
Annals of Tourism Research	1	0,83%
Corporate Ownership and Control	1	0,83%
Critical Perspectives on Accounting	1	0,83%
Decision Sciences	1	0,83%
Economists' Voice	1	0,83%
Financial Accountability and Management	1	0,83%
International Journal of Accounting	1	0,83%
International Journal of Precision Engineering and Manufacturing - Green Technology	1	0,83%
International Journal of Recent Technology and Engineering	1	0,83%
International Transactions in Operational Research	1	0,83%
Journal of Applied Business Research	1	0,83%
Journal of Economics/ Zeitschrift für Nationalokonomie	1	0,83%

continues on following page

Table 5. Continued

Journal	Number of articles published	%
Journal of Hospitality and Tourism Management	1	0,83%
Journal of Industrial Engineering and Management	1	0,83%
Journal of Self-Governance and Management Economics	1	0,83%
Journal of Sustainable Tourism	1	0,83%
Journal of Travel Research	1	0,83%
Management Science	1	0,83%
Pacific Accounting Review	1	0,83%
Production and Operations Management	1	0,83%
Schmalenbachs Zeitschrift für Betriebswirtschaftliche Forschung	1	0,83%
Tourism Economics	1	0,83%
Utilities Policy	1	0,83%
World Bank Economic Review	1	0,83%
TOTAL	121	100,00%

The CiteScore is an indicator for evaluating the impact of scientific journals, which determines the frequency with which an article is cited. This metric was first launched by Elsevier in 2016 as an alternative to the Journal Citation Reports (JCR) impact factors that are typically used in this analysis, based on Scopus-verified citations of articles published in the previous three years, rather than two or five (Silva and Memon, 2017). That is, the CiteScore of a scientific journal identifies the average number of times its published articles have been cited per year in the last three years.

Considering the CiteScore impact factor for the year 2019, it was verified that the Journal of Cleaner Production, with the highest number of published articles (48 articles) is one of the journals with the highest CiteScore, of 10.9 (Table 6). Like Omega (United Kingdom) and the Journal of Travel Research, however, interestingly the first journal published only 2 scientific articles and the second 1 article.

The second journal with the highest number of scientific publications on the subject "Carbon Tax", International Journal of Production Economics with 11 articles, also has the second highest CiteScore, of 10.5. On the contrary, the CiteScore of the National Tax Journal with 9 publications is only 2.2.

Table 6. Top five of the scientific journals with the highest impact factor CiteScore 2019

Journal	Number of articles published	%	CiteScore 2019
Journal of Cleaner Production	48	39,67%	10,9
Omega (United Kingdom)	2	1,65%	10,9
Journal of Travel Research	1	0,83%	10,9
International Journal of Production Economics	11	9,09%	10,5
Organization and Environment	2	1,65%	9,1

Finally, it is relevant to analyze the thematic associations of the articles that constitute the sample of this investigation, for this we resorted to the analysis of the most used keywords.

Taking into account Figure 10, it can be seen that the keyword “Carbon Tax” is mainly related to “cap-and-trade” (13 binding force), “carbon dioxide emissions” (14), “[6] climate change” (11), “Computable General Equilibrium (CGE) Model” (7) and “Emissions ” (6).

Figure 10. Keywords used by the authors in scientific articles
Source: Own elaboration through VOSviewer

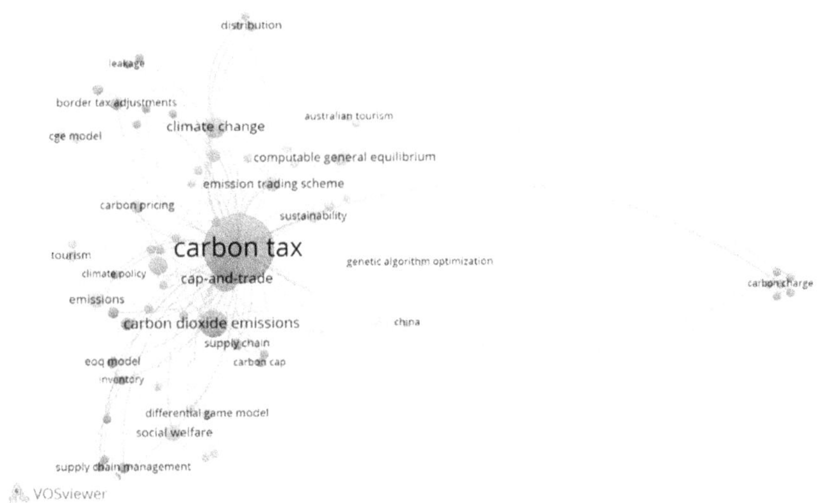

CONCLUSION

The growing environmental concerns on the part of society due to the increase in misunderstandings and imbalances that have been witnessed in recent years, such as global warming resulting from excess CO2 emissions into the atmosphere, requires governments to take appropriate measures to protect the environment. One of these measures is the carbon tax, which has proven to be quite effective in reducing CO2 emissions and which has been implemented over the years by several countries, thus increasing the number of scientific research related to this policy.

In this sense, it is useful to evaluate the scientific production in “Carbon Tax” so that researchers interested in this theme can follow the relevant literature and in addition, contributes to the recognition of researchers and educational institutions.

In the present study, the general objective was to characterize the research carried out on the theme “Carbon Tax” through articles published between 1995 and August 31, 2020. In order to achieve this goal, two specific objectives were defined. The first specific objective is the analysis of a set of characteristics related to the content of the published scientific articles and the second is the analysis of a set of characteristics related to the profile of the researchers of the respective articles.

The approach applied to carry out this study was bibliometrics that uses bibliometric methods, as was followed by Zhang et al. (2016) in a study on the same theme using the Web of Science database and without restrictions as to the area of study.

In the data collection, the Scopus database was used, delimiting the publications only to scientific articles, with the keyword "Carbon Tax" and the area of study Business, Management and Accounting. A total of 137 scientific articles were obtained, however, 16 articles were withdrawn because they were not available, thus resulting in a total of 121 scientific articles considered eligible for the sample of this study.

Subsequently, in Excel, a database was built with information taken from Scopus' CSV file, namely about the author, article title, year of publication, journal, keywords, number of citations and geographical affiliation. In addition, the VOSviewer software was used in the construction of the bibliometric maps.

Regarding the results of this study, regarding the first specific objective, it was found that the number of scientific publications on the theme "Carbon Tax" increased over the years, growing on average 14% per year in the period analyzed (1995-2020). Similarly, there was also a positive evolution in the number of citations received from these articles during the 25 years. This year, 2020, recorded the highest number of publications, with 26 articles published, and the highest number of citations received, with 701 citations. This means that the scientific community is increasingly interested in this theme, considering it relevant today. It was found that the most influential articles, that is, frequently used for the development of other investigations, are those of Sinn, H. -W. (2008) and Becken, S. (2007), with 350 and 234 citations, respectively. Using the H-Index metric that allows measuring the productivity and impact of a researcher, based on their most cited articles (Costa et al., 2012), it was concluded that the sample of this study has an index of 27, that is, of the 121 scientific articles, 27 of these have at least 27 citations each.

Regarding the sources of publication, the Journal of Cleaner Production was identified as dominant in the publication of scientific articles on the theme "Carbon Tax" in management and accounting, with 48 articles published. This is a journal focused essentially on research and practices of cleaner production, environment and sustainability, having presented in 2019 a CiteScore of 10.9, that is, the average number of annual citations of its articles in the last 3 years was 10.9. In addition, it was possible to conclude that most of the scientific publications inserted in Business, Management and Accounting are also associated with the areas of Engineering, Energy and Environmental Science.

The analysis of the keywords allowed to demonstrate that "Carbon Tax" is mainly related to environmental themes, such as "cap-and-trade", "carbon dioxide emissions", "climate change", "computable general equilibrium (cge) model" and "emissions" ". Thus, it is understood that these are the themes most referred to by most researchers in their scientific articles when addressing the topic of carbon taxation policy.

Regarding the second specific objective, it was found that Zhang, Y. is the author who contributed the most in the publication of scientific articles on "Carbon Tax" in management and accounting, with 6 articles published, followed by Cao, K. and Zhang, J.with 4 articles each. It was possible to notice that Zhang, Y. and Zhang, J. jointly produced 3 scientific articles. However, the majority of authors, about 82%, contributed to the scientific community with only 1 scientific article on this subject. In addition, it was observed that collective authorship is the predominant type of authorship in the analyzed articles, with a representation of 84%. Within these, the articles produced by 2 authors or by 4 or more authors have the highest representativeness, of 29.8%.

In addition, other researchers frequently cited in the research on "Carbon Tax" were identified, since they are equally relevant in research on this topic, such as Benjafaar, S. and Chen, X. especially focused on "Environmental Markets", "Pollution Tax" and "Trade Regulations".

Finally, it was found that the five existing continents are represented in this study, with the Asian continent being the most active, with 157 affiliated authors (55%). Within this, China stands out with 112 affiliated authors (38.9%). In addition, in the analysis of the relationship between the countries, it was noticed that the USA is the country that has more links to other countries (9 associations), having the strongest relationship with China, that is, with the largest number of shared articles (7 articles). Next is India, China and Australia with 7, 6 and 5 associations, respectively.

This study contributes to the knowledge of the state of current research on "Carbon Tax" in management and accounting, by identifying the latest trends on this topic, authors and influential journals in this area, as well as other related and relevant themes, thus serving as a guiding role for researchers in future research on carbon taxation policy.

Given the results obtained in this study, it is suggested the development of the same type of study by adding the keyword "cap-and-trade" to obtain a larger sample and better understand how this theme is so related to "Carbon Tax". In addition, the relationship with "Green Tax" can also be explored, since the carbon tax falls into the category of "green" or environmental taxes.

REFERENCES

Andrew, J., Kaidonis, M. A., & Andrew, B. (2010). Carbon tax: Challenging neoliberal solutions to climate change. *Critical Perspectives on Accounting*, *21*(7), 611–618. doi:10.1016/j.cpa.2010.03.009

APA. (2019). *European Emissions Trading (ETS)*. APA. https://apambiente.pt/index.php?ref=17&subref=295 (2019).

APA. (2020). *Green Taxation*. APA. https://apambiente.pt/index.php?ref=17&subref=1104 (2020).

Apleni, L., Mxunyelwa, S., & Vallabh, D. (2017). The implications of the proposed carbon tax in South Africa's tourism industry. *African Journal of Hospitality, Tourism and Leisure*, *6*(1), 1–12.

Baranzini, A., Goldemberg, J., & Speck, S. (2000). A future for carbon taxes. *Ecological Economics*, *32*(3), 395–412. doi:10.1016/S0921-8009(99)00122-6

Cao, K., Xu, B., He, Y., & Xu, Q. (2020a). Optimal carbon reduction level and ordering quantity under financial constraints. *International Transactions in Operational Research*, *27*(5), 2270–2293. doi:10.1111/itor.12606

Cao, K., Xu, B., & Wang, J. (2020b). Optimal trade-in and warranty period strategies for new and remanufactured products under carbon tax policy. *International Journal of Production Research*, *58*(1), 180–199. doi:10.1080/00207543.2019.1617448

Chang, N., & Han, C. (2020). Cost-push impact of taxing carbon in China: A price transmission Perspective. *Journal of Cleaner Production*, *248*, 1–9. doi:10.1016/j.jclepro.2019.119194

Chen, W., & Hu, Z.-H. (2018). Using evolutionary game theory to study governments and manufacturers' behavioral strategies under various carbon taxes and subsidies. *Journal of Cleaner Production*, *201*, 123–141. doi:10.1016/j.jclepro.2018.08.007

Choudhary, A., Sarkar, S., Settur, S., & Tiwari, M. K. (2015). A carbon market sensitive optimization model for integrated forward–reverse logistics. *International Journal of Production Economics, 164*, 433–444. doi:10.1016/j.ijpe.2014.08.015

Church, B., Jiang, W., Kuang, X., & Vitalis, A. (2019). A Dollar for a Tree or a Tree for a Dollar? The Behavioral Effects of Measurement Basis on Managers' CSR Investment Decision. *The Accounting Review, 94*(5), 117–137. doi:10.2308/accr-52332

Costa, T., Lopes, S., Fernández-Llimós, F., Amante, M., & Lopes, P. (2012). Bibliometrics and the Evaluation of Scientific Production: Indicators and tools. *ACTAS – National Congress of Bibliotecários. Archivists and Documentalists, 11*, 1–7.

Costello, K. W. (2019). Essay on climate apocalypse and a carbon tax. *The Electricity Journal, 32*(10), 1–6. doi:10.1016/j.tej.2019.106669

Datta, T. K., Nath, P., & Choudhury, K. (2020). A hybrid carbon policy inventory model with emission sourcebased green investments. *OPSEARCH, 57*(1), 202–220. doi:10.100712597-019-00430-y

Dwyer, L., Forsyth, P., & Spurr, R. (2012). Wither Australian tourism? Implications of the carbon tax. *Journal of Hospitality and Tourism Management, 19*(1), 15–30. doi:10.1017/jht.2012.18

Dwyer, L., Forsyth, P., Spurr, R., & Hoque, S. (2013). Economic Impacts of a Carbon Tax on the Australian Tourism Industry. *Journal of Travel Research, 52*(2), 143–155. doi:10.1177/0047287512461568

EEA. (2019). *Atmospheric greenhouse gas concentrations*. EEA. https://www.eea.europa.eu/data-and-maps/indicators/atmospheric-greenhouse-gas-concentrations-6/assessment-1

Gale, W., Brown, S., & Saltiel, F. (2013). Carbon taxes as part of the fiscal solution. Brookings Institution.

Gray, W. B., & Metcalf, G. (2017). Carbon tax competitiveness concerns: Assessing a best practices carbon credit. *National Tax Journal, 70*(2), 447–468. doi:10.17310/ntj.2017.2.08

Haites, E. (2018). Carbon taxes and greenhouse gas emissions trading systems: What have we learned? *Climate Policy, 18*(8), 955–966. doi:10.1080/14693062.2018.1492897

Hou, Y., Jia, M., Tian, X., Wei, F., & Wei, K. (2015). Optimal decisions of countries with carbon tax and carbon tariff. *Journal of Industrial Engineering and Management, 8*(3), 981–1001. doi:10.3926/jiem.1506

Hovelaque, V., & Bironneau, L. (2015). The carbon-constrained EOQ model with carbon emission dependent demand. *International Journal of Production Economics, 164*, 285–291. doi:10.1016/j.ijpe.2014.11.022

Hu, X., Yang, Z., Sun, J., & Zhang, Y. (2020). Carbon tax or cap-and-trade: Which is more viable for Chinese remanufacturing industry? *Journal of Cleaner Production, 243*, 1–16. doi:10.1016/j.jclepro.2019.118606

IPCC. (2005). *Carbon Dioxide Capture and Storage*. IPCC. https://www.ipcc.ch/site/assets/uploads/2018/03/srccs_wholereport-1.pdf (08/06/2020).

Jeffrey, C., & Perkins, J. D. (2015). The association between energy taxation, participation in an emissions trading system, and the intensity of carbon dioxide emissions in the European Union. *The International Journal of Accounting*, 1–21.

Jin, M., Shi, X., Emrouznejad, A., & Yang, F. (2018). Determining the optimal carbon tax rate based on data envelopment Analysis. *Journal of Cleaner Production*, *172*, 900–908. doi:10.1016/j.jclepro.2017.10.127

Kalita, U., & Barua, N. A. (2019). Determining a Carbon Tax Rate for India in the Context of Global Climate Change. *International Journal of Recent Technology and Engineering*, *8*(3), 8185–8191. doi:10.35940/ijrte.C6655.098319

Koks, E., Carrera, L., Jonkeren, O., Aerts, J., Husby, T., Thissen, M., Standardi, G., & Mysiak, J. (2016). Regional disaster impact analysis: Comparing input–output and computable general equilibrium models. *Natural Hazards and Earth System Sciences*, *16*(8), 1911–1924. doi:10.5194/nhess-16-1911-2016

Kortum, S., & Weisbach, D. (2017, June 01). Weisbach & The design of border adjustments for carbon prices. *National Tax Journal*, *70*(2), 421–446. doi:10.17310/ntj.2017.2.07

Kuo, T. C., Hong, I.-H., & Lin, S. C. (2016). Of carbon taxes work? Analysis of government policies and enterprise strategies in equilibrium. *Journal of Cleaner Production*, *139*, 337–346. doi:10.1016/j.jclepro.2016.07.164

Lee, J.-Y. (2019). Investing in carbon emissions reduction in the EOQ model. *The Journal of the Operational Research Society*, 1–12.

Li, G., & Masui, T. (2019). Assessing the impacts of China's environmental tax using a dynamic computable general equilibrium model. *Journal of Cleaner Production*, *208*, 316–324. doi:10.1016/j.jclepro.2018.10.016

Liang, Q.-M., Wang, T., & Xue, M.-M. (2016). Addressing the competitiveness effects of taxing carbon in China: Domestic tax cuts versus border tax adjustments. *Journal of Cleaner Production*, *112*, 1568–1581. doi:10.1016/j.jclepro.2015.02.092

Liao, H., Tang, M., Luo, L., Li, C., Chiclana, F., & Zeng, X.-J. (2018). A Bibliometric Analysis and Visualization of Medical Big Data Research. *Sustainability (Basel)*, *10*(1), 1–18. doi:10.3390u10010166

Lin, B., & Xu, D. M. (2019). Exploring the green total factor productivity of China's metallurgical industry under carbon tax: A perspective on factor substitution. *Journal of Cleaner Production*, *233*, 1322–1333. doi:10.1016/j.jclepro.2019.06.137

Liu, X., Wang, C., Niu, D., Suk, S., & Bao, C. (2015). An analysis of company choice preference to carbon tax policy in China. *Journal of Cleaner Production*, *103*, 393–400. doi:10.1016/j.jclepro.2014.12.084

Ma, N., Li, H., Wang, Y., Zhang, J., Li, Z., & Arif, A. (2020). The short-term roles of sectors during a carbon tax on Chinese economy based on complex network: An in-process analysis. *Journal of Cleaner Production*, *251*, 1–14. doi:10.1016/j.jclepro.2019.119560

Meng, X., Yao, Z., Nie, J., Zhao, Y., & Li, Z. (2018). Low-carbon product selection with carbon tax and competition: Effects of the power structure. *International Journal of Production Economics*, *200*, 224–230. doi:10.1016/j.ijpe.2018.03.029

Metcalf, G., & Weisbach, D. (2009). *The Design of a Carbon Tax.* Reg-Markets Center Working Paper, 1-63.

Metcalf, G. E. (2014). Using the tax system to address competition issues with a carbon tax. *National Tax Journal, 67*(4), 779–806. doi:10.17310/ntj.2014.4.02

Miller, R., & Blair, P. (2009). *Input-output analysis: Foundations and extensions* (2nd ed.). Cambridge University Press. doi:10.1017/CBO9780511626982

Mishra, U., Wu, J.-Z., & Sarkar, B. (2020). A sustainable production-inventory model for a controllable carbon emissions rate under shortages. *Journal of Cleaner Production, 256*, 1–19. doi:10.1016/j.jclepro.2020.120268

Moioli, S., & Pellegrini, L. A. (2020). Fixed and Capture Level Reduction operating modes for carbon dioxide removal in a Natural Gas Combined Cycle power plant. *Journal of Cleaner Production, 254*, 1–12. doi:10.1016/j.jclepro.2020.120016

Ndhaief, N., Nidhal, R., Hajji, A., & Bistorin, O. (2019). Environmental issue in an integrated production and maintenance control of unreliable manufacturing/remanufacturing systems. *International Journal of Production Research*, 1–19.

Nordhaus, W. D. (1992). An Optimal Transition Path for Controlling Greenhouse Gases. *Science, 258*(5086), 1315–1319. doi:10.1126cience.258.5086.1315 PMID:17778354

Nordhaus, W. D. (2010). Carbon taxes to move toward fiscal sustainability. *Economists' Voices*, 1-5.

Park, S. J., Cachon, G. Q., Lai, G., & Seshadri, S. (2015). Supply Chain Design and Carbon Penalty: Monopoly vs. Monopolistic Competition. *Production and Operations Management, 24*(9), 1494–1508.

Parry, I., Morris, A., & Williams, R. (2015). Implementing a US carbon tax: Challenges and debates. Routledge. doi:10.4324/9781315747682

Peng, P., Zhu, L., & Fan, Y. (2017). Performance evaluation of climate policies in China: A study based on an integrated assessment model. *Journal of Cleaner Production, 164*, 1068–1080. doi:10.1016/j.jclepro.2017.07.016

Pigou, A. C. (1920). The economics of welfare. Macmillan.

Pillay, S., & Buys, P. W. (2013). Carbon tax pricing and the social cost of carbon: The case in the South African motor vehicle manufacturing industry. *Journal of Applied Business Research, 29*(6), 1751–1762. doi:10.19030/jabr.v29i6.8212

Pillay, S., & Buys, P. W. (2015). Investigating the impact of carbon tax on socially responsible corporate governance: The case of south African motor vehicle manufacturers. *Corporate Ownership and Control, 12*(2), 128–134. doi:10.22495/cocv12i2p10

Saelim, S. (2019). Carbon tax incidence on household demand: Effects on welfare, income inequality and poverty incidence in Thailand. *Journal of Cleaner Production, 234*, 521–533. doi:10.1016/j.jclepro.2019.06.218

Shi, Q., Ren, H., Cai, W., & Gao, J. (2019). How to set the proper level of carbon tax in the context of Chinese construction sector? A CGE analysis. *Journal of Cleaner Production*, *240*, 1–12. doi:10.1016/j.jclepro.2019.117955

Shuang, Y., Diabat, A., & Liao, Y. (2019). A stochastic reverse logistics production routing model with emissions control policy selection. *International Journal of Production Economics*, *213*, 201–216. doi:10.1016/j.ijpe.2019.03.006

Smith, J., & Memon, A. (2017). CiteScore: A cite for sore eyes, or a valuable, transparent metric? *Scientometrics*, *111*(1), 553–556. doi:10.100711192-017-2250-0

van Heerden, J., Blignaut, J., Bohlmann, H., Cartwright, A., Diederichs, N., & Mander, M. (2016). The economic and environmental effects of a carbon tax in South Africa: A dynamic CGE modelling approach. *Suid-Afrikaanse Tydskrif vir Ekonomiese en Bestuurswetenskappe*, *19*(5), 714–732. doi:10.4102ajems.v19i5.1586

Wang, C., Wang, W., & Huang, R. (2017). Supply chain enterprise operations and government carbon tax decisions considering carbon emissions. *Journal of Cleaner Production*, *152*, 271–280. doi:10.1016/j.jclepro.2017.03.051

Williams, R. C. III, Gordon, H., Burtraw, D., Carbone, J. C., & Morgenstern, R. D. (2014). The initial incidence of a carbon tax across U.S. States. *National Tax Journal*, *67*(4), 807–830. doi:10.17310/ntj.2014.4.03

Williams, R. C. III, Gordon, H., Burtraw, D., Carbone, J. C., & Morgenstern, R. D. (2015). The initial incidence of a carbon tax across income groups. *National Tax Journal*, *68*(1), 195–214. doi:10.17310/ntj.2015.1.09

Xu, C., Wang, C., & Huang, D. R. (2020). Impacts of horizontal integration on social welfare under the interaction of carbon tax and green subsidies. *International Journal of Production Economics*, *222*, 1–13. doi:10.1016/j.ijpe.2019.09.027

Xu, X., Xu, X., & He, P. (2016). Joint production and pricing decisions for multiple products with cap-and-trade and carbon tax regulations. *Journal of Cleaner Production*, *112*, 4093–4106. doi:10.1016/j.jclepro.2015.08.081

Yang, H., & Chen, W. (2018). Retailer-driven carbon emission abatement with consumer environmental awareness and carbon tax: Revenue-sharing versus Cost-sharing. *Omega*, *78*, 179–191. doi:10.1016/j.omega.2017.06.012

Yang, H., Luo, J., & Wang, H. (2017). The role of revenue sharing and first-mover advantage in emission abatement with carbon tax and consumer environmental awareness. *International Journal of Production Economics*, *193*, 691–702. doi:10.1016/j.ijpe.2017.08.032

Yu, B., Wang, J., Lu, X., & Yang, H. (2020). Collaboration in a low-carbon supply chain with reference emission and cost learning effects: Cost sharing versus revenue sharing strategies. *Journal of Cleaner Production*, *250*, 1–26. doi:10.1016/j.jclepro.2019.119460

Yu, V. F., Maglasang, R., & Tsao, Y.-C. (2018). Shelf space allocation problem under carbon tax and emission trading Policies. *Journal of Cleaner Production*, *196*, 438–451. doi:10.1016/j.jclepro.2018.05.165

Zakeri, A., Dehghanian, F., Fahimnia, B., & Sarkis, J. (2015). Carbon pricing versus emissions trading: A supply chain planning perspective. *International Journal of Production Economics*, *164*, 197–205. doi:10.1016/j.ijpe.2014.11.012

Zhang, J., & Zhang, D. Y. (2020a). Chinese tourism economic change under carbon tax scenarios. *Current Issues in Tourism*, *23*(7), 836–851. doi:10.1080/13683500.2018.1551339

Zhang, J., & Zhang, D. Y. (2020b). Examining the economic and environmental effects of emissions policies in China: A Bayesian DSGE model. *Journal of Cleaner Production*, *266*, 1–9. doi:10.1016/j.jclepro.2020.122026

Zhang, K., Wang, Q., Liang, Q.-M., & Chen, H. (2016). A bibliometric analysis of research on carbon tax from 1989 to 2014. *Renewable & Sustainable Energy Reviews*, *58*, 297–310. doi:10.1016/j.rser.2015.12.089

Zhang, Z., & Baranzini, A. (2004). What do we know about carbon taxes? An inquiry into their impacts on competitiveness and distribution of income. *Energy Policy*, *32*(4), 507–518. doi:10.1016/S0301-4215(03)00152-6

Zhao, L., Yang, C., Su, B., & Zeng, S. (2020). Research on a single policy or policy mix in carbon emissions Reduction. *Journal of Cleaner Production*, *267*, 1–10. doi:10.1016/j.jclepro.2020.122030

Zhou, G., & Zhang, D. Y. (2017). Integration and consolidation in air freight shipment planning: An economic and environmental perspective. *Journal of Cleaner Production*, *166*, 1381–1394. doi:10.1016/j.jclepro.2017.07.145

Zhou, Y., Hu, F., & Zhou, Z. (2018). Pricing decisions and social welfare in a supply chain with multiple competing retailers and carbon tax policy. *Journal of Cleaner Production*, *190*, 752–777. doi:10.1016/j.jclepro.2018.04.162

Zhu, J., & Liu, W. (2020). A tale of two databases: The use of Web of Science and Scopus in academic papers. *Scientometrics*, *123*(1), 321–335. doi:10.100711192-020-03387-8

Zhu, L., Duan, H.-B., & Fan, Y. (2015). CO2 mitigation potential of CCS in China and an evaluation based on an integrated assessment model. *Journal of Cleaner Production*, *103*, 934–947. doi:10.1016/j.jclepro.2014.08.079

Zhu, N., Bu, Y., Jin, M., & Mbroh, N. (2020a). Green financial behavior and green development strategy of Chinese power companies in the context of carbon tax. *Journal of Cleaner Production*, *245*, 1–7. doi:10.1016/j.jclepro.2019.118908

Zhu, N., Qian, L., Jiang, D., & Mbroh, N. (2020b). A simulation study of China's imposing carbon tax against American carbon tariffs. *Journal of Cleaner Production*, *243*, 1–12. doi:10.1016/j.jclepro.2019.118467

Zuo, H., Du, H., Broadstock, D., Guo, J., Gong, Y., & Mao, G. (2016). China's future energy mix and emissions reduction potential: A scenario analysis incorporating technological learning curves. *Journal of Cleaner Production*, *112*, 1475–1485. doi:10.1016/j.jclepro.2015.08.012

KEY TERMS AND DEFINITIONS

Carbon Tax: A carbon tax is levied on the carbon content of fossil fuels. The term can also refer to taxing other types of greenhouse gas emissions, such as methane. A carbon tax puts a price on those emissions to encourage consumers, businesses, and governments to produce less of them (Tax Foundation).

Climate Changes: Climate change is a change in the statistical properties of the climate system that persists for several decades or longer—usually at least 30 years. These statistical properties include averages, variability and extremes. Climate change may be due to natural processes, such as changes in the Sun's radiation, volcanoes or internal variability in the climate system, or due to human influences such as changes in the composition of the atmosphere or land use (Australian Academy of Science).

CO2 Emissions: Carbon dioxide (CO2) is a colourless, odourless and non-poisonous gas formed by combustion of carbon and in the respiration of living organisms and is considered a greenhouse gas. Emissions means the release of greenhouse gases and/or their precursors into the atmosphere over a specified area and period of time. Carbon dioxide emissions or CO2 emissions are emissions stemming from the burning of fossil fuels and the manufacture of cement; they include carbon dioxide produced during consumption of solid, liquid, and gas fuels as well as gas flaring (Eurostat).

Global Warming: Changes in the surface-air temperature, referred to as the global temperature, brought about by the greenhouse effect which is induced by emission of greenhouse gases into the air (European Environment Agency).

Cap-and-trade: It's a system designed to reduce pollution in our atmosphere. The cap on greenhouse gas emissions that drive global warming is a firm limit on pollution. The cap gets stricter over time. The trade part is a market for companies to buy and sell allowances that let them emit only a certain amount, as supply and demand set the price. Trading gives companies a strong incentive to save money by cutting emissions in the most cost-effective ways (Environmental Defense Found).

ENDNOTES

1 The Kyoto Protocol introduced 3 flexible mechanisms: International Emissions Trading, Joint Implementation and the Clean Development Mechanism (APA, 2020; Mishra et., 2020; Choudhary et al., 2015).

2 The Paris Agreement aims to direct countries for a carbon-neutral economy, setting "as one of their long-term goals to limit the increase in global average temperature to levels well below 2°C above pre-industrial levels and pursue efforts to limit temperature rise to 1.5°C" (APA, 2020; Directive (EU) 2018/410).

3 This model is based on the general equilibrium theory of Walras, establishing a quantitative relationship between the various components of the national economy, making it possible to assess the impact of the disturbance of one part of the economy on the other part of the economy (Zhang and Zhang, 2020b; Kuo et al., 2016; Liang et al., 2016). Following Liang et al. (2016: 1570), the model *GCE* is "*good at describing the interactions among different agents and at assessing the direct, indirect and economy-wide impacts of a given policy*".

4 The breakdown structure of the model *I-O*, first proposed by Wassily Leontief In the 1930s, it consists of the description of the structural relations between industries within an economy, that is, it analyzes the flows of goods and services from one sector (producer) with others (consumer) (Koks et al., 2016; Miller and Blair, 2009).

5 Technology that extracts CO emissions, transports and stores in geological locations deep in the soil, in order to isolate emissions from the atmosphere and not to arise in harmful consequences to the environment (Zhu et al., 2015; Parry et al., 2015; IPCC, 2005).

6 The binding force (*link Strength*) concerns the frequency of co-occurrence.

Chapter 3
Corporate Climate Change-Related Accounting and Taxation:
A Comparative Analysis of Portuguese and Spanish Frameworks on GHG Emissions Allowances

Sónia Monteiro
Research Centre on Accounting and Taxation, Polytechnic Institute of Cávado and Ave, Portugal

Kátia Lemos
https://orcid.org/0000-0002-3961-5964
Research Centre on Accounting and Taxation, Polytechnic Institute of Cávado and Ave, Portugal

Jose Jurado Martin
https://orcid.org/0000-0002-7079-3291
Huelva University, Spain

María-Pilar Martín-Zamora
University of Huelva, Spain

ABSTRACT

This chapter aims to provide an overview of corporate climate change-related accounting and taxation practices in Portugal and Spain. The Portuguese Accounting and Financial Reporting Standard No. 26: Environmental Matters and the Spanish Royal Decree 602/2016 establish the current accounting treatment for greenhouse (GHG) emissions. While the initial measurement basis is the same in both countries (cost when licenses are acquired or fair value when obtained for free), the accounting regulations prescribe different accounting treatments for the initial recognition. In Spain, allowances are recognized as inventories, while in Portugal, they are recognized as intangible assets. Interestingly, the Portuguese regulation uses the FIFO inventory method to calculate the amortization of the intangible asset. Regarding corporate income tax, there is no specific treatment for emission allowances in either country's tax legislation. Additionally, a reverse-charge mechanism on CO2 emissions trading has been introduced in Europe, resulting in no differences between countries concerning VAT.

DOI: 10.4018/978-1-6684-8592-7.ch003

INTRODUCTION

Due to the rising levels of CO_2 and other greenhouse gases, the Earth's climate is undergoing changes that are causing significant economic, social, and environmental damage both in the present and for future generations (Arnold, 2016). As a result, global warming has become a pressing concern for countries and international organizations worldwide.

Since the Kyoto Protocol, European Union (EU) member states have committed to reducing atmospheric emissions. To this end, the creation of a European market for greenhouse gas (GHG) emissions transactions has been one of the most critical measures implemented, based on a cap-and-trade system. A limited number of licenses were granted to entities for a set period, and any excess licenses can be sold or purchased if they become insufficient (Ellerman et al, 2016).

One of the primary initiatives established by the European Emissions Trading Scheme (EU-ETS) was the allocation of free emission allowances to companies through grandfathering to help limit CO_2 emissions. Each member state allocates CO_2 emission allowances free of charge, which constitutes the initial allocation. It is also possible to auction 5% of CO_2 emission allowances during the first phase (2005-2008) and 10% in the second phase (2008-2012). With the introduction of the new Directive 2009/29/EC, there were significant changes, particularly with the start of the third phase (2013-2020), which involved allocating allowances through auctions, while free allocation continued on a marginal basis.

Regarding the accounting of CO_2 emission allowances, there is currently no dedicated international accounting standard at either the European or international level. In December 2004, the International Financial Reporting Interpretations Committee (IFRIC), a part of the International Accounting Standards Board (IASB), released an interpretive standard on accounting for emission rights called IFRIC No. 3 - Emission Rights. However, this standard was withdrawn in July 2005 due to opposition from the European Financial Reporting Advisory Group (EFRAG). Since then, no other standard or interpretation has been issued specifically addressing environmental issues.

The lack of a dedicated international accounting standard has resulted in varying accounting treatments adopted by different entities. As a result, some countries such as Portugal, Spain, Belgium, France, and the United Kingdom have developed specific regulations for the accounting treatment of emission licenses within their respective jurisdictions.

In Portugal, the Accounting Standardization System (Sistema de Normalização Contabilística, SNC) was approved, and the Accounting and Financial Reporting Standard (Norma Contabilística e de Relato Financeiro, NCRF) No. 26 - Environmental Matters was published in accordance with Decree Law (DL) No. 158/2009 on July 13th (amended by DL 98/2105). NCRF No. 26 includes an appendix specifically dedicated to the accounting treatment of CO_2 emission allowances.

In Spain, the Resolution of the Institute of Accounting and Auditing (ICAC) regarding greenhouse gas emission rights (published in 2006, revised in 2009 and 2010) clarified the accounting treatment of emission allowances received free of charge, purchased, or acquired through non-reciprocal transactions.

Regarding taxation, there is a lack of specific treatment in the tax regulations for operations related to emission allowances. Therefore, it is up to the accounting framework to determine their impact on the results of companies engaging in such operations, which in turn can affect the tax obligations of these companies.

The purpose of this paper is to present an overview of the corporate climate change-related accounting and taxation frameworks for GHG emission allowances in Portugal and Spain.

This chapter is structured as follows: after this introduction, the next section presents the European Carbon Trading market. A section regarding the background and empirical studies on accounting and taxation of GHG emission allowances is also presented. Next, the accounting regulations of greenhouse gas emission allowances are presented, both in Spain and Portugal. Subsequently, the tax regime for greenhouse gas emission allowances is presented, also in both countries. Finally, solutions and recommendations are presented, as well as suggestions for future research and the conclusions.

THE EUROPEAN CARBON TRADING MARKET

To meet the emissions reduction target set by the Kyoto Protocol, the European Union (EU) implemented cost-minimizing measures for its Member States, including the creation of the European Union Emissions Trading System (EU ETS), also known as the European Carbon Market. The EU ETS was established by Directive 2003/87/EC of the European Parliament and of the Council of 13 October 2003, which was later modified by Directive 2004/101/EC and Directive 2009/29/EC, to regulate more sectors and greenhouse gases (GHGs) in the post-2012 period (Santos, 2013). Later, on November 8, 2013, the Commission Regulation (EU) No 1123/2013 was enacted to establish international credit entitlements in accordance with Directive 2003/87/EC for the period spanning from 2008 to 2012. The revision of the EU ETS Directive, adopted in 2018, sets the total quantity of emission allowances for phase 4 (2021-2030), in line with the previous EU emission reduction target (40% reduction below 1990 levels by 2030). On july 2023, the Commission published a legislative proposal for a revision of the EU ETS, to align it with the target of a 55% reduction of EU net GHG emissions by 2030, compared to 1990 levels.

The CO2 emissions trading market started operating in 2005 and covers approximately 45% of the EU's GHG emissions. It is a revolutionary and unprecedented program within environmental policy instruments, designed to combat greenhouse gas emissions by incentivizing companies to reduce their emissions and encouraging the development of low-carbon technologies.

The EU Carbon Market employs a cap-and-trade mechanism, which involves setting a limit, or cap, on the total amount of greenhouse gas emissions that can be produced by certain industries. This cap is then divided into allowances, with each allowance representing the right to emit one tonne of CO2 or its equivalent. The allowances are allocated to the companies covered by the scheme, either for free or through auctioning, and they can be traded among them. Under the EU ETS, companies are allocated a specific number of allowances, which represent the right to emit a certain amount of greenhouse gases. If a company emits more than its allotted allowances, it must purchase additional allowances from companies that emit less than their allotted allowances (Fernandes, 2009).

The cap-and-trade mechanism of the EU Carbon Market is a crucial policy tool for reducing greenhouse gas emissions in a cost-effective manner and encouraging innovation in clean technologies.

Monast et al. (2017) identify three distinct stages in the implementation of the EU Carbon Market:

- The first phase of the EU Carbon Market, which took place from 2005 to 2007, was considered a pilot phase that preceded the adoption of the Kyoto Protocol. During this period, the market was in a learning phase, and emission allowances were allocated to companies free of charge.
- During the second phase (2008-2012), which coincided with the validity period of the Kyoto Protocol, the allocation of emission allowances for free was reduced by 10% compared to the previous phase.

- The third phase (2013-2020) reflected the EU's unilateral commitment to reducing greenhouse gas emissions, as negotiated under the Kyoto Protocol. The goal for this phase was to achieve a 20% reduction in emissions by 2020, compared to 2005.

Overall, these three phases reflect different stages in the evolution of the market, with changes in the allocation of emission allowances and an increasing commitment to reduce greenhouse gas emissions over time. In general, there was an extension of the scope with the introduction of new gases and new sectors. Additionally, the total quantity of allowances has been determined at the Community level, and there has been a shift towards the allocation of allowances through auctions, with some marginally free allocation remaining based on benchmarks defined at the Community level.

In November 2017, the EU reached an agreement on the regulations for Phase 4 (2021-2030) of its ETS. This new phase replaces the previous fixed cap with a cap that is determined based on market outcome. The Directive (EU) 2018/410 amends Directive 2003/87/EC to enhance cost-effective emission reductions and low-carbon investments. The revised EU ETS Directive entered into force on 8 April 2018 and sets the framework for the fourth trading period from 2021 to 2030. The revised EU ETS Directive, will enable the EU to achieve its overall greenhouse gas emissions reduction target for 2030. Under the 2030 framework, the EU aims to reduce emissions by at least 40% by 2030 compared to 1990 levels. To achieve this target, emissions within the EU ETS must fall by 43% by 2030 compared with 2005 levels.

On 14 July 2021, the European Commission proposed the revision of the EU ETS Directive under the "Fit for 55" package of legislative proposals, in view to achieve climate neutrality in the EU by 2050, including the intermediate target of an at least 55% net reduction in greenhouse gas emissions by 2030. The EU ETS is under revision from 2021 as part of the EU Green Deal (Bruninx & Ovaere, 2022). The Carbon Market Watch, an environmental non-governmental organization, has called for several changes to be made to the system, namely, to ensure that the EU meets its target of becoming carbon neutral by 2050 and for a strengthening of the market stability reserve (that is, to balance the supply and demand of emission allowances). Additionally, the Carbon Market Watch has called for an end to the free allocation of pollution permits and for all auctioning revenues from the EU ETS to be earmarked for delivering further climate action, industrial innovation, just transition, and international climate finance. Furthermore, a one-off reduction of allowances is necessary to ensure that the cap better reflects real emission levels.

The Directive (EU) 2023/959 - adopted by the co-legislators on 10 May 2023 - introduces amendments to Directive 2003/87/EC and Decision (EU) 2015/1814, establishing the Market Stability Reserve (MSR). The MSR is an add-on to the EU ETS legislated in 2015 that started operating in 2019. It postpones the issue date of allowances as a function of the number of unused allowances in the market ('the bank') and is an autonomous version of the backloading implemented in Phase 3 (Perino, 2018).

ACCOUNTING AND TAXATION OF CO2 EMISSION ALLOWANCES: THEORETICAL AND EMPIRICAL BACKGROUND

With the growing concern for sustainability and environmental protection, the number of companies seeking to disclose information about the measures they take and how they demonstrate their environmental concern is increasing. In this regard, research on the accounting and disclosure of carbon emissions has been the subject of study by an increasing number of researchers. Some studies are based on gathering information obtained through voluntary surveys (González-González & Zamora-Ramírez, 2016; Hörisch,

2013; Yunus et al., 2016), such as the survey conducted by CDP (formerly Carbon Disclosure Project). Focusing on studies regarding the disclosure of information related to CO2 allowances, there are some that analyze whether the imposition of specific and mandatory accounting standards in this area effectively influences greater disclosure (García-Torea et al., 2022; González-González & Zamora-Ramírez, 2016; Lovell et al., 2010; Reis & Lopes, 2011). Multiple approaches for the accounting treatment of these rights in different countries are also analyzed, caused by the absence of a European regulation (García-Torea et al., 2022; Boas & Monteiro, 2012; Steenkamp et al., 2011; Warwick & Ng, 2012; Haseeb Ayaz, 2017; Gallego-Alvarez et al, 2016; Black, 2013). Some other studies focus on the type of information that is disclosed (Black, 2013; Fernando et. al., 2010; Gonçalves & Góis, 2012; Jurado et al., 2007), and finally there are also studies that seek to identify the factors that influence the extent of disclosure (Florindo et al., 2020; Hermawan et al., 2018, Akbas & Canikli, 2018; Boas & Monteiro, 2012; Borghei et al., 2018; Gallego-Álvarez et al., 2016; Giorgios & Dimitrios, 2009; Hapsoro & Falih, 2020; Hermawan et al., 2018; Matsumura et al., 2011).

Considering the objective of this chapter, the studies that have focused on the consequences of the lack of standardisation of the accounting procedure for CO2 emission allowances in the context of the EU ETS stand out among those that have been carried out. The withdrawal in 2005 of IFRIC 3 - Emission Rights meant the elimination of common rules for the accounting of CO2 emissions by companies affected by the EU-ETS. In response to this situation, some Member States proceeded to issue rules applicable by companies located in their respective jurisdictions.

Thus, at present, at the EU level, there is a multiplicity of accounting practices for recording and disclosing information on emission allowances granted and purchased (Ayaz, 2017; Black, 2013; Gallego-Álvarez et al., 2016; García-Torea et al., 2022; Lovell et al., 2010; Warwick & Ng, 2012). In fact, alongside the accounting regimes established by mandatory rules, there are other procedures applied by companies where no accounting provisions have been established in this regard. The situation described above, which calls into question the comparability of the accounting information disclosed by companies operating in the EU-ETS, highlights the need for common rules for the accounting of CO2 emission allowances.

Ayaz (2017), applying a content analysis, analysed the different accounting approaches adopted by 28 large carbon producers in the EU-ETS. The main conclusion of the study is the lack of uniformity in the recognition and valuation of emission allowances obtained, acquired or generated. In fact, although a significant number of companies do not disclose information on allowances, most of those that include allowances in their accounting statements recognise them as intangibles, with a residual number of companies placing allowances among inventories. Differences in accounting recognition also extend to the initial valuation (cost versus fair value) or subsequent valuation of allowances.

In line with previous studies (Lovell et al., 2010; Warwick & Ng, 2012), Black (2013), through a content analysis of 62 CO2 emitting companies, identified three main policies in the treatment of allowances in the context of the EU ETS following the withdrawal of IFRIC 3: (i) net liability approach, based on the classification of emission allowances as intangibles and where a liability is only recognised when emissions exceed the free allocation; (ii) an approach broadly based on IFRIC 3, where allowances obtained for free are valued at fair value and a liability is recognised for the delivery obligation; and (iii) an approach based on inventory classification, where allowances allocated for free are valued at zero cost. The situation described above remains unchanged today, as the implementation of the auctioning system in 2013 did not change the practices that companies had been applying in relation to emission allowances (García-Torea et al., 2022).

In order to explain the lack of uniformity in the accounting treatment of emission allowances, Gallego-Álvarez et al. (2016) focus on the factors that condition the selection of a particular practice. In this regard, using a sample of 119 companies from different countries and sectors of activity, they conclude that firms operating in the EU ETS tend to record emission allowances through provisions, investments or as inventories. On the other hand, companies that disclose climate change indicators according to the Global Reporting Initiative model tend to recognise allowances as expenses, usually for research and development (I+D). However, companies located in countries that joined the Kyoto Protocol do not recognise allowances.

In short, the lack of uniformity in accounting practices adopted in relation to emission allowances not only hinders the comparability of financial statements, but also conditions the level of transparency of financial information on greenhouse gas emissions and emission allowances (Milanés et al., 2020), which will hinder the comparability of corporate performance (Karatzoglou and Karatzoglou, 2011).

Taking into account the geographical scope of this chapter, in Portugal, there are notable studies that specifically address the disclosure of information regarding CO2 emission allowances. Boas & Monteiro (2012), Dias et al., (2016) and Fernando et al. (2010), analyzed the information disclosed about emission allowances in Portuguese companies that are part of the National Allocation Plan for Emission Allowances (PNALE). Reis & Lopes (2011) conducted a similar analysis on companies listed on Euronext Lisbon, and Gonçalves & Góis (2012) also added the Top 100 Largest Portuguese Companies to their analysis. Florindo et al. (2020) expanded this approach by including the analysis of information related to derivative instruments on emission allowances as well.

In the Spanish context, we find articles that analyze the regulations applicable to emission allowances both at the national level (Martín & Jurado, 2005; Fernández Cuesta et al., Mateos, 2006; López & López, 2007; Giner, 2007; Gonzalo & Larriba, 2007; Alborg, 2015; Pérez & Soto, 2017a; Pérez & Soto 2017b) as well as in the international regulatory scope, emphasizing the need to reach a consensus for the treatment of this asset (Bilbao & Mateos, 2007; Bebbington & Larrinaga, 2008; Mateos, 2009). On the other hand, a series of articles have also been published that analyze specific aspects of emission allowances. For instance, Fernández & Fronti (2010) focus their study on the legal, technical, and environmental viability of clean development and joint implementation projects. In the field of management accounting, Jurado & Martín (2006a) center their research on the impact of emission rights on the business profit equation; Jurado & Martín (2006b) analyze the budgetary implications of CO2 emissions. Rabadan et al (2017) propose an indicator to measure the quality of information provided in the financial statements regarding emission allowances. Additionally, the study by Gonzalez-Gonzalez and Zamora Ramírez (2016) analyzed the factors that contribute to the decision of Spanish companies to disclose carbon information.

In a comparative study, Milanés et al. (2020) analyzed the transparency of financial reporting on emission allowances and GHG emission among companies registered with the Portuguese, Spanish, and French National Plans of Allocation.

In contrast to the existence of studies specifically dedicated to accounting and disclosure of CO2 emission allowances, only a limited number of studies focus on taxation. Among them, the analysis of the positive impact of the implementation of environmental taxes on the mitigation of CO2 emissions stands out (Ghazouani et al., 2020; Lin & Li, 2011; Ulucak et al., 2020). However, most of the existing literature on this topic has a strong theoretical character (Alonso Madrigal, 2012; Nunes, 2009; Tozer, 2007) or a macroeconomic perspective, focusing on the feasibility of environmental taxation as an instrument to reduce CO2 emissions (Haites, 2018; Metcalf, 2021; Shmelev & Speck, 2018). In the fiscal field, the works of Bilbao & Mateos (2006a), Bilbao & Mateos (2006b), and Madrigal (2012) stand out, analyzing the relationship of emission allowances with the corporate tax at the national level, while Bilbao & Mateos (2009) and Martínez (2016) study it at the European level. In this line of research, Bilbao (2010a) also stands out, analyzing taxes originating from CO2 emissions, as well as Martín & Jurado (2008), who explore the accounting treatment of so-called ecological taxes. The relationship between the Value Added Tax and emission allwances is examined in Bilbao (2010b) and Bilbao & Rodríguez (2010).

ACCOUNTING REGULATION OF GREENHOUSE GAS EMISSION ALLOWANCES IN SPAIN

Evolution of Accounting Standardization of Greenhouse Gas Emission Allowances in Spain

The first regulation to account for greenhouse gas emission rights in Spain was a Resolution of the Institute of Accounting and Auditing (ICAC) in 2006, which approves rules for the registration, valuation, and reporting of greenhouse gas emission allowances. This regulation, which only addressed the treatment of rights that would be consumed in the production process, established that these rights, in accordance with the guidelines of IFRIC 3, were to be considered accounting assets - specifically, intangible assets. In fact, they had to be integrated into the company's assets on the date of the rights' origination, appearing in a specific asset item on the balance sheet under the name "Greenhouse Gas Emission Rights."

In accordance with this provision, at the time of initial recognition, the rights had to be valued at their acquisition price if they had been purchased on the market, or at their fair value (as determined by this regulation, the market value "will generally be used as the fair value of the emission right) if the certificates had been obtained for free (generally, through the national allocation plan). In the latter case, the company was also required to recognize, for the same amount, a deferred income that would be recognized in the results as expenses related to emissions were incurred. Finally, when the company did not have enough rights to meet the delivery obligation for the emissions recorded, it had to recognize a provision for the best estimate to cover the deficit.

The accounting treatment of emission rights acquired for speculative purposes was not incorporated into Spanish law until 2013, when the rules applicable to intangible assets were developed (Resolution of May 28, 2013, of the ICAC, which sets forth rules for registration, valuation, and information to be included in the intangible assets note). Specifically, this last provision, following the business model criteria introduced by the Autorité des Normes Comptables (Regulation No. 2012-03, of October 4, 2012, on the accounting of greenhouse gas emission rights and similar units, later repealed by Regulation ANC No. 2014-03), determined the guidelines to be followed in accounting for the rights in three different situations. First, if the rights were allocated or acquired to be consumed in the production process, they had to be classified as intangible assets, with the registration and valuation rules applicable to these assets. On the other hand, if the rights were acquired with the intention of being sold, the rules applicable to inventories had to be applied. Finally, the criteria provided for financial instruments - specifically, advances or purchase or sale commitments - would apply in the case of contracts formalized for the purpose of receiving or delivering emission rights.

Finally, since January 1, 2016 (the change was introduced by Royal Decree 602/2016, of December 2nd, whose origin is accounting harmonization in the EU), the criteria applicable to inventories have been applied to all emission rights, whether acquired for the purpose of cancelling the delivery obligation or for obtaining a gain from their sale. This regulatory change, which required the accounting reclassification of rights that were listed among intangible assets on that date to their new location, is justified by the legislature by stating that, although the classification of rights as intangible assets or inventories has been a widely discussed issue in the international context, "the practical impact is not excessively relevant. Therefore, given arguments in favour of their treatment as inventories, such as the fact that we are dealing with an asset whose presence in the company's assets, in general, will not be long-lasting, and whose systematic depreciation is not evident, the change in its accounting classification is configured as a practical solution for accounting for an economic fact that had a difficult fit in the new regulatory framework for intangible assets" (Section IV of the introduction of Royal Decree 602/2016, of December 2nd).

In short, as shown in Figure 1, since the emergence of the emissions market, Spanish law has incorporated rules for the registration and valuation of rights, although in a period of ten years, three different approaches have been adopted in defining their accounting nature, which undoubtedly affects the comparability of the information disclosed on this asset.

Figure 1. Evolution of the recognition of emission allowances according to the approved regulations in Spain.
Source: Own elaboration

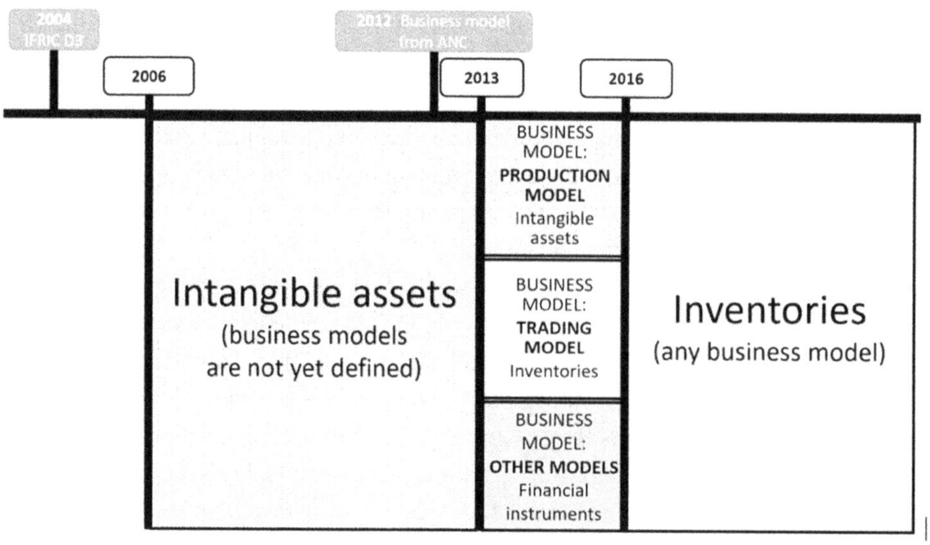

The Accounting Regime of Greenhouse Gas Emission Allowances in Spain

The approval of Royal Decree 602/2016 involved the modification of the accounting regime for greenhouse gas emission allowances, altering their accounting classification. However, there were no significant changes in the valuation criteria that had been previously applied, nor in the information to be disclosed in the financial statements in relation to this asset. In fact, the introduction of this regulation states that the previous criteria contained in the ICAC Resolution of 2013 remain in force, except for the change in the accounting classification.

In the following sections, the current regulations in Spain on the accounting treatment of emission allowances will be analysed.

Accounting Nature of Emission Allowances

Regardless of the company's purpose in acquiring greenhouse gas emission allowances, the allowances will be recorded as inventory. Therefore, they will appear in the current assets section of the balance sheet. However, if the company estimates that some of the allowances will be consumed over a period longer than one year, the amount of the allowances estimated to be consumed before and after that period will be separately shown in the balance sheet (Specifically, the first additional provision of Royal Decree 602/2016 determines the breakdown of the item "2. Raw materials and other supplies" of section B.II of the current assets in the event that the indicated situation occurs).

Initial Valuation of Emission Allowances

Greenhouse gas emission allowances will be initially valued according to the method by which they have come into the company. In this sense, the current accounting standard provides for different valuation criteria depending on whether the allowances come from a purchase on the market, have been allocated by the government, or have been produced through a certified reduction of emissions or in an emissions reduction unit.

When the allowances are purchased on the market, whether for use in the production process or for subsequent sale, they will be recognized at their acquisition cost (Section 8 of the sixth rule of the ICAC Resolution of May 28, 2013), which will coincide with the amount paid. Indirect taxes levied on the purchase will be included in the initial valuation, provided they cannot be directly recovered from the Tax Authority.

In the case of allowances obtained through allocation by the State, either without consideration or for a substantially lower amount than their market value, they will be initially valued at fair value (Section 1.2 of the 18th Accounting and Valuation Standard (NRV) of the General Accounting Plan (PGC). If this situation arises, the company must recognize a deferred income at the beginning of the natural year to which the allowances correspond. The deferred income, whose valuation will be the difference between the fair value of the emission allowance and, where applicable, the consideration paid, will be transferred to income as the expense for emissions associated with the allowances received without consideration or for a substantially lower amount than their market value is recognized.

Finally, when the allowances originate from a certified reduction of emissions or an emissions reduction unit, they will be recognized at their production cost, which will include (Section 2 of the third standard of the ICAC Resolution of May 28, 2013):

a. The cost of materials and services used or consumed in the generation of the allowances.
b. The cost of employee remuneration derived from the generation of the allowances.
c. Fees accrued for the registration of the allowances.
d. The amortization of patents and licenses used in the generation of the allowances.

Expenditures related to personnel training or costs attributable to administrative or commercial functions will not be included in the cost of the allowances unless they can be directly attributed to their acquisition (Section 3 of the second norm of the Resolution of ICAC of May 28, 2013).

Subsequent Valuation of Emission Allowances

The accounting placement of emission allowances among inventory requires that they be shown in the balance sheet at their acquisition cost or production cost unless their net realizable value is lower. If the latter circumstance arises, "the necessary valuation adjustments shall be made, recognizing them as an expense in the profit and loss account" (NRV 10th of the PGC -Plan General de Contabilidad). Therefore, allowances are subject to impairment adjustments, which recognize "the estimated loss in the value of the asset that represents the difficulty of recovering, through its use, sale, or other form of disposal, the entirety of its book value." (Section 1 of the second norm of the ICAC Resolution of May 28, 2013).

Regarding emission rights intended to be consumed in the production process, we believe that recognition of impairment loss is not appropriate when the estimated value of the delivery obligation exceeds the carrying amount of the rights needed for its cancellation (Since the emission rights will be consumed in the production process by being delivered, the application by analogy of NRV 10th of PGC leads to not making a value adjustment regarding the rights that will be delivered to cancel said obligation, even if their net realizable value is lower. Specifically, that standard states that "in the case of raw materials and other consumable materials in the production process, no value adjustment will be made as long as it is expected that the finished products to which they are incorporated will be sold above cost"). Conversely, when the carrying amount of emission rights exceeds the estimated amount needed to cancel the delivery obligation, an impairment loss may be recognized if the carrying amount of the excess rights exceeds their net realizable value, which would be considered fully or partially irreversible if they were assigned for free or below market value (NRV 18th of the PGC).

Deregistration of Emission Rights

Deregistration of emission rights occurs in the following situations (Section 8.1.2.2 of the ICAC Resolution of May 28, 2013):

a. transfer to third parties through sale.
b. delivery to the administration to comply with the obligation to transfer a number of rights equivalent to the total of verified emissions.

c. expiration, as valid emission rights for each period automatically expire when four months have elapsed since the end of the period.

In the case of the sale of rights to third parties, the company recognizes revenue on the transaction date for the agreed-upon selling price, net of any discounts, price reductions, or similar items that the company may grant (NRV 14th of the PGC). On the other hand, when the rights are delivered to the administration to comply with the obligation contracted for verified emissions, the deregistration occurs in the following year, when the cancellation of the rights is communicated to the corresponding administrative register (Section 8.1.3.2 of the ICAC Resolution of May 28, 2013).

Recognition and Valuation of Expenses Associated With Greenhouse Gas Emissions

The emission of greenhouse gases requires the recognition of an operating expense, the counterpart of which is a provision, as it is an expense of the period that is clearly specified in terms of its nature but indeterminate in terms of its exact amount at the closing date. Since the obligation will be fulfilled by delivering the corresponding rights for verified emissions before April 30 of the following year, the provision will be classified as a current liability on the balance sheet.

The valuation of the expense for emissions and the associated provision, which when significant will require the creation of specific items in the financial statements (specifically, the item called "Expenses for greenhouse gas emissions" will be created in the "Other operating expenses" section of the income statement. Additionally, the item "Provision for greenhouse gas emission rights" will be enabled in the "Short-term provisions" heading of the current liabilities section of the balance sheet), is conditioned by the order of precedence of the rights established by the ICAC (Section 8.1.3.2 of the ICAC Resolution of May 28, 2013) for the purpose of fulfilling the obligation. First, the emission rights received through an allocation under the emission scheme will be considered. These rights will be allocated to emissions made in proportion to the total emissions expected for the entire period for which they were allocated. The expense corresponding to this portion of the obligation will be quantified according to the book value of the transferred emission rights. In addition, if deferred revenue was recorded because the rights were acquired for free or below market value, revenue will be recognized in the income statement in proportion to the recognized expense.

Second, the remaining emission rights shown in the company's balance sheet, acquired in the market, or generated through other clean development mechanisms, will be considered. The expense corresponding to this part of the obligation will generally be quantified according to the weighted average cost method of the emission rights. The application of this valuation criterion requires that they be homogeneous assets, so the possible restrictions on the use of the rights must be considered.

When emissions made by companies exceed those that can be cancelled with the rights shown in the company's balance sheet, the expense corresponding to the rights deficit must also be considered. To this end, the expense for emissions will be quantified for this part according to the best possible estimate of the amount necessary to cover the deficit.

Emissions Allowances in the Financial Statements

The treatment of emission rights in Spanish accounting standards, despite their classification as inventory, produces different effects on financial statements depending on the company's business model regarding rights (see Figure 2). However, apart from the impacts on the balance sheet and income statement, the notes to the accounts will include all significant information related to emission rights. In particular, the following aspects will be disclosed:

Figure 2. Impact on balance sheet and income statement of emission allowances according to the business model
Source: Own elaboration

	Balance sheet	Income statement
CO_2 emitter	**Assets** Inventories **Liabilities** Provision for liabilities and charges **Equity** Deferred income	**Other operating revenues** (-) Purchase of CO_2 emission rights [Cost / Fair value] (-) Cost of CO_2 emissions (+) CO_2 emission rights activated (+) Recognition of the deferred income *= Estimated cost of CO_2 emissions*
Trading business	**Assets** Inventories (-) Impairment of inventories	**Turnover** (-) Purchase of CO_2 emission rights [Cost / Fair value] (+) Income from selling of CO_2 emission rights (+) CO_2 emission rights activated *= Profit/loss from selling CO_2 emission rights*

a. The "Valuation Standards" note will indicate the valuation criteria for emission rights, as well as the criteria for charging expenses related to emissions to results.
b. A specific note is created, limited to the normal model of the report, under the name "Information on greenhouse gas emission rights", where, in a concise manner, information will be provided on issues such as:
    i. Quantity of assigned rights and their annual distribution, as well as whether the assignment was free or paid.
    ii. Analysis of the movement during the financial year of emission rights and the corresponding impairment losses that could adjust their value, indicating the following: initial balance, entries or acquisitions, disposals and other reductions, and final balance.
    iii. Expenses for the year derived from emissions, specifying their calculation.
    iv. Amount of the "Provision for greenhouse gas emission rights", specifying, if applicable, the amount due to a deficit in emission rights.
    v. Information on contracts related to emission rights.
    vi. Subsidies received for emission rights and the amount charged to results as income for the year.
    vii. Contingencies related to penalties or provisional measures, in accordance with the Law.
    viii. If applicable, being part of an installation group.

ACCOUNTING REGULATION OF GREENHOUSE GAS EMISSION ALLOWANCES IN PORTUGAL

Portugal was not indifferent to the issue of CO2 emission allowances and took steps to establish appropriate accounting practices following the establishment of the EU carbon market. In 2002, an accounting directive (Diretiva Contabilística - DC) called DC 29 - Environmental Matters was created in Portugal specifically for environmental issues. However, DC 29 did not correspond directly with any standard issued by the IASB and was inadequate regarding the specific issue of the accounting treatment of CO2 emission allowances. Consequently, the Portuguese Accounting Standardization Committee issued Technical Interpretation (IT) No. 4 in 2006 - Greenhouse Gas Emission Rights - Accounting for Emission Allowances.

On January 1, 2010, the Accounting Standardization System came into effect, leading to the repeal of both IT4 and DC 29, which were replaced by NCRF 26 - Environmental Matters. This standard not only prescribes the treatment of environmental issues but also has an appendix that resembles the content of IT No. 4.

The appendix of NCRF 26, which is fully dedicated to greenhouse gas emission allowances, advocates for the recognition of these allowances as intangible assets, whether they have been acquired in the market or granted for free (§5), with initial measurement at fair value (§10). Greenhouse gas emission allowances granted for free should lead to the simultaneous recognition of a subsidy (§6).

According to NCRF 22 - Grants and Other Assistance from Government Entities (§12a), this type of subsidy should be classified as an asset-related subsidy, and its recognition should be made in equity. Paragraph 12a of NCRF 22 states that "Non-refundable subsidies related to intangible assets from public entities should be initially recognized in Equity and subsequently (...) systematically recognized as income during the periods necessary to balance them with the related expenses they are intended to compensate." This is the major difference between the current accounting standards and IT no. 4, as the latter treated the recognition of the subsidy as deferred income.

Initial Recognition of Emission Allowances

To initially recognize CO2 emission allowances, the basis of measurement to be adopted for both free grants and those acquired for a fee is fair value. It is assumed that fair value in the latter case coincides with the acquisition cost, which is consistent with the treatment provided in NCRF 6 - Intangible Assets and NCRF 22 - Grants and Other Support from Public Entities.

Subsequent Recognition of Emission Allowances

NCRF 26 does not provide guidance on subsequent measurement of emission allowances, but NCFR 6 offers a choice between the cost model and revaluation model (§70) for intangible assets. Under the cost model, subsequent measurement of emission allowances is based on cost less accumulated amortization and impairment losses (NCRF 6, §72). Under the revaluation model, subsequent measurement is based on the revalued amount less accumulated amortization and impairment losses (NCRF 6, §73). Fair value for revaluation purposes should be determined with reference to an active market (such as the European carbon market for CO2 emission allowances). Any increase in value is recognized directly in equity while any decrease exceeding the value previously recognized in equity is recognized in the income statement for the period (NCRF 6, §83 and §84).

Regarding subsequent measurement, NCRF 26 only addresses the recognition of amortization when greenhouse gases are emitted and the corresponding expense recognition based on the cost of the allowances held using FIFO costing formula (§7, §8, and §11). If the entity emits gases above the allowances held, a provision must be recognized in accordance with NCRF 21, and the measurement is based on the best possible estimate of the price for their acquisition (§9, §11).

Emissions Allowances in the Financial Statements

The Portuguese regulatory framework recognizes CO2 emission allowances as intangible assets, which should be presented under the non-current assets section of the balance sheet. In the income statement, it is necessary to recognize the amortization and any provisions related to the intangible asset. Additionally, the standard requires certain disclosures in the notes to the financial statements, including:

a) the allowances allocated for the period.
b) Greenhouse gas emissions in tonnes of carbon dioxide equivalent.
c) Allowances sold during the period in tone-carbon dioxide and its price.
d) Allowances acquired during the period, in tonnes of carbon dioxide and their price.
e) Fines, and related ancillary sanctions associated with greenhouse gases emissions. and
f) fair value of the licenses held.

Table 1 summarizes the main aspects provided in the national accounting standard for CO2 emission allowances.

THE TAX REGIME FOR GREENHOUSE GAS EMISSION ALLOWANCES IN SPAIN

The following lines briefly analyse the tax implications of operations related to emission rights both within the framework of the Corporate Income Tax and the Value Added Tax.

Corporate Income Tax

Corporate Income Tax (CIT) levies on the income of companies and other legal entities residing in Spanish territory. Its taxable base is calculated by adjusting the accounting result according to the provisions established in the tax law. In the absence of specific provisions, the tax regime coincides with the treatment established by accounting standards, and generally no adjustment is necessary. This situation applies to emissions trading rights, where applicable tax legislation does not contemplate any uniqueness. However, to analyse the tax implications of emissions trading rights in the CIT, the following three issues will be considered:

a. Initial and subsequent valuation of emissions trading rights.
b. Expense derived from emissions trading rights.
c. Transfer of emissions trading rights.

Table 1. Accounting treatment of GHG emission allowances (NCRF 26 and its appendix)

Recognition	**GHG Emission Allowances**	**Recognition of an intangible asset.**
	Licenses acquired free of charge	**Recognition of a subsidy evidenced in equity. The allowance will be charged to income on the basis of the depreciation recorded (NCFF 22).**
Measurement	Initial measurement of licenses acquired free of charge	Measurement at Fair Value.
	Initial measurement of licenses acquired for a fee	Measurement at fair value, which is assumed to coincide with the acquisition cost.
Subsequent Measurement	Subsequent Measurement	Period expense (amortization of intangible assets) measured by the cost of the licenses, on a FIFO basis.
	Subsequent Measurement	Recognition of a provision and an expense for the best estimate of expenditure (NCRF 21)
	Cost Model vs. Revaluation Model	Missing (however, these models are set out in IFRS 6 - Intangible assets)
Derecognition	Delivery of licenses	Derecognition of the carrying amount of the IIA (accumulated amortization and intangible assets).
	Alienation of Licenses	Derecognition of Intangible assets, and recognition of other income and gains / expenses and losses, and recognition of net financial debt. Derecognizing the allowance against a revenue and earnings account (if it is a free allocation)
	Cancellation of unused licenses (purchased free of charge)	Derecognition of the subsidy and the intangible asset.
	Cancellation of unused licenses (purchased for a fee)	Recognition of the expense or loss and derecognition of the intangible asset.
	Cancellation of licenses that you do not have	Recognition of new licenses acquired, the provision, as well as the IA, are derecognized for the licenses issued (not available at the date of issue)
Disclosure	Disclosure	- Emissions of greenhouse gases in tonnes of carbon dioxide equivalent; - EA sold in the year in tonnes of carbon dioxide and their price; - EA acquired in the year in tonnes of carbon dioxide and their price; - Fines, and ancillary sanctions related to the emission of greenhouse gases; - Fair value of the licenses held.

Source: Adapted from Boas and Monteiro (2012)

From a tax perspective, the initial accounting valuation of emissions trading rights is accepted, which will be determined by their cost - acquisition or production. Even when rights are obtained free of charge or at a price below market value, the fair value accounting is admitted, and the deferred income quantified and recognized as income for the year as expenses associated with the rights obtained free of charge are recorded. Accounting-wise, a value adjustment will be made in the subsequent valuation of the rights when their book value is higher than their net realizable value. The impairment loss recognized in the income statement is tax-deductible when, at the end of the fiscal year, it is justified that the market value of the rights is lower than their purchase price or production cost.

At the end of the fiscal year, the company must recognize an expense for emissions during the period, with a counterpart to a short-term provision for risks and expenses. The current legislation does not consider this provision as deductible in the CIT. However, we understand that the expense associated with the provision will be deductible since it is an accrued expense that corresponds to a duly justified payment, although it is uncertain in amount since the emissions for the year are not yet verified. On the other hand, expenses arising from administrative sanctions that may result from non-compliance with regulations governing emissions trading rights will not be deductible.

Finally, when the company sells emissions trading allowances, it recognizes revenue for the sales price. Since the Spanish accounting standards have opted for the inventory division method, the profit or loss derived from the sale will not be known until the end of the fiscal year when the account representing the emissions trading rights is regularized. This income will form part of the taxable base of the CIT. Similarly, in the event that the rights were acquired in whole or in part free of charge, it will be necessary to recognize in income the corresponding portion of deferred income, which will also be integrated into the taxable base of the tax.

Value Added Tax

In order to understand the implications of emission rights in Value Added Tax (VAT), two different operations will be considered:

a. The allocation of emission rights by the administration.
b. Subsequent transfers of emission rights.

Generally, the transfer of emission rights is considered a taxable service subject to VAT. However, in the case of the free allocation of rights by the State, it must be understood that the State is not acting as an entrepreneur or professional but in the exercise of a public function and, therefore, the free allocation is not subject to the tax. However, when the allocation is made in exchange for a price, this action is also outside the scope of VAT, but would be subject to the Tax on Onerous Asset Transfers (Alonso Madrigal, 2012).

On the other hand, subsequent transfers of emission rights are subject to VAT, so entrepreneurs who sell these rights are required to charge and pay the corresponding VAT. However, the main difficulty in this type of operation lies in determining where the service provision is located, which is especially relevant when dealing with transactions between cross-border operators (Bilbao Estrada & Rodríguez Márquez, 2010). In order to prevent fraud, in 2009, the list of situations where the reverse charge mechanism applies was extended, including the provision of services related to emission rights, certified emission reductions, and greenhouse gas emission reduction units. Therefore, since that date, in sales of emission rights, whenever the transfer has been made by an operator not established in Spain, the taxable person for VAT purposes required to present and declare the transaction will be the buyer.

THE TAX REGIME FOR GREENHOUSE GAS EMISSION ALLOWANCES IN PORTUGAL

The Portuguese tax regime regarding **Corporate Income Tax** is similar to the Spanish legislation in that it aligns with the accounting treatment when there is no specific tax legislation for emission allowances. As previously mentioned, emission allowances are considered intangible assets and are amortized based on the FIFO method (§5 and 11, appendix NCRF 26). These amortizations are recognized as deductible expenses for tax purposes in the period (art. 29º, nº 1, a), Corporate Income Tax Code), therefore no further tax adjustment is needed.

However, NCRF 26 (§5, appendix) also requires the recognition of a provision for GHG emissions that exceed the allowances held (NCRF 21), but this provision is not tax-deductible (art. 34º Corporate Income Tax Code).

Regarding VAT, the Portuguese government has recognized the importance of the European carbon market for achieving emission targets and since 2010 has implemented a reverse-charge rule for CO2 allowances. Under this rule, the buyer of the licenses is responsible for paying and delivering the VAT to the state. In other words, it is the purchaser's own responsibility to settle and deliver to the State the VAT levied on the transmission of CO2 allowances.

SOLUTIONS AND RECOMMENDATIONS

Considering that, in both Spain and Portugal, listed companies are required to adopt the IASB standards, which lack regulation regarding the accounting treatment of emission allowances, an inconsistency is observed in the accounting treatment of these allowances by these entities compared to non-listed entities. Therefore, it is urgent to define a standard on this matter by the IASB, which should be taken into account by both countries in order to promote accounting harmonization.

It is also important that these two countries incorporate into national standardization the disclosure requirements that may be published following the publication of the exposure draft being developed by the International Sustainability Standards Board, regarding Climate-related disclosures.

FUTURE RESEARCH DIRECTIONS

In Portugal and Spain, non-listed companies have a harmonized accounting treatment and disclosure requirements for CO2 emission trading. However, there is currently a lack of accounting standards for listed companies that follow IASB standards and hold CO2 emission licenses. To address this, the International Sustainability Standards Board (ISSB), as part of the IFRS Foundation, is finalizing requirements for climate-related disclosures. In March 2022, the ISSB published Exposure Draft IFRS S2 Climate-related Disclosures, which builds on the Task Force on Climate-Related Financial Disclosures (TCFD) recommendations and incorporates industry-based disclosure requirements from SASB Standards. The aim of Draft IFRS S2 is to require entities to disclose information about their exposure to significant climate-related risks and opportunities, as well as to improve processes for measuring and disclosing Scope 3 GHG emissions (i.e. indirect emissions from sources outside an organization's control). Future research could focus on developing frameworks or models for estimating Scope 3 emissions across different industries and exploring how these estimates could be integrated into GHG reporting frameworks.

Although there is a growing trend towards voluntary GHG disclosure, there is limited understanding of the factors that encourage firms to disclose their emissions. Research could examine how external factors (e.g. regulatory pressure, stakeholder activism, reputation concerns) influence firms' decisions to disclose GHG emissions.

Furthermore, there is currently a lack of research on the impact of GHG taxation. Future research could analyze this impact on different sectors, including those that are sensitive to taxation. The results of such research could inform policymakers in designing more effective GHG taxation policies that take into account sector-specific challenges and opportunities.

CONCLUSION

The absence of a specific international accounting standard for GHG emission allowances has led to the development of specific regulations in some European Member States, such as Portugal and Spain, for the accounting treatment of these licenses. In Portugal, the Accounting Standardization System's Accounting and Financial Reporting Standard (NCRF) nº. 26 - Environmental Matters has an appendix dedicated to the accounting of CO2 emission allowances. Generally, emission allowances are treated as intangible assets and amortized as they are used up, calculated using the FIFO method (§5 and 11, appendix AFRS 26).

In Spain, the first version of the Resolution of Instituto de Contabilidad y Auditoría de Cuentas (ICAC, 22nd February 2006) stipulated that emission allowances should be classified in the balance sheet as intangible assets. Contrary to Portuguese legislation, emission allowances were not subject to systematic amortization, without prejudice to the company's obligation to provision the value of the emission allowances if, at year-end, their book value is lower than their market value. However, after the approval of Royal Decret 602/2016, emission rights are recognized as inventories since 2016.

Both regulations establish the initial measurement basis as either cost (when licenses are acquired) or fair value (when licenses are obtained free of charge). If licenses are obtained for free, both regulations require the recognition of a subsidy. However, there is a difference between NCRF 26 and Spanish regulations in terms of how the subsidy is recognized. According to NCRF 26, the subsidy is recognized in equity, while Spanish regulations require the subsidy to be recognized as deferred income.

Although the initial measurement basis is identical, as stated before, the accounting regulations prescribe a completely different accounting treatment in the initial recognition. In Spain, the allowances are recognized as inventories and in Portugal as intangible assets. Interestingly, in the subsequent measurement of the licenses, the Portuguese regulation uses a specific inventory criterion, FIFO, to calculate the amortization of the intangible asset.

In both regulations, the emission of greenhouse gases must be recognized as an expense. In Portugal, this expense corresponds to the amortization of emission allowances, whereas in Spain, the expense corresponds to a provision. The differentiated treatment of the type of assets to be recognized (inventories vs intangible assets, in Spain and Portugal, respectively) leads to an impairment test being carried out in the Spanish case, comparing the cost of the allowances with their net realizable value, which may result in the recognition of an impairment loss in inventories.

Both regulations recognize an expense (provision) for the responsibility of the operator for the emission of gases above the licenses held, whose measurement is made by the best possible price estimate.

Regarding taxation, there is no specific tax treatment for GHG emissions in either country. In Portugal, the corresponding amortizations are accepted as fiscal expenses in the period as they are recognized as intangible assets. However, the provision recognized in the case of a company exceeding the emission allowances is not fiscally accepted. In Spain, as the company makes GHG emissions, an obligation arises to surrender emission allowances to the government at the end of the emission period.

As stated before, this obligation is recognized as provision for liabilities and charges. This provision is a duly justified outstanding payment, even though it may be uncertain in amount, and should therefore be considered as a deductible expense for societies tax purposes. Also, the impairment loss recognized in the income statement, in case of the book value is higher than the realizable value of the allowance, shall be tax deductible.

In terms of income tax, the absence of specific treatment in the tax regulations for operations relating to emission licenses leaves it to the accounting framework to determine their effect on the results of the companies that carry them out and, consequently, on the tax. Thus, the tax treatment of emission allowances for the purposes of corporate income tax is completely inseparable from their accounting treatment.

Regarding VAT, in Europe it has been introduced a reverse-charge mechanism on CO2 emissions trading, seeking to reassure markets and avoid disparities in treatment. Therefore, there is no differences between both countries in this matter.

ACKNOWLEDGMENT

The authors acknowledge the financial support of FCT - Foundation for Science and Technology, I.P., within the scope of multi-annual funding UIDB/04043/2020

REFERENCES

Akbaş, H. E., & Canikli, S. (2019). Determinants of voluntary greenhouse gas emission disclosure: An empirical investigation on Turkish firms. *Sustainability (Basel)*, *11*(1), 107. doi:10.3390u11010107

Alborg, I. R. (2015). *Contabilidad de los derechos de emisión de carbono*. Instituto de Contabilidad y Auditoría de Cuentas.

Alonso Madrigal, J. (2012). Los derechos de emisión de CO2 y la fiscalidad española. ICADE. *Revista de la Facultad de Derecho (Montevideo)*, *86*, 55–90.

Arnold, D. (2016). Corporate Responsibility, Democracy, and Climate Change. *Midwest Studies in Philosophy*, XL.

Ayaz, H. (2017). Analysis of carbon emission accounting practices of leading carbon emitting European Union companies. *Athens Journal of Business & Economics*, *3*(4), 463–486. doi:10.30958/ajbe.3.4.5

Bebbington, J., & Larrinaga-González, C. (2008). Carbon Trading: Accounting and Reporting Issues. *European Accounting Review*, *17*(4), 697–717. doi:10.1080/09638180802489162

Bilbao Estrada, I. (2010a). Los impuestos mundiales sobre el CO2. *Revista Aranzadi de derecho ambiental*, *17*, 115-133.

Bilbao Estrada, I. (2010b). La tributación de los derechos de emisión en el IVA y en el IS. *Partida doble*, *226*, 84-97.

Bilbao Estrada, I. & Mateos Ansótegui, A. (2006a). El impuesto sobre sociedades y los derechos de emisión de gases de efecto invenadero. Retos y oportunidades. *Revista de contabilidad y tributación, 285*, 83-128.

Bilbao Estrada, I. & Mateos Ansótegui, A. (2006b). La planificación fiscal en el Impuesto sobre Sociedades a través de los derechos de CO2. *Tribuna fiscal, 194*, 16-31.

Bilbao Estrada, I., & Mateos Ansótegui, A. (2009). Los derechos de emisión y su contabilización: especial referencia a la provisión por derechos de emisión tras el nuevo Plan General Contable y la reforma del Impuesto sobre Sociedades, en La Fiscalidad de los Derechos de Emisión: Estado de Situación y Perspectivas de Futuro, Instituto de Estudios Fiscales, Madrid.

Bilbao Estrada, I., & Mateos Ansótegui, A. I. (2007). Planificación contable y derechos de emisión de gases de efecto invernadero. *Revista Universo Contábil, 3*(3), 101–122.

Bilbao Estrada, I., & Rodríguez Márquez, J. (2010). IVA y derechos de emisión: Especial referencia a su utilización como instrumento de fraude. *Crónica Tributaria, 136*, 37–84.

Bilbao Estrada, I., & Rodríguez Márquez, J. (2010). IVA y derechos de emisión: Especial referencia a su utilización como instrumento de fraude. *Crónica Tributaria, 136*, 37–84.

Black, C. M. (2013). Accounting for carbon emission allowances in the European Union: In search of consistency. *Accounting in Europe, 10*(2), 223–239. doi:10.1080/17449480.2013.834730

Boas, S., & Monteiro, S. (2012). O relato sobre licenças de emissão de CO2: o caso das empresas portuguesas do PNALE II, *XV Encontro AECA - Novos caminhos para a Europa: O papel das empresas e dos governos*, Ofir, 20-21 de Setembro.

Borghei, Z., Leung, P., & Guthrie, J. (2018). Voluntary greenhouse gas emission disclosure impacts on accounting-based performance: Australian evidence. *Australasian Journal of Environmental Management, 25*(3), 321–338. doi:10.1080/14486563.2018.1466204

Bruninx, K., & Ovaere, M. (2022). COVID-19, Green Deal and recovery plan permanently change emissions and prices in EU ETS Phase IV. *Nature Communications, 13*(1), 1165. doi:10.103841467-022-28398-2 PMID:35246534

Cazorla Prieto, L. M., & Cazorla González-Serrano, L. (2009). *El tratamiento en el ordenamiento jurídico español del incremento de los precios de la energía eléctrica como consecuencia de la internalización de los derechos de emisión de CO2. Tratado de regulación del sector eléctrico* (Vol. 1). Aranzadi, Navarra.

Real Decreto 1514/2007, de 16 de noviembre, por el que se aprueba el Plan General de Contabilidad (PGC).

Dias, A., Montero, P., & Quirós, J. (2016). O relato financeiro das licenças de emissão atribuídas gratuitamente no período 2008-2012. *XVII Encuentro AECA*, Bragança, 22-23 setembro.

Ellerman, D., Marcantonini, C. & Zaklan, A. (2015). The European Union Emissions Trading System: Ten Years and Counting. *Review of Environmental Economics and Policy. 10*(1), 89–107.

Fabra, N., & Reguant, M. (2014). Pass-Through of Emissions Costs in Electricity Markets. *The American Economic Review*, *104*(9), 9, 2872–2899. doi:10.1257/aer.104.9.2872

Fernandes, M. T. (2009). *O mercado de direitos de emissão de CO2: um estudo experimental*. [PhD Thesis in Economic Sciences, Universidade do Minho].

Fernández Cuesta, C., & Fronti de García, L. (2010). El cambio climático y su control económico en la empresa. *Partida Doble*, *226*, 54–69.

Fernández Cuesta, C., Larrinaga González, C., & Moneva Abadía, J. M. (2006). Derechos de emisión de gases efecto invernadero: Registro, valoración e información. *Partida Doble*, *182*, 88–97.

Fernando, A., Albuquerque, F., Almeida, M., & Delgado, S. (2010). Tratamento contabilístico das licenças de emissão de gases com efeito de estufa: o caso português, *XIV Encuentro AECA*. Coimbra.

Florindo, A., Lemos, K., Monteiro, S., & Ribeiro, V. (2020). Corporate social responsibility (CSR) reporting and climate change: evidence of carbon emissions disclosures by Portuguese companies, in Book Conceptual and Theoretical Approaches to Corporate Social Responsibility, Entrepreneurial Orientation and Financial Performance. IGI Global.

Gallego-Alvarez, I., Martínez-Ferrero, J., & Cuadrado-Ballesteros, B. (2016). Accounting Treatment for Carbon Emission Rights. *Systems*, *4*(1), 12. doi:10.3390ystems4010012

Garcia-Torea, N., Giordano-Spring, S., Larrinaga, C., & Rivière-Giordano, G. (2022). Accounting for Carbon Emission Allowances: An Empirical Analysis in the EU ETS Phase 3. *Social and Environmental Accountability Journal*, *42*(1–2), 93–115. doi:10.1080/0969160X.2021.2012496

Ghazouani, A., Xia, W., Jebli, M., & Shahzad, U. (2020). Exploring the Role of Carbon Taxation Policies on CO2 Emissions: Contextual Evidence from Tax Implementation and Non-Implementation European Countries. *Sustainability (Basel)*, *12*(20), 8680. doi:10.3390u12208680

Giner Inchausti, B. (2007). La contabilidad de los derechos de emisión: Una perspectiva internacional. *Revista Española de Financiación y Contabilidad*, *36*(133), 175–193. doi:10.1080/02102412.2007.10779619

Giorgios, X., & Dimitrios, X. (2009). *Accounting treatment of emission credits*. International Hellenic University.

Gonçalves, A., & Góis, C. (2012). Reconhecimento contabilístico das licenças de emissão de CO2: o caso português, *XV Encontro AECA - Novos caminhos para a Europa: O papel das empresas e dos governos*. Ofir.

Gonzalez-Gonzalez, J., & Zamora Ramírez, C. (2016). Voluntary carbon disclosure by Spanish companies: An empirical analysis. *International Journal of Climate Change Strategies and Management*, *8*(1), 57–79. doi:10.1108/IJCCSM-09-2014-0114

Gonzalo Angulo, J.A. & Larriba Díaz-Azorita, A. (2007). Contabilización de los derechos de emisión de gases de efecto invernadero. *Cuadernos de sostenibilidad y patrimonio natural*, 12, 54-66.

Haites, E. (2018). Carbon taxes and greenhouse gas emissions trading systems: What have we learned? *Climate Policy*, *18*(8), 1–12. doi:10.1080/14693062.2018.1492897

Hapsoro, D., & Falih, Z. (2020). The Effect of Firm Size, Profitability, and Liquidity on The Firm Value Moderated by Carbon Emission Disclosure. *Journal of Accounting and Investment*, *1*(2), 240–257. doi:10.18196/jai.2102147

Hermawan, A., Aisyah, I., Gunardi, A., & Putri, W. (2018). Going Green: Determinants of Carbon Emission Disclosure in Manufacturing Companies in Indonesia. *International Journal of Energy Economics and Policy*, *8*(1), 55–61.

Hörisch, J. (2013). Combating climate change through organisational innovation: An empirical analysis of internal emission trading schemes. *Corporate Governance (Bradford)*, *13*(5), 569–582. doi:10.1108/CG-06-2013-0077

Jurado Martín, J., Martín Zamora, M. & Cabello Léon, C. (2007b). *Tratamiento contable de los derechos de emisión: la aplicación de la resolución de 8 de febrero de 2006 del ICAC en el sector eléctrico español. VI Reunión en Investigación Social y Medioambiental*. Navarra.

Jurado Martín, J. A. (2016). *La información contable y los derechos de emisión: presupuesto de emisiones y la utilidad de la información revelada en las cuentas anuales*. Universidad de Huelva.

Jurado Martín, J. A., & Martín Zamora, M. P. (2006a). El Mercado de Derechos de Emisiones y su Incidencia en el Resultado Empresarial, Decisiones Basadas en el Conocimiento y el Papel Social de la Empresa. *XX Congreso Nacional y XVI Congreso Hispano Francés de AEDEM*, Palma de Mallorca (España.

Jurado Martín, J. A., & Martín Zamora, M. P. (2006b). *El Proceso Presupuestario y el Protocolo de Kioto: el Presupuesto de Emisiones de CO2. XVI Jornadas Luso-Espanholas de Gestao Científica, "Inovação e Empreendedorismo para o Séc*. XXI.

Jurado Martín, J. A., Martín Zamora, M. P., & Cabello León, C. (2007a). *Los Derechos de Emisión y su Repercusión en la Cuenta de Resultados en las Empresas del Sector Eléctrico Español. XI Jornada de Trabajo Sobre Contabilidad de Costes y de Gestión. X Jornada de Trabajo Sobre Contabilidad de Costes y Gestión*. ASEPUC.

Karatzoglou, B., & Karatzoglou, O. (2011). Carbon accounting in Greek companies participating in the European Union's Emissions Trading Scheme: Current practice and projected financial implications. In R. Burritt, S. Schaltegger, M. Bennett, T. Pohjola, & M. Csutora (Eds.), *Environmental management accounting and supply chain management. Eco-Efficiency in industry and science, 27*. Springer. doi:10.1007/978-94-007-1390-1_10

Lin, B., & Li, X. (2011). The effect of carbón tax on per capita CO2 emissions. *Energy Policy*, *39*(9), 5137–5146. doi:10.1016/j.enpol.2011.05.050

López Gordo, M. G., & López Gordo, J. F. (2007). Normativa sobre el registro contable de los derechos de emisión. [Revista Científica y Técnica de Ecología y Medio Ambiente]. *Ecosistemas (Madrid)*, *16*(1), 90–96.

Lovell, H., Aguiar, T., Bebbington, J., & Larrinaga-Gonzalez, C. (2010). Accounting for Carbon. *Certified Accountants Educational Trust for the Association of Chartered Certified Accountants. Research Reports (Montgomery)*, 122.

Madrigal, J. A. (2012). Los derechos de emisión de CO2 y la fiscalidad española. icade. *Revista de la Facultad de Derecho (Montevideo)*, (86), 55–90.

Martín Zamora, M., & Jurado Martín, J. A. (2005). Nuevos Procedimientos Contables: La Contabilidad del Mercado de Emisiones de CO2. *Técnica Contable*, *680*, 5–36.

Martín Zamora, M. P., & Jurado Martín, J. A. (2008). *La Contabilización de los Impuestos Ecológicos. X Seminario Hispano-Luso de Economía Empresarial*. Servicio de Publicaciones de la Universidad de Huelva.

Martínez, M. (2016). Derechos de emisión y tributos en la unión Europea. *Jurídicas Cuc*, *12*(1), 139–180.

Mateos Ansótegui, A. (2006). La contabilidad de los derechos de emisión. *New Challeges of Accounting Ethics, Independece and Critical Thought*. XI Accounting Congress, Coimbra.

Matsumura, E., Prakash, R., & Vera-Muñoz, S. (2011). *Voluntary Disclosures and the Firm-Value Effects of Carbon Emissions*. University of Notre Dame.

Metcalf, G. E. (2021). Carbon taxes in theory and practice. *Annual Review of Resource Economics*, *13*(1), 245–265. doi:10.1146/annurev-resource-102519-113630

Milanés, P., Pérez-Calderón, E., & Dias, A. (2020). Transparency of Financial Reporting on Greenhouse Gas Emission Allowances: The Influence of Regulation. *International Journal of Environmental Research and Public Health*, *17*(3), 893. doi:10.3390/ijerph17030893 PMID:32023942

Monast, J., Anda, J., & Profeta, T. (2009). *U.S. Carbon Market Design: regulating emission allowances as financial instruments.* Working paper, Nicholas Institute for Environmental Policy Solutions, Duke University.

Monast, J., Murray, B., & Wiener, J. (2017). On morals, markets and climate change: Exploring Pope Francis' challenge. *Law and Contemporary Problems*, *80*, 135–162.

Montero, P., Calderón, E., & Dias, A. (2020). Transparency of Financial Reporting on Greenhouse Gas Emission Allowances: The Influence of Regulation. *International Journal of Environmental Research and Public Health*, *17*(3). PMID:32023942

Nunes, C. (2009). Fiscalidade vs. Licenças de emissão de CO2 no combate à mitigação. *da Sphera*, julho, 1-40.

Pérez, A., & Soto, R. (2017a). Real Decreto 602/2016: Aplicaciones prácticas de las modificaciones contables (I). *Revista de Contabilidad*, (53), 14–57.

Pérez, A., & Soto, R. (2017b). Real Decreto 602/2016: Aplicaciones prácticas de las modificaciones contables (II y última parte). *Revista de Contabilidad*, (54), 8–37.

Perino, G. (2018). New EU ETS Phase 4 rules temporarily puncture waterbed. *Nature Climate Change*, *8*(4), 262–264. doi:10.103841558-018-0120-2

Portuguese Corporate Income Tax Code Rabadán Martín, I., Jurado Martín, J. A., & Martín Zamora, M. P. (2017). Análisis de la utilidad de la información contable relativa a los derechos de emisión: Propuesta de un indicador. *Revista International Business and Economics Review.*

Real Decreto 602/2016, de 2 de diciembre, por el que se modifican el Plan General de Contabilidad (PGC)

Reis, C., & Lopes, A. (2011). Fatores económicos e contabilísticos que influenciam a qualidade da informação ambiental: um estudo com empresas admitidas à cotação na Euronext Lisbon que tenham títulos de Emissão de gases com efeito de estufa. *XVI Encuentro AECA, Granada.*

Resolución de 28 de mayo de 2013, del Instituto de Contabilidad y Auditoría de Cuentas (ICAC), por la que se dictan normas de registro, valoración e información a incluir en la memoria del inmovilizado intangible.

Resolución de 8 de febrero de 2006 del Instituto de Contabilidad y Auditoría de Cuentas (ICAC), por la que se aprueban normas para el registro, valoración e información de los derechos de emisión de gases de efecto invernadero.

Santos, A. (2013). O comércio de licenças de emissão em Portugal. *International Business and Economics Review: Revista Internacional de Gestão e Comunicação, 4*, 62–77.

Shmelev, S. E., & Speck, S. U. (2018). Green fiscal reform in Sweden: Econometric assessment of the carbon and energy taxation scheme. *Renewable & Sustainable Energy Reviews, 90*, 969–981. doi:10.1016/j.rser.2018.03.032

Spanish Corporate Income Tax Law, Steenkamp, N., Rahman, A., & Kashyap, V. (2011). Recognition, measurement and disclosure of carbon emission allowances under the EU ETS: an exploratory study. *10th Australasian CSEAR Conference, Session 15: Climate change*, University of Tasmania.

Tozer, R. (2007). Fiscalidad de los derechos de emisión de CO2. *Cuadernos de Sostenibilidad y Patrimonio Natural*, 12.

Ulucak, R., & Kassouri, Y. (2020). An assessment of the environmental sustainability corridor: Investigating the non-linear effects of environmental taxation on CO2 emissions. *Sustainable Development (Bradford), 28*(4), 1010–1018. doi:10.1002d.2057

Warwick, P., & Ng, C. (2012). The 'Cost' of climate change: How carbon emissions allowances are accounted for amongst European Union companies. *Australian Accounting Review, 60*(22), 54–67. doi:10.1111/j.1835-2561.2011.00158.x

Yunus, S., Elijido-Ten, E., & Abhayawansa, S. (2016). Determinants of carbon management strategy adoption. *Managerial Auditing Journal, 31*(2), 156–179. doi:10.1108/MAJ-09-2014-1087

ADDITIONAL READING

Monteiro, S., Lemos, K., Ribeiro, V., & Florindo, A. (2022). Derivative instruments on emission allowances: Factors influencing disclosure in Portuguese companies. *International Journal of Professional Business Review*, *7*(4), 1–30. doi:10.26668/businessreview/2022.v7i4.e389

Prado-Lorenzo, J.-M., Rodríguez-Domínguez, L., Gallego-Álvarez, I., & García-Sánchez, I.-M. (2009). Factors influencing the disclosure of greenhouse gas emissions in companies world-wide. *Management Decision*, *47*(7), 1133–1157. doi:10.1108/00251740910978340

Prentice, J. (2018). The Revision of the European Union's Emissions Trading System Ahead of the Fourth Trading Period, 2021–2030. *Climate Law*, *8*(3-4), 338–348. doi:10.1163/18786561-00803013

KEY TERMS AND DEFINITIONS

Emission Allowance: the right to emit one tonne of carbon dioxide (CO2) or any other greenhouse gas (GHG) having an equivalent effect for a given period.

EU Emission Trading Scheme (EU ETS): Market established by Directive 2003/87 / EC of 13 October, which allows entities that exceed established targets to purchase pollution rights from entities that have achieved reductions above the limits set.

Expenses for GHG Emissions: refer to the costs or financial obligations associated with the release or reduction of GHG into the atmosphere.

Greenhouse Gas (GHG) emissions: are gases released into the Earth's atmosphere that trap heat and contribute to the greenhouse effect.

Intangible Assets: are assets that do not have a physical presence but are valuable to a business because of their economic benefits.

Inventories: refer to the assets that a business entity holds for the purpose of resale or for use in its production process.

Taxation: is the process by which governments impose charges or levies on individuals, businesses, or other entities to generate revenue for public expenditures and fund various government functions and services.

Chapter 4
Environmental Non-Governmental Organizations' Tax Regime:
The Portuguese Case

Ana Paula Rocha
CEOS.PP, ISCAP, Polytechnic of Porto, Portugal

ABSTRACT

In the most recent years, the Environmental Non-Governmental Organizations (commonly known as Environmental NGO's) had assumed a very important role in the defense and enhancement of the environment and natural heritage. In this sense, the tax regime applicable to those entities can be framed as a green tax regime, to the extent that it may constitute an instrument for the State' intervention towards the protection of the environment. Therefore, in this chapter, the author addresses the main topics concerning the Portuguese tax regime applicable to the Environmental NGO's, not only in what concerns to the framework specifically applicable to those organizations under the main Portuguese' taxes but, also, in what concerns to the tax benefits that, in respect for the fundamental principle of tax justice, can be recognized to them.

INTRODUCTION

The exponential growth of climate' changes and environmental problems require the approval of public policies that, underlying a paradigm shift, promote the adoption of environmentally friendly behaviors by the different members of our society. Among such public policies, a special mention shall be made to the so-called *green tax regimes* that, once harmonized with the fundamental principle of tax justice or tax equality (in Portuguese, "princípio da justiça" ou "princípio da igualdade fiscal"), clearly constitute a green tax policy, e.g., an instrument for the State' intervention in the economy and in the society towards the protection of the environment.

DOI: 10.4018/978-1-6684-8592-7.ch004

It is precisely in this context that the study of the tax regime applicable to the Environmental NGO's is mandatory, bearing in mind the very important role that these entities have taken, in the most recent years, as legal persons that exclusively aim to defend and enhance the environment and the natural heritage, as well as the nature' conservation, instead of pursuing lucrative purposes for themselves or for their members. And we will try to develop such analysis by focusing our attention on the Portuguese tax regime, whose complexity and recent developments represent a fertile field to our analysis, allowing us to question ourselves whether that regime is designed and qualified to protect and promote the important role carried out by these organizations in what concerns to the protection and the defense of the environment without harming, as far as possible, the aforementioned principle of tax justice.

To do so, we will begin to understand the principal features of the Portuguese legal regime applicable to the Environmental NGO's (in Portuguese, Organizações Não-Governamentais de Ambiente – ONGA's), which is a key point to analyze and understand the tax regime currently applicable to those entities in Portugal.

Then, we will decompose this tax regime in what concerns to the main Portuguese taxes, namely the corporate income taxation ("Imposto sobre o Rendimento das Pessoas Coletivas" – IRC), the value added taxation ("Imposto sobre o Valor Acrescentado" – IVA) and the real estate taxation ("Imposto Municipal sobre Imóveis" – IMI and "Imposto Municipal sobre as Transmissões Onerosas de Imóveis" – IMT). In this context, we objectively intend to:

- Explain and problematize the major aspects and particularities that the Environmental NGO's face and encounter regarding the tax' settlement in Portugal;
- Understand and problematize the application of the rules concerning the access to Portuguese tax benefits by the Environmental NGO's; and
- Identify and problematize the potential tax advantages that can be granted to the benefactors of the Environmental NGO's.

Please note that our aim does not focus on the development of the Portuguese legal regime of Environmental NGO's nor the Portuguese tax regime applicable to the non-profit entities in general terms, once that the vast number of subjects related to these topics would not suit the concrete purpose of our work. Therefore, those general regimes will be considered in our *Background Section*, to the extent that they are suitable to reach our objective: present and problematize the tax specificities applicable to the Environmental NGO's, as previously mentioned.

Knowing that there are still few scientific studies regarding the Portuguese tax regime applicable to the non-profit sector in general (Fernandes, 2022) – none of which specifically addressed, to the extent that we propose, to the Environmental NGO's –, we hope to make a serious and objective contribution on the comprehension and discussion of the Portuguese tax regime currently applicable to the Environmental NGO's and, to that extent, on the reflection regarding the role played by the Portuguese' green tax policies in the actual protection of the environment.

BACKGROUND

Portuguese Legal Framework Applicable to the Environmental NGO'S: Main Considerations

The Portuguese legislator provides a legal framework for the organizations dedicated to the defense and the enhancement of the environment since 1987, when it was approved the Law number 10/87, of April 4th, regarding the Associations for the Defense of the Environment (in Portuguese, "Associações de Defesa do Ambiente").

Later, this legal regime was abrogated by the Law number 35/98, of July 18th, which established the Portuguese legal framework for the Environmental NGO's in an attempt to accompany the evolution verified in the international law in this subject and to confer a renewed effectiveness to the action of the organizations dedicated to the defense of the environment (Agência para a Modernização Administrativa, 2023). Being approved in 1998, the Portuguese regime for the Environmental NGO's already has 2 legislative amendments: the significative amendment introduced by the Law number 82-D/2014, of December 31st, known as Portuguese Green Tax Reform, and the recent amendment introduced by the Law number 36/2021, of June 14th, which approved the legal regime regarding the Public Utility Status.

Under the terms of the Law number 35/98 currently in force (specifically, articles 2 and 3), the legal form of an Environmental NGO may be granted to entities endowed with legal personality that, requesting their inscription on the Agência Portuguesa do Ambiente, IP (hereinafter, APA)[1]-[2], do not pursue lucrative purposes for themselves neither for their members, since that they must exclusively aim to defend and enhance the environment, the natural and constructed heritage as well as the nature' conservation. As such, the overall result of these organizations' activity generally has the nature of a public good, grounded in the protection of the environment (Franco, 2015).

In this sense, and given that the Environmental NGO's chase purposes that benefit the community at large and which, to that extent, are deemed as altruistic or disinterested purposes with social relevance, these entities must be considered as included in the "Social Economy Sector" (Martins, 2021) as established in the articles 2 and 4 of the *Lei de Bases da Economia Social* (LBES – Law number 30/2013, May 8th) whenever they:

1. Are endowed with legal personality,
2. Comply with the guiding principles provided in the article 5 of that law which, in their most relevant aspects, are the following:
    a) Primacy of people and social objectives;
    b) Free and voluntary adhesion and participation;
    c) Democratic control by its members;
    d) Conciliation between the members' interests, the beneficiaries' interest and the general interest;
    e) Respect for the values of solidarity, equality and non-discrimination, social cohesion, justice and equity, transparency, shared responsibility and subsidiarity;
    f) Autonomous and independent management from public authorities and any other entities outside the social economy;
    g) Allocation of the surpluses on the pursuit of the Entity' purposes, in accordance with the general interest, and

3. Appear in the Portuguese database regarding the Social Economy Sector (https://basededadossocial.pt/).

Notwithstanding, and as noticed by Meira (2013), the concept of social economy is "an open concept under construction", being evidently difficult for the doctrine to "arrive to a precise and safe definition". Therefore, we agree with Meira (2013) when the author states that "the legislator did well by opting, regarding the delimitation of the concept of social economy, for a combined technique, that is, complementing the definition of social economy contained in the article 2 by an open enumeration of social economy entities (article 4) and by the enunciation of its guiding principles (article 5)".

It is precisely on the basis of this legal framework that the Environmental NGO's can benefit from a set of rights suitable to their purposes, as the right to participate in the definition of the policies and the main legislative guidelines regarding the environment, as well as in the filing of legal actions deemed as necessary for the prevention, the correction, the suspension or the cessation of acts or omissions committed by public or private entities that constitute – or are able to constitute – a step towards the degradation of the environment, as deemed in the articles 5 to 10 of the Law number 35/98 (Pereira, 2021). Likewise, is that legal framework that, as we understand, justifies *i)* the financial support that may be granted to this organizations under the Portuguese Regulation on Sustainability and Efficiency in the Resources' Use (Ordinance number 57-B/2015, of February 27th), which has the specific objective of promoting the conservation and enhancement of cultural and natural heritage, as instruments of sustainability of the territories, namely through its tourism valorization, and *ii)* the specific tax regime foreseen to these entities, which is fundamental for their subsistence and for the effective accomplishment of their purposes.

It is just to the tax support granted to the Environmental NGO's registered in Portugal that we will dedicate our attention in the next chapter, not without first addressing one aspect that, as we will see, is crucial in what concerns to the determination and utilization of that regime, in line with the assignments and missions of the Environmental NGO's: the attribution of the public utility' status, currently established in the Law number 36/2021, of June 14th.

In general terms, and as foreseen in the articles 4 and 5 of the Law number 36/2021, the public utility' status can be assigned to legal persons that pursue diffuse and generalized goals (such as the protection and development of the environment, natural heritage and life' quality – as pursued by Environmental NGO's) and that cooperate, to that extent, with the central, regional or local public administration of the Portuguese State. Furthermore, and within the scope of their activities, the public utility' entities must act in accordance with the aforementioned guiding principles provided by the Portuguese LBES.

The acknowledgment of the public utility' status is not automatic, once that it must be required by the Environmental NGO's and depends on the fulfillment of several specific requirements foreseen in the Law number 36/2021, such as the following ones, that are specifically highlighted due to their importance in what concerns to the use of tax benefits:

1. The entity must have three years of effective and relevant activity, as well as an equal period of uninterrupted registration in the APA;
2. The organization must also be duly constituted, have the necessary personnel, infrastructure, facilities and equipment to ensure the pursuit of its objectives, maintain an updated list of its members and notify the General Secretariat of the Portuguese Council of Ministers' Presidency of any changes to its statutes or to its internal regulations within three months of that amendment, among other obligations foreseen in the articles 6, 7, 8 and 12 of the Law number 36/2021;

3. The granting of the status by the Portuguese Prime Minister (with the possibility of delegation) depends on a prior opinion prepared by the APA.

In an analysis undertaken to the list of Environmental NGO's and legally assimilated entities with an active registration in the National Registry until December 31st, 2022 (Notice number 4195/2023 from APA) and to the Portuguese Government's platform that aggregates information on the entities with public utility' status regarding June, 2023 (https://eportugal.gov.pt/fichas-de-enquadramento/fundacoes-e-pessoas-coletivas-de-utilidade-publica/pesquisa), it is possible to understand that only 38.5% of the Environmental NGO currently registered as such do benefit from the referred status, as outlined in the following table.

Table 1. Environmental NGO's with public utility' status

Type of Entity	Number of Registered Entities	Number of Entities with Public Utility Status	Percentage
National' NGO	14	9	64,3%
Regional' NGO	15	7	46,6%
Local' NGO	29	7	24,1%
No scope' NGO	20	10	50%
Entity assimilated to NGO	26	7	26,9%
TOTAL	**104**	**40**	**38,5%**

The public utility' status is important because, once recognized as public utility' organizations, Environmental NGO's become eligible for certain tax benefits specifically provided for in the Portuguese law (articles 11, 30 and Annex III of the Law number 36/2021), provided that they meet the requirements that we will adress hereafter. And, as we will see in the next chapter, this circumstance proves to be essential in what concerns to the definition of the tax regime applicable to the NGO's.

SOCIAL ECONOMY SECTOR AND TAXES: INITIAL APPROACH

As referred above, the Environmental NGO's endowed with legal personality that chase purposes that benefit the community at large and which, to that extent, are deemed as altruistic or disinterested purposes with social relevance take part of the Social Economy Sector, whenever they observe the guiding principles provided in the LBES and appear in the Portuguese database regarding that sector. In this sense, those entities are not included in the general Portuguese tax regime but, alternatively, in the special Portuguese tax regime applicable to the Social Economy Sector. But beware: as noticed by Taborda & Martins (2023), it is presumed that these entities "do not primarily engage in a commercial, industrial or agricultural activity", once that "if they do so as a main objective, moving away from their structural nature, they will be taxed by profit (like companies)" and shall be included in the general Portuguese tax regime.

The Portuguese legislator does not foresee a legal tax diploma specifically and solely applicable to the Social Economy Sector nor, as such, for the tax regime applicable to the Environmental NGO's. In fact, and contrary to what it happens, for example, in Spain (in the *Ley 49/2002, de 23 de diciembre, de régimen fiscal de las entidades sin fines lucrativos y de los incentivos fiscales al mecenazgo*, whose last update was published on 12.10.2021), the Portuguese tax regime applicable to the social economy' organizations is dispersed throughout the legislation, circumstance that accentuates the difficulties regarding the interpretation, integration and application of the tax law on this subject.

Notwithstanding, and as we will develop further up, there is one common aspect that covers most of the tax rules specifically applicable to the social economy' entities, namely the fact that the tax benefits' provisions are closely linked to the effective pursuit of the social object defined in the bylaws of those entities and the effective performance of activities targeted to accomplish such object.

This aspect is undeniably relevant since, as it is well known, tax benefits constitute a derogation from the general rules of taxation, as tax advantages with an exceptional nature. As foreseen in the article 2 of the Portuguese Statute of Tax Benefits (in Portuguese, Estatuto dos Benefícios Fiscais), tax benefits are measures of an exceptional nature, such as total or partial tax exemptions, instituted to protect relevant extra-fiscal public interests that are greater than those of the actual taxation that they prohibit. Without underlying objectives, such advantages could not truly be called tax benefits, rather constituting "a privilege that, given the principle of equality, have no reason to exist", which is why it is not possible to conceive "a tax benefit that do not constitute a mean to achieve a hierarchically higher value purpose than" the tax justice and the equality of all the taxpayers (Pereira, 2014). This happens because the principle of fiscal equality is embedded in the idea of generality and universality, according to which all the citizens and legal persons are bound to the duty to pay taxes, and uniformity, requiring that this duty shall be measured by the same criteria – the ability to pay' criteria (Nabais, 2021).

Therefore, and as foreseen in the article 11 of the LBES, the establishment of a more favourable tax status can only be interpreted as a path to encourage and promote the activities that are developed by the social economy' entities, to the extent that those activities do benefit, directly or diffusely, all the society' members. In fact, it is only on the basis of an objective criteria that it is possible to overcome the tight scrutiny of the European Union state' aid regime and to ensure that the granting of the tax benefits won't be used as a vehicle towards abusive tax planning (Aguiar, 2016).

Likewise, and given the heterogeneity of the entities that are included in the social economy sector, the tax regime shall be differentiated among them, considering the degree to which they pursue general interest objectives (Meira, 2013). And it is precisely for this reason that it is important to analyse the essential features of the tax rules specifically applicable to Environmental NGO's in Portugal in what concerns to the main Portuguese taxes, questioning whether the Portuguese legislator has managed to create a tax regime that fulfills the stated purpose without harming, as far as possible, the aforementioned principle of tax justice.

PORTUGUESE TAX REGIME APPLICABLE TO THE ENVIRONMENTAL NGO'S

Corporate Income Taxation (IRC)

In general terms, IRC is a direct tax levied on the global income obtained in each tax period by legal persons, with the main objective of encumbering the actual income of those taxpayers, as foreseen in the

Article 1 of the IRC' Code and in the article 104 of the Portuguese Constitution (Amorim & Azevedo, 2016).

Legal persons with their headquarters or effective management in Portugal are considered as resident taxable persons and are subject to taxation on their *worldwide income*, e.g., on all the results obtained in a given tax year, whether their source is located inside or outside Portugal. Legal persons that do not have their headquarters or their effective management in Portugal can also be considered as non-resident taxable persons ***if*** they earn income in the Portuguese territory. However, in this case, the worldwide income principle does not apply in Portugal, reason why the only income that can be subject to IRC is, precisely, the income obtained in the Portuguese territory (articles 2 and 4 of the IRC' Code).

Broadly speaking, and regarding IRC, the resident taxable persons can be framed into two different regimes (article 3 of the IRC' Code):

a. In the regime applicable to the entities that carry out, as their main purpose, a commercial, industrial or agricultural activity – *e.g.*, "trading entities". As a matter of default, these entities shall be taxed on their profit under the rules provided in the articles 17 to 52 of the IRC' Code, which is applicable to the enterprises in general;
b. In the regime applicable to the entities that do not carry out, as their main purpose, a commercial, industrial or agricultural activity – *e.g.*, "non-profit entities" or "non-traders". It is precisely in this regime that the Environmental NGO's that take part of the Social Economy' Sector and who are not mainly engaged in a lucrative activity are framed, which requires the analysis of this tax regime with greater attention.

The non-profit entities are IRC' taxpayers that shall be taxed on their global income, which corresponds to the algebraic sum of the net income of the several categories provided in the Portuguese Personal Income Tax' Code, namely capital income, property income, business income obtained incidentally, capital gains and asset increases, including the asset increases that, being gratuitously' obtained, are not intended for the direct and immediate accomplishment of the entity's statutory purposes (Silva & Soares, 2015). This calculation excludes, however, the dues paid by the members and the subsidies granted in accordance with the bylaws of the organization, once that they are considered as untaxed income (Articles 2, 3, 15, 53 and 54 of the IRC' Code and the Circular Letter number 20219/2020, of March 16th, from the CIT' Directorate of Services).

Once the global income of the organization has been determined in these terms, it is then important to calculate its taxable income (in Portuguese, "matéria coletável"), which is obtained by deducting to the global income the expenses concretely authorized by the legislator (Ribeiro & Santos, 2013). The taxable income will then be subject to the IRC rate applicable to the non-profit entities which, from 2016 on, is the same as the rate applicable to trading entities: 21% (article 87 of the IRC' Code).

Notwithstanding, Environmental NGO's can be partially or totally exempt from the Portuguese IRC if they comply with the requirements provided in the article 10 of the IRC' Code, which establishes a tax exemption for *i)* the Private Institutions of Social Solidarity (in Portuguese, Instituições Particulares de Solidariedade Social – IPSS) and, also, for *ii)* the public utility persons that pursue, exclusively or predominantly, one of the following purposes: scientific, cultural, charity, assistance, beneficence, social solidarity, *defense of the environment* or agro-food interprofessionalism (in harmony with the provisions of the article 11 of the Law number 36/2021).

Hence, the importance of the attribution of the public utility' status to the Environmental NGO's is axiomatic, since such status allows these entities to request, to the Portuguese Minister of Finance, the use of the IRC' exemption provided for in the referred article 10 of the IRC' Code. Please note that this request must be submitted no later than the deadline to present the organization' income tax return and must be duly accompanied by an adequate proof of the exemption' conditions (as provided in the article 65 of the Portuguese Tax Process and Procedure Code and in the Circular Letter number 20140 issued by the IRC Directory Services on August 4th, 2009), namely:

- The proof of the legal status of the corporate entity which, in what concerns to an Environmental NGO, refers to its registration as such type of entity. Notwithstanding, please note that, in our opinion, this obligation has recently lost some of its relevance, since the information regarding the granting, renewal and revocation of the public utility status – including the NGO's corporate name, its legal form, its legal entity identification number, its sector of activity, the duration of the status and the date it takes effect – must be directly provided to the Portuguese Tax Authorities by the General Secretariat of the Portuguese Council of Ministers' Presidency (articles 24 and 33 of the Law number 36/2021 and article 5 of the Ordinance number 138-A/2021, of June 30th);
- Updated copy of the organization' statutes (simple photocopy);
- The activity' report and the accounts of the last two tax years; and
- A declaration of the non-existence of any direct or indirect interest of the members of the statutory bodies, either by themselves or through an intermediary, in the results of the entity's economic activities.

Due to its relevance, it is important to underline that the IRC' exemption won't be recognized to an Environmental NGO that, despite benefiting from the public utility' status, has failed to pay any income, value added or property taxation, or does not have its contributory situation regularized before the Portuguese Institute of Social Solidarity (under the terms specifically provided in the article 13 of the Estatuto dos Benefícios Fiscais). However, if this circumstance does not occur and all the requirements foreseen for the recognition of the tax benefit are fulfilled, the Portuguese Minister of Finance will prepare an order that will define the scope of the exemption in accordance with the activities carried out by the organization and will be duly published in the Portuguese official newspaper.

In any case, the exemption does not cover business income derived from the exercise of commercial or industrial activities that may be incidentally carried outside the scope of the statutory purposes of the entity, as well as the income derived from bearer securities not registered or deposited, which are excluded from the scope of the exemption in order to guarantee the underlying reason that supports the attribution of the tax benefit, as referred above. And it is also for this reason that the exemption will be conditioned on the continued compliance of the following requirements:

1. Effective exercise of activities associated to the environmental purposes that justified the exemption. Such activities must be pursued by the organization in an exclusively or predominantly manner;
2. Imputation, to the purposes referred in the previous paragraph, of at least 50% of the overall net income that would be subject to taxation under general terms, until the end of the 4th tax period subsequent to the one in which it was obtained (except in the event of a fair impediment to meet the assignment deadline, that should be notified to the Portuguese General-Director of the Tax

Authority, duly accompanied by a written justification, until the last working day of the 1st month following the deadline);

3. Absence of any direct or indirect interest of the members of the statutory bodies, by themselves or through an intermediary, in the results of the exploitation of the economic activities pursued by the Environmental NGO's.

REAL ESTATE TAXATION (IMI AND IMT)

IMI and IMT are framed as property taxes since they are levied on the ownership and on the transfer of real estate located in the Portuguese territory. They revert to the vaults of the municipalities where those properties are located (Pires, 2016).

In general terms, IMI is levied on the taxable value of rural and urban properties located in the Portuguese territory and IMT is levied on the acquisition of the ownership' right over those properties. In this sense, the Environmental NGO's will be subject to IMI whenever they figure in the building booklet as its owner at the last day of each year and will be subject to IMT whenever they acquire a building' property through the celebration of an onerous contract (articles 1 and 8 of IMI' Code and articles 2 and 4 of IMT' Code).

Although they are subject to IMI and IMT in this very linear terms, the Environmental NGO's may be exempt from these taxes to the extent that they benefit from the public utility' status (which, once again, highlights the importance associated to this condition), once that the Law number 36/2021 establishes in its article 11 that public utility' entities may benefit from IMI and IMT' exemptions in what concerns to the real estate property that is directly and immediately used to carry out their statutory purposes.

The article 6 of the IMT' Code goes precisely in this direction, by providing that the legal persons with public utility status are exempt from taxation regarding the "goods destined, directly and immediately, to carry out their statutory purposes". The exemption operates automatically whenever the property' value is equal to or less than EUR 300,000; in the remaining cases, the Environmental NGO must present to the Portuguese Tax Authority, before the acquisition' date, an application that shall contain the identification, the description and the purpose of the assets and that must be accompanied by *i)* a document proving the legal status of the entity and *ii)* a certificate or certified copy of the resolution regarding the asset acquisition, which must expressly mention the property' destination (Article 10 of the IMT' Code). In either case (this is, whether it operates automatically or is dependent on recognition), the exemption ceases to apply whenever the assets are sold or are given another destination without a prior authorization from the Portuguese Minister of Finance (Article 11 of the IMT' Code).

However, the IMT' clarity does not extend, unfortunately, to the IMI. In fact, the Law number 36/2021 did not introduced any changes to the IMI' Code or to the Estatuto dos Beneficios Fiscais in what concerns to the public utility status which, in our opinion, disturbs the interpretation and the application of the tax legislation in this context, clearly reducing the legal certainty of the tax law in this field. In fact, and within Estatuto dos Beneficios Fiscais, two possibilities remain in force:

- On one hand, we have to consider the article 44 of the Estatuto dos Beneficios Fiscais in its number 1 – paragraph e), which establishes that "legal persons with public administrative utility and the ones with mere public utility" are exempt from IMI regarding all the properties that are aimed to be directly used to carry out their purposes – which, although has not accompanied the legislative evolution at the level of the legal status of the public utility' entities, goes with the spirit of the Law number 36/2021' provisions;

- On the other hand, the number 1 – paragraph m) of the referred article 44 also remains in force. This rule establishes that culture and recreation communities, *non-governmental organizations* and other types of non-profit associations are exempt from IMI, but only in what concerns to the buildings used as their headquarters – which is less comprehensive than the Law number 36/2021, leaving out from the exemption any properties that, in addition to the headquarters, can be used by the entity to pursuit the organization' purposes.

For the organizations that do benefit from the public utility' status, an interpretation according with the Law number 36/2021 suggests that these entities can take advantage of the IMI exemption under the terms provided in the referred number 1 – paragraph e) of the article 44, e.g., regarding all the properties that are aimed to be directly used to carry out their purpose. Nonetheless, we believe that a legislative clarification regarding this topic would be important, since the lack of updating of the Estatuto dos Benefícios Fiscais following the approval of the Law number 36/2021 creates increased difficulties in what concerns to the integration and the application of the tax law in a topic that cannot be subject to an analogical interpretation, but only to extensive interpretation (as it is the case of tax benefits, as provided in the Article 10 of the Estatuto dos Benefícios Fiscais).

For the NGO's that do not benefit from the public utility' status, it seems that they can only take advantage of the IMI exemption under the terms provided in the referred number 1 – paragraph m) of the article 44, e.g., in what concerns to the buildings used as their headquarters. Which means that for the eventual properties excluded from the scope of the exemption, the taxation shall be levied under the general terms foreseen in the IMI' Code.

VALUE ADDED TAXATION (IVA)

In general terms, IVA is a general tax on the consumption tendentially levied on all the expensive transactions that result from the exercise of an economic activity (Ribeiro & Santos, 2013). Harmonized at the European Community' level, IVA is configured as a multiphase tax based on the indirect subtractive method, which "is nothing more than the technique of assessing and deducting the tax at each stage of an economic circuit" (Palma, 2015), if the rules laid down for such assessment and deduction are respected. As economic operators, Environmental NGO's are IVA' taxable persons, whose expensive transfer of goods and provision of services constitute operations subject to taxation, under the general terms (articles 1 to 4 of the IVA' Code).

Despite the broad scope of IVA' incidence and, consequently, the vast number of operations that are subject to taxation, Environmental NGO's may benefit from IVA' exemptions in the so-called *domestic or internal operations* in the Portuguese territory (please note that in the context of this study, we are not addressing, naturally, to the IVA' exemptions that arises from the application of the destination principle in the international and intra-EU transactions practiced by economic operators in general terms). Once that we are dealing with incomplete exemptions, please note that the organizations won't charge IVA on their transfer of goods and provision of services but are, however, forbidden to deduct the IVA supported upstream, e.g., the IVA paid on eventual purchases of goods and services (Tracana & Dias, 2020). In the opposite direction, for all the economic operations subject to tax and not exempt from it, the organization must charge IVA on its active operations (using, for this purpose, the general rates that are provided for in the article 18 of the IVA' Code and in the Lists attached to this code), being able to deduct the IVA supported upstream.

In this context, it is important to start by stressing out that the activities carried out by the Environmental NGO's may take advantage of the incomplete tax exemptions provided in the Article 9 of the IVA' Code ***if*** the legal conditions regarding the legal nature of the entity and the nature of the economic operation in question are met. Conditions that, in our opinion, are not easy to meet – even if the Environmental NGO benefits from the public utility' status.

In fact, in order to benefit from a tax exemption in the domestic operations, it is mandatory that the specific goods sold, or the services provided by the organization are foreseen in the list provided by the Article 9 of the IVA' Code. However, it turns out that, although such list includes 38 different types of goods and services exempt from IVA in domestic transactions, the items theoretically applicable to the activity developed by the NGO's are (very) reduced. In fact:

- We understand that the congresses, colloquiums, conferences, seminars, courses and similar events that *i)* are held by Environmental NGO's, *i)* have a scientific, cultural, educational or technical nature, and *iii)* are referent to the protection and the defense of the environment, can benefit from the IVA' exemption foreseen in the paragraph 14 of the article 9 of the IVA' Code, provided that the Environmental NGO is qualified as a *non-profit organization* as a result of the compliance with the cumulative requirements foreseen in the article 10 of the IVA' Code. Please note that pursuant to this article 10, and for the purposes of the exemption, "only the entities that, simultaneously, fulfill the following conditions shall be considered as non-profit organizations:
    a) They shall never distribute profits, and their governing bodies shall not have, either by themselves or through an intermediary, any direct or indirect interest in the results of the entity' operations;
    b) They shall maintain accounts that cover all their activities, making them available to the tax authorities, namely to verify the condition mentioned in the preceding paragraph;
    c) They shall apply prices approved by the public authorities or, for the operations not subject to approval, prices lower than those charged by commercial enterprises in similar operations;
    d) They shall not engage in direct competition with taxable persons".
- We understand that it is difficult to consider that other sales or services provided by Environmental NGO's are able to take advantage of some of the other exemptions provided in the article 9 of the IVA' Code, even if those operations are exclusively carried out in the interest of the NGO' members and there is no amount spent by them to access to those activities (in addition to the fees/quotas paid to the organization as their associates). In fact, paragraph 19 of the article 9 of the IVA' Code omits any express reference to environmental purposes, only granting a tax exemption to the activities carried out in the collective interest of its associates by non-profit organizations with political, syndical [e.g., trade union' activity], religious, humanitarian, philanthropic, recreational, sporty, cultural or civic' objectives or regarding the representation of economic interests, with no fee other than the quota fixed under the terms of the entity' statutes. Which does not include, expressly, the reference to the protection of the environment.

Considering *i)* that an analogical interpretation of tax benefits is forbidden (as referred above and as foreseen in the article 10 of the Estatuto dos Benefícios Fiscais), *ii)* that the Court of Justice of the

European Union has "insisted on a narrow interpretation of the exceptional situations falling within the scope of these exemptions" (Silva & Soares, 2015 and Ribeiro & Santos, 2013) and *iii)* that the Law number 35/98 no longer makes a generic legal reference to the tax exemptions provided for non-profit organizations since the approval of the Law number 36/2021, we believe that it is very difficult to argument that the exemptions provided for in the paragraph 19 of the article 9 of the IVA' Code may apply to the Environmental NGO's. And this circumstance is especially important as it entails that any occasional fundraisings organized by the Environmental NGO's for their exclusive benefit cannot be considered as tax exempt, due to the inability to apply the provisions of the paragraph 20 of the article 9 of the IVA' Code (pursuant to which it can only be considered as exempt a fundraising that is carried out by organizations whose usual activities are exempt from IVA).

Therefore, and even if the environmental purposes stays excluded of the paragraph 19 of article 9 of the IVA' Code, we understand that it is strongly advisable to reflect on a law' amendment that expressly includes the fundraising' activities intended to protect and promote the environment as activities worthy of a IVA' exemption, to the extent that those activities are, naturally, carried out by non-profit entities, as defined in the article 10 of the IVA' Code.

Notwithstanding, and considering the current difficulties related to the application of the IVA' domestic exemptions by the Environmental NGO's, we understand that this organizations may consider the application of the *i)* special exemption' regime foreseen in the articles 53 to 58-A of the IVA' Code whenever, in the previous calendar year, the organization had not a turnover greater than EUR 13,500 (which will be actualized for EUR 14,500 in 2024 and for EUR 15,000 in 2025 and following years)or *ii)* in the *Cash IVA Regime* introduced by the Decree-Law no. 71/2013, May 30th whenever they have not reached, in the previous calendar year, a turnover for IVA purposes exceeding EUR 500,000 (and provided that all the additional requirements are met).

GENERAL ASPECTS ON THE PORTUGUESE TAX REGIME REGARDING DONATIONS

At this point, we intend to analyze whether the Environmental NGO's can take advantage of the tax mechanisms and benefits created by the Portuguese tax legislator in order to promote the attribution of donations to Social Economy Organizations, not only in what concerns to the contributions made by individuals but also to the contributions made by corporations (which are called, in Portugal, "mecenato empresarial"), once that such tax benefits can constitute an important and significant boost to the financial survival of the Environmental NGO's.

IRS' ALLOCATION

Every individual who is a taxpayer of Personal Income Tax (in Portuguese, Imposto sobre o Rendimento das Pessoas Singulates – IRS) can assign a quota equivalent to 0,5% of the tax settled on their annual income tax return (referred, in Portugal, as Declaração Modelo 3) to a duly registered religious organization or to a non-profit entity, indicating in the referred tax return the specific beneficiary entity to whom they want to allocate the assigned amounts (Ribeiro & Santos, 2013). Therefore, it is our understanding that this tax mechanism does not grant to the individual taxpayers a real tax benefit, as it only allows them to make a direct choice on the destination or allocation of their taxes.

Environmental NGO's can be beneficiaries of this allocation, according to the regime that, since the Portuguese Green Tax Reform that took place in 2014, are provided for in paragraphs 5 to 14 of the article 14 Law number 35/98. The essential legal requirements of this regime are the following:

- Only the Environmental NGO's that have been granted with the public utility' status can benefit from the allocation of the IRS;
- Those entities must expressly require to be considered as beneficiary' entities;
- The Environmental NGO's eligible for the tax' allocation will be referred as beneficiary entities in a list created by APA and validated by the Portuguese Tax Authority. This list will be duly published in an annual basis on the institutional electronic website of the Portuguese Tax Authority;
- The amounts assigned by individuals must be transferred to the beneficiaries by March 31st of the year following the submission of the IRS' declaration; and
- The beneficiary must submit an annual report to the Portuguese Tax Authority explaining the use given to the amounts received.

As it can be seen, the IRS' allocation may constitute a very important measure in what concerns to the financing of the Environmental NGO's. However, we have verified that only a moderate percentage of organizations are enjoying of this regime.

In fact, and in an analysis undertaken to the list of Environmental NGO's and legally assimilated entities with an active registration in the National Registry until December 31st, 2022 (Notice number 4195/2023 from APA) and to the list of entities that can benefit from the IRS' allocation regarding 2022 (duly published on the electronic website of the Portuguese Tax Authority), we had verified that out of a total of 104 Environmental NGO's existing in Portugal (including legally assimilated organizations), only 29 entities took advantage of the IRS consignment in that year, which represents a percentage of only 27,9% of the total universe of entities.

These global values can be disaggregated as follows:

Table 2. Environmental NGO's entitled to benefit from IRS' allocation in 2022

Type of Entity	Number of Registered Entities	Number of Entities able to receive IRS' consignment	Percentage
National' NGO	14	9	64,3%
Regional' NGO	15	4	26,7%
Local' NGO	29	5	17,2%
No scope' NGO	20	6	30%
Entity assimilated to NGO	26	5	19,2%
TOTAL	**104**	**29**	**27,9%**

Bearing in mind the results provided in the Table 1, there are strong indicators that suggest that this data results from the lack of compliance with the requirements legally established for the access to the

IRS' allocation (namely, the grant of the public utility' status, as demonstrated in the Table 1) and, additionally, from the lack of fiscal literacy of the Environmental NGO's in what concerns to the access to this tax measure. In this context, we understand that, in the near future, it is important to undertake a study that allows us to understand the reasons that do motivate the low use of the IRS' allocation by the Environmental NGO's once that the obtention and organization of such data will be, in our opinion, very important for a better global implementation of the Portuguese Green Tax Reform.

CORPORATE PATRONAGE

The Portuguese tax legislation foresees specific provisions to encourage corporate patronage, i.e., the distribution of donations with a liberality' character by legal persons and individual entrepeneurs to the Social Economy Organizations (Ribeiro & Santos, 2013). For tax purposes, and in accordance with the article 61 of the Estatuto dos Benefícios Fiscais, these donations constitute deliveries in cash or in kind that, without counter obligations of pecuniary or commercial' nature, are granted to public or private entities whose activity consists predominantly in carrying out initiatives in the social, cultural, *environmental*, sports or educational areas.

Focusing specifically on the IRC' rules, it is possible to verify that corporate patronage constitutes a true tax benefit once that *i)* regarding the Social Economy Entities, the donations are tax exempt if they are used for the direct and immediate fulfillment of the statutory purposes of the organization (article 54 of the IRC' Code), and *ii)* regarding the patrons, it is possible not only to accept the expense for tax purposes but also, in some situations, theoretically increase it for the same purposes (in an operation called as *majoração*). Let's focus our attention on the possible application of this tax benefits to the Environmental NGO's, knowing that the corporate patronage is "one of the main financing instruments" of the non-profit sector (Silva & Soares, 2015).

Regarding the donations granted to Environmental NGO's, it is important to stress the two most relevant possibilities that, in our opinion, are currently in force.

First, the donations can be framed in the so-called regime of **associative patronage**, since that NGO' associates may accept as a tax expenditure the total amount assigned to the accomplishment of the organization' statutory purposes (paragraph 8 of the article 62 of the Estatuto dos Benefícios Fiscais). In this context, the amounts accepted by the associates for tax purposes may not exceed the limit of 1/1000 of the respective volume of sales or services provided on their normal commercial, industrial or agricultural activity. In this case, there is no possibility of increasing the donation' amount for tax purposes.

It is also important to address the **environmental patronage regime**, as it is referred to in the Estatuto dos Benefícios Fiscais. Under this regime, patrons *i)* can accept, as a tax expense, the total value of the donations granted to an Environmental NGO (even if they are not its members or associates) and *ii)* can theoretically increase, for tax purposes, the value of the donation by 20%, which will also be accepted as a tax expense on their IRC (as foreseen in paragraphs 6 and 7 of the article 62 of the Statute of Tax Benefits). However, the total value of the donation that is accepted for tax purposes cannot exceed the limit of 6/1000 of the volume of sales or services provided by the patron in the financial year in which the donations are granted.

But beware: in addition to the concrete and specific limits provided for each of the patronage' types mentioned above, it is also important to pay attention to two limitations of a global (or aggregate) nature. Therefore:

- Whenever the patron grants several donations provided in paragraphs 3 to 8 of the article 62 of the Estatuto dos Benefícios Fiscais – as it is the case of the donations analyzed above – the overall tax deduction may not exceed 8/1000 of the volume of sales or services provided by the company in the year (paragraph 12 of the article 62 of the Estatuto dos Benefícios Fiscais);
- The patron must ensure the compliance with the so-called "result of the assessment" (in Portuguese, resultado da liquidação) provided for in the Article 92 of the IRC' Code. In this context, he must "ensure that the tax paid, net of deductions related to international double taxation and tax benefits, is not less than 90% of the amount that would be determined if [he had not] enjoyed" tax benefits and other specific regimes (Ribeiro & Santos, 2013).

In this context, we cannot fail to underline the need for a legislative update on the article 13 of the Law number 35/98, which links the environmental patronage to "the rules foreseen for the cultural patronage provided for in the IRS and IRC' Codes". This circumstance brings significant uncertainty to the interpretation and application of the tax law, once that, actually, *i)* the environmental patronage regime is expressly provided for in the Estatuto dos Benefícios Fiscais, *ii)* the cultural patronage is also established in the Estatuto dos Benefícios Fiscais, with different rules than the ones established for the environmental patronage, *iii)* the IRS and IRC' Codes do not currently establish any rules related to the cultural patronage and *iv)* the legal provision under analysis still refers to public institutes that are no longer in activity. Therefore, and regarding these problems, a legislative amendment is strongly recommended in order to revoke the rule or in order to update its terms (namely referring to the rules of environmental or cultural patronage currently in force and to the public institutes currently in activity).

At last, please note that the tax regime regarding the corporate patronage can only be applicable if the Environmental NGO comply with the several obligations listed in the article 66 of the Estatuto dos Benefícios Fiscais, namely:

- Issue a document proving the amount received in each donation, which must include:
    - The legal status of the beneficiary entity and the corresponding legal provisions;
    - The framework of the donation within the scope of the Estatuto dos Benefícios Fiscais;
    - The amount of cash or, alternatively, the identification of the goods received; and
    - The mention that the donation is granted without counter obligations of pecuniary or commercial' nature.
- Have a complete and up-to-date record of its patrons with the identification of their name or designation, the tax identification number and the date and value of each donation awarded, among other additional references that are deemed as necessary; and
- Submit the so-called "Model 25 Declaration" by the end of February of each year, declaring in it all the donations with tax relevance that were received in the previous year.

In our opinion, these obligations are fundamental to allow a greater control over the use of the tax benefits under consideration and, consequently, *i)* to ensure that such benefits are not subject to abusive tax planning and *ii)* that they *effectively* are a path to encourage and promote the activities solely developed by the Environmental NGO's in benefit of all the society.

SOLUTIONS AND RECOMMENDATIONS

Regarding the issues previously identified, the author objectively understands and recommends that:

- The establishment of a close link between the attribution of tax benefits to Environmental NGO's and their public utility' status should be maintained in the future. In fact, such a link makes it possible to justify the attribution of a special tax regime to the Environmental NGO's as an effective path to encourage and promote the activities that are developed by these organizations, to the extent that those activities do benefit, directly or diffusely, all the society' members;
- The legal' requirements that impose the compliance with formal or declarative conditions regarding the use of tax benefits are also important, as a way of ensuring that those tax benefits won't be used as a vehicle towards abusive tax planning;
- We found, in the course of our analysis, the existence of some legislative discrepancies towards the tax regime applicable to the Environmental NGO's: namely, in what concerns to the IMI' exemption applicable to the organizations with public utility' status, the tax regime of environmental patronage and to the proof of the organization' legal status deemed as necessary to grant the application of the IRC' exemption. Noticing that this circumstance accentuates the difficulties regarding the interpretation, integration and application of the tax law, it is advisable to consider the introduction of clear and precise legislative amendments that clarifies the tax rules applicable to Environmental NGO's and that, as far as possible, avoid the dispersion of those rules through the Portuguese legislation;
- Even if the environmental purposes stay excluded of the paragraph 19 of article 9 of the IVA' Code, we understand that it is strongly advisable to reflect on a law' amendment that expressly includes the fundraising' activities intended to protect and promote the environment as activities worthy of a IVA' exemption, to the extent that those activities are, naturally, carried out by non-profit entities, as defined in the article 10 of the IVA' Code.

FUTURE RESEARCH DIRECTIONS

In the near future, the author believes that it will be important:

- To undertake a broad analysis of some specific issues regarding the Portuguese tax regime applicable to the Environmental NGO's, as the reflection on the applicability of the corporate patronage' regime to the organizations that have their headquarters or effective management outside the Portuguese territory and on the tax regime applicable to the organizations that are only legally assimilated to Environmental NGO's;
- To deeply study the tax regime applicable to the Environmental NGO's that do not fulfil the requirements to be considered as an entity of the Social Economy' Sector; and
- To better understand, as referred above, the reasons that do motivate the low use of the IRS' allocation by the Environmental NGO's.

The author believes that these studies may have a relevant impact on the global comprehension and evolution of the tax policies regarding the Environmental NGO's and, subsequently, the tax purposes concerning the environment and the sustainability.

CONCLUSION

As a result of the exponential growth of climate' changes and environmental problems, the Portuguese legislator began, in the most recent decades, to dedicate a greater attention to the activity carried out by Environmental NGO's, not only in what concerns to the legal definition of their legal status but also to the establishment of specific tax rules that are consistent with their purposes and attributions.

Specifically addressing the Environmental NGO's that took part of the Social Economy Sector (as legal persons that, in general terms, exclusively aim to defend and enhance the environment and the natural heritage instead of pursuing lucrative purposes for themselves or for their members), the Portuguese legislator allows these organizations to be integrated in the Portuguese tax regime applicable to the non-profit entities. Among other implications, this circumstance affords the Environmental NGO's and their patrons to access to some specific tax benefits, as it happens with the tax regime of corporate patronage, which is a very important measure toward the financing of any non-profit entity.

Notwithstanding, and as we had noticed during this chapter, this more favorable tax regime is, in general, more expressive when there is an official attribution of the public utility' status to the Environmental NGO's. As we understand, this circumstance is an important measure to guarantee that the attribution of the tax benefits under analysis do not harm the fundamental principle of tax justice once that, as foreseen in the article 11 of the LBES, the establishment of a more favorable tax status to the Social Economy Sector can only be applied as a path to encourage and promote the activities that are developed by these entities, to the extent that those activities and the corresponding purposes effectively benefit all the society' members. Concomitantly, the legal' requirements that impose the compliance with formal or declarative conditions regarding the use of tax benefits are also important, as a way of ensuring that those tax benefits won't be used as a vehicle towards abusive tax planning.

Additionally, we have found, in the course of our analysis, the existence of some legislative discrepancies towards the tax regime applicable to the Environmental NGO's: namely, in what concerns to the IMI' exemption applicable to the organizations with public utility' status, the tax regime of environmental patronage, the IVA regime regarding the fundraising' activities intended to protect and promote the environment and to the proof of the organization' legal status deemed as necessary to grant the application of the IRC' exemption. In our opinion, this circumstance can expressively accentuate the difficulties regarding the interpretation, integration and application of the tax law, which advises the introduction of some legislative amendments that *i)* in an objective and precise manner clarify the tax rules applicable to the Environmental NGO's and *ii)* as far as possible, avoid the dispersion of those rules through the Portuguese legislation.

In fact, this topic is very important, once that the clarity and objectivity of the tax legislation undoubtedly contribute to a more effective application of the tax rules, to a better application of the tax justice' principle and, ultimately, to the fulfilment of one of the main objectives on which the tax policies regarding the Environmental NGO's are focused: the protection and conservation of the environment against climate' changes and environmental problems.

REFERENCES

Agência para a Modernização Administrativa, IP (2023). *Obter informações sobre Organizações Não Governamentais de Ambiente (ONGA)*. ePortugal.https://eportugal.gov.pt/pt/servicos/informar-sobre-organizacoes-nao-governamentais-de-ambiente-onga-

Aguiar, N. (2016). O Problema da Tributação do Rendimento das Cooperativas – reflexão a partir do direito português. *Revista Cooperativismo e Economía Social, 28*, 163-190.

Amorim, J. C., & Azevedo, P. A. (2016). *Lições de Direito Fiscal*. Calendário de Letras.

Fernandes, F. V. (2022). *Direito Fiscal do Terceiro Setor*. AAFDL Editora.

Franco, R. C. (2015). *Coord.. Diagnóstico das ONG em Portugal*. Fundação Calouste Gulbenkian.

Martins, C. Q. (2021). *ESNL – Entidades do Setor Não Lucrativo. Abordagem Contabilística e Fiscal*. APECA.

Meira, D. (2013). A Lei de Bases da Economia Social Portuguesa: do projeto ao texto final. *CIRIEC-España, Revista jurídica de economía social y cooperativa* (number 24/2013).

Nabais, J. C. (2021). *Direito Fiscal* (11th ed.). Almedina.

Palma, C. C. (2015). *Introdução ao Imposto Sobre o Valor Acrescentado* (6th ed.). Almedina.

Pereira, M. F. (2014). *Fiscalidade* (5th ed.). Almedina.

Pereira, M. G. (2021). *Environmental Law and Practice in Portugal: Overview*. Practical Law. https://uk.practicallaw.thomsonreuters.com/3-500-2889?navId=AE5DEF8415A6B6C9866CE77

Pires, J. M. F. (2016). *Lições de Impostos sobre o Património e do Selo* (3rd ed.). Almedina.

Ribeiro, F. M., & Santos, S. F. (2013). *A Fiscalidade e as Organizações de Economia Social*. Vida Económica.

Silva, S. T., & Soares, C. D. (2015). *Regime Fiscal das Entidades da Economia Social e Civil*. Vida Económica.

Taborda, D., & Martins, A. (2023). As isenções em IRC das atividades exercidas pelas entidades do setor não lucrativo. *Revista Jurídica Portucalense, n.º 33*. https://revistas.rcaap.pt/juridica/article/view/28483

Tracana, D., & Dias, C. C. (2020). Tributação do Setor Cooperativista em Portugal: análise dos principais aspetos e questões em aberto. *Revista Eletrónica de Fiscalidade da Associação Fiscal Portuguesa, II:2*. https://www.afp.pt/revista/edicoes/708-revista-electronica-de-fiscalidade-da-afp-ano-ii-n-2-2020?start=4

ADDITIONAL READING

Santos, L. M. (2022). *Donations and Tax: The Corporate Conundrum*. IBFD Doctoral Series.

KEY TERMS AND DEFINITIONS

Corporate Patronage: Donations granted, with a liberality' character, by legal persons and individual entrepeneurs to the Social Economy Organizations.

Environmental NGO's: legal persons that exclusively aim to defend and enhance the environment and the natural heritage, as well as the nature' conservation, instead of pursuing lucrative purposes for themselves or for their members.

Green Tax' Policies: The use of the taxes as an instrument for the State' intervention in the economy and in the society towards the protection of the environment.

IRS' Allocation: Assignment made by individual' taxpayers, equal to a quota of 0,5% of the tax settled on their annual income tax return, in favor of a duly registered religious organization or a non-profit entity.

Lei de Bases da Economia Social **(LBES):** Law that establishes the fundamental political options of the Portuguese legal regime applicable to the social economy, that should be implemented by the Portuguese government through development decrees.

Public Utility' Status: Status assigned to legal persons that pursue diffuse and generalized goals, act in accordance with the guiding principles provided by the Portuguese LBES and cooperate, to that extent, with the public administration of the Portuguese State.

Social Economy Sector: Although it is an open concept under permanent construction (Meira, 2013), the social economy sector integrates the organizations that, being endowed with legal personality, chase purposes that benefit the community at large and which, to that extent, are deemed as altruistic or disinterested purposes with social relevance. They also comply with the guiding principles provided in the article 5 of the LBES and appear in the Portuguese database regarding the Social Economy Sector.

Tax Benefits: Measures of an exceptional nature that are established to protect relevant and public interests that are greater than the receipt of taxes.

Tax Justice or Tax Equality: tax principles embedded in the idea of generality or universality, according to which all citizens (and legal persons) are bound to the duty to pay taxes, and the idea of uniformity, which requires that this duty shall be measured by one only criteria – the ability to pay' criteria.

ENDNOTES

1. The rules regarding the registration of an Environmental NGO's in Portugal are currently provided in the articles 17 to 20 of the Law number 35/98 and, as well, in the Statute for the National Registry of Environmental NGO's and Legally Assimilated Companies (Ordinance number 478/99, of June 29th, with the changes introduced by the Ordinance number 71/2003, of January 1st and by the Ordinance number 771/2009, of July 20th).
2. APA is the Portuguese environment agency, a public institute integrated in the State' indirect administration, under the supervision of the Environment' Ministry. APA is endowed with administrative and financial autonomy and its own assets, having the mission to propose, develop and monitor the integrated and participatory management of environmental and sustainable development policies.

Chapter 5
Environmental Tax Reform in Western Balkan Countries

Enriko Ceko
https://orcid.org/0000-0002-3372-2785
Canadian Institute of Technology, Albania

Edmira Cakrani
https://orcid.org/0000-0001-6259-1895
Canadian Institute of Technology, Albania

ABSTRACT

This chapter's goal is to provide a comprehensive description of the environmental tax reform that has been implemented in Western Balkan nations and to examine how it has been carried out. The analysis of the environmental tax's effectiveness and its contribution to the promotion of green initiatives is another goal of the chapter. Since environmental taxation has the potential to address all facets of environmental protection, the goal of environmental taxation is, in theory, to factor environmental harm or negative externalities into prices to influence production and consumption choices in an eco-friendly direction.

INTRODUCTION

The objective of the paper is to verify the state of environmental taxation in the Western Balkans. The paper is organized into several parts starting with the Background, followed by Environmental taxation and application, with its parts of Environmental taxes, Issues, controversies, problems, Literature on environmental taxation, Brief WB6 countries' economy and country profiles (Albania, Bosnia and Herzegovina, Kosovo, Montenegro, North Macedonia, and Serbia), EU on environmental tax reform, EU green deal, Why environmental tax reform?, The environmental tax reform in WB Countries, ISO 14001as an international standard for environmental protection, Western Balkans approach.

Western Balkan countries aspire to join the European Union (EU), but they face numerous obstacles that limit their ability to develop economically. They are considered as transitional economies because they underwent the transition from a centrally planned economy to a market economy in the early 1990s,

DOI: 10.4018/978-1-6684-8592-7.ch005

and the process of economic restructuring is still ongoing. Croatia and Serbia, which are both specialized in producing goods requiring low and middle technologies, are named as the region's two largest economies. The remaining countries, including Albania, Bosnia and Herzegovina, North Macedonia, Montenegro, and Kosovo, are seen as having modest, open economies with labor-intensive sectors including agriculture, textiles, and services. There is currently a dearth of examination of the economy-environment interaction in the Western Balkans in the scientific and professional literature, despite the countries' shared geographic location and economic history. Mining, industry, agriculture, and forestry all contribute significantly to the economy of the Western Balkans and are big environmental polluters. According to the Health and Environment Alliance (HEA) report on Chronic Coal Pollution (2019), more Sulphur dioxide pollution was produced in 2016 by 16 outdated coal-fired power stations in the Western Balkans than by all 250 coal-fired power plants in Europe put together. Particularly in the case of Serbia, according to the Ministry of Mining and Energy, around 75% of Serbia's current energy comes from fossil fuels. According to the OECD, each of the Western Balkan countries can be considered to have a diverse and competitive economy in this region, but they also share many environmental issues, such as exposure to environmental dangers, high carbon intensity and low energy efficiency, heavy reliance on coal, excessive pollution levels in the air, and inadequate handling of wastewater and waste. The European Union agreed to a financial investment plan for the Western Balkans worth up to 9 billion euros at the end of 2020 which was based on: climate action with a focus on decarbonization, the circular economy with emphasis on recycling and sustainable production, biodiversity, tackling air, water and soil pollution and the development of sustainable food systems and rural areas. All Western Balkan countries are committed to upholding the Paris Agreement and achieving the EU2020 and EU2030 goals for reducing greenhouse gas (GHGE) emissions as EU candidates or potential candidates. Due to a lack of human and financial resources, the burden of the effort associated to the climate change issues that each economy is required to accomplish is acknowledged as a significant setback, which promotes regional cooperation and development. (Vuković & Vujadinović, 2018) The environmental tax plays an important role in addressing issues related to climate change and has significant advantages such as economic efficiency and the ability to raise public revenue.

BACKGROUND

Environmental taxation advocacy dates back to early 1970, especially with the British political agenda by the 'Pearce Report' to the Department of the Environment in 1989 (Pearce et al.,1989). Environmental taxes were presented in the report as a means of meeting environmental objectives more cheaply than through regulation. Later, the purpose of environmental taxes was set as to oblige polluters to pay a tax to safeguard the environment. Environmental taxes have many important advantages, such as environmental effectiveness, economic efficiency, the ability to raise public revenue, and transparency. In the course of the issue, many environment-related taxes were worldwide presented, e.g., emission tax, energy tax, carbon tax, sulfur tax, etc., with the idea of further generating positive environmental impact by changing the consumption behavior of the public in related fields and increasing revenue for the government. In the 1990s, green tax, ecological tax, and environmental tax were proposed respectively, the idea behind which are the same, i.e., sustainable development. Since then, the term environmental taxation suggested by OECD has gradually been accepted and many developed countries started building their environmental taxation system while developing countries started their attempts at the implementation of environmental taxation. Some further progress on the introduction of ecological tax reform measures came during early 2000, with the introduction of a carbon tax based on carbon content or CO2 emission amount of fossil fuel, widely implemented in developed countries while energy tax based on the price of energy was adopted in many developing countries.

In a report by Larry Lohmann (2006), he describes how climate change politics has unraveled throughout the years. For example, when the UN delegations assembled in Kyoto in 1997, there was a big sense of urgency in relation to global warming. Southern countries "had been insisting that the industrialized North take responsibility for a global crisis of which it had been the main cause" (Lohmann, 2006, p.296). In Kyoto, the USA offered to commit to emission reductions in exchange for international commitments to carbon trading. Exhausted by the long process of negotiations, the southern nations embraced the move and, almost unwittingly, furthered the techno-political process that led to the numerous defeats that we have seen in the recent years. While many countries in the Global South were preoccupied with getting the USA and other countries in the Global North to commit to "legally-binding emission cuts", the carbon trading arrangements were used on the other hand only to assert the political position of the countries involved. The "binding cuts" were undermined through the "widespread use of the offsets that carbon trading made possible, as well as the growing 'offshoring' of emissions" (Peters et al., 2011). In another UN conference in 2007, an action plan was created. Its reach would have expanded to encompass developing economies like China, India, and Brazil as well as rising demands for wealthier nations to decrease their carbon emissions. The latter category, which is labeled as "evolving," has so far been exempt from limitation. The new strategy aimed to slow the emissions' rapid increase. Plans were for the revised agreement to be finalized during the COP15 conference in Copenhagen, Denmark, two years from then. The Conference's conclusion was very awful, even tragic. A purely political agreement without legal ramifications or specific objectives was established. Perhaps the most significant treaty related to climate change is the Paris agreement in 2015 which resulted in a landmark global agreement to combat climate change, where 196 countries took part. Following voluntary pledges to cut emissions known as NDCs—which are currently nowhere near sufficient to achieve this goal—nearly the entire international community chose to commit to keeping the increase in average world temperatures well below 2 degrees Celsius. Another important pledge that was established is the Green Climate Fund, the world's largest climate fund. It was established to assist developing countries in achieving their Nationally Determined Contributions (NDC) goals of adopting low-emission technologies and climate-resilient pathways. The idea of this fund is to invest across four transitions: transformational planning and programming, catalyzing climate innovation, de-risking investment to mobilize finance at scale, mainstreaming climate risks and opportunities into investment decision-making to align finance with sustainable development. (GCF) The fund is mandated to invest 50% of its resources to mitigation and 50% to adaptation in grant equivalent. (Nationally Determined Contributions, or NDCs, are countries' self-defined national climate pledges under the Paris Agreement, detailing what they will do to help meet the global goal to pursue 1.5°C, adapt to climate impacts and ensure sufficient finance to support these efforts. NDCs represent short- to medium-term plans and are required to be updated every five years with increasingly higher ambition, based on each country's capabilities and capacities. NDCs represent politically backed commitments by countries. If used right, they could be our way out of tackling the world's current crises – not just the climate crisis, but other systemic problems like biodiversity loss and energy security as well.

With the increasing importance of achieving the climate change goals at a global scale, the concept of green economy has been trending more in the recent years. Green economy has been defined by UNEP (2011a) as one that results in improved "well-being and social equity, while significantly reducing environmental risks and ecological scarcities". A green economy can be easily defined as low carbon, resource efficient and socially inclusive (UNEP, 2011a). UNEP emphasizes the protection of natural capital, including ecosystems and natural resources. In addition to, or sometimes synonymously with, green economy, the term green growth is often used (EEA, 2014). Green growth is qualitative growth that uses natural resources efficiently, is clean because it minimizes pollution and environmental damage, and is resilient because it accounts for natural disasters (World Bank, 2012). The idea of a "Green Economy" as a means of promoting economic growth and development while protecting natural resources was verified and accepted at the UN Conference on Sustainable Development (Rio+20), which was held in 2012. In order to pull people out of poverty and foster more social cohesion, growth must be inclusive, particularly in the form of job creation and skills development.

If we look at the implementation of the climate change policies and green economy strategies in the Western Balkans, Knez, et al (2022) conclude that the execution of specified actions is at a low level, despite the fact that Croatia, an EU member, has enacted the required legislation and plans for combating climate change, and the other five countries in the Western Balkans are in the process of doing the same. This is frequently due to decision-makers' lack of commitment to making meaningful adjustments in the transition from climate change. The Western Balkans are currently under the Green Recovery and Support Program (2022) in partnership with the European Union. The main goals of this program are to design: policy and regulatory instruments that promote more sustainable growth paths and integrated approaches to air quality management and climate change mitigation. Achieving these objectives also requires the development of green financing instruments. UNEP defines green financing as financial flows towards the fulfillment of the objectives of sustainable development. Despite some progress being made in terms of climate policy implementation, a report by the Stockholm Environment Institute (2022) states that climate finance is still lacking in the Western Balkans mainly due to the region's huge reliance on fossil fuels.

For green finance to have an effect, countries must encourage the use of green financial instruments through appropriate legal changes. One of these changes should be the taxation of activities that have a negative impact on the environment. Environmental taxes are considered as instruments for reducing pollution and promoting green activities. The purpose of these taxes is not only the highest taxation of activities that burden the environment, but the use of tax revenues from these activities to promote other activities that preserve the environment.

An ecotax has been enacted in Germany by means of three laws in 1998, 1999 and 2002. The first introduced a tax on electricity and petroleum, at variable rates based on environmental considerations; renewable sources of electricity were not taxed; the second adjusted the taxes to favor efficient conventional power plants; the third increased the tax on petroleum. At the same time, income taxes were reduced proportionally so that the total tax burden remained constant. A variety of ecotaxes (often called "severance taxes") have been enacted by various states in the United States. The Supreme Court of the United States held in Commonwealth Edison Co. v. Montana, 453 U.S. 609 (1981), that in the absence of federal law to the contrary, states may set ecotaxes as high as they wish without violating the Commerce Clause or the Supremacy Clause of the United States Constitution (Elison & Snyder, 2001). The Netherlands, Portugal, Canada, Spain and Finland have introduced differentiations into their car registration taxes to encourage car buyers to opt for the cleanest car models. Austria has had a registration tax based on fuel consumption for several years.

In the UK, starting from 1993, fuel taxes have been introduced, stating that they are not fundamentally there to raise revenue; Changing people's behavior is only achieved by market forces and price signals, and, as our understanding of climate change increases, it is clear more needs to be done (BBC,2006).

The French government shared its intentions to establish a new fee on plane tickets with the purpose to fund environment-friendly alternatives, such as eco-friendly transport infrastructure, including rail. The proposed tax would range between 1.50 euros ($1.7) and 18 euros ($20) and apply to most flights departing in France. The French government expected the new tax to raise over 180 million euros ($200 million) from 2020 (Adamson & Jordans,2019). The carbon tax was implemented in 2014 based on the "polluter pays" principle, "according to which all persons must contribute to the repair of the damage they cause to the environment". Nearly 4 million tons of CO2 were avoided by France in 2018 thanks to its carbon tax, according to an OECD study, which represents a 5% reduction in emissions from the manufacturing sector between 2014 and 2018 (Dussaux, 2020).

The preservation of the ozone layer, local air quality, acidification and eutrophication, water supply and water quality, waste management, and biodiversity losses are just a few of the environmental issues that countries around the world must deal with. How to cut greenhouse gas emissions is a topic that is currently high on the agenda in the majority of countries. Economic tools have become increasingly important in global environmental strategies over the past ten years. The increasing significance of taxes related to the environment in this context stands out. To differing degrees, all countries have enacted environmental levies, and a growing number are putting comprehensive reforms into place while others are just thinking about it.

Environmental taxes may or may not support the "polluter pays principle," which states that the expenses of pollution prevention and control should be reflected in the cost and output of goods and services that produce or consume pollution.

In Europe, taxes on the environment generate an average of 2% of the GDP. These taxes are imposed in support of a number of policy goals, where selecting the best course of action frequently requires balancing competing objectives. When it comes to environmental issues, environmental taxes give a price signal that enables polluters to consider the consequences of pollution on the environment when they decide how much to produce and consume.

Taxes are a flexible tool for policymaking that can reduce the cost of pollution management while also encouraging technical advancement and future emissions reductions. The majority of money collected from taxes related to the environment comes from taxes on the purchase or use of motor vehicles and fuels, including taxes on gasoline and diesel. Taxes are also utilized in certain countries to address a variety of other environmental issues. However, there is room to increase the usage of environmental taxes in many nations. The effectiveness of environmental levies in EU/OECD countries as a way to lessen environmental harm is becoming more and more clear.

There are also hints that if tax rates on energy products are re-modulated in accordance with environmental requirements, significant fuel switching may result from modifications to tax rates for competing energy sources.

Green tax reforms can be accomplished by a number of complementary measures, such as the enactment of new taxes or the restructuring of already-existing ones, such as those on transportation or energy, to better represent the damaging properties of the activities or goods that are subject to the taxes. Additionally, it is crucial to eliminate or modify fiscal measures that negatively impact the environment, such as tax breaks or subsidies, while considering the non-environmental goals they were intended to support.

Pre-announcement of the implementation of environmental taxes and tax rate hikes, along with a progressive reduction or phase-out of rebates and exemptions, are two policy alternatives that could simplify implementation, boost the efficacy of environmental taxes, and address concerns about competitiveness. There may be room to enhance the architecture of tax laws in some nations so that any remaining exemptions and refund mechanisms are effectively targeted to meet their stated goals. The financial expenses of accomplishing a certain environmental goal could be decreased with careful planning and targeting.

Sharing information, experiences, and best practices with regard to potential options and opportunities for broadening the application of ecologically linked taxes is one approach for countries to address concerns about their ability to compete internationally.

ENVIRONMENTAL TAXES AND THEIR APPLICATIONS

Issues, Controversies, Problems

An environmental tax, also known as an eco-tax or a green tax, is a tax imposed on activities that are deemed to be environmentally damaging with the goal of encouraging environmentally favorable behaviors through financial incentives. A carbon tax is one such instance. Such a policy can supplement regulatory (command and control) techniques or eliminate the necessity for them. A green tax shift toward ecological taxation is when a policy proposal for an ecotax tries to retain overall tax revenue by correspondingly cutting other taxes (such as taxes on renewable resources and labor). The lack of consideration for environmental effects by free markets is addressed through eco-taxes (OECD, 2015). Pigouvian taxes, which aim to make the engaged private parties feel the social cost of their actions, include ecotaxes. The Global Resources Dividend presented by philosopher Thomas W. Pogge (2001) serves as an illustration. Examples of taxes that a change to a green tax could reduce or abolish include:

- Taxes on earnings, income, and, to a lesser extent, sales.
- Corporate taxes (investment and business taxes).
- Taxes on real estate for infrastructure (buildings).

Examples of ecotaxes that could be introduced or raised include:

- Carbon taxes on the production of greenhouse gases due to the usage of fossil fuels. Production of greenhouse gases (GHGs) is not penalized under the old hydrocarbon levies.
- Severance taxes on the extraction of mineral, energy, and forestry products.
- Duties on imported items containing significant non-ecological energy input (to a level necessary to treat local manufacturers fairly).
- The cost of camping, hiking, fishing, and hunting licenses as well as any equipment required.
- Specific taxes on goods and technology that have significant negative externalities.
- Refundable fees and taxes for waste removal.
- Imposing charges on pollutants, hazardous wastes, and effluents.
- Site value taxes based on the value of the land as-is.

A green tax shift frequently aims to achieve "full cost accounting" or "true cost accounting," which uses fiscal policy to internalize externalities that distort markets and generate sustainable prosperity. The more extensive actions needed to achieve this are also sometimes referred to as ecological fiscal reform, particularly in Canada (Beauregard-Tellier, 2006), where the government has frequently used this phrase. Eco-social market economies are known as such in several countries.

In order to preserve overall progressiveness and revenue neutrality for the government, tax shifting typically entails balancing taxing levels. Additionally, it typically includes safeguards for the most vulnerable, such as raising the required minimum income for filing income taxes or raising the amounts of pension and social support to balance rising fuel costs.

The presence of externalities and their potential for harm are acknowledged in fundamental economic theory. Green taxes align with conventional economic theory insofar as they account for externalities like pollution. Setting the proper taxation amount or the necessary tax collecting mechanism, however, is challenging in practice and may have unforeseen or additional negative effects.

LITERATURE ON ENVIRONMENTAL TAXATION

Currently the literature for environmental issues, and especially for environmental taxes is expanding. A paper on reducing EU emissions stresses that the EU aims to reduce emissions by at least 55% by 2030 (EC, 2020). To achieve this, the European Commission launched proposals to introduce price signals through the modification or extension of the Emission Trading System (ETS) and the Energy Tax Directive (ETD). These instruments could have a negative impact on households and companies' competitiveness. This research focuses on the one hand, on the role of energy taxation in the achievement of climate objectives and budget stability. And, on the other, in analyzing the role of taxation to develop compensatory measures so that the transition is efficient, socially fair and inclusive. Carbon pricing instruments, such as the ETS or energy taxes, can result in negative distributional impacts on the most vulnerable sectors. Likewise, the adoption of these measures must consider their impact on the loss of competitiveness of EU companies and, therefore, the risks of "carbon leakage" and relocation to jurisdictions with less ambitious environmental standards than those introduced. Therefore, it is necessary to consider not only the impact on the current legal system and the current system of the proposals to modify the ETD or the ETS, but also the possible impacts in social, distributive and equity terms. Therefore, Commission's proposals should not be insensitive to the energy and social crisis caused by the pandemic first and the Ukraine war later. A gradual application with temporary regimes for some sectors would increase social acceptance and facilitate the adaptation of the affected sectors. The previous position, in turn, could be a factor that would allow obtaining the favorable vote of all the Member States to approve the measures proposed by the European Commission. The accelerated deterioration of environmental quality and the need to reduce the energy dependency from countries like Russia could discourage the adoption of transitory measures and gradual adaptation. In these cases, the adoption of initiatives such as the ETS or the ETD will require the configuration of compensatory packages that deal with the distributional impacts on business competitiveness. These mechanisms should be designed considering, in addition, the need to increase the acceptance of the Commission's proposals both from the social point of view and from the point of view of MS. The tax system, both from the income and spending side, has the potential to complement transition strategies that are designed to deal with the social, distributive and labor impacts of the change in the production model (Antón,2022).

In a paper investigating introduction of carbon emission tax in Europe (Ghazouani et al, 2020) is mentioned that during the past decades, environmental related taxes, energy, and carbon taxes has been recommended by environmental scientists as a policy tool to mitigate pollutant emissions in developed and developing economies. Among developed nations, Denmark, Finland, Sweden, the Netherlands, and Norway are the first regions to adopt a tax on carbon dioxide (CO2) emissions and research into the impacts of carbon tax on carbon emissions brings significant implications. The prime objective and goal of this work is to explore the role of carbon tax reforms for environmental quality in European economies. This is probably the first comparative study in European context for carbon-tax implementation and non-implementation policies. To this end, the study reports new conclusions and implications regarding the effectiveness of environmental regulations and policies for climate change and sustainability. Using the propensity score matching method, the results of the estimation of the different matching methods allows to observe a positive and significant impact of the adoption of the carbon-tax on stimulating the reduction of carbon emissions (Ghazouani et al, 2020).

Guo et al (2022), in a paper regarding the implementation of environmental protection tax law in China mention that air pollution significantly impacts sustainable development and public health. Taking the implementation of China's Environmental Protection Tax Law as a quasi-natural experiment, authors employ the difference-in-differences (DID) and spatial DID models to evaluate the effects of environmental tax reform on urban air pollution. The findings are as follows: (1) Environmental tax reform can significantly reduce urban air pollution, and a series of robustness tests are conducted to provide further evidence; (2) Green technology innovation and industrial structure upgrading form a vital transmission mechanism for environmental tax reform to improve air quality; (3) Environmental tax reform significantly inhibits urban air pollution in cities located north of the Qinling-Huaihe line and big cities; (4) Environmental tax reform not only promotes the improvement of local air quality but also has a significant negative spatial spillover effect, reducing air pollution in neighboring cities. The research conclusions provide theoretical support and policy suggestions for promoting sustainable economic development, rationally optimizing environmental protection tax policies and improving urban air quality (Guo et al, 2022).

In a paper exploring the relationship between environmentally related taxes and the economic growth rate, Hassan et al (2020) analyze also whether this relationship differs between countries that have implemented environmental tax reforms (ETRs) and those that have not. Using panel data from 31 OECD countries over the period 1994-2013, the paper finds that when we allow environmentally related tax revenues to interact with an initial level of real GDP per capita, the overall revenues of these taxes are negatively associated with the economic growth rate in the short and long term. Furthermore, it was shown that the higher the initial level of GDP per capita, the more environmentally related tax revenues can promote the economic growth rate. The analysis also reveals that the relationship between environmentally related tax revenues and economic growth varies between countries that have a mechanism to redistribute environmentally related tax revenues and those that do not (Hassan et al, 2020).

Jiang et al (2020) investigate the role of taxes on iron ore in China. Authors adopt a computable general equilibrium model to simulate the socio-economic and environmental impacts of different iron ore resource tax reforms from an economy-wide perspective. The results show that this reform has had a limited influence on reducing the tax burden in the iron ore industry. However, it is found that adopting a 1% ad valorem tax rate would enhance the competitiveness of domestic iron ore enterprises and reduce dependence on iron ore imports. In addition, if the resource tax is halved or a 1% ad valorem tax rate in implemented, GDP and household welfare would both rise; however, a 3% or 6% ad valorem tax rate would have a negative impact on GDP and household welfare. Finally, a lower (higher) ad valorem tax rate would increase (decrease) carbon dioxide emissions and carbon intensity. However, if a carbon tax policy is adopted under the low ad valorem tax rate policy, the disadvantageous environmental impact could be significantly improved and the tax burden of iron ore enterprises would also fall (Jiang et al, 2020).

Li and Zhou (2022), using a GMM model, analyze the impact of fiscal and tax policies on China's carbon emissions. The research shows that there is an inverted U-shaped relationship between financial expenditure and carbon emissions, that is, with the increase of financial expenditure, carbon emissions show a trend of first rising and then declining. Tax revenue has the effect of restraining carbon emissions, but the effect is limited. Based on this, authors suggest the improvement of fiscal and tax policies and the expansion of the implementation path from the perspective of "dual carbon" (Li & Zhou, 2022).

In a paper exploring how carbon taxation and economic growth affect environment hazards in Malaysia, using time series data over the period 1974–2010, applying cointegration and causality approaches to determine the long-term cointegration and direction of causal relationship between these variables, Loganathan et al (2014) find a cointegration relationship between the variables. Furthermore, it is noted that although Kuznets' theory, i.e. inverted-U shaped curve between economic growth and CO2 emissions is valid for Malaysia, the carbon taxation policy is ineffective to control CO2 emissions. The causality analysis reveals that there is a bidirectional relationship between carbon tax and CO2 emissions. Granger Economic growth causes CO2 emissions and Granger carbon tax causes economic growth. To enhance the awareness concerning pollution issues, governments should rely on alternative instruments that may give benefit to the taxpayers and reduce pollution, which is the pivotal issue to be tackled globally (Loganathan et al, 2014).

Mardones (2022), in a paper investigating the mining taxes, finds that in several countries, mining taxes are applied as a mechanism to collect resources, but without adequately evaluating their economic impacts. From a theoretical perspective, the implementation of royalties can be justified in the non-renewable nature of mining resources. In contrast, Pigouvian taxes aim to make productive activities internalize their environmental damages. This study illustrates the general equilibrium effects on the Chilean economy of replacing the current royalty on copper mining with Pigouvian taxes, considering various environmental damages. For the above, a computable general equilibrium model is developed and calibrated with a Social Accounting Matrix base year 2016. The results show that the internalization of all environmental damages would generate a significant drop in mining activity and electricity generation, contributing to reducing greenhouse gas emissions. Also, it is observed that the fall in GDP and increase in the real exchange rate would modify the production of other tradable and non-tradable sectors, benefiting non-copper export sectors. The fall in economic activity reduces the payment to productive factors, although the impact is greater on capital and skilled labor. The fall in income is more remarkable in the richest households than in the poorest households, which improves income distribution but increases poverty. Finally, the results on poverty reflect the need to implement monetary transfers to the poorest households (Mardones, 2022).

The environmental extension from the Leontief pricing model is used to simulate the economic and environmental effects of setting different CO_2 tax rates in three Latin American countries (Brazil, Mexico and Chile), considering first a regulatory scenario in which the tax is applied only to emissions from the electricity sector and another scenario in which tax is applied to emissions from all economic sectors. The results show that the effects of a CO_2 tax on sectoral prices and emissions in each country are very different for the same tax rate, but these effects are similar in each country when comparing both regulatory scenarios. This is explained by the composition (carbonization) of the electricity generation (and energy matrix) of each of the countries, which significantly affects the effectiveness of CO_2 taxes to reduce emissions. In addition, the economic sectors most affected by the different regulatory scenarios are similar in all countries (Mardones, & Baeza, 2018).

In a paper considering the source of CO_2 emissions in Europe, is found that the transport sector is the biggest source of CO_2 emissions in Europe (Meireles et al, 2021). It is responsible for over a quarter of all greenhouse gas emissions. Passenger vehicles, alone, account for nearly 41% of these emissions, resulting in human health impacts. To meet the Paris climate commitments, cars and vans should be decarbonized until 2050. Such a transformation requires general changes, such as how the vehicles are owned, taxed, and driven. The European Federation for Transport and Environment revealed that Mediterranean countries tend to emit less per vehicle compared to the northern and central Europeans. Intriguingly, this does not necessarily correspond to motorization rates. In this article it is assessed whether the observed reductions in CO_2 emissions in the Mediterranean countries can be attributed to vehicle taxation on CO_2 emissions. The authors applied panel data econometric techniques using data on annual registrations from 2008 to 2018 and model the demand for new-vehicle purchases and their responsiveness to changes in both CO_2-based taxation and circulation tax, and the results show that fiscal policies can have an important role in reducing the emission in the Mediterranean countries (Meireles et al, 2021).

Mitic et al (2023), in a paper investigating the balance between economic growth and environmental preservation, find that the implementation of sustainable policies is necessary. As such, it is imperative to have a comprehensive understanding of the intricate relationship between the economy and the environment. This study examines the relationship between CO_2 emissions and GDP per capita in five WB countries during 1960-2018, using a methodological framework that includes panel unit root testing, cointegration testing, and a Vector Error Correction Model. The study provides robust evidence of a long-term cointegration between CO_2 emissions and GDP per capita. Furthermore, the study also reveals a short-run bidirectional causal relationship between CO_2 emissions and GDP per capita. In the long run, no statistically significant causality exists from GDP per capita to CO_2 emissions, but it is statistically significant from CO_2 emissions to GDP per capita. These findings offer valuable insights for policymakers to develop comprehensive policies that promote economic growth and environmental sustainability, such as investing in clean energy, implementing stronger environmental regulations, and encouraging environmentally sound management practices (Mitic et al, 2023).

Olja, and Mirjana (2012) provide an overview of key economic issues in the use of taxation as an instrument of environmental policy. In a paper investigating National Strategy for Sustainable Development of Serbia, where a clear guidelines for the direction of the economy, respect for ecological principles, development and implementation of better social and institutional policies is given, it is shown that environmental taxes and all other taxes, the fiscal revenue and its application has special significance, because in addition to economic, it is ecological, institutional and social character, and as such it is an instrument of sustainable development. With this work the authors suggest moving the value of revenues from environmental taxes in the EU in the period 2005-2010. As this is a period characterized by economic problems caused by the crisis in all countries, there has been a decline in revenues from environmental taxes even in the most developed EU countries. On average, the EU has achieved the highest revenues from environmental taxes in 2008. The largest revenue from environmental taxes is reached in Germany (EUR 56.031,00 million) at the beginning of the period, while Malta has the lowest income (138,21 million EUR) (Olja, & Mirjana, 2012).

In a paper considering the implications of environmental tax reform and public spending policy for growth and welfare, using a two-sector endogenous growth model where the interactions between health, education, and the environment are taken into account, Oueslati (2015) shows that revenue-positive tax reforms combined with a change in the public spending structure may improve long-run growth and welfare. However, this outcome incurs relatively high welfare cost during the transition phase. This is particularly the case when the spending policy favors education spending (Oueslati, 2015).

Oueslati et al (2017), analyze the macroeconomic relationship between taxes on energy products and income inequality. They also investigate whether this relationship is different in cases where explicit mechanisms to shift tax burden from labor and income to environmentally harmful activities have been established. In contrast to earlier empirical studies which analyze the distributional effects of energy taxes on the uses of household income, this paper focuses on their macroeconomic relationship with income sources. Using a panel of 34 OECD countries from 1995 to 2011, the paper shows that in the absence of explicit tax revenue recycling mechanisms there is a positive, although modest, relationship between the share of revenues from energy taxes in GDP and the Gini coefficient, which is the adopted measure of income inequality. In contrast, where such mechanisms have been implemented, there is an inverse and relatively stronger relationship between the share of energy tax revenues in GDP and inequality in income sources (Oueslati et al, 2017).

In a paper regarding Western Balkan countries, Sanfey et al (2016) find that Western Balkans – Albania, Bosnia and Herzegovina, FYR Macedonia, Kosovo, Montenegro and Serbia – all aspire to membership in the European Union, but they face a major convergence challenge in terms of living standards. The main reason behind this prosperity gap lies in the failure over the years of Western Balkans to be competitive, meaning that they lack the appropriate factors and institutions needed for high levels of long-term productivity. The key issue is whether the Western Balkans can narrow the gap in the coming decade and, if so, what do they need to do to achieve this. The vital requirement for catching up with the rest of the European Union is a boost to investment in the region. The Western Balkans have a number of attractive features for investors. The long-term EU perspective is a major plus and a unique quality of the region compared with other emerging markets, as it helps to anchor market-oriented reforms and European standards. Strong macroeconomic stability, strategic geographic location, diverse economies, favorable tax regimes and low unit labor costs, combined with a relatively well-educated population, are common attributes throughout the region. Looking ahead, sources of growth include trade integration, within the region and with the rest of the world, exploitation of the region's energy resources, improvement of the transport infrastructure and technological innovation. However, long-term challenges remain, such as the possibility of a slowdown in reforms, risks posed by financial instability, adverse demographic trends and climate change. But provided governments in the region remain committed to reforms and regional cooperation, these challenges can be mitigated, if not overcome (Sanfey et al, 2016).

Skjærseth (2021), in a paper regarding decarbonization finds that decarbonization cannot be achieved with single instruments like carbon pricing alone. A broader mix of reinforcing policies is required. Applying the policy mix literature to the European Union (EU), this article examines the development of climate and energy policies from separate and narrow initiatives to coordinated policy packages to achieve increasingly ambitious climate targets, culminating with the European Green Deal. The starting point to explain this policy development is that EU policies will reflect the positions of the 'least ambitious' actors when unanimity is required. Examination of different policy phases shows that EU policy mixes are not only needed to fulfil different transition functions—they also provide opportunities to combine different actor interests to raise climate ambitions. The EU institutions have been instrumental in crafting policy packages that exempt and compensate the least climate-ambitious actors. The Paris Agreement has also provided an enabling context for higher EU ambitions. Looking towards the future, the corona-induced recession has so far mainly been used by the EU as an opportunity to strengthen climate ambitions and the European Green Deal (Skjærseth, 2021).

Wang et al (2023), in a paper regarding China's objectives on carbon peak and neutrality find that to achieve its carbon peak and neutrality targets, China needs balancing environmental and economic sustainability. Based on the data of A-share listed companies in Shanghai and Shenzhen from 2012 to 2021, this study takes China's environmental fee-to-tax as a quasi-natural experiment to evaluate the impact of environmental tax on enterprises' total factor productivity (ETFP). The results show that the environmental fee-to-tax significantly hinders the growth of ETFP. Further, this negative impact mainly stems from the fact that the reform prompts enterprises to choose strategic innovation and conduct greenwashing behavior. Furthermore, when considering enterprises' heterogeneity, authors find that enterprises with large scale, high investment efficiency, state ownership, high market competitiveness and located in western regions are more resistant to policy shocks (Wang et al, 2023).

Research on environmental taxation has captured significant attention in recent years. The 590 literature is retrieved from the Scopus database (2001–2022 March) with the keywords Environmental* Tax* OR Carbon Tax*, after limiting the subject area to economics, business, finance, and social science. The analysis is confined to the most impactful journals, organizations, countries, keywords, international collaboration, and the direction of recent research areas. It is found that the "Journal of Cleaner Production" has the highest growth rate due to the most productive (43 publications) and highly cited (1115 total citations) journal. The 'North China Power University' has published the highest 26 documents, and China is the most productive (148 publications) and highly cited (2598 total citation) country. The USA has the highest international collaboration in 118 documents with 17 countries. The study found that over half of the top 30 keywords are about policymaking, reform, regulation, and environmental tax impact. Analysis of keywords and network analysis directs that the Impact of environmental and carbon tax, Double dividend, Effect on macroeconomic variables, Emission trading system and Public acceptability are the research hotspots in the published literature. It is suggested from the analytical results that research collaboration among social science and environmental science subjects and developed and developing nations will bridge the knowledge gap and bring new insights to the environmental tax area (Patel, & Jhalani, 2023).

BRIEF WB6 COUNTRIES' ECONOMY

The current geopolitical situation is confronting Western Balkans with unprecedented challenges. After Russia's unprovoked, barbaric invasion of Ukraine, Europeans are called to counter this aggression and to defend freedom and values together. In times like these, solidarity and close international cooperation are more important than ever. This includes the integration of the countries of the Western Balkans into the EU, where important steps forward were made in 2022: the beginning of EU accession talks with Albania and North Macedonia in July and the granting of the EU candidate status to Bosnia and Herzegovina in December. Also, the substantial progress on visa liberalization for Kosovo with a clear timeline is a very positive development. These steps revitalize the accession process and bring new dynamics to the Western Balkans.

As for the current energy crisis, the need for a green transition of all energy supplies to renewable energies is clearer than ever. Consequently, this issue was at the heart of the Berlin Process summit on November 3, 2022, where leaders of the Western Balkans signed a joint declaration on energy security and green transition. At the summit, Germany announced its support for the governments of the Western Balkans for the development of a regional energy and climate plan for the region (EC, 2022). Of crucial importance in this regard is the Green Agenda for the Western Balkans, adopted by the countries at the Sofia Summit in November 2020, which aims at spurring the long-term economic recovery of the Western Balkans and their convergence with the EU. It is financed through EU funds amounting almost to 6 billion EUR and additional loans from international financial institutions.

The transition from coal toward renewables is not only increasing the political and economic independence but also improving the life and health of citizens. Making use of the huge potential for decarbonization and green transition is a political priority for all governments both in the region and the EU.

Albania, Bosnia and Herzegovina, Kosovo, Montenegro, North Macedonia, and Serbia are all part of the Western Balkans. These nations work to maintain stability while dealing with the problems posed by climate change. They accomplish this while having limited finances to deal with environmental protection issues, inadequate environmental regulation or implementation, little public involvement, and political difficulties. EU integration and EU membership are the primary objectives of the region's nations. All nations are driven to uphold the Paris Agreement and accomplish the EU2020 and EU2030 goals of enhancing energy efficiency, cutting GHG emissions, and producing energy from renewable sources as potential candidates and candidates for EU membership. In the decades to come, climatic and environmental policies, legislation, and practices will be significantly impacted by the EU's integration (Knez et al., 2022).

The Western Balkan economy continues to be challenged by an unstable external environment, which puts people's homes, businesses, and governments under extreme stress. The war in Ukraine, together with high energy prices and slowing global growth, is negatively impacting the economies of all six countries. Inflation has reached previously unheard-of levels as a result of rising food and energy costs. Over predictions, the first half of 2022 had strong economic growth. Investment and private consumption acted as the main stimuli for growth. Private consumption has been boosted by rising earnings, remittances, and private credit growth. North Macedonia, Serbia, and Bosnia and Herzegovina saw a lot of investment.

By mid-2022, some countries saw employment levels hit all-time highs, building on the results from 2021. All countries had an improvement in the employment rate of 3 percentage points since mid-2021. The job market recovered across all industries, but services—including tourism—played a particularly significant contribution. By mid-2022, the Western Balkans' unemployment rate had fallen to a record-low 13.5 percent, or 151,000 fewer people were unemployed. While poverty has continued to decrease in 2022, the future of poverty reduction is at stake due to dramatically rising inflation.

The intensity of the energy and food price shocks might lead to a 13% increase in the region's disadvantaged population in the absence of government assistance. As a result of the spikes in energy and food prices, fiscal spending pressures have countered the revenue boost from high inflation and prevented fiscal consolidation.

Excessive inflation has aided in increasing nominal revenue, particularly for indirect taxes like the VAT. On the other hand, as a result of the government's actions in reaction to the energy crisis and growing inflation, public spending has increased dramatically. During 2022, the Western Balkans' inflation rate skyrocketed as a result of a confluence of supply- and demand-side variables. For 2022, all Western

Balkan countries forecast inflation to remain in double digits, with Albania being the sole exception. In North Macedonia, Montenegro, and Bosnia and Herzegovina, the price of food increased by up to 25%. Price pressures are becoming more widespread despite the fact that rises in commodity prices have historically been the main causes of inflation.

Although the banking sector has remained stable, a slowdown in GDP and greater inflation will unavoidably put it to the test. Non-performing loans are still declining (at 4.4 percent on average in March 2022), and the war in Ukraine has had no overall direct effect on the region's banking industry. However, the region's financial industry will undoubtedly be impacted by tighter global financial conditions, a slowdown in both domestic and international demand, and declining corporate and consumer confidence.

The region's export boom has started to slow down, and imports have climbed significantly as a result of rising food and energy prices, leading to wider current account deficits. All of the Western Balkans' deficits are growing, and in some, like Kosovo, Montenegro, and Serbia, they have even reached double digits. The current account deficit for the area is anticipated to increase, rising from 54.9 percent of GDP in 2021 to 8.7 percent of GDP in 2022. The region's exports of both commodities and services have remained robust, although the rate of expansion has started to decline.

Although the first half of 2022 saw growth that was very solid, it is obvious that the area is now about to experience another storm. The region's future is anticipated to be severely hampered by a convergence of indirect supply and demand shocks that will keep prices high and undermine investor and consumer confidence. A winter COVID-19 flare-up remains a concern, lending conditions are becoming more restrictive, and supply chains around the world are still under pressure. The Western Balkans' growth rate for 2023 has been further reduced (by 0.3 percentage points) to 2.8 percent.

Governments should prioritize policy support for the most vulnerable in the short term, making sure that actions are focused and time-limited to reduce fiscal risks. Policymakers should prioritize reforms that will increase medium-term growth at low fiscal costs. This would entail taking steps to boost market competition, lower business entry barriers, increase foreign investor retention and reinvestment, lower barriers to women entering the workforce, improve education quality, and raise governance standards, including digitalization (WB, 2022).

WHY ENVIRONMENTAL TAX REFORM?

The European Union is currently facing challenges related to climate change, biodiversity loss, ecosystem degradation, human health effects of chemical pollution, growing material resource scarcity, worries about the security of food, energy, and water, as well as national budget deficits and an aging population. The cost-effectiveness of integrated packages of policy measures is being pointed out at the same time as there is a greater awareness of how many environmental, economic, and social problems are interconnected (EEA, 2010). As suggested in "Europe 2020" a policy that transfers a portion of the tax base to activities that harm the environment can be a crucial component of a comprehensive policy package intended to address these various issues.

A tax change of this kind is frequently referred to as an Environmental Tax Reform (ETR). The results of ETR implementation in numerous EU member states in the 1990s and early 2000s are largely favorable. By reducing environmentally harmful subsidies, Environmental Fiscal Reform (EFR) expands on the ETR concept and frees up limited financial resources for more effective use elsewhere.

Table 1. Western Balkans key performance indicators, 2021

Wb6 Key Economic Indicators (2021)	Real Gdp Frowth (%)	Consumer Price Inflation	Public Expend. &Gdp	Public Revenues (%Gdp)	Fiscal Balance (%Gdp)	Public Debt (%Gdp)	Goods Exports (%Gdp)	Trade Balance (%Gdp)	Current Account Balance (%Gdp)	External Debt (%Gdp)	Unemployment Rate (%)
Albania	8.5	2.6	31.6	27	-4.5	74	8.2	-13.3	-7.7	62.7	11.5
B&H	7.5	2	43.3	43	-0.3	37.6	34.1	-12.3	-2.6	65.4	17.4
Kosovo	10.5	3.4	29.3	27.9	-1.3	21.9	9.3	-30.7	-8.3	37.3	18
RNM	4	3.2	37.7	32.3	-5.4	60.8	51.1	-16	-3.5	81.4	15.7
MNE	13	2.4	45.9	44	-1.9	86.8	10.6	-19.4	-9.2	214.9	16.6
Serbia	7.4	4	47.4	43.3	-4.1	57.1	38.9	-8.5	-4.4	68.5	11
WB6	8.48	3.3	39.2	36.3	-2.9	56.4	31.8	-12.7	-4.9	88.4	15.03

Source: Western Balkans Regular Economic Report. No.22. Fall 2022

Table 2. Country profiles

Albania	• Following robust growth in early 2022, GDP is likely to decelerate in the remaining part of the year, as rising inflation affects real disposable income, and a slowdown in the global economy translates into tighter financing conditions and lower foreign demand. • High food and energy inflation have prompted additional government support to households and SMEs, which are already benefitting from regulated electricity prices. This adds to the fiscal pressures. • Medium-term prospects hinge on the global recovery and structural reforms, and on the launch of fiscal consolidation. • Poverty is expected to continue declining but the continuation of inflationary pressures will decelerate the recent gains.
Bosnia & Herzegovina	• Robust economic activity continued in the first half of 2022. Nevertheless, employment improved only marginally; unemployment remains elevated and is especially high among youths. • Headline inflation accelerated, fueled by food and transport prices, raising inflation to 12 percent by July 2022. Fiscal revenues benefited from strong growth and inflation, with the deficit expected to reach 1 percent of GDP in 2022, on election year. • Real output growth is expected to decelerate in the second half of 2022 as private consumption slows due to the erosion of real disposable income caused by high inflation, and a deterioration in net exports. • Elections took place on October 2, 2022, potentially setting the scene for a return to much needed and delayed structural reforms to boost potential growth in the medium term.
Kosovo	• Kosovo's economic growth moderated in early 2022, with activity affected by broad-based price increases. Risks to the outlook remain high as the country continues to grapple with rising inflation pressures. • The price shock will inevitably continue affecting demand and economic activity in 2022, and growth is expected to moderate to 3.1 percent, mainly on account of higher real exports. • Strong tax revenue collection continues to favor the fiscal position, supported by higher inflation and tax compliance measures. The fiscal deficit is expected to reach 0.8 percent of GDP in 2022, amid significant capital underspending. • In a context of high uncertainty, maintaining buffers to respond to the changing macroeconomic environment, particularly in the context of an ongoing energy crisis, is vital. Over the medium term, it is imperative to advance structural reforms to enhance competitiveness and private sector development in support of the current export growth momentum.
Montenegro	• While still recovering from the pandemic, the economy is facing renewed headwinds. Growth remains very strong, estimated at 6.9 percent in 2022, led by private consumption and tourism recovery. • Inflation surged to new highs, but its adverse impact on the cost of living was largely mitigated by an increase in real disposable income. • The fiscal deficit is estimated to widen to 4.9 percent of GDP in 2022, due to the forgone revenues of the recent tax reform and increased social spending. • High public debt and the deteriorating global environment require near-term fiscal consolidation.
Republic of North Macedonia	• As the war in Ukraine and the energy crisis dim growth prospects, inflation is racing toward an all-time high, disproportionately eroding the real income of the poor. • With limited fiscal space, elevated public debt, and increased cost of financing, fiscal support needs to target the most vulnerable households and firms. At the same time, monetary policy tightening needs to strike a balance between containing inflation and avoiding stifling economic activity. • Disruptions related to the war in Ukraine, overstretched global supply chains, mounting inflationary and wage pressures, and the intensifying energy supply crisis continue to weigh on the outlook and the economic prospects of the country.
Serbia	• Growth continued to be strong in H1 2022, despite major domestic and external challenges. • Inflation accelerated more rapidly than projected, and in line with developments in the region, reaching 13.2 percent y-o-y in August, driven by food and energy prices. • The fiscal deficit turned lower than anticipated, thanks to a strong performance of revenues, while public debt is around 57% of GDP. • While growth projections over the medium term (2022-2024) remain unchanged, risks to the outlook are clearly tilted to the downside. • The most significant deterioration is expected on the external side, with the CAD widening to around 10% of GDP due to the major increase in imports.

Source: Western Balkans Regular Economic Report No.22. Fall 2022

When debating the justification for an ETR/EFR, it is important to keep in mind that the poorest segments of a community typically bear a heavier share of the weight of much current environmental pollution and deterioration and frequently have less access to green spaces.

As a result, labor tax reductions (possibly targeted at young or unskilled workers) as part of an ETR can help offset the greater negative effects of some environmental taxes on poorer households, such as domestic energy taxes (Blobel et al., 2009). Progressive social security reforms and progressive levies on energy and water supplies can also partially counteract the negative effects of a large ETR/EFR on equity.

Environmental fiscal reform (EFR) entails legislative changes that change the way money is raised from the use of labor and capital to the pollution of resources. Environmental fiscal reform may also involve the introduction of full-cost pricing (with user fees for water supply, waste disposal, and sewage), auctioned permits in emissions trading schemes, and the gradual elimination of environmentally harmful subsidies, in addition to taxes related to the environment (EEA,2010).

Environmental fiscal reform has frequently been planned to be fully or partially revenue neutral in those nations that have previously implemented such tax policies (for example, Estonia, Sweden, Denmark, Finland, Netherlands, Slovenia, Germany, and the UK), with the goal of relieving pressure on labor as a tax base.

Environmental fiscal reform can result in five benefits:

(1) increased employment;
(2) increased resource productivity and eco-innovation;
(3) improved environmental and human health;
(4) a more effective tax system;
(5) sharing the financial burdens of an aging population also according to consumption.

EU on Environment Tax Reform

The European Environment Agency (EEA) issued four papers on market-based mechanisms and the possibilities for environmental tax reform in the EU between 1996 and 2005 (EEA, 1996, 2000, 2005, and 2006). Some studies (Bassi et al., 2009, Blobel et al., 2009, Gehr et al., 2009) on the effects of Environmental Tax Reform (ETR) on eco-innovation, equity, and the political viability of considerable ETR were commissioned by the government. Other organizations like the OECD (2001, 2006, 2010) and Green Budget Europe, as well as a wide range of national and EU research initiatives (such as Clinch & Dunne, 2006; Wissema & Dellink; 2007; Convery et al.; 2009; Andersen & Ekins; 2010; Green Fiscal Commission; Ekins & Speck, 2011) have successfully focused on addressing questions about the goal, validity, and effectiveness of many ETR instruments. Although extensive research into the many advantages of ETR has encouraged some modest ETR in some EU and other countries, it is now understood that the main obstacles to ETR are largely political and institutional, necessitating agreement among many stakeholders and appreciation of the various national contexts in which ETR can work.

The fact that the base on which environmental taxes and fees are assessed will decrease once they go into effect is a crucial consideration when introducing them. Making reliable projections of revenue requires some comprehension of the connection between an increase in a tax and its effects on the environment. However, environmental taxes and fees can promote innovation and increase resource efficiency in terms of use of energy, water, and other resources (OECD, 2010).

Experience has also shown that as resource efficiencies increase, some of the income gains that result are spent on increased consumption, such as traveling longer in vehicles with higher fuel efficiency. As a result, advances in eco-efficiency may cause an increase in the overall consumption of energy and resources.

The "rebound effect" is what causes this (see for example Sorrell, 2007; Polimeni et al., 2008; Barker et al., 2009).

By gradually boosting the tax in accordance with the advances in eco-efficiency, it is possible to counteract both a declining environmental tax base and the rebound effect. Such progressive increases in environmental taxes are supported by growing awareness of the true costs (as with tobacco and increasingly other forms of pollution), ongoing resource depletion, and resource scarcity. Therefore, the EEA has made the decision to support the implementation of ETR in EU Member States by planning ETR awareness-raising activities with senior civil servants and members of civil society in collaboration with country organizations.

EU Green Deal

The EU Green Deal aims to achieve "no net emissions of greenhouse gases by 2050, economic growth decoupled from resource use, no person and no place left behind." The Green Agenda is a similar roadmap for the countries of the Western Balkan to adapt to these EU climate targets. Albania, Bosnia and Herzegovina, Kosovo, Montenegro, North Macedonia, and Serbia all committed themselves to the Green Agenda at the Sofia Summit in November 2020. The Green Agenda is structured along five pillars: (1) cleaning energy sources and protecting the climate; (2) moving to a circular economy; (3) depolluting air, water, and soil; (4) building sustainable agriculture and food systems; and (5) protecting biodiversity and ecosystems.

However, implementation is oftentimes lacking due to the (perceived) financial, economic, and social costs of the transformation processes, diverging interests of societal actors, information gaps among decision-makers, and insufficient regional cooperation. A green transition, especially in the field of energy, is inevitable in order to tackle the climate crisis, mitigate its effects, and ensure future economic development and competitiveness as well as social justice and cohesion (AI, 2022).

- Income from environmental taxes could assist in fulfilling the goals of the European Green Deal, which calls for a net reduction of greenhouse gas emissions of 55% by 2030, and the transition to a climate-neutral economy by 2050. The case for environmental taxes has never been greater due to Europe's aging population and its effects on labor tax receipts and social system expenditures.
- At the national, European, and international levels, there have been requests for additional environmental taxes, but implementation has been exceedingly slow. Environmental taxes make up 5.9% of all taxes in the EU, down from 6.6% over 20 years ago. However, this varies greatly from country to country.
- The European Green Deal recognizes that taxes play a critical role in the shift to a greener and more sustainable economy.
- The European Green Deal and the European Commission's 'Fit for 55' package both include crucial elements that help the transition to a climate-neutral economy. Increased income will also result from reviewing and adjusting the EU's present energy taxes and carbon pricing policies, particularly the EU Emission Trading System.

- However, as present energy taxes and carbon pricing schemes mainly rely on non-renewable energy supplies, decarbonizing the European economy would unavoidably diminish the revenue base. In order to future-proof tax systems with new revenue streams, it will be crucial to strike the correct balance between achieving transition objectives and ensuring revenue stability.
- In order to promote sustainability goals, the long-standing proposal of moving taxes from labor to the environment has mostly not been accomplished. The concept might be repackaged as sustainable fiscal reform, embracing taxes from additional revenue sources like financial transactions, land, and wealth, given the multiple fiscal challenges the EU would face in the future decades (Speck, 2022).

THE ENVIRONMENTAL TAX REFORM IN WESTERN BALKAN COUNTRIES

The Western Balkans are particularly susceptible to expensive natural catastrophes including floods, wildfires, earthquakes, and droughts; climate change is only predicted to increase the frequency and severity of weather-related risks in these countries. Fighting climate change has its own unique set of difficulties because shifting to less carbon-intensive economic activity can disrupt some areas of the economy and necessitate additional investments. The region must continue to be on the lookout for how climate change will affect its financial sectors. Affected industries like mining, energy, or agriculture may experience revenue losses, increased credit risk, or reduced funding sources from customers due to the physical and transition-related dangers of climate change (UNDP, 2015).

At the same time, financial resources have been exhausted throughout the Western Balkans as a result of economic shocks like the COVID-19 outbreak and the aftermath of Russia's invasion of Ukraine. Domestic banking sectors are viewed as crucial players to mobilize private funds for green projects because the governmental sector has limited resources for investments in climate adaptation and mitigation and because capital markets are small.

Although progress in the Western Balkans is still in its early stages, green finance—in which financial institutions' loans and investments contribute to climate reduction, climate adaptation and resilience, and mitigating environment risks—is crucial. Governments throughout the region lack coherent strategies and roadmaps to address climate and environmental risks, despite the substantial physical and transition concerns as well as their close proximity to and economic ties to the European market (Hwang et al, 2023).

Green taxonomies, a crucial component for the development of a green loan and bond market, have not yet been created, and huge corporations and other economic agents do not yet have to comply with disclosure rules. And banks regulators are just now starting to look more closely at how exposed financial institutions are to climate concerns. The misalignment of fiscal, economic, and environmental policies would be addressed as a first step. For instance, governmental subsidies are still provided to industries with huge carbon footprints, and there is little to no effective carbon taxes. The development of sustainable economic ventures may also be hindered by uncertainty regarding long-term government policies, and financial institutions' abilities to identify and source green assets and manage climate risks may be constrained by data gaps.

Furthermore, the real and perceived riskiness of green projects as a whole may grow due to the high upfront financing costs, transaction expenses, lack of track record of new technologies, and lengthy payback periods for some green initiatives. The appetite of investors may be adversely affected by these circumstances, which may have an effect on lending and investment decisions. WB6 is under the way of integration into the EU, and should follow EU environmental policy, while according to European Environment Agency (EEA):

- Revenues from environmental taxes could support the transition to a climate-neutral economy by 2050 and achieve the objectives of the European Green Deal reaching a net reduction of greenhouse gas emissions by 55% by 2030. Given Europe's aging population and the implications of this for labor tax revenues and welfare system costs, the arguments in favor of environmental taxes have never been stronger.
- Despite calls for more environmental taxes at the national, European, and global levels, implementation has been very slow. In the EU, environmental taxes account for 5.9% of total taxes, less than the 6.6% almost 20 years ago. This varies considerably by country, however.
- The European Green Deal acknowledges the crucial role of taxation in the transition to a greener and more sustainable economy, and this may drive the implementation of environmental taxes in coming years.
- Carbon pricing measures are essential components of the European Green Deal and also the European Commission's 'Fit for 55' package, supporting the transition to a climate-neutral economy.
- The amendment and revision of current EU energy taxation and carbon pricing schemes, including of the EU Emission Trading System, will also lead to higher revenues in the coming decade.
- However, decarbonizing the European economy will inevitably erode the tax base, as current energy taxation and carbon pricing schemes rely heavily on non-renewable energy products. It will be essential to strike the right balance between achieving transition objectives and maintaining revenue stability and to future-proof tax systems with new revenue sources.
- The long-standing idea of shifting taxes from labor to environment in support of sustainability objectives has largely not been realized. With the EU facing multiple fiscal pressures in coming decades, the idea could be reformulated as sustainable fiscal reform, embracing taxes from other revenue sources e.g. financial transactions, land, and wealth (EEA,2022).

BW6 should cope with these approaches as soon as possible.

WB6 Environmental Performance

Environmental Performance Index (EPI) offers a data-rich sustainability scorecard that translates cutting-edge scientific findings into policy insights. Using 40 performance indicators, the EPI ranks 180 countries on their national efforts to protect environmental health, enhance ecosystem vitality, and mitigate climate change. These indicators measure how close countries are to meeting internationally established sustainability targets for specific environmental issues. While the overall EPI scores provide a way to spotlight sustainability leaders and call out laggards, the accompanying disaggregated data offers a more refined tool for identifying policy weaknesses and anomalies, as well as proven programs that countries can adopt from their top performing peers. With a methodology refined over two decades that builds on the most recent data, the EPI enables decision-makers to recognize the drivers of top-tier performance. Analysis of the EPI data demonstrates that financial resources, good governance, human development, and regulatory quality matter for elevating a country's sustainability. Highlighting these connections, the EPI helps to promote sustainable development in support of a more environmentally secure and equitable future. High-scoring countries exhibit longstanding and continuing investments in policies that protect environmental health, preserve biodiversity and habitat, conserve natural resources, and decouple greenhouse gas emissions from economic growth.

Many bottom-tier countries face war and other sources of unrest as well as a lack of financial resources to invest in environmental infrastructure. Deteriorating air quality and rapidly rising greenhouse gas emissions pose especially urgent challenges.

With an EPI Mean 43.10333, based on EPI score, in Balkan countries, Slovenia (67.3), Croatia (60.2), Romania (56), NR Macedonia (54.3), Bulgaria (51.9), Albania (47.1), Montenegro (46.9), and Serbia (43.9), have a higher score than mean 43.1, while Bosnia & Herzegovina (39.4), and Turkey (26.3) have a score below mean, but considering the highest score of 77.9 (Denmark), Balkan countries in general are far away of European agenda related to environment performance, while WB6 countries are somehow in middle of the rank, so they have a hard way forward towards EU integration compare with Slovenia, Croatia, etc.

Figure 1. The 2022 EPI framework

Source: EPI Report 2022. EPI Framework includes three policy objectives and 11 issue categories. A total of 40 individual environmental metrics contributes to a country's overall EPI score

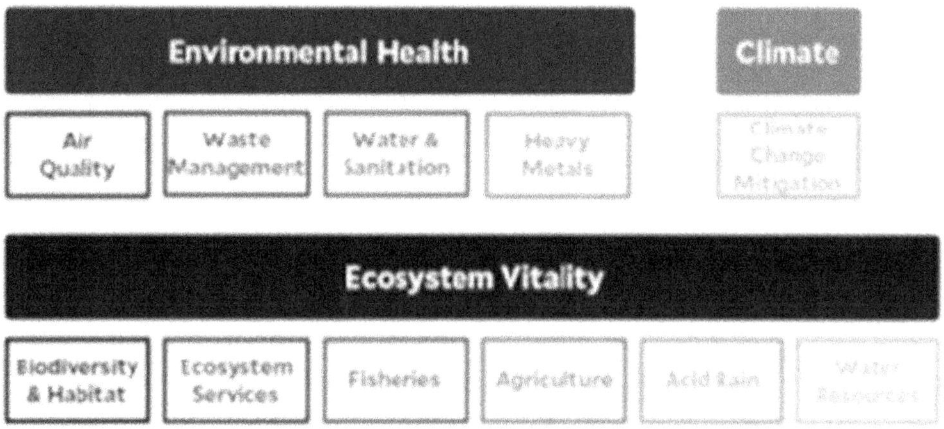

WB6 Green Agenda

The Economic and Investment Plan's (EIP) key goal continues to be the Green Agenda for the Western Balkans (GAWB). The five-pillar strategy depends on regulatory changes in the area and necessary investments to bring it into line with the goals of the European Green Deal. Out of the six EIP priorities, four are fundamental tenants of the GAWB, including (1) Green power, (2) Environment and climate, (3) Private sector growth, and (4) Sustainable transportation.

Table 3. Environmental performance index, EPI score 2022

No	Country	EPI Score	No	Country	EPI Score	No	Country	EPI Score	No	Country	EPI Score
1	Denmark	77.9	46	Poland	50.6	91	Cabo Verde	41.9	136	Senegal	33.9
2	UK	77.7	47	Panama	50.5	92	Argentina	41.1	137	Qatar	33
3	Finland	76.5	48	Portugal	50.4	93	Kazakhstan	40.9	138	Cote d'Ivoire	32.8
4	Malta	75.2	49	Belize	50	94	Paraguay	40.9	139	Rwanda	32.8
5	Sweden	72.7	50	Canada	50	95	El Salvador	40.8	140	Sierra Leone	32.7
6	Luxembourg	72.3	51	Gabon	49.7	96	Tunisia	40.7	141	Lesotho	32.3
7	Slovenia	67.3	52	Ukraine	49.6	97	Malawi	40.6	142	Lebanon	32.2
8	Austria	66.5	53	S. Lucia	49.4	98	Gn-Bissau	40.2	143	Ethiopia	31.8
9	Switzerland	65.9	54	Kiribati	49	99	Bolivia	40.1	144	Eritrea	31.7
10	Iceland	62.8	55	Belarus	48.5	100	Congo	40.1	145	Mozambique	31.7
11	Netherlands	62.6	56	Armenia	48.3	101	Peru	39.8	146	Guinea	31.6
12	France	62.5	57	Israel	48.2	102	B&H	39.4	147	Fiji	31.3
13	Germany	62.4	58	Grenada	47.9	103	Georgia	39.1	148	Kenya	30.8
14	Estonia	61.4	59	Trnd&Tbg	47.8	104	Azerbaijan	38.6	149	Laos	30.7
15	Latvia	61.1	60	Cuba	47.5	105	Guyana	38.5	150	Oman	30.7
16	Croatia	60.2	61	Djibouti	47.5	106	Zambia	38.4	151	Angola	30.5
17	Australia	60.1	62	Albania	47.1	107	Uzbekistan	38.2	152	Burundi	30.5
18	Slovakia	60	63	S. Korea	46.9	108	Thailand	38.1	153	Cameroon	30.2
19	Czech Rep	59.9	64	Montenegro	46.9	109	S. Arabia	37.9	154	Cambodia	30.1
20	Norway	59.3	65	Chile	46.7	110	Niger	37.7	155	Benin	29.6
21	Belgium	58.2	66	Ecuador	46.5	111	Nicaragua	37.7	156	Algeria	29.6
22	Cyprus	58	67	Venezuela	46.4	112	Russia	37.5	157	Mongolia	29.6
23	Italy	57.7	68	Costa Rica	46.3	113	Micronesia	37.4	158	Philippines	28.9
24	Ireland	57.4	69	Zimbabwe	46.2	114	Maldives	37.4	159	Mali	28.5
25	Japan	57.2	70	Suriname	45.9	115	Uruguay	37.4	160	China	28.4
26	N Zealand	56.7	71	Brunei Drsl	45.7	116	S. Africa	37.2	161	Morocco	28.4
27	Spain	56.6	72	Jamaica	45.6	117	Tajikistan	37.1	162	Nigeria	28.3
28	Bahamas	56.2	73	Mexico	45.5	118	Turkmenistan	37	163	Nepal	28.3
29	Greece	56.2	74	Taiwan	45.3	119	Congo	36.9	164	Indonesia	28.2
30	Romania	56	75	Cntrl Africa	44.9	120	Vanuatu	36.9	165	Mauritania	28.1
31	Lithuania	55.9	76	Eswatini	44.9	121	Honduras	36.5	166	Chad	28.1
32	Seychelles	55.6	77	Eqtr Guinea	44.8	122	Gambia	36.4	167	Guatemala	28
33	Hungary	55.1	78	Mauritius	44.8	123	Samoa	36.4	168	Madagascar	28
34	NR Mcdnia	54.3	79	Serbia	43.9	124	Mrsh Islands	36.2	169	Iraq	27.8
35	Botswana	54	80	Tonga	43.8	125	Uganda	35.8	170	Ghana	27.7
36	Barbados	53.2	81	Afghanistan	43.6	126	Kyrgyzstan	35.7	171	Sudan	27.6
37	S.V&Grnd	53.2	82	Brazil	43.6	127	Burkina Faso	35.5	172	Turkey	26.3

continues on following page

Table 3. Continued

No	Country	EPI Score	No	Country	EPI Score	No	Country	EPI Score	No	Country	EPI Score
38	S. T&Prncp	52.9	83	Jordan	43.6	128	Egypt	35.5	173	Haiti	26.1
39	UAE	52.4	84	Moldova	42.7	129	Timor-Leste	35.1	174	Liberia	24.9
40	Antig&Brbd	52.4	85	Bhutan	42.5	130	Malaysia	35	175	PN Guinea	24.8
41	Bulgaria	51.9	86	Comoros	42.5	131	Slmn Islands	35	176	Pakistan	24.6
42	Dominica	51.2	87	Colombia	42.4	132	Sri Lanka	34.7	177	Bangladesh	23.1
43	USA	51.1	88	Kuwait	42.4	133	Iran	34.5	178	Viet Nam	20.1
44	Namibia	50.9	89	DominicanR	42.2	134	Tanzania	34.2	179	Myanmar	19.4
45	Singapore	50.9	90	Bahrain	42	135	Togo	34	180	India	18.9

Source: EPI Report 2022

Table 4. Environmentally related tax revenue as a share of total tax revenue and per capita

No	Country	Compared to GDP, %, 2020	As a share of total tax revenue, %, 2020	Per capita, USD, 2020	No	Country	Compared to GDP, %, 2020	As a share of total tax revenue, %, 2020	Per capita, USD, 2020
1	Seychelles	4.66	13.6	1 208.4	65	Japan	1.25	6.2	505.74
2	Serbia	4.21	11.6	521.01	66	Senegal	1.25	6.78	42.07
3	Croatia	3.91	10.53	991.04	67	India	1.22	17.47	79.42
4	Greece	3.77	9.68	964.74	68	Ireland	1.2	5.96	1 057.3
5	Solomon Isl	3.77	20.12	91.02	69	Pakistan	1.2	8.65	55.76
6	Slovenia	3.61	9.8	1 254	70	Burkina Faso	1.12	6.52	21.61
7	Netherlands	3.55	8.96	1 832.72	71	Canada	1.11	3.56	497.11
8	Latvia	3.31	10.4	931.26	72	New Zealand	1.07	3.18	416.44
9	Denmark	3.17	6.72	1 639.95	73	Bolivia	0.99	4.43	69.43
10	Italy	3.08	7.2	1 103.29	74	Philippines	0.96	5.39	75.44
11	Bulgaria	3.03	9.88	623.29	75	Viet Nam	0.96	4.25	74.89
12	Estonia	2.99	8.68	992.47	76	Egypt	0.89	6.18	113.59
13	Montenegro	2.96	0	401.09	77	Paraguay	0.89	6.65	107.55
14	South Africa	2.92	10.11	355.53	78	Kazakhstan	0.86	6.1	207.66
15	Guyana	2.77	12.99	514.89	79	China	0.84	4.2	137.18
16	Mauritius	2.75	12.17	515.73	80	Lesotho	0.84	3.94	24.58
17	Finland	2.73	6.52	1 217.91	81	OECDAO	0.8	3.43	333.44
18	Korea	2.66	11.38	986.17	82	Guatemala	0.78	6.24	61.84
19	Türkiye	2.63	11.01	748.13	83	Saint Lucia	0.76	3.64	87.85
20	Israel	2.55	8.27	966.79	84	Bhutan	0.73	7.9	72.06
21	Cyprus	2.48	7.2	888.86	85	Brazil	0.72	2.28	96.46
22	Hungary	2.46	6.92	743.86	86	Barbados	0.72	2.14	92.14

continues on following page

Table 4. Continued

No	Country	Compared to GDP, %, 2020	As a share of total tax revenue, %, 2020	Per capita, USD, 2020	No	Country	Compared to GDP, %, 2020	As a share of total tax revenue, %, 2020	Per capita, USD, 2020
23	Poland	2.41	6.71	745.89	87	Togo	0.71	4.76	11.02
24	France	2.38	5.18	949.8	88	Jamaica	0.7	2.52	56.77
25	Portugal	2.38	6.77	725.55	89	Malawi	0.7	5.84	7.53
26	Costa Rica	2.29	9.79	440.35	90	OECDAM	0.69	2.74	292.48
27	Czech Rep	2.28	6.64	824.13	91	Namibia	0.67	3.31	65.09
28	Malta	2.27	7.67	863.42	92	Maldives	0.67	3.52	82.97
29	OECDE	2.27	6.17	888.7	93	United States	0.66	2.58	384.16
30	Austria	2.1	5	1 028.25	94	Malaysia	0.65	5.67	171.05
31	Kenya	2.1	13.39	84.46	95	Cameroon	0.6	4.31	21.27
32	OAVG_A	2.09	6.18	851.35	96	Morocco	0.58	2.06	44.07
33	Sweden	2.07	4.87	1 018.39	97	Australia	0.56	2.12	265.54
34	Honduras	2.04	10.64	95.79	98	Colombia	0.56	2.99	69.98
35	United Kingdom	2.03	6.29	824.52	99	Peru	0.5	3.2	53.44
36	Belgium	2.01	4.69	916.7	100	El Salvador	0.46	2.12	34.5
37	Norway	1.98	5.13	1 208.03	101	Albania	0.41	1.75	33.42
38	Lithuania	1.98	6.41	707.64	102	Fiji	0.39	2.4	44.38
39	Eswatini	1.97	10.76	180.89	103	Cabo Verde	0.38	1.78	25.96
40	Ukraine	1.94	5.98	215.02	104	Equatorial Guinea	0.38	3.89	59.55
41	Slovak Republic	1.9	5.49	604.38	105	Panama	0.37	2.79	84.53
42	Uganda	1.84	15.04	41.27	106	Botswana	0.33	2.44	53.46
43	Sierra Leone	1.83	15.57	32.15	107	Chad	0.28	3.57	4.33
44	Iceland	1.82	5.04	899.63	108	Belize	0.27	1	16.35
45	Spain	1.75	4.77	590.93	109	Ecuador	0.27	1.43	26.31
46	Argentina	1.75	5.84	289.03	110	Niger	0.25	2.45	3.15
47	Uruguay	1.75	6.59	341.84	111	Singapore	0.25	1.88	314.47
48	Germany	1.71	4.45	816.72	112	Antg&Brbda	0.22	1.09	38.03
49	Ghana	1.66	12.55	99.8	113	Liechtenstein	0.19	0.93	–
50	Romania	1.59	5.89	416.66	114	DRC	0.18	2.37	1.7
51	Bahamas	1.58	7.46	460.49	115	Samoa	0.16	0.82	10.09
52	Mongolia	1.57	7.42	178.5	116	Trnd&Tbg	0.16	0.73	34.87
53	Madagascar	1.55	13.59	25.36	117	Nauru	0.15	0.32	19.84
54	Mali	1.48	8.45	32.59	118	Kyrgyzstan	0.13	0.75	5.47
55	Mexico	1.43	8.03	244.15	119	Côte d'Ivoire	0.1	0.79	5.71
56	Luxembourg	1.41	3.68	1 511.6	120	Nigeria	0.02	0.35	1.05

continues on following page

Table 4. Continued

No	Country	Compared to GDP, %, 2020	As a share of total tax revenue, %, 2020	Per capita, USD, 2020	No	Country	Compared to GDP, %, 2020	As a share of total tax revenue, %, 2020	Per capita, USD, 2020
57	Domncn Rp	1.38	10.98	225.45	121	Congo	0.02	0.28	0.78
58	Switzerland	1.37	4.94	914.21	122	P. N. Guinea	0.01	0.05	0.23
59	Rwanda	1.37	7.9	29.07	123	Bangladesh	0.01	0.11	0.57
60	Tunisia	1.37	4.27	152.57	124	Mozambique	0	0	0
61	Chile	1.35	7	287.89	125	Venezuela	0	0	0
62	OECD	1.35	4.56	551.89	126	Thailand	0	0	0
63	Nicaragua	1.3	5.11	64.41	127	Tokelau	0	0	0
64	Mauritania	1.29	9.38	57.81	128	Cambodia	0	0	0

Source: OECD 2020

Table 5. Descriptive statistics for (1) EPI Score, (2) environmentally related tax revenue compared to GDP, (3) as a share of total tax revenue and (4) per capita

EPI Score		Compared to GDP, %, 2020		As a share of total tax revenue, %, 2020		Per capita, USD, 2020	
Mean	43.10333	Mean	1.412188	Mean	5.725078	Mean	371.118
Standard Error	0.916613	Standard Error	0.094493	Standard Error	0.355534	Standard Error	37.62252
Median	41.95	Median	1.27	Median	5.285	Median	144.875
Mode	43.6	Mode	0	Mode	0	Mode	0
Standard Deviation	12.29765	Standard Deviation	1.069068	Standard Deviation	4.022406	Standard Deviation	425.6503
Sample Variance	151.2323	Sample Variance	1.142907	Sample Variance	16.17975	Sample Variance	181178.2
Kurtosis	0.029505	Kurtosis	-0.07066	Kurtosis	0.800663	Kurtosis	0.512038
Skewness	0.565688	Skewness	0.712734	Skewness	0.837132	Skewness	1.147651
Range	59	Range	4.66	Range	20.12	Range	1832.72
Minimum	18.9	Minimum	0	Minimum	0	Minimum	0
Maximum	77.9	Maximum	4.66	Maximum	20.12	Maximum	1832.72
Sum	7758.6	Sum	180.76	Sum	732.81	Sum	47503.11
Count	180	Count	128	Count	128	Count	128
Largest(1)	77.9	Largest(1)	4.66	Largest(1)	20.12	Largest(1)	1832.72
Smallest(1)	18.9	Smallest(1)	0	Smallest(1)	0	Smallest(1)	0
Confidence Level(95.0%)	1.808757	Confidence Level(95.0%)	0.186985	Confidence Level(95.0%)	0.703537	Confidence Level(95.0%)	74.44819

Source: Authors calculations based on data from OECD

Table 6. Balkan region EPI score and 10 years change

No	Country	Rank	Epi score	10-year change
1	Albania	62	47.1	9.9
2	B&H	102	39.4	10
3	Bulgaria	41	51.9	4.6
4	Croatia	16	60.2	17.2
5	Montenegro	63	46.9	8.5
6	North Macedonia	34	54.3	3.6
7	Romania	30	56	5.3
8	Serbia	79	43.9	8.5
9	Slovenia	7	67.3	8.6
10	Turkey	172	26.3	-0.5

Source: EPI 2022

Figure 2. Balkan region EPI score and ten years' change
Source: EPI 2022

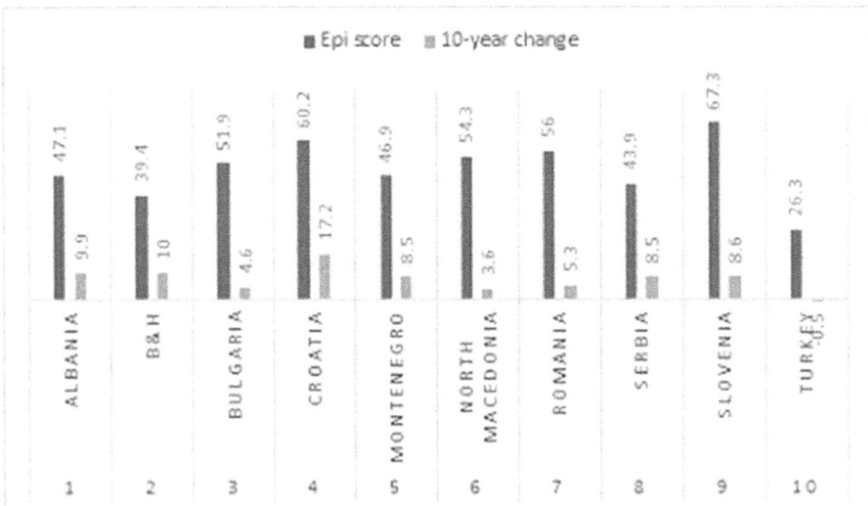

Implementation of the Environmental Tax Reform in WB6

The Western Balkan countries are implementing certain initiatives at a low level, but they are working to develop the appropriate laws and strategies to mitigate climate change.

Switching to power plants that use cleaner energy sources is one of Albania's measures for reducing climate change in the energy industry. Agriculture-related initiatives are concentrated on enhancing crop rotation, grazing management, and manure management. In order to lower emissions associated to garbage, new landfills with methane recovery and solid waste incinerators are also being built. The creation of sustainable forestry, the development of forest management techniques, and the restoration of damaged forest land are all strategies related to land-use change and forestry.

Table 7. Issues of WB6 green agenda

Decarbonization and climate resilience	The Western Balkans must commit to decarbonization goals by the end of 2022 in order to comply with the Energy Community Treaty. After that, they'll create integrated energy and climate plans. The EU has been funding the platform for coal regions in transition in the Western Balkans to aid in the region's transition away from coal. According to the potential of the region, the EU is helping the Western Balkans boost their production of renewable energy, improve their energy efficiency, make the transition to a low-carbon economy, and build efficient electricity transmission lines. The Sustainable and Smart Mobility Strategy is being implemented in the Western Balkans. This includes support from the EU for the renovation and development of new resilient and intelligent transportation infrastructure, including rail and public transportation, as well as environmentally friendly multimodal transportation options. The EU has helped the area manage its flood risk in an effort to promote climate resilience.
Moving toward the circular economy	The Western Balkans are obligated by the Green Agenda to transition to a circular economy by enhancing waste management, raising resource productivity, establishing new, more effective methods of production and consumption, and incorporating circularity into business models. A collaborative statement on the prevention of plastic pollution, especially marine litter, which is a serious transboundary concern, is being developed throughout the region. Through the EU4Circular economy project, a €30 million initiative, the EU supports the development of a legal framework for the circular economy, of circular business models, and of waste management infrastructure in Albania.
Depolluting air, water and soil	The Western Balkans have some of the worst air pollution in all of Europe. The EU is helping the area comply with EU standards for water, waste water management, and air quality. This entails upgrading the systems for monitoring the air and water as well as making further investments in waste-water management. Through EU4Environment, the EU assists Kosovo* in tackling environmental risks brought on by waste, water, and air pollution.
Building agriculture and food sustainable systems	The EU is funding IPARD (IPA Rural Development) initiatives in Albania, North Macedonia, Montenegro, and Serbia with up to €500 million. This will aid farms and agribusinesses in implementing more environmentally friendly and digitally advanced food production strategies. The EU4Prespa project in North Macedonia supports the implementation of the Green Agenda in the transboundary Prespa lake area, concentrating on, among other things, biodiversity and the greening of local businesses and agriculture.
Protection and restoration of biodiversity	Nature not only provides food, fresh water, clean air, and protection from storms and floods, but also helps to regulate the climate, fight land degradation, and improve food security by reducing the impact of natural disasters, pests, and illnesses. The area is aiming to define a post-2020 biodiversity framework and create a long-term strategy for reversing the loss of biodiversity, safeguarding ecosystems, and restoring extraordinary biological richness (EU,2022).

Source: European Union

A national mitigation policy connected to national economic and development plans, a list of the actions and measures required for implementation, and provisions for direct government involvement at the entity level should all be part of the creation of Bosnia and Herzegovina's climate change mitigation strategy.

Kosovo hopes to become energy self-sufficient, with a focus on enhancing waste disposal and the recycling sector while safeguarding a number of geographical areas.

The Republic of North Macedonia's plan to cut emissions calls for using natural gas to generate electricity, utilizing hydropower to its fullest extent, and increasing energy efficiency. Enhancing car energy efficiency as well as public urban and interurban transportation are examples of transportation measures.

The waste strategy offers strategies for waste selection and recycling, composting, and reducing garbage that is dumped on the ground. Manure management and afforestation are the main objectives of forestry and agriculture, respectively.

Each sector's emission reduction plans have been identified by Montenegro. Increasing the efficiency of industrial boilers, switching to liquefied petroleum gas in industrial boiler rooms, and creating high-temperature heat are all strategies for the energy sector. Plans for the transportation industry call for the creation of a more effective transportation system as well as the substitution of fossil fuels with alternative fuels. Agriculture strategies include promoting organic farming and enhancing manure management.

Montenegro plans to raise the proportion of highly productive forests, repair damaged forests, and maintain and enhance forest habitats in the land use and forestry sector.

The Republic of Serbia's goals for energy sector mitigation include increasing renewable energy sources and boosting energy efficiency. The major methods for lowering emissions in the realm of industrial operations include automation, monitoring, and the use of waste energy. The Republic of Serbia aims to end the production of leaded gasoline, restore roads, raise the level and effectiveness of river transportation, and reestablish an effective international railway system. An important method for lowering agricultural emissions is the use of biogas in the generation of heat and power for local use, and afforestation is a desirable technique in the sector of land-use change and forestry.

Republic of Serbia intends to build regional landfills and boost recycling rates in the waste management industry. The level of specific activities being implemented is minimal. The main cause of this is the lack of commitment on the part of decision-makers to undertake substantial adjustments in the area of transitioning from the effects of climate change (lower level of economic development, lack of investment, preservation of social peace) (Knez et al, 2022).

CO2 Taxation

CO2 taxation charges a fee for every ton of CO2 released in an effort to persuade customers to avoid buying or acting in ways that produce a lot of greenhouse gas emissions. To consume less fossil fuels, in other terms. It is a Pigouvian tax that encourages calculating the price of products and service negative externalities. The domestic consumption tax on energy products (TICPE), natural gas (TICGN), and coal (TICC) are all combined into a single, more comprehensive computation that includes a "carbon component" that represents the carbon tax. It serves as a climate reference for investment decisions by public and private economic participants and is indexed to the carbon price, which is expressed in euros per ton of CO2. It is more of a fee than a tax. Households (on gasoline or gas, for instance), businesses, and government agencies all pay the tax. There are numerous exemptions, though: the most polluting large industries are subject to European regulations; air and sea transportation are exempt due to international agreements and European directives; national river transportation, taxis, road freight transportation, public transportation, as well as agricultural uses are also exempt from this tax.

A plan for CO2 taxation in contracting parties has been developed by the Energy Community Secretariat. It calls for the immediate implementation of internal carbon pricing, the development of national systems by 2025, the unification of those systems into a single regional system by 2030, and the achievement of the EU's carbon price by 2040. Free carbon allowances would be provided as part of the gradual introduction of carbon pricing.

The EU plans to implement the Carbon Border Adjustment Mechanism (CBAM) starting in 2026 on imports of cement, iron, steel, fertilizers, aluminum, and electricity from nations without a carbon tax. The Western Balkan countries are under pressure to enact carbon pricing, but WB6 authorities have noted that the Western Balkans and the EU must show understanding for the CBAM.

In order to speed up decarbonization and avoid paying the EU's (European Union) carbon tax on imports, which the bloc plans to enact starting in 2026, Bosnia and Herzegovina, Montenegro, and North Macedonia have taken the first concrete steps to introduce carbon dioxide (CO2) taxation. Currently, Serbia, Kosovo, and Albania are falling behind.

Internal carbon pricing was implemented by the energy company Elektroprivreda Bosne i Hercegovine (EPBiH) in January 2021. The money is saved in a special fund that, in the previous year, had 10 million euros and is only to be used for investments in renewable energy sources.

The BiH government has created a legal framework for the implementation of a CO2 tax. Given that around 72% of BiH's exports go to the EU, the carbon border tax imposed by the EU will have an impact on the entire country's economy. In May 2021, BiH and the EU agreed a memorandum of understanding to implement the EU ETS starting in 2026 as part of the NECP. The mechanism will be used to decarbonize the energy industry and EPBiH by 2040. The internal carbon price was implemented by EPBiH in January 2021, and it immediately resulted in the creation of a special fund that, in 2021, had a value of around 10 million euros.

About 95% of the company's emissions were taxed internally in 2021, and the BiH Council of Ministers had also created a legal framework for carbon pricing. The money is only designated for investments in renewable sources. The implementation of CO2 taxes would also strengthen EPBiH's position in the electricity export market (Spasić,2022).

By implementing a CO2 tax in 2020 in accordance with the EU ETS regulations, Montenegro has made the most progress.

Elektrani na Severna Makedonija (ESM), the power company in North Macedonia, also proposed internal CO2 pricing in 2021, with the goal of implementing CO2 taxes and bringing carbon pricing in accordance with the EU ETS by 2030.

In line with the Paris Agreement, the country's nationally determined contributions (NDCs) call for emissions to be reduced by 51% by 2030, or 80% from 1990 levels. One of the largest hydroelectric facilities in the area, Ebren, is set to be built in North Macedonia, which is also diversifying its supply sources and making it easier for private enterprises to invest in renewable energy. A new climate change law is now being prepared with the goal of implementing carbon pricing initially as a temporary fix and bringing it into compliance with EU ETS regulations by 2030 (Spasić,2022).

In Figure 3, CO2 emission in Western Balkan countries is shown. As a matter of fact, Western Balkan countries don't belong to the largest and large contributors of CO2 worldwide and in Europe, and as the figure shows, their levels of CO2 emission are in a stable direction, while, Albania remains at the lowest level, followed by Montenegro, and Croatia. Serbia and Bosnia & Herzegovina have the highest level are around 7metric tons per capita, which still is a low emission compared with developed and emerging economies worldwide.

In Figure 4, the total greenhouse emission in Western Balkan countries is shown. As it can be noticed from the graph, their level of total greenhouse emission is in a stable direction, while, Albania remains in the lowest level, followed by Bosnia & Herzegovina and Montenegro, while Serbia has the highest level around 7 kt of CO@ equivalent, but still below levels of developed and emerging economies worldwide.

Figure 5 shows environmental taxes for 2019, where Croatia has the highest level of them as a percentage of GDP, around 4.2%, followed by Serbia and Albania, while the OECD average is 2.4%.

Figure 3. CO2 emissions in Western Balkans
Source: Knez et al,2022

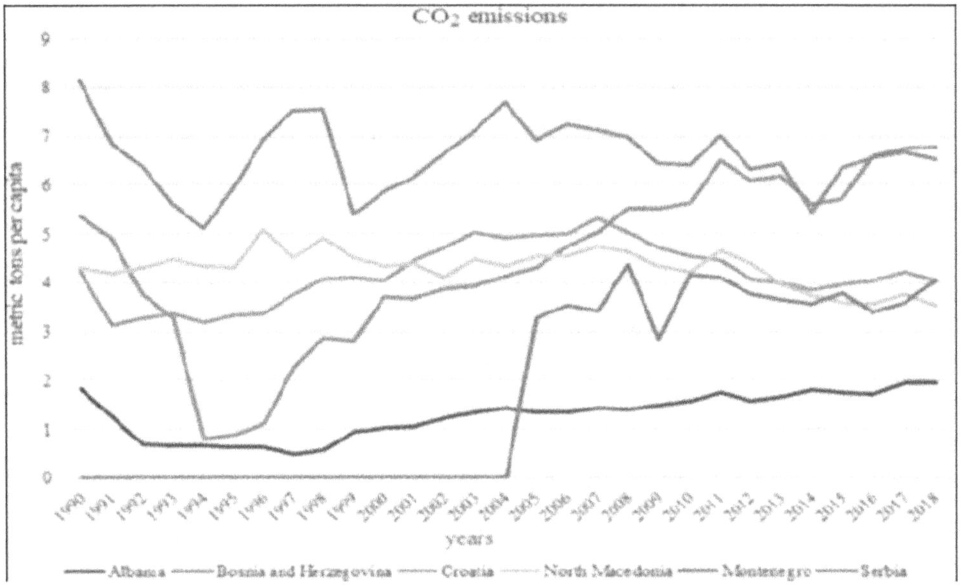

Figure 4. Total greenhouse emission in Western Balkans
Source: Knez et al,2022

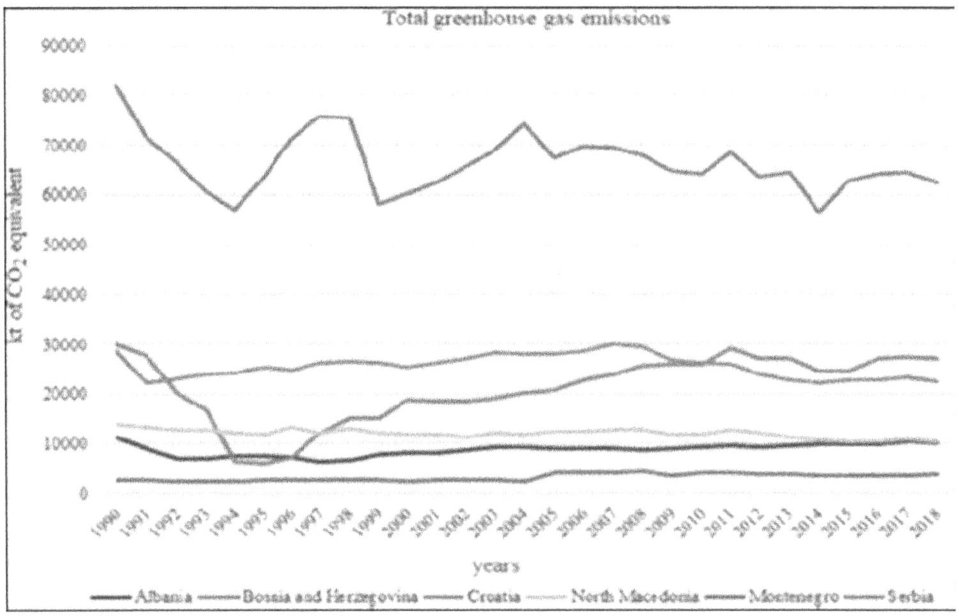

In Table 8, Western Balkans environmental taxes as a percentage of GDP, as a share of total tax revenue and per capita have been shown, where Serbia, Greece Slovenia, Bulgaria, and Montenegro have the highest values (as percentage of GDP), while Slovenia, Croatia, Greece, and Turkiye, have the highest tax per capita, between 1245 – 748 Euro. The rest of Western Balkans has a lower tax per capita.

Figure 5. Environmental taxes, 2019 (% of GDP)
Source: IMF,2022

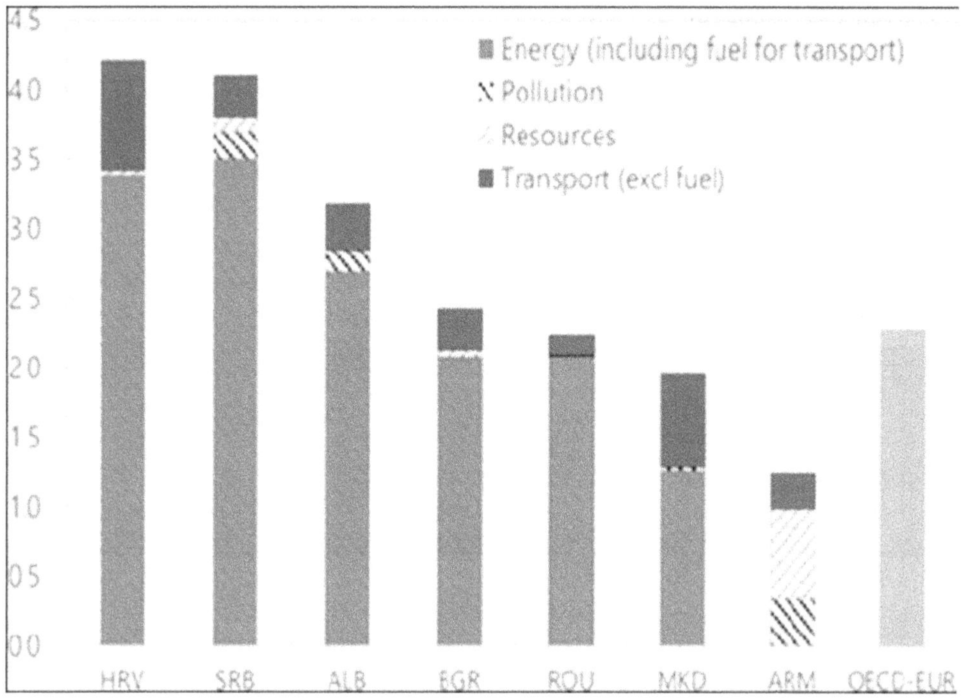

Table 8. WB6 environmental taxes (% of GDP)

Country	Compared to GDP, %, 2020	As a share of total tax revenue, %, 2020	Per capita, USD, 2020
Serbia	4.21	11.6	521.01
Croatia	3.91	10.53	991.04
Greece	3.77	9.68	964.74
Slovenia	3.61	9.8	1 254
Bulgaria	3.03	9.88	623.29
Montenegro	2.96	0	401.09
Türkiye	2.63	11.01	748.13
Romania	1.59	5.89	416.66
OECD	1.35	4.56	551.89
Albania	0.41	1.75	33.42

Source: OECD,2020.

The mean of environmental taxes (as % of GDP) is around 1.41, with Serbia (4.41), Croatia (3.91), Greece (3.77, Slovenia (3.61), Bulgaria 3.03), Montenegro (2.96), Turkiye (2.63), and Romania (1.61) above it, while Albania (0.41) is below the mean of 1.41.

The mean of environmental taxes as a share of total tax revenue is around 5.72, and above it is Serbia (11.6), Croatia (10.53), Greece (9.86), Slovenia (9.8), Bulgaria (9.88), Turkiye (11.01), while Romania (5.89), and Albania (1.75) are below the mean.

The mean of environmental tax per capita is 371.118 $, with Serbia (521.01), Croatia (991.04), Greece (964.74), Slovenia (1.254,00), Bulgaria (623.29), Montenegro (401.09), Turkiye (748,13), Romania (416.66), while Albania (33.42) is below the mean.

In Figure 6, environmental-related tax revenues as a percentage of GDP have been shown, where Albania has the lowest level of around 0.35, while Greece, Croatia, and Serbia have the highest level between 3.8 – 4.3%.

If we look at the data for environmental-related tax revenues as a percentage of total related tax revenue (Figure 7) we notice that Albania has the lowest level of around 1.85%, while all the other countries are above the OECD level of 4.3%. Data for Montenegro are not available.

Figure 6. Environmentally related tax revenue compared to GDP
Source: OECD, 2020

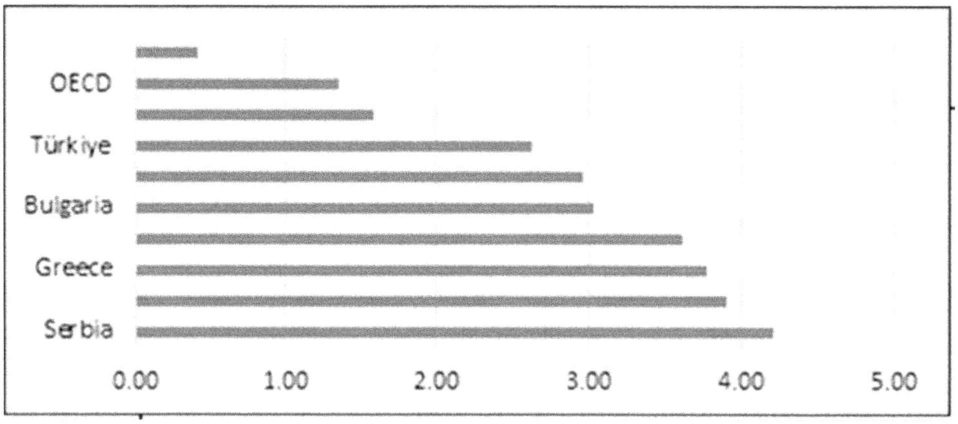

Figure 7. Environmental-related tax revenue as % of total tax revenue
Source: OECD, 2020

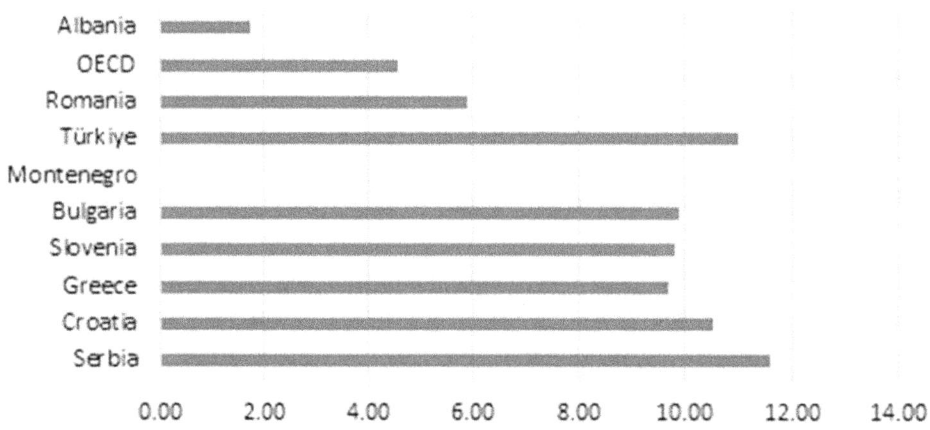

In Figure 8, environmental-related tax revenues per capita have been shown, where Albania has the lowest level, followed by Romania, Turkiye, and Serbia, that are below the OECD level, while other countries are above the OECD level.

Figure 8. Per capita environmental tax revenue
Source: OECD, 2020

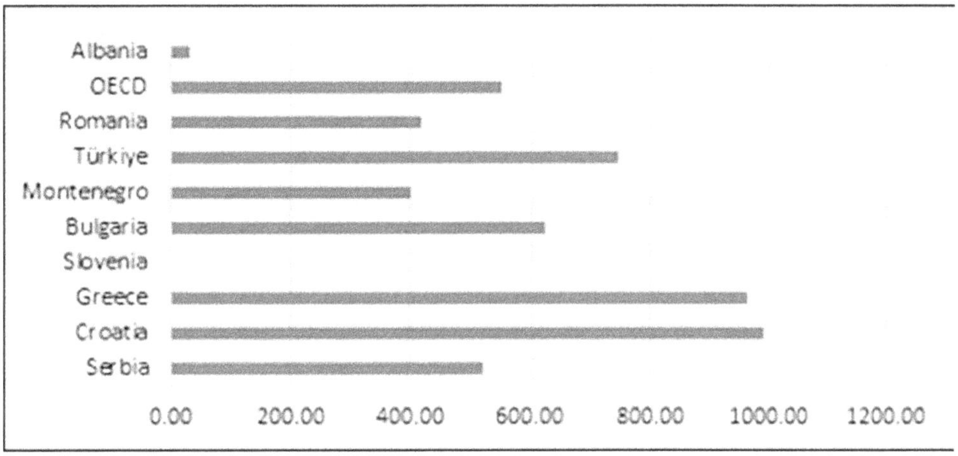

PERFORMANCE OF THE ENVIRONMENTAL TAX IN WB6

Based on the European Green Deal (2019) and the Recovery Plan for Europe (2021), the European Commission (EC) plans to allocate 30% of the budget to climate-related projects, initiatives, and programs, demonstrating its commitment to becoming the first climate-neutral region by 2050. Activities are also being planned for non-EU nations that need complex adjustments to legislation, strategic planning, implementation, and monitoring (Knez et al., 2022).

The Western Balkan countries are characterized by "low" consumer environmental awareness levels, which are linked to a weak market for sustainable consumer goods. Financial resources are commonly cited as a hindrance to the purchase of sustainable consumer goods throughout the entire region. Because consumers have limited purchasing power, the primary factor influencing consumption decisions is product price. In different ways across the region, sustainable consumer behavior and status are related.

The human resources needed for professional development in the green industries present difficulties for all countries. These issues are characterized by two things: either a large number of people leaving the area who are skilled workers, or a mismatch between labor supply and demand that already exists due to gaps between educational institutions and the industry.

Different policies and regulations are in place in the region to encourage the growth of green businesses, but their implementation and enforcement are not guaranteed for a number of reasons, including lack of supporting sub-legal frameworks, improper oversight, dearth of inspectors, shortage of experienced and qualified workers, corruption, and insufficient financial resources. Although there are just few, there are entities in Western Balkans that are in charge of integrating and controlling policy compliance. Businesses are operating without environmental licenses in various countries in this region despite existing laws. Albania and North Macedonia have begun talks on EU Chapter 27 on Environment and Climate Change as part of the WB6 countries' current integration process toward EU membership.

In this region, the definition, application, and integration of the circular economy idea and its guiding principles in strategies and action plans are still in their infancy. The majority of the earliest advancements have been made in the field of waste management, typically on the basis of EU directives and policies, where various (national) policies concentrate on the enhancement of waste management through CE principles. However, the current state of waste management in these countries still faces many difficulties. Different strategies are used in the region to apply CE concepts to other sectors. Outside of waste management, concrete references to CE in policies and action plans have mostly been lacking from the Western Balkan countries.

However, a number of plans and initiatives in action now could spur the integration of CE across a wider spectrum. The CE Roadmap in Montenegro and UNDP projects in Bosnia and Herzegovina are two examples. WB countries have expanded their CE initiatives to the energy sector, focusing on boosting energy efficiency techniques. Environmental-related tax policies appear to be inadequately integrated in the Western Balkan countries. Comparatively to the 39 OECD economies, Israel and Turkey generate above-average revenue through environmental tax measures. In this area, there are almost no subsidies or financial benefits for the development of green or CE businesses.

Ecolabelling is a concept that has been implemented through a number of techniques in the region, typically linked to the food industry and organic goods. At this point, attempts to expand ecolabelling into other industries have been negligible or failed in the Western Balkan countries.

The region's countries have expanded producer responsibility or polluter pays laws, although their implementation and enforcement may be better or still leave something to be desired. Although there are policies to raise the proportion of green public procurement across the area, these incentives' regulatory and practical implementation varies and appears to be lacking.

There are no data on the percentage of green public procurement in the entire region, which suggests inadequate monitoring. The current systems for public procurement are usually corruptible.

Boosting Green Finance for WB6

The countries of the Western Balkan are highly vulnerable to costly natural disasters such as floods, wildfires, earthquakes and droughts, and climate change is only expected to exacerbate the frequency and intensity of weather-related hazards. Grappling with climate change also brings its own substantial challenges, as transitioning towards less carbon-intensive economic activities can lead to transformation in some economic sectors that may require new investments.

The region needs to remain vigilant about the impact of climate change on its financial sectors. Both the physical and transition-related risks of climate change can cause revenue losses, heightened credit risks and lower funding sources from customers in affected sectors such as mining, energy or agriculture. At the same time, economic shocks such as the COVID-19 pandemic and the fallout of Russia's invasion of Ukraine have depleted fiscal space across the Western Balkans. With limited public sector resources for investment in climate adaptation and mitigation and shallow capital markets, domestic banking sectors are seen as key players to mobilize private funds for green investments.

The Western Balkans still has some way to go in greening finance. Green finance—whereby financial institutions' lending and investments contribute to climate mitigation, climate adaption and resilience and mitigating environment risks—is critical, but progress remains at the initial stages in the Western

Balkans. Despite the significant physical and transition risks as well as their proximity and economic linkages to the European market, governments across the region lack comprehensive strategies and roadmaps to tackle climate and environmental risks. Green taxonomies, a key enabler for the creation of a green loan and bond market, are yet to be developed, while disclosure requirements for large corporates and other economic agents have not been introduced. And banking supervisors are only beginning to take a closer look at the exposure of financial institutions to climate risks.

An initial step would be to address misalignment of fiscal, economic and environmental policies. For example, public subsidies are still being offered to economic activities with a high carbon footprint, and effective carbon taxation is largely missing. Uncertainty on long-term government policies may also inhibit the development of sustainable economic projects, while data gaps can limit the capacity of financial institutions to identify and originate green assets and manage climate risks. Furthermore, as outlined in a recent World Bank toolkit, the high upfront financing costs, transactions costs, lack of track-record of new technologies, and long payback periods for some green projects could increase the real and perceived riskiness of green projects overall. These factors can indirectly influence investors' appetite, negatively impacting investment and lending decisions.

Ways Forward on Green Finance in the Western Balkans

To green the region's financial sector, governments in the Western Balkans need to consider:

Prioritizing Sustainability Strategies and Green Taxonomies

Central banks and banking supervisors in the region are taking their first steps towards preparing and approving sustainability strategies and roadmaps. These strategies are being developed by the central banks in Albania, Kosovo, Montenegro and North Macedonia, and the banking agencies in Bosnia and Herzegovina.

In parallel, ministries of finance should take the initiative for developing green taxonomies, which are key to defining green investment and finance that will enable banks, insurers and capital market institutions to develop new instruments, central banks and supervisors to rely on standardized and comparable data, and corporates to make appropriate disclosures.

Promoting Prudential Management of Climate and Environmental Risks

To support prudential management of climate-related and environmental risks, financial authorities need to develop an approach to climate and environmental risk assessment. Regulators should have a comprehensive understanding of the situation of the banking sector's climate and environmental risks. Albania, Bosnia and Herzegovina, Kosovo and North Macedonia are in the initial stages of issuing surveys of environmental risks to commercial banks, while Bosnia and Herzegovina and North Macedonia have begun drafting a Strategy for Green Finance.

Figure 9. Green finance and the role of the financial sector
Source: World Bank FinSAC, 2022

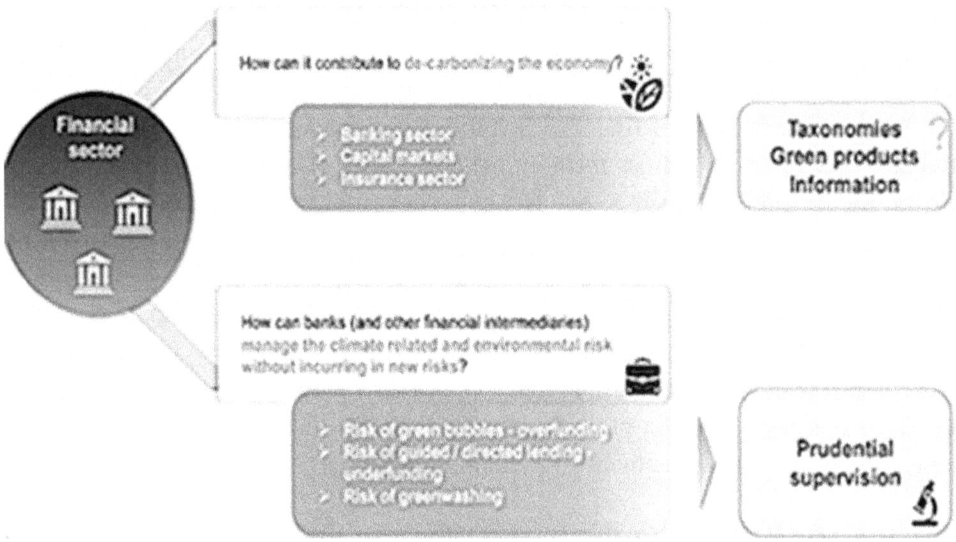

Developing Green Finance Instruments

Western Balkan governments could explore opportunities to issue sovereign green and blue bonds; Serbia issued sovereign green bonds in 2021. To help mobilize private capital toward climate mitigation and adaptation, financial authorities could cooperate with corporates and financial market participants to develop green finance guidelines for issuances as well as establish minimum entry requirements for independent assessors, ratings, or verification services to ensure quality external reviews and enhance comparability. Because the corporate bond market is nascent amongst the Western Balkans financial sector, governments may want to consider introducing incentive mechanisms to reduce transaction costs such as for external reviews of green bonds. Moreover, working with institutional investors to integrate sustainability considerations into the investment decision-making process could foster demand for green assets (Hwang et al. 2023). Ministries of finance may also explore options to introduce industry-level risk pooling mechanisms for natural disaster insurance like agriculture insurance. Existing national development banks and credit guarantee schemes could also play a leading role in greening the financial sector. Finally, for capacity building and knowledge exchange purposes, Western Balkans governments might also consider joining international platforms such as the Coalition of Finance Ministers for Climate Action.

Table 9. Green agenda for WB6 local, national, and regional approach

	Local	National	Regional
Implementing the Green Agenda and Ensuring a Just Transition	Strengthen local governance capacities to im prove the ability of local governments in the Western Balkans to conduct proper environmental impact assessments, and decentralize the implementation of the Green Agenda. Raise awareness about the importance of clean energy and environmental protection both among the population and at the policy level, including by sensitizing youth to the severity of the issue and the potential for the future of the energy transition through education. More public campaigns are needed to place climate change more squarely on the political agenda. Counter public apprehensions of being confronted with unsightly renewables technologies in their immediate vicinity through decentralized energy grids, strategic landscaping, educational campaigns, and positive messaging.	Decentralize the implementation of the Green Agenda by incorporating local municipalities ward more renewables most urgently; energy efficiency could best be improved in the construction sector. Respect emission limits that governments have committed to and avoid subsidies absorbing energy costs and funding black and brown energy. Look beyond short-term plans that would secure re-election and envisage a long-term transition strategy that goes beyond procuring sufficient heating for voters during winter. Work to find a sustainable balance between economic and environmental interests instead of posing the question which of the two should be prioritized. Ensure a "just" transition by assuring that the social impact of the Green Agenda finds con- sideration in all implementation mechanisms. Democracy and rule of law are critical components to assure enforcement of existing laws, strengthen business' position and security, and empowering communal approaches.	Strengthen regional cooperation to ease the green transition by lowering costs and avoiding inefficient parallel structures in implementing the Green Agenda. Regional alliances are crucial to the transition process, both in the form of stronger partnerships between the Western Balkans (WB) and the EU on the one hand, as well as among WB countries on the other. Maintain momentum for the Green Agenda by aligning Western Balkans markets more closely with European Union (EU) schemes and recognize that the implementation of the Green Agenda is a crucial step in the EU accession process of the WB.
Energy Transition and Energy Security	Decentralize energy supplies to counteract the vested interests of brown and black industries and allow private households to produce energy and contribute to the country's grid. Ensure that future energy strategies are transparent and enable functional and meaningful public participation. Encourage scaling up of smart strategies that include local populations. Measures to implement the energy transition (e.g. building power stations) must avoid causing harm to biodiversity or wasting financial and technical resources.	Move faster and more decisively to implement an energy transition and reaffirm commitments already made. Where needed, governments should be reminded of their commitments. Recognize that renewable energies are the only reasonable way to secure energy independence. Shift away from short-term plans which favor the reliance on brown and black energy sup plies towards long-term visions which favor sustainable renewables even in cases where renewables may have higher initial costs. Develop more profitable incentives to encourage companies and private investors to install and operate renewable capacities. Adapt ad-hoc efforts to mitigate the current energy crisis to be more complementary to the overall vision of the Green Agenda. Ensure that energy transition strategies are coordinated and multidisciplinary, involving social, legal, economic, technical, and educational considerations. Stop government subsidies for dirty energy sources that hurt the competitiveness of energy companies that focus on green energy. Move past the idea that coal is a cheap energy source. Seek out expert-informed recommendations during the policy-making process rather than trying to draft solutions to the energy crisis without specific expertise. Foster innovation more rigorously and bring about policy reforms. This would include incentivizing private energy production as well as improving property rights to prevent companies from leaving the country.	Develop robust platforms for regional cooperation and exchange while critically examining current arrangements. Facilitate pragmatic regional integration of energy markets to reduce volatility in electricity prices and allow for higher political acceptance of less regulated electricity markets. Use the energy crisis exacerbated by Russia's war in Ukraine as a catalyst for a green transition.

continues on following page

Table 9. Continued

	Local	National	Regional
Sustainable Agriculture and Food Security	Focus on the practical implementation of steps at a more granular, local level rather than on strategic documents and declarations of intent. Reduce food waste along the supply chain, from farm to fork, to improve food security. Support small farms to increase the region's international competitiveness. This includes urgent land reform to prevent further land fragmentation.	Develop and adopt national food-security strategies to have contingency plans in place to ensure the sufficient and acute provision of food to domestic markets in critical situations. Make available large expanses of unused agricultural land to produce strategic crops. In this regard, a long-term solution for the management of state land is needed, perhaps the establishment of national agencies or funds for agricultural land. This management strategy should carefully weigh industrial interests against bio-diversity and conservation needs. Provide support for the industrialization of food production and distribution and ensure employment opportunities. Improve the rule of law and combat corruption to avoid unnecessary barriers to the implementation of the Green Agenda. Give civil society a more prominent role to steer the discussion and help apply more pressure on the government. Expert committees that provide implementation recommendations and subsequently track governments' progress have proven useful to this end in the past. Promote EU integration as this is an indisputable condition for farmers' global competitiveness.	Improve university curricula and enhance regional cooperation to promote research and in novation in the agricultural sector through a regional research collaboration framework. Curricula need to be adapted to teach innovative agricultural techniques that take climate adaptation into consideration and expand green and sustainable agricultural technologies. While universities are responsible for developing the appropriate curricula, they require in creased funding to do so. Promote the exchange of knowledge and resources on a regional level. This could prove particularly useful for cross-border practices of responding to the climate crisis. Shift the research focus from technology generation to technology transfer.
Circular economy	Promote circular economy (CE) approaches at all levels of the region and across the entire value chain. Raise awareness for CE to boost public support through campaigns that introduce the full prices of products, such as the "Pay as You Throw" scheme. Counteract the public misconception of CE as a form of waste management with awareness and education campaigns that explain the complexity of the issue and emphasize the immense potential benefits inherent in CE.	Include CE in legislation and strategic documents since funds cannot be properly allocated if circular economy is not explicitly named in the relevant decision papers. Make relevant data on CE approaches more readily available. Increase funds set aside for CE, especially in the form of environmental schemes. Seek out a wider exchange with businesses and non-governmental organizations aimed at building compromise and partnerships in se curing energy and raw resources, as well as protecting employment.	

Source: Aspen Institute, 2022

Table 10. Western Balkans' strategic and legislative frameworks on energy and climate

	Energy Strategy	Low-carbon Development Strategy	Climate-change Law	Energy Efficiency Strategy	Renewables Development Strategy
Albania	National Energy Strategy [illegible]	National Climate Change Strategy (adopted in 2019)	Law on Climate Change (adopted in December 2020)	National Energy Efficiency Action Plan expired in 2020	National Action Plan for Renewable Energy Resources in Albania 2019-2021
Bosnia and Herzegovina	Framework Energy Strategy 2035	Climate change adaptation and Low Emissions Growth Strategy 2025	-	Action Plan for Energy Efficiency of Bosnia and Herzegovina 2019-2021 (NEEAP BiH) (final draft)	National Renewable Energy Action Plan 2020
Kosovo	Energy Strategy 2017-2028	Climate Change Strategy 2019-2028 and Action Plan on Climate Change 2019-2021 (approved)	-	National Energy Efficiency Action Plan (NEEAP) 2019-2021 (draft)	National Renewable Energy Action Plan (NREAP 2011-2020)
North Macedonia	Energy Development Strategy 2040	Long-term Strategy on Climate Action and National Action Plan on Climate Change (drafts)	Law on Climate Action (draft)	Fourth National Energy Efficiency Action Plan (NEEAP) (adopted)	Renewable Energy Action Plan [illegible] 2025
Serbia	Energy Sector Development Strategy for the Period until 2025; Energy Development Strategy 2040 (draft ongoing)	Draft low-carbon development strategy	Law on Climate Change (adopted in 2021)	Fourth National Energy Efficiency Action Plan (NEEAP) (until 2021) (adopted)	National Renewable Energy Action Plan 2020 (adopted in 2013)

Source: Adapted from OECD, Multi-dimensional Review of the Western Balkans: From Analysis to Action, April 2021, section 14.2

Note: Green: document approved and valid. Black: document still valid but requires revision. Blue: draft document exists, but has not yet been approved. Red: document expired.

ISO 14001, AN INTERNATIONAL STANDARD FOR ENVIRONMENTAL PROTECTION: WESTERN BALKANS APPROACH

According to studies, there is an indication that ISO 14001 certification increases compliance with environmental regulations, and this effect persists after implementing the appropriate controls for endogeneity (McGuire,2014). Manufacturers have relied on environmental management systems to meet government environmental regulations, enhance their environmental performance, and reduce their impact on the environment. In a research investigating the impact of ISO 14001 certification on manufacturers' toxic release levels, applying the censored quantile instrumental variable estimator (CQIV) to data on U.S. transportation equipment manufacturing subsector facilities, resulted that for large firms, encouraging voluntary adoption of ISO 14001 might be an effective government strategy to reduce on-site pollution, while for small firms and for the purpose of reducing off-site pollution, other economic incentives or regulations might be warranted (Nemati et al, 2019). In another research, Camilleri (2022) suggests that the use of this certifiable standard may result in operational efficiencies through better utilization of resources and waste management systems. It provides opportunities for practitioners to re-conceive their license to operate and to enhance their credentials with stakeholders (Camilleri, 2022). Facilities that integrate ISO 14001 standards into their daily operations are more likely to report improvements in environmental performance. Environmental improvements are also more likely to occur in facilities that include performance management elements in their ISO 14001 standards (Yin & Schmeidler. 2008).

A globally recognized standard, ISO 14001 outlines the specifications for an environmental management system. Through more effective resource utilization and waste reduction, it aids organizations in improving their environmental performance, giving them a competitive advantage and the confidence of stakeholders. Like all ISO management system standards, ISO 14001 stresses the need for ongoing development of a company's environmental policies and procedures. The standard has recently undergone revisions that include significant advancements like elevated status of environmental management within the organization's strategic planning procedures, enhanced leadership involvement, and a stronger dedication to proactive actions that improve environmental performance.

ISO 14001 helps:

- Demonstrate compliance with current and future statutory and regulatory requirements
- Increase leadership involvement and engagement of employees
- Improve company reputation and the confidence of stakeholders through strategic communication
- Achieve strategic business aims by incorporating environmental issues into business management
- Provide a competitive and financial advantage through improved efficiencies and reduced costs
- Encourage better environmental performance of suppliers by integrating them into the organization's business systems.

ISO 14001:2015 now requires:

- Environmental management to be more prominent within the organization's strategic direction
- A greater commitment from leadership
- The implementation of proactive initiatives to protect the environment from harm and degradation, such as sustainable resource use and climate change mitigation
- A focus on life-cycle thinking to ensure consideration of environmental aspects from development to end-of-life
- The addition of a stakeholder-focused communication strategy It also allows for easier integration into other management systems thanks to the same structure, terms, and definitions.

In Table 11, the number of ISO 14001: 2015 certificates issued in Balkan countries between 2007 – 2021 period have been shown. It is clear, the number of ISO 14001: 2015 certificates issued in this territory is increasing, and this is because of the private sector awareness about environment protection as well as per business improvement processes requirements, especially for those companies which are intended to do business with foreign companies, and/or operating abroad. In total, from only 2857 certificates issued in 2007, the number is increased to 17813 in 2021, or about 6 times higher. Bulgaria, Romania, and Greece, as EU member states, have the largest number.

SOLUTIONS AND RECOMMENDATIONS

The Western Balkan countries aspire to join the European Union (EU), but they encounter many challenges that prevent them from growing economically, especially because of moderate economies with labor-intensive industries like textiles, agriculture, and services, mostly based on mining, industry, agriculture, and forestry, all of which are important environmental polluters. As EU candidates or potential candidates, all countries in the Western Balkans are dedicated to respecting the Paris Agreement and meeting the EU2020 and EU2030 goals for decreasing greenhouse gas (GHG) emissions.

Fighting climate change has its own unique set of difficulties because shifting to less carbon-intensive economic activity can disrupt some areas of the economy and necessitate additional investments.

Table 11. ISO 14001 certificates issued during 2007 – 2021 in some countries

Country	2007	2008	2009	2010	2011	2012	2013	2014	2015	2016	2017	2018	2019	2020	2021
Albania					9	12	49	38	118	153	68	144	151	165	163
B&H			50	55	86	136	66	109	100	124	202	291	742	272	319
Bulgaria			335		559	717	893	1279	1183	1376	1605	1946	1993	1941	2166
Croatia			297	392	448	598	561	762	707	735	878	1027	1182	1138	1297
Greece	51	285	584	450	321	458	944	779	863	1023	1348	1415	1542	1696	1891
Montenegro	4	11	11	6	2	4	10	14	14	17	15	63	91	178	255
RNM	2	4	6	13	13	41	91	108	95	96	208	306	292	381	434
Romania	2222	3519	4221	2005	5267	7041	1083	6146	3123	3417	5284	4553	4658	5221	6174
Serbia	82	81	124	172	193	560	428	666	1190	1027	792	1169	1275	1629	1794
Slovenia	74	249	184	190	204	600	303	269	195	424	418	432	468	494	557
Turkey	422	559	862	474	1122	996	1251	1750	1634	2009	1598	2521	2360	2954	2763
TOTAL	2857	4708	6674	3757	8224	11163	5679	11920	9222	10401	12416	13867	14754	16069	17813

Source: ISO survey, 2021

Although progress in the Western Balkans is still in its early stages, green finance—in which financial institutions' loans and investments contribute to climate reduction, climate adaptation and resilience, and mitigating environmental risks—is crucial. The Western Balkans must commit to decarbonization goals by the end of 2023 in order to comply with the Energy Community Treaty. After that, they'll create integrated energy and climate plans.

The Western Balkan countries are characterized by "low" consumer environmental awareness levels, which are linked to a weak market for sustainable consumer goods. The human resources needed for professional development in the green industries present difficulties for all nations.

Different policies and regulations are in place in the region to encourage the growth of green businesses, but their implementation and enforcement are not guaranteed for a number of reasons, including a lack of supporting sub-legal frameworks, improper oversight, a dearth of inspectors, a shortage of experienced and qualified workers, corruption, and insufficient financial resources.

In this region, the definition, application, and integration of the circular economy idea and its guiding principles in strategies and action plans are still in their infancy. Concrete references to CE in policies and action plans have mostly been lacking in the Western Balkan countries.

Environmental-related tax policies appear to be inadequately integrated in the Western Balkan countries. Attempts to expand eco-labeling into other industries have been negligible or failed in the Western Balkan countries, and even though the region's counties have expanded producer responsibility or polluter pays laws, their implementation and enforcement may be better or still leave something to be desired. The incentives' regulatory and practical implementation varies and appears to be lacking.

There are no data on the percentage of green public procurement in the entire region, which suggests inadequate monitoring. The current systems for public procurement are usually corruptible.

The increasing number of ISO 14001 standard certificate issues in the Balkan region shows private sector is moving forward with environment protection and environmental management practices, a good sign; The legal framework regarding environmental taxes shall be implemented in the near future.

There is no any transparent information in the Western Balkan countries where environmental taxes collected are spent. Additionally, it should be clear explicitly in development papers of WB6 governments how green taxation can address the specific environmental challenges in the Western Balkans, such as pollution from mining, industry, agriculture, and forestry, as well as how environmental tax measures can contribute to the Western Balkan countries' commitment to the Paris Agreement and their goals for reducing greenhouse gas emissions.

Besides the fact environmental taxation aligns with the objectives of the European Union's (EU's) financial investment plan for the region, it is not clear yet how environmental tax measures can support the countries in meeting their climate targets and the goals set by the EU.

There is a need for a more reflective link between green taxation and tax reform in the countries under review. This connection should be further developed to highlight the specific ways in which environmental tax reform can address the economic and environmental challenges faced by these countries.

FUTURE RESEARCH DIRECTIONS

The future research directions for Western Balkan countries is not difficult to match, since the way towards integration path to EU already paved the way. The most important think is connecting the WB6 practicalities with EU legislative and strategic approach generally and related to environment, green agenda and

CO_2 emissions, etc. Saying this, besides there is a promising climate between WB6 governments to be in accordance with EU towards the integration path, in reality things are moving very slow, considering here 30 years of transition period from collapse of communist regime in these countries in early 1990.

An important issue in WB6 remain the question how green taxation can address the specific environmental challenges in the Western Balkans, such as pollution from mining, industry, agriculture, and forestry, since there is no transparency how environment taxes are been spent, and how environment is being considered towards investing and protecting it with state budget money expenditures.

It is clear that environmental tax reform is relevant and necessary in the context of Western Balkan countries, considering here how environmental tax measures can contribute to the Western Balkan countries' commitment to the Paris Agreement and their goals for reducing greenhouse gas emissions, as well as aligns the environmental tax reform with the objectives of the European Union's (EU's) financial investment plan for the region.

It is clear that green taxation should address the specific environmental issues faced by the Western Balkan countries and contribute to their economic development and achievement of climate goals too.

The preservation of the ozone layer, local air quality, acidification and eutrophication, water supply and water quality, waste management, and biodiversity losses are just a few of the environmental issues that WB6 are suffering, and must deal with, and the economic tools have become increasingly important for WB6, making sure there is a link between environmental taxes and "polluter pays principle", which seems to be weak in WB6.

In WB6 the majority of money collected from taxes related to the environment comes from taxes on the purchase or use of motor vehicles and fuels, including taxes on gasoline and diesel, as well as taxation in business level for their activity, while all taxable collections should be distributed carefully.

Green tax reforms can be accomplished by a number of complementary measures, such as the enactment of new taxes or the restructuring of already-existing ones, and additionally, it is crucial to eliminate or modify fiscal measures that negatively impact the environment, such as tax breaks or subsidies, while considering the non-environmental goals they were intended to support.

Sharing information, experiences, and best practices with regard to potential options and opportunities for broadening the application of ecologically linked taxes is one approach for WB6 to address concerns about their ability to compete internationally.

WB6 should come soon close to the approach based on the European Green Deal (2019) and the Recovery Plan for Europe (2021), the European Commission (EC) plans to allocate 30% of the budget to climate-related projects, initiatives, and programs, demonstrating its commitment to becoming the first climate-neutral region by 2050, and the activities for this already planned, need complex adjustments to legislation, strategic planning, implementation, and monitoring.

The countries of the Western Balkan are highly vulnerable to costly natural disasters such as floods, wildfires, earthquakes and droughts, and climate change is only expected to exacerbate the frequency and intensity of weather-related hazards. The region needs to remain vigilant about the impact of climate change on its financial sectors. Both the physical and transition-related risks of climate change can cause revenue losses, heightened credit risks and lower funding sources from customers in affected sectors such as mining, energy or agriculture.

Despite the significant physical and transition risks as well as their proximity and economic linkages to the European market, governments across the region lack comprehensive strategies and roadmaps to tackle climate and environmental risks. Furthermore, the high upfront financing costs, transactions costs, lack of track-record of new technologies, and long payback periods for some green projects could increase the real and perceived riskiness of green projects overall. These factors can indirectly influence investors' appetite, negatively impacting investment and lending decisions.

To green the region's financial sector, governments in the Western Balkans need to consider (1) *prioritizing sustainability strategies and green taxonomies, and (2) promoting prudential management of climate and environment risks*

Western Balkan governments could explore opportunities to issue sovereign green and blue bonds. To help mobilize private capital toward climate mitigation and adaptation, financial authorities could cooperate with corporates and financial market participants to develop green finance guidelines for issuances as well as establish minimum entry requirements for independent assessors, ratings, or verification services to ensure quality external reviews and enhance comparability. Ministries of finance may also explore options to introduce industry-level risk pooling mechanisms for natural disaster insurance like agriculture insurance. Existing national development banks and credit guarantee schemes could also play a leading role in greening the financial sector.

Finally, for capacity building and knowledge exchange purposes, Western Balkan governments might also consider joining international platforms such as the Coalition of Finance Ministers for Climate Action.

REFERENCES

Adamson, Th., & Jordans, F. (2019). France to slap new 'ecotax' on plane tickets from 2020. *Houston Chronicle. Associated Press.* https://apnews.com/article/be0605b81b214502b2fce419869ea09e

Antón, Á. (2022). EU energy taxation framework revision in the context of the green deal: Towards a green, efficient, socially just, and inclusive economy transition. *VISUAL REVIEW. International Visual Culture Review / Revista Internacional de Cultura Visual, 9,* 1–13. https://journals.eagora.org/revVISUAL/article/view/3670

Aspen Institute. (2022). *Green Agenda for the Western Balkans. The Road Toward Effective and Sustainable Implementation.* Aspen Institute. https://www.aspeninstitute.de/wp-content/uploads/Green-Agenda-for-the-Western-Balkans_2023.pdf

BBC. (2006). Miliband Draws Up Green Tax Plan: Environment Secretary David Miliband Has Confirmed the Government Is Holding Discussions On Tackling Climate Change Using Green Taxes. http://news.bbc.co.uk/2/hi/uk_news/6095680.stm

Beauregard-Tellier, F. (2006). Ecological fiscal reform (EFR), *Library of Parliament Canada.* https://publications.gc.ca/site/eng/9.560939/publication.html

Blau, A. & Janssen, C. (2020). *Regional Baseline Assessment. Region III: Albania, Bosnia and Herzegovina, Israel, Montenegro and Turkey.* SwitchMed. https://switchmed.eu/wp-content/uploads/2020/12/GCB_Regional-Baseline-Assessment-03.pdf

Camilleri, M. A. (2022). The rationale for ISO 14001 certification: A systematic review and a cost–benefit analysis. *Corporate Social Responsibility and Environmental Management, 29*(4), 1067–1083. doi:10.1002/csr.2254

Dussaux, D. (2020), "The joint effects of energy prices and carbon taxes on environmental and economic performance: Evidence from the French manufacturing sector", *OECD Environment Working Papers*, No. 154, OECD Publishing, Paris, https://doi.org/ doi:10.1787/19970900

EEA. (2010). *EEA Briefing Note for ETR Workshop, Dublin.* EEA.

EEA. (2013). *EEA Briefing Note for ETR Workshop, Lisbon.* EEA.

EEA. (2022). *The role of (environmental) taxation in supporting sustainability transitions.* EEA. https://www.eea.europa.eu/publications/the-role-of-environmental-taxation

Elison, L., & Snyder, F. (2001). *The Montana State Constitution: A Reference Guide.* Greenwood Publishing Group.

EPI. (2022). *Environmental Performance Index.* Yale Center for Environmental Law & Policy. https://epi.yale.edu/

European Commission. (2020). *Stepping up Europe's 2030 climate ambition Investing in a climate-neutral future for the benefit of our people.* EC. https://eur-lex.europa.eu/legal-content/EN/TXT/PDF/?uri=CELEX:52020DC0562

European Commission. (2022). *Energy Support package for the Western Balkans.* EC. https://www.wbif.eu/storage/app/media/Library/12.%20Economic%20and%20Investment%20Plan/WBIF%20Energy%20Support%20Package%20%20for%20the%20Western%20Balkans%202022.pdf

European Union. (2022). *Implementing Green Agenda for the Western Balkans.* EU. https://neighbourhood-enlargement.ec.europa.eu/system/files/2022-11/factsheet%20green%20agenda%20nov2022.pdf

Ghazouani, A., Xia, W., Ben Jebli, M., & Shahzad, U. (2020). Exploring the Role of Carbon Taxation Policies on CO2 Emissions: Contextual Evidence from Tax Implementation and Non-Implementation European Countries. *Sustainability (Basel)*, *12*(20), 8680. https://pdfs.semanticscholar.org/f387/0b63757560571461417d40e5f20b6e87f3f7.pdf. doi:10.3390u12208680

Guo, B., Wang, Y., Feng, Y., Liang, C., Tang, L., Yao, X., & Hu, F. (2022). The effects of environmental tax reform on urban air pollution: A quasi-natural experiment based on the Environmental Protection Tax Law. *Frontiers in Public Health*, *10*, 967524. https://www.frontiersin.org/articles/10.3389/fpubh.2022.967524/full. doi:10.3389/fpubh.2022.967524 PMID:36033767

Hassan, M., Oueslati, W., & Rousselière, D. (2020). Environmental taxes, reforms and economic growth: An empirical analysis of panel data. *Economic Systems*, *44*(3), 100806. https://www.sciencedirect.com/science/article/abs/pii/S093936251830400X?via%3Dihub. doi:10.1016/j.ecosys.2020.100806

Huang, L., & Xiao, E. (2021). Peer effects in public support for Pigouvian taxation. *Journal of Economic Behavior & Organization*, *187*, 192–204. https://www.sciencedirect.com/science/article/abs/pii/S0167268121001608?via%3Dihub. doi:10.1016/j.jebo.2021.04.019

Hwang, J., Fontan, I. A., Oguz, A., & Vardanyan, V. (2023). *How can the Western Balkans boost green finance?* https://blogs.worldbank.org/europeandcentralasia/how-can-western-balkans-boost-green-finance

IMF. (2022). *IMF Staff Country Reports 2022.* IMF. https://www.elibrary.imf.org/view/journals/002/2022/363/article-A001-en.xml

Jiang, H.-D., Hao, W.-T., Xu, Q.-Y., & Liang, Q.-M. (2020). Socio-economic and environmental impacts of the iron ore resource tax reform in China: A CGE-based analysis. *Resources Policy*, *68*, 101775. https://www.sciencedirect.com/science/article/abs/pii/S0301420719307251?via%3Dihub. doi:10.1016/j.resourpol.2020.101775

Knez, S., Štrbac, S., & Podbregar, I. (2022). Climate change in the Western Balkans and EU Green Deal: Status, mitigation and challenges. *Energy, Sustainability and Society*, *12*(1), 1. https://energsustainsoc.biomedcentral.com/articles/10.1186/s13705-021-00328-y. doi:10.118613705-021-00328-y

Li, Z., & Zhou, Z. (2022). Research on the Impact Mechanism of Fiscal and Tax Policies on Carbon Emissions under the Dual Carbon Goals. *Frontiers in Business. Economics and Management*, *6*(2), 32–37. doi:10.54097/fbem.v6i2.2637

Loganathan, N., Shahbaz, M., & Taha, R. (2014). The link between green taxation and economic growth on CO2 emissions: Fresh evidence from Malaysia. *Renewable & Sustainable Energy Reviews*, *38*, 1083–1091. doi:10.1016/j.rser.2014.07.057

Mardones, C. (2022). Pigouvian taxes to internalize environmental damages from Chilean mining-A computable general equilibrium analysis. *Journal of Cleaner Production*, *362*, 132359. doi:10.1016/j.jclepro.2022.132359

Mardones, C., & Baeza, N. (2018). Economic and environmental effects of a CO2 tax in Latin American countries. *Energy Policy*, *114*, 262–273. doi:10.1016/j.enpol.2017.12.001

McGuire, W. (2014). The effect of ISO 14001 on environmental regulatory compliance in China. *Ecological Economics*, *105*(September), 254–264. doi:10.1016/j.ecolecon.2014.06.007

Meireles, M., Robaina, M., & Magueta, D. (2021). The Effectiveness of Environmental Taxes in Reducing CO2 Emissions in Passenger Vehicles: The Case of Mediterranean Countries. *International Journal of Environmental Research and Public Health*, *18*(10), 5442. doi:10.3390/ijerph18105442 PMID:34069643

Mitic, P., Fedajev, A., & Kojic, M. (2023). *Exploring the Economy -Environment Interactions in the Western Balkans: A Panel Data Analysis*. http://dx.doi.org/ doi:10.28934/ea.23.56.1.pp43-56

Nemati, M., Zheng, Y., & Hu, W. (2019, May). The ISO 14001 Standard and Firms' Environmental Performance. *Journal of Agricultural and Resource Economics*, *44*(2), 422–438. https://www.jstor.org/stable/26797565

OECD. (2015). *Environmental Taxation A Guide for Policy Makers.* OECD. https://www.oecd.org/env/tools-evaluation/48164926.pdf

Olja, M. I., & Mirjana, G. (2012). Environmental Taxation as a Tool for Sustainable Development Policy-State Comparison of Serbia and Application of Ecological Taxation Reform in European Union. *Economic Analysis*, *45*(1), 32–44. https://www.semanticscholar.org/paper/Environmental-Taxation-as-a-Tool-for-Sustainable-of-Ivanovi%C4%87-Golu%C5%A1in/0924238c7e174f17307587f592002f3076e8ff21

Oueslati, W. (2015). Growth and welfare effects of environmental tax reform and public spending policy. *Economic Modelling*, *45*, 1–13. doi:10.1016/j.econmod.2014.10.040

Oueslati, W., Zipperer, V., Rousselière, D., & Dimitropoulos, A. (2017). Energy taxes, reforms and income inequality: An empirical cross-country analysis. *Inter Economics*, *150*, 80–95. doi:10.1016/j.inteco.2017.01.002

Patel, S. K., & Jhalani, P. (2023). Formulation of variables of environmental taxation: A bibliometric analysis of Scopus database (2001–2022). *Environment, Development and Sustainability*. https://link.springer.com/article/10.1007/s10668-023-03027-0

Pogge, T. W. (2001). Eradicating Systemic Poverty: Brief for a global resources' dividend. *Journal of Human Development*, *2*(1), 59–77. doi:10.1080/14649880120050246

Sanfey, P., Milatovic, J., & Kresic, A. (2016). How the Western Balkans Can Catch Up. *EBRD Working Paper No.185.* https://papers.ssrn.com/sol3/papers.cfm?abstract_id=3119685

Skjærseth, J. B. (2021). Towards a European Green Deal: The evolution of EU climate and energy policy mixes. *International Environmental Agreement: Politics, Law and Economics*, *21*(1), 25–41. doi:10.100710784-021-09529-4

Spasić, V. (2022). Which Western Balkan countries intend to introduce carbon tax? *Balkan Green Engineering News.* https://balkangreenenergynews.com/which-western-balkan-countries-intend-to-introduce-carbon-tax/

Speck, S. (2022). *The role of environmental taxation in supporting sustainability transitions.* EEA. https://www.eea.europa.eu/publications/the-role-of-environmental-taxation

UNDP. (2016). *Human Development Report 2016 Risk-Proofing the Western Balkans: Empowering People to Prevent Disaster*s. UNDP. https://www.undp.org/albania/publications/human-development-report-2016-risk-proofing-western-balkans-empowering-people-prevent-disasters

Wang, L., Ma, P., Song, Y., & Zhang, M. (2023). How does environmental tax affect enterprises' total factor productivity? Evidence from the reform of environmental fee-to-tax in China. *Journal of Cleaner Production*, *413*, 137441. doi:10.1016/j.jclepro.2023.137441

WB. (2022). Western Balkans Regular Economic Report, No.22, Fall 2022. *Western Balkans Regular Economic Report*, (22). doi:10.1596/38189

Yin, H., & Peter, J. (2009, November). Schmeidler.(2008). Why do standardized ISO 14001 environmental management systems lead to heterogeneous environmental outcomes? *Business Strategy and the Environment*, *18*(7), 469–486. https://onlinelibrary.wiley.com/doi/10.1002/bse.629. doi:10.1002/bse.629

Zhou, S., Shi, M., Li, N., & Yuan, Y. (2011). Impacts of Carbon Tax Policy on CO2 Mitigation and Economic Growth in China. *Advances in Climate Change Research*, *2*(3), 124–133. doi:10.3724/SP.J.1248.2011.00124

APPENDIX

Figure 10. EBRD scores on green quality of a sustainable market economy, 2021

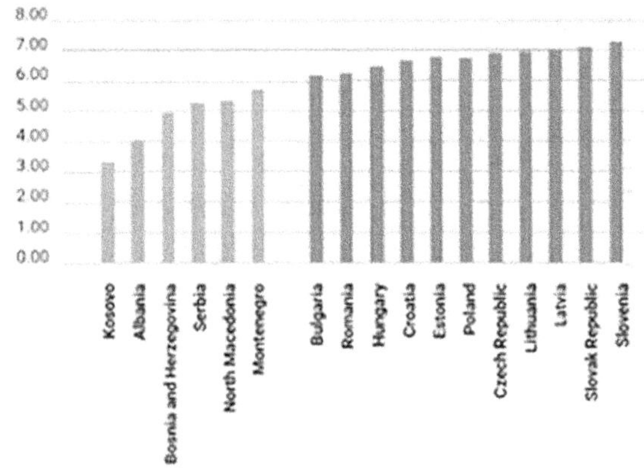

Source: Prepared on the basis of data provided in the EBRD Transition Report 2021-22. EBRD Transition Report 2021-22 (accessed December 20, 2022).

Note: Blue color bars: EU new member states; green color bars: Western Balkan countries. Scores range from 1 to 10, where 10 represents a synthetic frontier corresponding to the standards of a sustainable market economy.

Figure 11. Energy Intensity of GDP in EU and Western Balkan countries, 2020

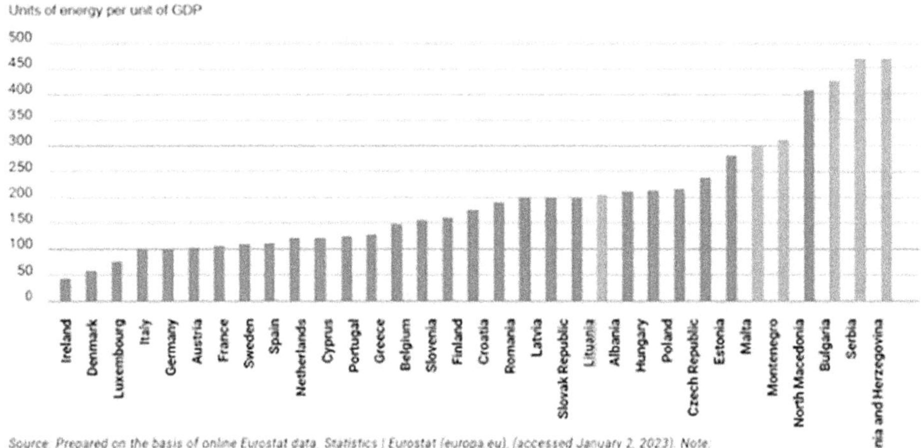

Source: Prepared on the basis of online Eurostat data. Statistics | Eurostat (europa.eu). (accessed January 2, 2023). Note: Energy intensity is calculated as units of energy per unit of GDP. The unit of measure is kilograms of oil equivalent (KGOE) per thousand euro in purchasing power standards (PPS).

Note: Blue color bars: EU member states; green color bars: Western Balkan countries.

Figure 12. Energy and Carbon Intensities of GDP; Energy intensity of GDP (total energy supply per GDP [GJ/USD 1000 2015 PPP])

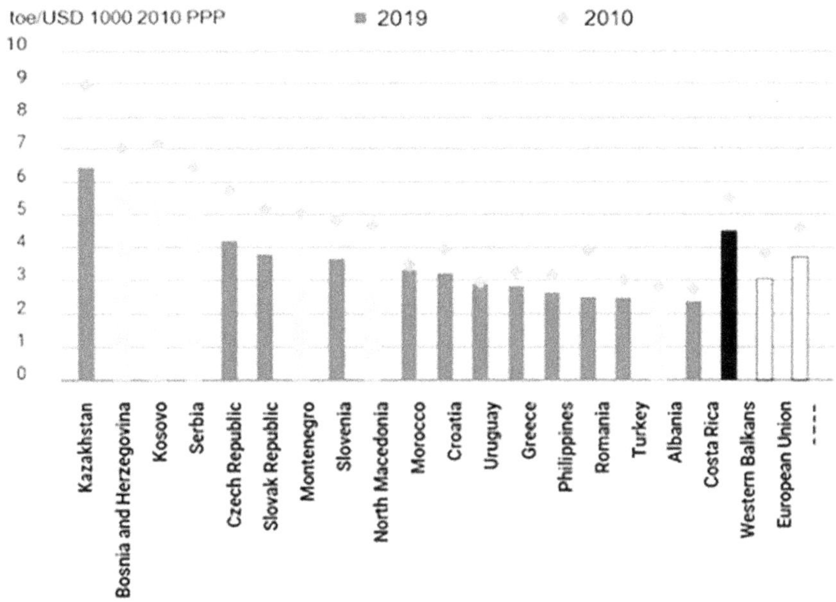

Figure 13. Carbon Dioxide Emissions (metric tons per capita), 2019

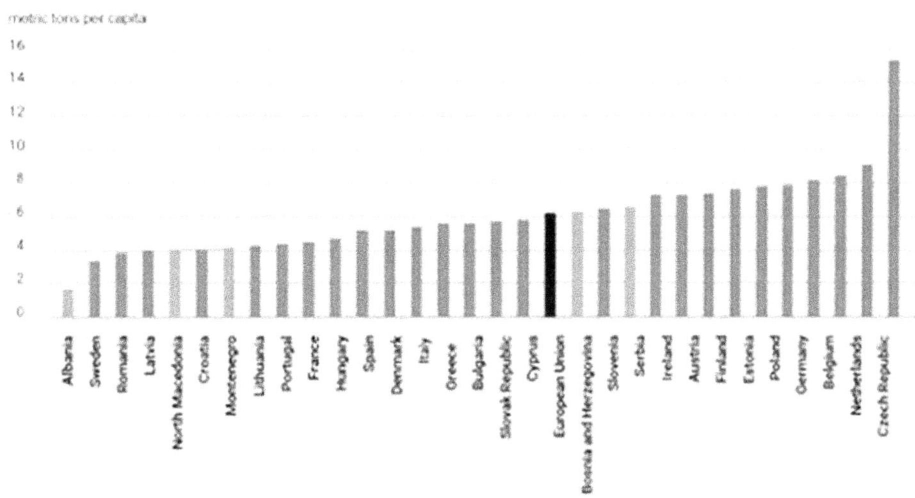

Source: World Bank, World Development Indicators. CO2 emissions (metric tons per capita) | Data (worldbank.org) (accessed January 3, 2023)

Note: Blue color bars: EU member states; green color bars: Western Balkan countries. No data is available for Kosovo. Carbon dioxide emissions are those stemming from the burning of fossil fuels and the manufacture of cement. They include carbon dioxide produced during consumption of solid, liquid, and gas fuels and gas flaring.

Figure 14. (a) Electricity generation mix (%), 2019; (b) CO2 Emissions by sector (Mt Co2), 2019

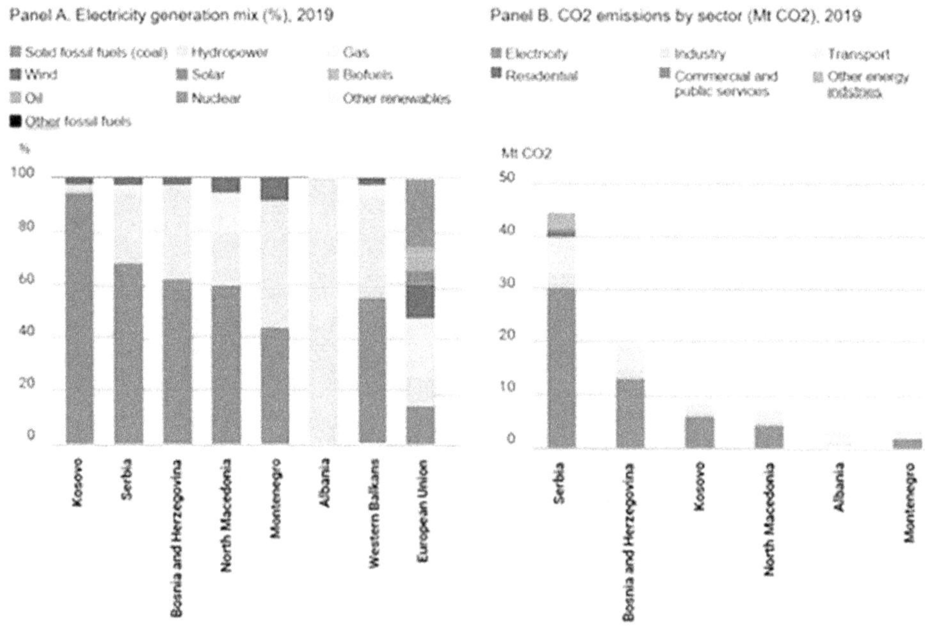

Figure 15. Carbon intensity of GDP (CO2 emissions per unit of GDP [kg/USD 2015 PPP])

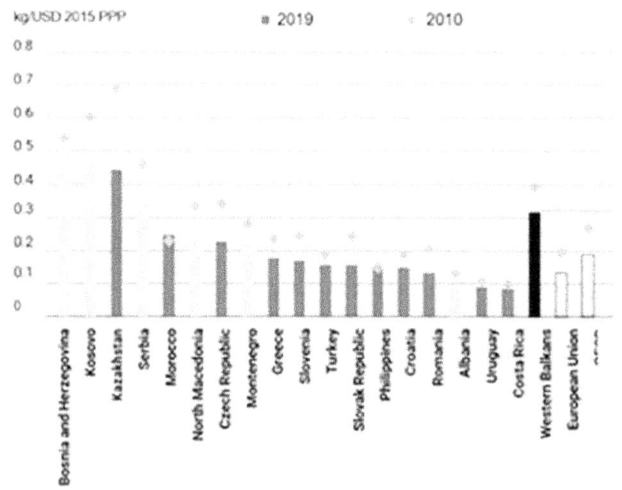

Source: Adapted from OECD, Multi-dimensional Review of the Western Balkans: From Analysis to Action, April 2022, 368. OECD Development Pathways, OECD Publishing, Paris, https://doi.org/10.1787/8824c5db-en (accessed December 20, 2022).

Figure 16. Share of energy frm renewable sources in EU and Western Balkan countries, 2020

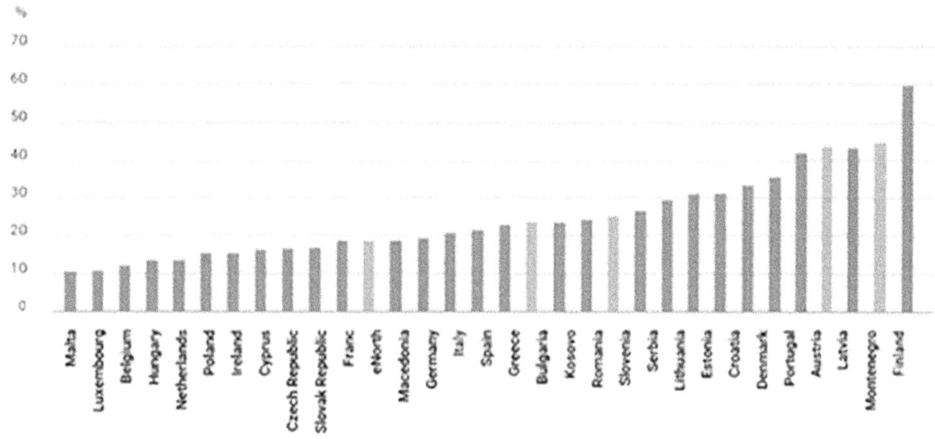

Source: Prepared on the basis of online Eurostat data. Statistics | Eurostat (europa.eu) (accessed October 1, 2022).

Note: Blue color bars: EU member states, green color bars: Western Balkan countries

Figure 17. Use of renewables for heating and cooling, 2020

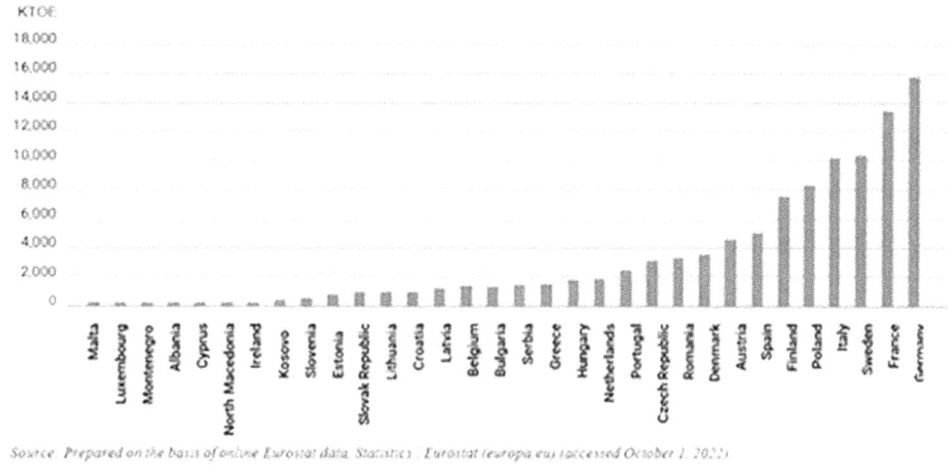

Source: Prepared on the basis of online Eurostat data. Statistics | Eurostat (europa.eu) (accessed October 1, 2022).

Note: Unit of measure: Thousand tonnes of oil equivalent [KTOE]

Figure 18. Use of renewables for transport, 2020

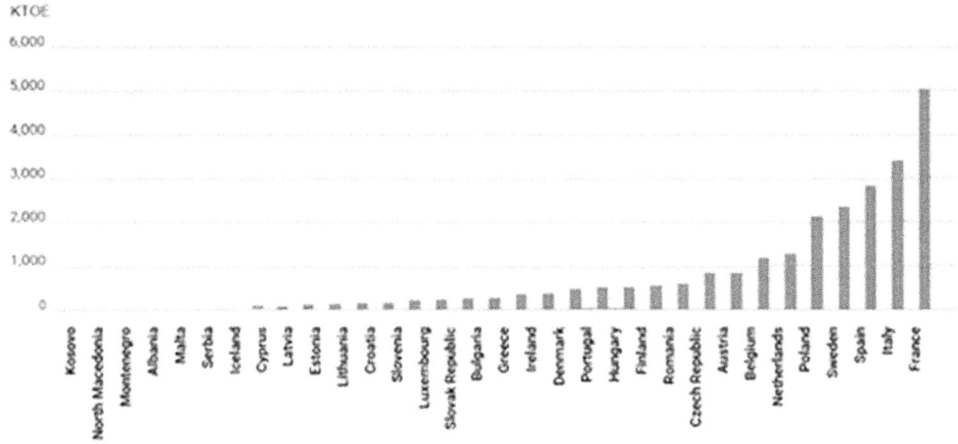

Source: Prepared on the basis of online Eurostat data. Statistics | Eurostat (europa.eu) (accessed October 1, 2022).

Note: Unit of measure: Thousand tonnes of oil equivalent [KTOE]

Figure 19. Years of life lost per 100,000 inhabitants attributable to exposure to pollution in 2016

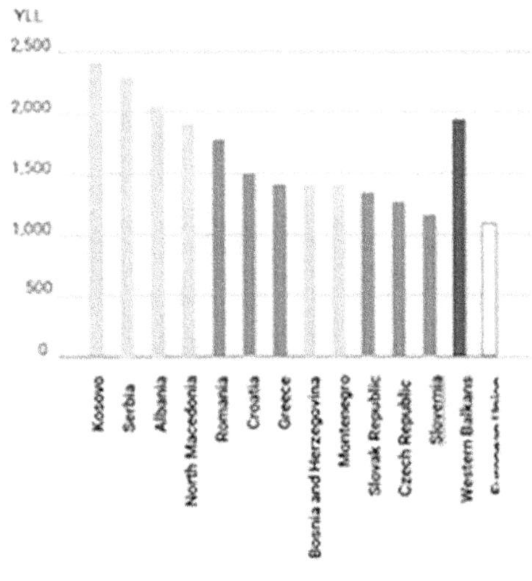

Source: European Environment Agency, Air Quality in Europe: 2020 Report, as reported in OECD, Multi-dimensional Review of the Western Balkans. From Analysis to Action, April 2022, 367, OECD Development Pathways, OECD Publishing, Paris, https://doi.org/10.1787/8824c5db-en (accessed December 20, 2022).

ACRONYMS

1. B&H – Bosnia and Herzegovina
2. CBAM - Carbon Border Adjustment Mechanism
3. CFMCA - Coalition of finance ministers for climate action.
4. COP 15 – United Nations by diversity Conference. 15th Meeting
5. EC – European Commission
6. EEA – European Economic Area

7. EEA - European Environment Agency
8. EFTE - European Federation for Transport and Environment
9. EIP - Economic and Investment Plan
10. EPBiH - Public Enterprise Electric Utility of Bosnia and Herzegovina
11. EPI – Environment Performance index
12. ETD - Energy Tax Directive
13. ETR - Environmental Tax Reforms
14. ETS - Emission Trading System
15. EU – European Union
16. EUETS - EU Emission Trading System
17. EUGD - EU Green Deal
18. GCF - Green Climate Fund
19. GE – Green economy
20. GDP – Gross Domestic Product
21. GHGE – Greenhouse Gas Emission
22. GAWB - Green Agenda for the Western Balkans
23. HEA - Health and Environment Alliance
24. IPARD - IPA Rural Development
25. MNE - Montenegro
26. NDC - Nationally Determined Contributions
27. NECP - The national energy and climate plans
28. RNM – Republic of North Macedonia
29. NSSD - National Strategy for Sustainable Development
30. OECD – Organization for Economic Cooperation and Development
31. SAM - Social Accounting Matrix
32. TFEU – Treaty of Functioning of European Union
33. TICC - The Domestic Consumption Tax on Natural Gas
34. TICGN - The Internal Consumption Tax on Natural Gas
35. TICPE - The domestic consumption tax on energy products
36. UN – United Nations
37. USA – United States of America
38. UNEP – United Nations Environment Program
39. WB – World Bank
40. WB6 – Western Balkans 6 economies, Albania, Bosnia & Herzegovina, Kosovo, Montenegro, Republic of North Macedonia, and Serbia

Chapter 6
Green Tax:
A Bibliometric Analysis

Paulo Dias
https://orcid.org/0000-0002-0351-9751
Instituto Universitário de Lisboa (ISCTE-IUL), Portugal

Iara Cardoso Santos
Instituto Universitário de Lisboa (ISCTE-IUL), Portugal

ABSTRACT

In this study the authors provide a bibliometric overview regarding available articles in SCOPUS database about green tax and double dividend. The results show that the theme has recently become more popular, with most articles affiliated with European countries. Regarding the sources of publication, the authors identified that the most influential journals are environmental, resources economics, and energy economics, where most of the articles are published. Finally, through keyword analysis the authors concluded that green tax focus areas are often related to double dividends, sustainable development, environmental fiscal reform and market-based instruments, while double dividend focus areas are related to Environmental tax reform, environmental policy, economic growth, employment, optimal taxation, and carbon tax demonstrating that these are the most influential topics on the subject.

INTRODUCTION

The theme on the environment is not new, in recent years there have been some concerns to protect the environment and nature. This awareness began to evolve with the industrial revolution and, later, an increased concern arose with the harmful effects of human action on nature and the repercussions caused by world wars.

These negative impacts that modern society entails are vast and require urgent measures so that the destructive tendency is changed and the pressure on resources and the natural system is alleviated. Specific legislation then began to be created and goals were established for the community with a view to preservation. It is at this moment that the concept of environmental tax and its derivatives arises.

DOI: 10.4018/978-1-6684-8592-7.ch006

Today, climate action has gained another boost thanks, in large part, to the expression of environmental activists and to the goals and commitments of the international community through the Paris Agreement. (OECD, 2018; UNFCCC, 2015) With this impulse also increases the number of investigations and scientific opinions making it difficult to monitor scientific production.

The objective of this study is to develop a significant reference for researchers with an interest in the theme of *Green tax* and *Double dividend*, as we provide an evaluation and perspective of the scientific literature with greater potential, contributing to increase the rigor of research specialties in this theme.

This research also discusses the elements on which environmental taxation is based and the basis for a green tax reform. This results in the importance of deepening the knowledge about what is meant by *Green Tax* and the *Double Dividend*.

Internationally, the studies developed in the context of environmental taxation are mostly concerned with the evaluation of the implementation of the Environmental Tax Reforms (ETR), as well as the "dividends" obtained, these are mostly studies from Europe.

A green tax reform seeks to fulfill two purposes of taxes: raising revenue and changing the behavior of taxpayers, focusing on social and environmental sustainability.

These objectives are achieved by using taxes as instruments of environmental policy, sending signals to the market, placing the tax burden on elements harmful to the environment and alleviating ordinary taxation. In this way the replacement will allow to achieve a multiplicity of fiscal and social objectives, the so-called double dividend.

The possibility of this fiscal policy being able to provide, simultaneously, the balance of public accounts and sustainable growth has caused the volume of scientific research on the subject to gradually increase, contributing to the wealth of data.

With this increase, it becomes important to evaluate the scientific production, not only for the recognition of researchers, but also to allow academics to follow the relevant literature.

The choice of the empirical study is related to the scarcity of bibliometric analysis methodologies, in *Green Tax* and *Double Dividend,* with the use of *VOSviewer* considered an advanced instrument of these analyses and relationships.

In this sense, and to carry out this study, we proceeded to the collection and processing of reliable and adequate information using the content provider *SCOPUS*. From which we collected published articles delimited by the criterion of keyword, area of study and type of scientific production, considering for this study only scientific articles, without any temporal limitation to be able to analyze all the articles available on these themes, comprising a period of 25 years from 1994 to 2019.

LITERATURE REVIEW

The year 2018 was a year marked by numerous ecological disasters, referred to in chapter four, and species extinction. With these current disasters it is necessary to impose some social changes and for this it is necessary that people and governments change consumer behaviors.

Since the concern about global warming, environmental taxes have been playing an important role in the EU's environmental policies. In the Europe 2020 strategy, environmental tax reform is an important and inevitable element in achieving a smart, competitive, sustainable and inclusive economy. (European Commission, 2010)

This chapter aims to provide the necessary bases for the understanding of this theme and for this it is necessary to use a framework of the types of management and economic instruments to introduce the concept of environmental tax. From this concept we were able to introduce the origin of environmental tax reforms that, in turn, generate the double dividend hypothesis.

Management Instruments and Economic Instruments (EI's)

Conventionally, regulation and control management instruments, such as laws and regulation, have been used to deal with externalities. However, this type of management called *Regulatory or Command and Control (CAC) Approach* was preferentially used until the 70's, when economic instruments began to gain importance due to the fact that the regulation and control regime is inefficient, because (i) it requires the same compliance with standards by all sources of pollution, Regardless of the marginal costs, (ii) high demands in technical and bureaucratic terms both in their definition, implementation and even in control and analysis, (iii) they are expensive and slow to apply instruments and (iv) promote small technological incentives. (Hawkins, 2000)

In mid 1972 OECD members were recommended to implement the polluter pays and user pays principles which were subsequently published in the Treaty on European Union Article 174(2). These principles aim, regardless of the existence of fault or intent, to make the polluter responsible by reflecting in the cost of goods and services the cost of production, environmental and social costs, making activities harmful to the environment less advantageous from an economic point of view.

To apply them it is necessary to resort to political instruments and, since the instruments of regulation and control alone are not efficient, the use of economic instruments that directly address the disadvantages of the previous ones arises, allowing to promote the reconciliation of environmental and social concerns, to value natural resources and to stimulate behavioral change.

However, to be effective, they must affect the costs and benefits of the choices of economic agents to boost the improvement of the environmental situation. The Economic Instruments must be different and appropriate for different countries, being subdivided into five groups (Hawkins, 2000): Fees and taxes; Subsidies; *Deposit-refund systems*; Creation of markets and financial incentives.

Definition of Environmental Tax

The relevant literature on this subject emerged during 1980, however the main reference work in environmental economics was that of *Arthur Pigou* who in 1920 suggested taxing pollution to internalize the negative externalities of economic activities in the environment. This proved to be an efficient way to correct some market failures and help economic agents to take into account the cost of pollution when making their decisions. [1] (Bovenberg & De Mooij, 1994)

This "new" instrument of taxation was dubbed *Pigouvian Taxes* by Bovenberg & De Mooij (1994) and, in accordance with what was said earlier, contributes to the optimization of social welfare not only by confronting the polluter with a price that reflects the true social marginal cost but also by creating an environmental tax that will generate revenues allowing to reduce or eliminate other distorting taxes, thus changing the weight of the factors considered as growth (capital and labor) to elements of reduction of well-being, as the case of the exhaustion of resources and pollution. (A Lans Bovenberg & de Mooij, 1994; Gouveia & Palma, 2018; Hjøllund & Svendsen, 2001; Sandmo, 2011);

However, and for the tax to be efficient, *Pigou* recommends that it be applied directly and uniformly per unit of pollutant emission. The existence of a uniform tax means that companies would reduce emissions to the point where the marginal cost of reducing an extra unit is equal to the tax rate. If the tax rate were not uniform each company would face a different rate. (Hjøllund & Svendsen, 2001)

The particularity that, in the event of no action, the costs of environmental damage are potentially high suggests that the stance of environmental policy, in relation to global warming, should be quite aggressive. (Klok, Larsen, Hansen, & Dahl, 2006; Pearce, 1991)

As such, this tax is often seen as an instrument that improves the alignment of tax rates because they are also used to confront the user with the social cost of consuming polluting environmental services or with a polluting process, reflecting it in prices, in order to encourage agents to minimize the level of consumption harmful to the environment and a more efficient use of available natural resources, complying with the extrafiscal character of the tax figure. (Jackson, 2010; Klok et al., 2006; Pearce, 1991; Sandmo, 2011)

We can therefore see that, in order to promote sustainable development and adopt the precautionary principle, environmental taxes are subdivided into taxes on energy, transport, pollution and taxes on resources, since these are the goods and services that are likely to degrade the environment through their production, consumption and/or disposal. (Gouveia & Palma, 2018)

However, the main difficulty encountered is the quantification of the value of environmental goods so that it is possible to have an idea of the cost-benefit and the estimation of future impacts. (Klok et al., 2006)

It was then that in 1990 some European countries decided to implement environmental tax reforms that we will discuss below. (Klok et al., 2006; Pearce, 1991)

Definition of Environmental Tax Reform (ETR)

Since, as stated earlier, environmental taxes increase public revenue, this can be used to reduce other taxes or fees thereby alleviating the increase in environmental costs for businesses and consumers. In this way, a policy of combining the tax burden was introduced, in which the revenue is neutral (fiscal neutrality), with the name of Environmental Tax Reform.

However, the definition is also the subject of discussion, and this is directly related to the nature of the use of revenues (recycling) from new environmental taxes or increase of existing taxes, the approaches are divided into: (i) Recycling revenues through tax reduction; (ii) Support environmental projects and initiatives using part of the revenues; (iii) Use the proceeds for compensation.

Clinch, Dunne, & Dresner, (2006) present various ways of using recipes as well as their economic, environmental, and political and social acceptance.

One of the ways presented is to resort to the recycling of revenue by dividing a portion for the financing of environmental projects and the other portion for the reduction of the tax on labor. This approach makes it possible to focus on more obvious environmental measures to increase the acceptance of the reform and lower costs if these are efficient projects.

This balance between project financing and tax relief should consider and depend on labor market conditions, the success of projects and the government. Being seen as a socially and politically accepted instrument.

Another form of recycling is to apply the revenue in order to reduce the regressive aspects of energy taxes by preparing a support system to protect those most vulnerable to the results of the reform. Clinch et al., (2006) present suggestions such as: reduce or change indirect taxes; return a global amount equal to each household, corresponding to the average value of the energy tax or make the structure of the tariff, the energy tax, progressive, among others.

For these reasons, these reforms have become an important tool for taxation or *pricing* as they improve the tax systems used while reducing environmental damage and achieving social gains. [2] (Pearce, 1991) Thus providing a justification for a progressive increase in revenues generated through environment-related taxes to reduce taxes derived from other, often distorting, sources, such as income, profits, and employment.

However, Hawkins (2000) states that the main particularity of these is the dependence and interaction of the market. For other authors, the FRG is characterized as a process by which taxation, on the consumption of resources, pollution or energy, is increased while (by recycling) the burden on labor, capital or personal income is decreased to promote fiscal neutrality, conferring social advantages (double dividend) such as the optimization of the level of employment, increased savings and investment. (Ekins & Barker, 2001; Silva & Chaves, 2015)

In this way and to assist in the design of policies, the scientific community has been developing mathematical models. These have been an essential tool to identify the best policy or mix of policies for governments to plan and make decisions taking into account a better balance between environmental and economic objectives and thus avoid substantial costs to the economy and estimate the possible benefits that may result in the scope of environmental taxation. (Gouveia & Palma, 2018; Mota & Palma, 2015)

The most used types of mathematical models are:

i. *Input-Output*: to analyze the interdependence of industries in the economy;
ii. *Computable General Equilibrium (*CGE): group of economic models that are essentially empirical versions of the *Walsarian General Equilibrium System* that apply the assumptions of neoclassical theory, used by (Pereira & Pereira, 2014);
iii. *Linear Programming Models;*
iv. *Non-Linear Programming Models;*
v. *Macroeconomic Models.*

In addition, it is important to take into account that the results should be interpreted as relative and be complemented with empirical studies and over-regulation should be taken into account, that is, overlapping law over law without thinking about the cumulative effect of the measures hindering the execution of taxation. (Hawkins, 2000)

In short, an ETR involves the use of environmental fiscal policy instruments to reflect the costs of environmental damage in the prices faced by polluters and users of polluting products/services and in turn this increase in public revenue is implemented in a way that is useful to society. This implementation is called the Double Dividend. (Clinch et al., 2006)

It is also important to analyse the real costs of implementing this reform because, although it has many advantages, it has some political-economic, legal and social weaknesses that may hinder its implementation. (Borges & Merlin, 2018)

Environmental impacts can be analysed in two ways: using ex *ante* analyses, predicting the possible effects, or *ex post* analyses, analyzing the real facts, as well as achieving the economic, political and social effects of energy taxation. (Clinch et al., 2006)

However, in addition to there being numerous studies on ETR, there is no clear tendency to emphasize the environmental impacts known as the first dividend. This disadvantage coupled with the difficulty in estimating the efficient rate and monetarily quantifying the value of environmental damage or benefit, underlying the fact that it is technically difficult and expensive, makes it difficult to anticipate this same impact. (Clinch et al., 2006)

However, according to Clinch et al. (2006), assuming that environmental taxes are based on scarce resources or substances with polluting emissions, these will have a higher price than other resources or substances, so we can get some answers estimating the elasticity of prices.

Although it should be borne in mind that if, at the same time, the tax rate is set at a relatively low level, and the demand for this resource is relatively insensitive to price changes, and if the revenue is not used to subsidise renewable energy, invest in energy saving or research and development, it will be difficult to achieve the environmental objective determined by jeopardising environmental efficiency. (Baranzini, Goldemberg, & Speck, 2000; Clinch et al., 2006)

There are also factors that must be taken into account that can reduce environmental efficiency, such as inflation and the entry of new pollutants into the market. (Baranzini et al., 2000)

However, the macroeconomic environment, impacts on competitiveness, inequalities between sectors and regressivity entail various political and economic costs. (Clinch et al., 2006)

Definition of Double Dividend

In theory, the ETR is associated with two principles: revenue neutrality and the achievement of the double dividend through governments' use of environmental tax revenues to lower other taxes. (A Lans Bovenberg & de Mooij, 1994; Mota & Palma, 2015) But the big question discussed by several authors is whether the implementation of an environmental tax reform is capable of giving rise to the double dividend; what will be the type of application of the revenue; and what sign (positive or negative) it assumes, thus contributing to a vast although controversial literature. (Borrego, 2016)

According to Clinch et al., (2006), it is first important to distinguish the effect of "revenue recycling" from "revenue increase". Revenue recycling consists of using the revenue from environmental taxes to reduce existing taxes, mostly distorting rather than setting a *lump-sum*.

Second, it is important to present a definition of the first and second dividend. Apparently, this does not exist in the literature, however the most usual and adopted by researchers is the first dividend, known as the Environmental Double Dividend, cover the negative externality of economic activities in the environment thus assuming a benefit or gain of well-being resulting from the improvement in the environment while the second dividend, known as the Double Economic Dividend, encompass the non-environmental welfare benefits.

Fullerton & Metcalf, (1998) contains, in the article, a literature review of studies and models of the double dividend that conclude with 3 main points:

1. The validity of the double dividend hypothesis cannot be resolved as a general question. Under certain circumstances the exchange of environmental taxes can improve the environment and reduce the burden on the system.

2. The emphasis in the literature on the importance of revenue is misguided. Three types of policies can have equivalent effects on the environment and labour supply. One of the policies increases revenue from the environmental component, another loss of revenue, and the other with no associated revenue.
3. Different regulatory approaches are available and not all regulations create scarcity rents.

With regard to the literature on the application of revenue and the associated economic gains, which will be seen in more detail below, it is interesting to note that they assume an indefinite amount of forms all these associated with tax benefits, economic growth (L.H. Goulder, 1995; Peng et al., 2019; Shi, Qiao, Shao, & Wang, 2019), economic well-being (Baranzini et al., 2000; Clinch et al., 2006; Koskela, Schöb, & Sinn, 1998) and employability (A Lans Bovenberg & De Mooij, 1994; Chaturvedi, Saluja, Banerjee, & Arora, 2014; Clinch et al., 2006; Pearce, 1991)

Since the European authors argue that the double dividend leads to a decrease in unemployment and consequently an increase in competitiveness (A.L. Bovenberg & van der Ploeg, 1994; A.L. Bovenberg & Van Der Ploeg, 1998), while the American authors assume the double dividend as a gain in reducing the negative impact of the tax system on the economy, making it more efficient and competitive by reducing taxation on capital, income, among others. (Borrego, 2016; Lawrence H. Goulder, 2013; Mota & Palma, 2015)

Regarding the question of which sign (positive or negative) assumes the double dividend in the implementation of a Federal Jar, the studies that defend that this presents a positive sign claim to be associated with the decrease in unemployment, because as the revenues from environmental taxation increase, so does the taxation on labor and social security contributions, thus providing an incentive for employability. (Clinch et al., 2006; Ekins & Barker, 2001; L.H. Goulder, 1995)

In turn, the negative sign is associated with the distortion of the purchasing power of households through taxation on the most polluting products and the insufficient compensation of the decrease in taxation on income, and there is also the possibility that, in the presence of distorting taxes, environmental taxation does not fully internalize environmental damage. (A Lans Bovenberg & De Mooij, 1994)

As a rule, the double dividend hypothesis assumes that the two dividends are independent. The introduction of interdependence can generate different policy implications, leading to instruments that do not generate revenue being more efficient than instruments that do. (Lai, 2018)

Given that many studies are focused on evaluating the implementation of an ETR and the associated double dividend, and this is seen to keep up with economic challenges, it is important to detail the economic gains presented earlier and highlight the behavior of the economy when introducing a reform.

The consolidation and balance of public accounts as well as sustainable growth are very important in a Federal Republic of Germany, as such *Goulder* (1995) in his study defined three types of dividend based on efficient costs being these dividends weak, intermediate and strong, however this definition does not have many supporters. (A Lans Bovenberg & De Mooij, 1994) assume that the conditions imposed by *Goulder* are too strict and prove that one cannot achieve the strong form of the double dividend.

Additionally, they discover the effect of tax interaction while searching for the validity of the double dividend. They concluded that in the presence of distorting taxes the optimum point of the environmental tax should be below the social damage of pollution.

The study conducted by Shi et al., (2019) concludes that if taxes on pollution increase the elasticity of capital production and decrease that of labor and increase the efficiency of resource use that, by convergence, promotes economic growth.

In dissent, Peng et al., (2019), says that energy taxes lead to resource efficiency but if this is excessive it leads to economic decline.

While, according to Freire-González & Ho, (2018) efficiency is achieved whether the revenue from the implementation of a carbon tax is used to reduce taxes on capital, on labor or on value added.

As for competitiveness, it can be defined at the business level as the company's ability to maintain or increase national and/or international market share and profit. (Baranzini et al., 2000) This is influenced by micro-factors such as the cost structure, quality, and stock market, and by macro-factors such as taxes, fees, subsidies, and regime accordingly.

Thus, through an open economy model, we can conclude that the implementation of a moderate environmental tax rate will induce the replacement of energy use by labor, leading to a reduction in costs that, consequently, can lead to a reduction in price thus making the economy more competitive. (Koskela et al., 1998; Metcalf, 2014)

Accordingly, Clinch et al., (2006), conclude that there are only impacts on competitiveness if environmental policy imposes different levels of costs or if countries have different policies and regulations between companies.

Clinch et al. (2006) explores three approaches to ETR and concludes that the most common is the reduction of taxes at work to promote employability, however it will depend greatly on macroeconomic conditions and the labor market (reasonable or perfect competitiveness) in the country in question.

Additionally, the study by Maxim, Zander, & Patuelli, (2019) compares the performance of an ETR that generates double employment dividends in European and non-European countries. The results show that revenue recycling policies and taxes play an important role in employability, however the optimal policy is not identical in European and non-European countries. That is why there is a need for each country to design its policy in such a way as to create an optimal effect on employment, considering its needs.

In short, if the recycling option is to use revenues from environmental taxes to reduce taxes on wages and others that hinder employment, the problem of unemployment can be solved. (Bosello & Carraro, 2001; Chiroleu-Assouline & Fodha, 2009; Kuralbayeva, 2019; Pearce, 1991)

Definition of Triple Dividend

In practice, ETR's can lead to multiple dividends, satisfying other macroeconomic objectives, such as improving the distribution of wealth and reducing public debt, taking a form of fiscal sustainability. (Pereira & Pereira, 2014)

In the study by Pereira and Pereira (2014) the first dividend is always achieved, and the second dividend can be in two ways: increase in GDP or in employability. The increase in GDP occurs if the reform adopts public or private investment policies while the increase in employability occurs if the policies fall on the reduction of social security contributions, by companies, and the personal income tax.

Liu, (2013) suggests that tax evasion may play a potential role in calculating the costs of reform, with certain environmental taxes having unique properties that make them difficult to circumvent. When we consider a green tax change, by changing the tax base from easy to circumvent tax ones to difficult ones we can decrease the total amount of tax evasion.

By using revenues from environmental taxes to reduce pre-existing taxes rather than directing revenues through non-productive transfers. By reducing marginal rates on easy-to-avoid taxes, policymakers can realize the substantial additional benefit of lowering tax evasion overall.

METHODOLOGY

Our research began, firstly, with the collection of scientific articles published in the various scientific journals about *Green Tax* until September 30, 2019. To this end, and in order to collect only reliable and adequate information to provide an objective answer, we used the *SCOPUS* database as a content provider, as it is considered the largest, and most common, bibliometric database. (Mongeon & Paul-Hus, 2016; Zupic & Čater, 2015)

In the elaboration of the criteria, we tried to include all the existing articles on the referred theme, so the first restriction to be made was that of all the articles presenting, in a comprehensive way, the keyword Green tax in order to include articles that have placed only the keyword *Green Tax* or *Green Tax Reform*.

Then, and to make the investigation more complete, the keyword *Double Dividend*, which as previously presented, is one of the principles of environmental tax reforms, was included in the research, like the previous criteria. Totaling, with the two keywords, a sample of 162 articles. Among these 162 articles there were 14 areas of study, from which we chose [REMOVED REF FIELD]*Economic, Econometrics and Finance* and *Business, Management and Accounting,* having reduced the sample to 107 articles presented in Table 1.

Table 1. Articles considered in the study, divided into author(s), title, and year of publication

Author(s)	Article title	Year of Publication
Shi H., Qiao Y., Shao X., Wang P.,	The effect of pollutant charges on economic and environmental performances: Evidence from Shandong Province in China	2019
Mardones C., Cabello M.,	Effectiveness of local air pollution and GHG taxes: The case of Chilean industrial sources	2019
Andreoni V.,	Environmental taxes: Drivers behind the revenue collected	2019
Weiss J.F., Anisimova T.,	The innovation and performance effects of well-designed environmental regulation: evidence from Sweden	2019
Jacobs B., van der Ploeg F.,	Redistribution and pollution taxes with non-linear Engel curves	2019
Freire-Gonzalez J., Ho M.S.,	Carbon taxes and the double dividend hypothesis in a recursive-dynamic CGE model for Spain	2019
Kuralbayeva K.,	Environmental Taxation, Employment and Public Spending in Developing Countries	2019
Meeks R., Sims K.R.E., Thompson H.,	Waste Not: Can Household Biogas Deliver Sustainable Development?	2019
Peng J.-T., Wang Y., Zhang X., He Y., Taketani M., Shi R., Zhu X.-D.,	Economic and welfare influences of an energy excise tax in Jiangsu province of China: A computable general equilibrium approach	2019
Maxim M.R., Zander K.K., Patuelli R.,	Green tax reform and employment double dividend in European and non-European countries: A meta-regression assessment	2019
Maxim M.R., Zander K.,	Can a green tax reform entail employment double dividend in European and non-European countries? A survey of the empirical evidence	2019
Tikoudis I.,	Second-Best Road Taxes in Polycentric Networks with Distorted Labor Markets	2019
Skolrud T.D., Galinato G.I.,	Revenue-neutral pollution taxes in the presence of a renewable fuel standard	2019
Klenert D., Schwerhoff G., Edenhofer O., Mattauch L.,	Environmental Taxation, Inequality and Engel's Law: The Double Dividend of Redistribution	2018
Ma Z., Zhao J., Ni J.,	Green Tax Legislation For Sustainable Development In China	2018
Perrier Q., Quirion P.,	How shifting investment towards low-carbon sectors impacts employment: Three determinants under scrutiny	2018
Yan S., Eskeland G.S.,	Greening the vehicle fleet: Norway's CO 2 -Differentiated registration tax	2018

continues on following page

Table 1. Continued

Author(s)	Article title	Year of Publication
Liu A.A., Yamagami H.,	Environmental Policy in the Presence of Induced Technological Change	2018
Silva S., Smith I., Pine C.,	Renewable energy subsidies versus carbon capture and sequestration support	2018
Singh S., Haldar N., Bhattacharya A.,	Offshore manufacturing contract design based on transfer price considering green tax: a bilevel programming approach	2018
S. Rausch, Yonezawa H.,	The Intergenerational Incidence Of Green Tax Reform	2018
Caron J., Cole J., Goettle R., Wave C., McFarland J., Woollacott J.,	Distributional Implications Of A National Co2 Tax In The U.S. Across Income Classes And Regions: A Multi-Model Overview	2018
Caron J., Cohen S.M., Brown M., Reilly J.M.,	Exploring The Impacts Of A National U.S. Co2 Tax And Revenue Recycling Options With A Coupled Electricity-Economy Model	2018
Lai Y.-B.,	The Feasibility of the Double-Dividend Hypothesis in a Democratic Economy	2018
Freire-Gonzalez J.,	Environmental taxation and the double dividend hypothesis in CGE modelling literature: A critical review	2018
Hong C.-Y., Huang C.-H., Li J.-F., Tsai Y.-C.,	Environmental Tax Reform, R&D Subsidies and CO2 Emissions: View Double Dividend Hypothesis	2018
Chloupkova J., Svendsen G.T., Zdechovsky T.,	A global meat tax: from big data to a double dividend	2018
Mathieu-Bolh N.,	Can tax reforms help achieve sustainable development?	2017
Pui K.L., Othman J.,	Economics and environmental implications of fuel efficiency improvement in Malaysia: A computable general equilibrium approach	2017
Islam A.K.M.N., Mitu J.A.,	Green fiscal reform in the context of Bangladesh: A scoping exercise	2017
Alexeev A., Good D.H., Krutilla K.,	Environmental taxation and the double dividend in decentralized jurisdictions	2016
Lai C.-F.,	Examining the double dividend effect of energy tax with the overlapping generations model	2016
Bostan I., Onofrei M., Dascălu E.-D., Fîrţescu B., Toderaşcu C.,	Impact of sustainable environmental expenditures policy on air pollution reduction, during European integration framework	2016
Zárate-Marco A., Vallés-Giménez J.,	Environmental tax and productivity in a decentralized context: new findings on the Porter hypothesis	2015
Tikoudis I., Verhoef E.T., van Ommeren J.N.,	On revenue recycling and the welfare effects of second-best congestion pricing in a monocentric city	2015
Jacobs B., of Mooij R.A.,	Pigou meets Mirrlees: On the irrelevance of tax distortions for the second-best Pigouvian tax	2015
Tuladhar S.D., Montgomery W.D., Kaufman N.,	Environmental policy for fiscal reform: Can a carbon tax play a role?	2015
Nerudova D., Dobranschi M.,	Double dividend hypothesis: Can it be validated by carbon taxation swap with payroll taxes?	2015
Gonand F., Jouvet P.-A.,	The "second dividend" and the demographic structure	2015
Metcalf G.E.,	Using the tax system to address competition issues with a carbon tax	2014
Figge F., Young W., Barkemeyer R.,	Sufficiency or efficiency to achieve lower resource consumption and emissions? The role of the rebound effect	2014
Allan G., Lecca P., McGregor P., Swales K.,	The economic and environmental impact of a carbon tax for Scotland: A computable general equilibrium analysis	2014
Orlov A., Grethe H.,	Introducing carbon taxes in Russia: The relevance of tax-interaction effects	2014
Chaturvedi A., Saluja M.S., Banerjee A., Arora R.,	Environmental fiscal reforms	2014
Andersson D., Nässén J., Larsson J., Holmberg J.,	Greenhouse gas emissions and subjective well-being: An analysis of Swedish households	2014
Markandya A., Gonzalez-Eguino M., Escapa M.,	From shadow to green: Linking environmental fiscal reforms and the informal economy	2013

continues on following page

Table 1. Continued

Author(s)	Article title	Year of Publication
Goulder L.H.,	Climate change policy's interactions with the tax system	2013
Liu A.A.,	Tax evasion and optimal environmental taxes	2013
Bjertnæs G.H., Tsygankova M., Martinsen T.,	Norwegian climate policy reforms in the presence of an international quota market	2013
Orlov A., Grethe H., McDonald S.,	Carbon taxation in Russia: Prospects for a double dividend and improved energy efficiency	2013
Böhringer C., Keller A., van der Werf E.,	Are green hopes too rosy? Employment and welfare impacts of renewable energy promotion	2013
Sun A.,	The establishment of the green tax policy in China - To accelerate the construction of circular economy experimental zone in Qaidam basin of Qinghai province as an example	2013
Fraser I., Waschik R.,	The Double Dividend hypothesis in a CGE model: Specific factors and the carbon base	2013
Rivers N., Groves S.,	The Welfare Impact of Self-supplied Water Pricing in Canada: A Computable General Equilibrium Assessment	2013
By Miguel C., Manzano B.,	Gradual green tax reforms	2011
Pang A., Shaw D.,	Optimal emission tax with pre-existing distortions	2011
By Miguel C., Manzano B.,	Green tax reforms and habits	2011
Jaeger W.K.,	The Welfare Effects of Environmental Taxation	2011
Nakabayashi M.,	Optimal tax rules and public sector efficiency with an externality in an overlapping generations model	2010
Aidt T.S.,	Green taxes: Refunding rules and lobbying	2010
Fernandez E., Perez R., Ruiz J.,	Double dividend, dynamic Laffer effects and public abatement	2010
Kronenberg T.,	Energy conservation, unemployment and the direction of technical change	2010
Chiroleu-Assouline M., Fodha M.,	Double dividend and distribution of welfare: Advanced results and empirical considerations	2009
Andersson J.O.,	Basic income from an ecological perspective	2009
Chen J.-h., Shieh J.-y., Chang J.-j., Lai C.-c.,	Growth, welfare and transitional dynamics in an endogenously growing economy with abatement labor	2009
Beladi H., Chao C.-C., Hazari B.R., Laffargue J.-P.,	Tourism and the environment	2009
Gomez C.M., Lozano J., Rey-Maquieira J.,	Environmental policy and long-term welfare in a tourism economy	2008
Glomm G., Kawaguchi D., Sepulveda F.,	Green taxes and double dividends in a dynamic economy	2008
Takeda S.,	The double dividend from carbon regulations in Japan	2007
Benedict A.M., Jacobsen M.,	Ricardian rents, environmental policy and the 'double-dividend' hypothesis	2007
Røpke I., Godskesen M.,	Leisure activities, time and environment	2007
Albrecht J.,	The use of consumption taxes to re-launch green tax reforms	2006
Heijdra B.J., Kooiman J.P., Ligthart J.E.,	Environmental quality, the macroeconomy, and intergenerational distribution	2006
Patuelli R., Nijkamp P., Pels E.,	Environmental tax reform and the double dividend: A meta-analytical performance assessment	2005
Bayindir-Upmann T., Raith M.G.,	Unemployment and pollution: Is one policy suited for two problems?	2005
Sugeta H., Matsumoto S.,	Green tax reform in an oligopolistic industry	2005
Bovenberg A.L., by Mooij R.A.,	Does money illusion rescue the double dividend?	2005
Bayındır-Upmann T.,	On the double dividend under imperfect competition	2004
Welsch H., Ehrenheim V.,	Environmental fiscal reform in Germany: a computable general equilibrium analysis	2004
Ng Y.-K.,	Optimal environmental charges/taxes: Easy to estimate and surplus yielding	2004

continues on following page

Table 1. Continued

Author(s)	Article title	Year of Publication
Kumbaroğlu G.S.,	Environmental taxation and economic effects: A computable general equilibrium analysis for Turkey	2003
Babiker M.H., Metcalf G.E., Reilly J.,	Tax distortions and global climate policy	2003
Böhringer C., Wiegard W., Starkweather C., Ruocco A.,	Green Tax Reforms and Computational Economics A Do-it-yourself Approach	2003
Goodstein E.,	The death of the Pigovian tax? Policy implications from the double-dividend debate	2003
Szomolányiová J.,	Environmental Taxation Benefits - Theorethical Analysis [Teoretická analýza přínosů environmentálních daní]	2002
Håkonsen L.,	A Note on Green Taxes and Double Dividends	2001
Ekins P., Barker T.,	Carbon taxes and carbon emissions trading	2001
Bosello F., Carraro C.,	Recycling energy taxes: Impacts on a disaggregated labour market	2001
Aronsson T.,	Green taxes and uncertain timing of technological change	2001
Mayeres I., Proost S.,	Marginal tax reform, externalities and income distribution	2001
Chia N.-C., Phang S.-Y.,	Motor vehicle taxes as an environmental management instrument: the case of Singapore	2001
Jansen H., Klaassen G.,	Economic impacts of the 1997 EU energy tax: Simulations with three EU-wide models	2000
Bosquet B.,	Environmental tax reform: Does it work? A survey of the empirical evidence	2000
Ligthart J.E., Van Der Ploeg F.,	Environmental policy, tax incidence, and the cost of public funds	1999
Bruvoll A., Ibenholt K.,	Green throughput taxation: Environmental and economic consequences	1998
Alden D.M., Proops J.L.R., Gay P.W.,	Industrial hemp's double dividend: A study for the USA	1998
Pezzey J.C.V., Park A.,	Reflections on the Double Dividend Debate: The Importance of Interest Groups and Information Costs	1998
Koskela E., Schöb R., Sinn H.-W.,	Pollution, Factor Taxation and Unemployment	1998
By Mooij R.A., Bovenberg A.L.,	Environmental Taxes, International Capital Mobility and Inefficient Tax Systems: Tax Burden vs. Tax Shifting	1998
Bovenberg A.L., Van Der Ploeg F.,	Consequences of environmental tax reform for unemployment and welfare	1998
Schöb R.,	Environmental Taxes and Pre-Existing Distortions: The Normalization Trap	1997
Håkonsen L., Mathiesen L.,	CO2-stabilization may be a 'no-regrets' policy: A general equilibrium analysis of the norwegian economy	1997
Majocchi A.,	Green fiscal reform and employment: A survey	1996
Kuper G.H.,	The effects of energy taxes on productivity and employment: The case of the Netherlands	1996
Proost S., Van Regemorter D.,	The double dividend and the role of inequality aversion and macroeconomic regimes	1995
Lans Bovenberg A., by Mooij R.A.,	Environmental taxes and labor-market distortions	1994
Bovenberg A.L., van der Ploeg F.,	Environmental policy, public finance and the labour market in a second-best world	1994

After data collection, it was necessary to evaluate the scientific production, through the application of several bibliometric indicators to introduce quantitative rigor to the evaluation of the literature. These indicators are divided into performance indicators, such as quality, importance and scientific impact indicators, and *science mapping* indicators, such as thematic associations and structure. (Costa, Fernández-Llimós, & Lopes, 2012; Zupic & Čater, 2015)

In the indicators of scientific activity we used, to measure the patterns of publication and authorship, the historical evolution, number of published works, geographical affiliation, and productivity of the authors.

Regarding scientific impact indicators to measure the standards of publication and use of scientific articles we have divided these into two types: the impact of the works on the community and the impact of the sources of publication.

Finally, to measure the indicator of thematic associations, we used the analysis of the *keywords* of the articles, references, and the collaborations in the authorship of the articles.

FINDINGS

The objective of this study is to analyze and evaluate the scientific production considered by articles published in scientific journals, on the theme *Green Tax and Double dividend*, being made an application of the various bibliometric indicators of scientific activity, importance, and impact.

The 107 scientific articles selected were distributed by three areas of study: *Economics, Econometrics and Finance* and *Business Management and Accounting,* as evidenced in the figure 1.

Figure 1. Distribution of articles analyzed by area of study

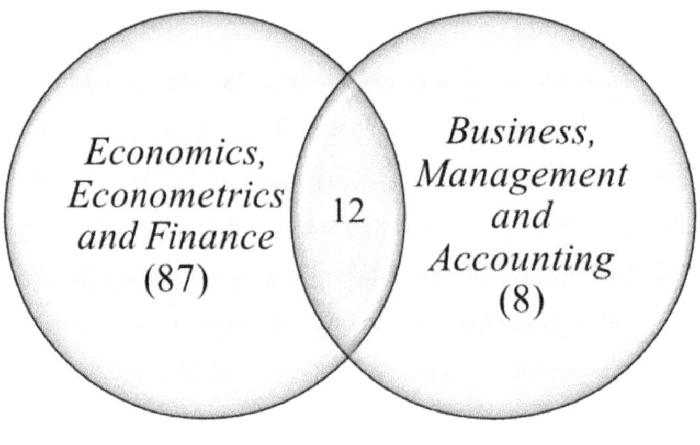

To evaluate the historical evolution of the publications, we rely on the analysis of Figure 2, where it is possible to verify that this theme has been presenting a positive general growth trend. The year in which the highest number of publications was registered was 2018, with 14 articles, and the years 1995 and 1999 were the years in which there were fewer publications (1 article).

In addition, we can observe that in the last 8 years more than half of the articles have been published, and the intervals with the most publications, since the beginning of the study, those of 1998-2001, 2009-2015 and 2018-2019. Interestingly, this interval intersects with important environmental and economic events, namely:

- The first Conference of the members of the United Nations Framework Convention on Climate Change (better known by its acronym UNFCCC) in 1995 with the aim of stabilizing or reducing greenhouse gas emissions;

- The adoption of the Kyoto Protocol in 1997 with more emphasis on the first conference;
- In 2001 the *Marrakech* Agreement defined the national system of emission inventories in order to emphasize sustainable development and climate change;
- In 2008 a UNFCCC conference agreed to fund a fund to help poorer countries with the effects of climate change;
- Approval in 2012 of the understanding of the validity of the Kyoto Protocol until 2020;
- In 2015 the Paris Agreement was approved where it was established that global warming should be contained below 2ºC;
- The year 2018 was a year marked by multiple natural disasters around the world with the forest fire in California, Spain and Portugal, Hurricane *Florence* and *Michael*, earthquakes in the state of Alaska, Peru, Taiwan, Iran and Japan, the eruption of the Fire Volcano in Guatemala, Kilauea Volcano in Hawaii and Soputan Volcano in Indonesia, the *tsunami* in Indonesia, area storms in India, heavy rains in Thailand that caused the blockage of a cave with 12 children and an adult, and heat waves in Canada and Japan.
- Finally, 2018 was also a year of great expression for environmental activists and *Greta Thunberg* was nominated for the Children's Climate Award.

Figure 2. Historical evolution of the publication of scientific articles

After analyzing the periodicity of the publication of the articles, we proceed to the analysis of the productivity of the authors, verifying which of these was the one that contributed the most to the scientific community with their knowledge. Thus, through Table 2 we can see that, in a total of 211 authors, the author who contributed most to the research on the subject of *Green Tax* was Bovenberg A. L., with a total of 5 articles, and second, with a total of 4 articles, Mooji R.A and van der Ploeg F.. We can also see that about 80% of the authors only contributed to the scientific community with one scientific article, and the remaining 18 authors contributed with 2 articles.

Table 2. Number of articles published by author

Authors	Number of articles per author	%
Bovenberg A.L.	5	2.092%
by Mooij R.A.	4	1.674%
van der Ploeg F.	4	1.674%
Bayindir-Upmann T.	2	0.837%
Böhringer C.	2	0.837%
Caron J.	2	0.837%
By Michael C.	2	0.837%
Freire-Gonzalez, J.	2	0.837%
Grethe H.	2	0.837%
Håkonsen L.	2	0.837%
Jacobs B.	2	0.837%
Ligthart J.E.	2	0.837%
Liu A.A.	2	0.837%
Manzano B.	2	0.837%
Maxim M.R.	2	0.837%
Metcalf G.E.	2	0.837%
Orlov A.	2	0.837%
Patuelli, R.	2	0.837%
Proost S.	2	0.837%
Schöb R.	2	0.837%
Tikoudis I.	2	0.837%
Other authors	1	79.498%
Total	**239**	**100%**

Through the geographical link we can check which country the article was affiliated with at the time of publication.

The best method to visualize the behavior of this indicator is through Figure 3, where we can see that it is divided into four continents, namely: Oceania, Europe, Asia and America. We can also observe that the vast majority of scientific articles on the subject are affiliated with Europe, in more detail, looking at Table 3, the Netherlands, the United Kingdom and Germany.

Nevertheless, we can state that the country with the most affiliations is the United States of America with about 27 affiliated articles.

Through the density map Figure 4 we can see that the authors with the most citations are: *Goulder L.H.*, with the article "The *effects of energy taxes on productivity and employment: The case of the Netherlands",* totaling 506 citations from a sample of 2303, *Bovenberg A.L.* and *van der Ploeg F.* and the article "*Environmental policy, public finance and the labour market in a second-best world"* with 152 citations, *Bosquet B.* with 128 citations in his article "*Marginal tax reform, externalities and income distribution*" and, finally, *Barket T. and Ekins P.* with 103 citations in "*Environmental Taxation Benefits - Theoretical Analysis [Teoretická analýza přínosů environmentálních daní]".*

Figure 3. Distribution of affiliated authors by continent

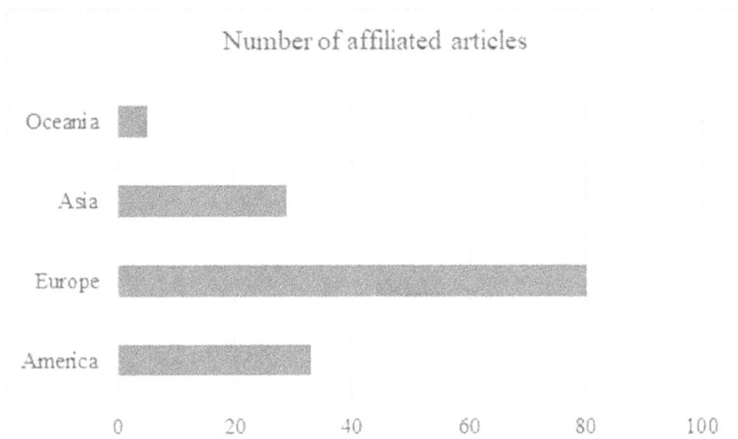

Table 3. Distribution of affiliated authors by country/territory

Continent/Country/Territory	Number of affiliated authors
America	**33**
United States of America	27
Canada	4
Chile	2
Europe	**80**
Holland	14
United Kingdom	13
Germany	12
Spain	8
France	6
Italy	5
Norway	5
Belgium	4
Sweden	3
Finland	2
Czech Republic	2
Denmark	1
Portugal	1
Switzerland	1
Austria	1
Romania	1
Turkey	1
Asia	**29**
Japan	7
China	5
Taiwan	5
Thailand	5
India	2
Malaysia	1
Hong Kong	1
Kuwait	1
Singapore	1
Bangladesh	1
Oceania	**5**
Australia	5
Total	**147**

Figure 4. Map of densities cited by scientific article

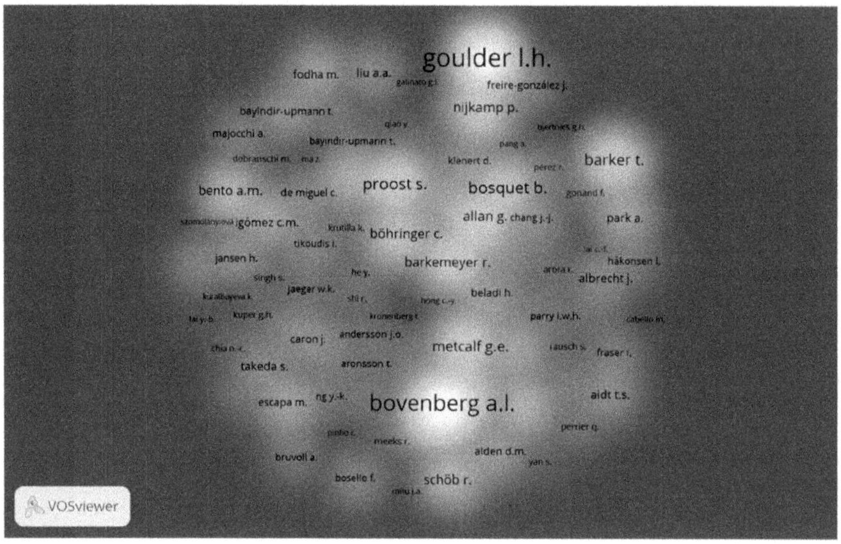

This analysis can also be done using the *H-Index* metric, created by the American physicist *JE Hirsch*, to measure the productivity and impact of researchers based on their most cited articles. (Costa et al., 2012)

Given that this metric can be applied to an individual publication or to a research group, institution, or country, we will use it to analyze the group of publications in this study. Through the information provided by *SCOPUS* and represented in Figure 5, we can affirm that this study has an H index of 24, that is, that of all the articles published (107 articles), 24 of these articles have at least 24 citations each.

Figure 5. H-Index of the articles under study

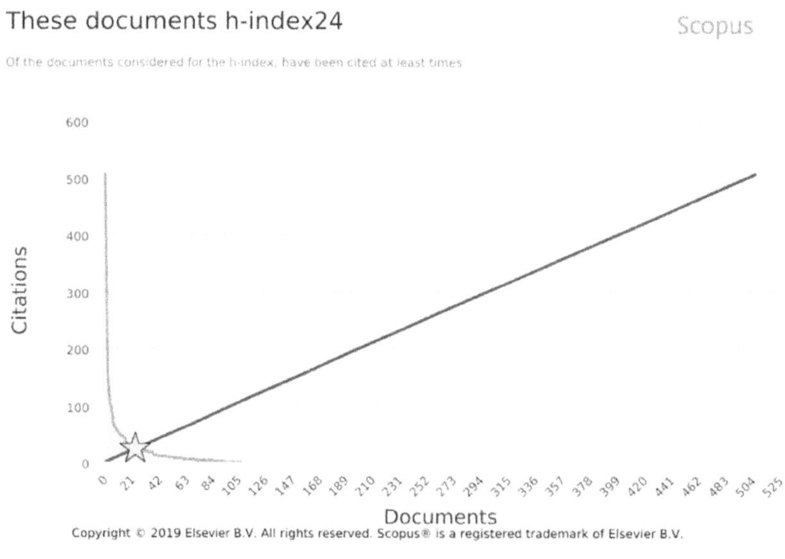

The analysis of Table 4 shows that of the 46 scientific journals analyzed, the one with the most publications is *Environmental and Resource Economics* with a total of 16 articles out of 107, corresponding to 15% of the analysis. We can also observe that about 55% of the journals have less than 6 articles published, of which 30% only have 1 article.

Table 4. Distribution of articles published by scientific journal

Magazine	Number of published articles	%
Environmental and Resource Economics	16	14.95%
Energy Economics	10	9.35%
Journal of Environmental Economics and Management	8	7.48%
Ecological Economics	6	5.61%
Resource and Energy Economics	6	5.61%
International Tax and Public Finance	5	4.67%
Journal of Cleaner Production	5	4.67%
Climate Change Economics	3	2.80%
Environmental Economics and Policy Studies	3	2.80%
International Journal of Energy Economics and Policy	3	2.80%
Journal of Policy Modeling	3	2.80%
Journal of Public Economics	3	2.80%
National Tax Journal	2	1.87%
Scandinavian Journal of Economics	2	1.87%
Agricultural Economics (Czech Republic)	1	0.93%
Amfiteatru Economic	1	0.93%
Asian Social Science	1	0.93%
B.E. Journal of Economic Analysis and Policy	1	0.93%
Basic Income Studies	1	0.93%
Computational Economics	1	0.93%
Economic Modelling	1	0.93%
Economic Record	1	0.93%
Economic Systems Research	1	0.93%
Economie Internationale	1	0.93%
Engineering Economics	1	0.93%
Environment, Development and Sustainability	1	0.93%
European Journal of Law and Economics	1	0.93%
European Journal of Political Economy	1	0.93%
Finance a Uver - Czech Journal of Economics and Finance	1	0.93%
German Economic Review	1	0.93%
IIMB Management Review	1	0.93%
Industry and Innovation	1	0.93%

continues on following page

Table 4. Continued

Magazine	Number of published articles	%
International Journal of Energy Economics and Policy,	1	0.93%
International Journal of Green Economics	1	0.93%
International Journal of Innovation and Sustainable Development	1	0.93%
International Journal of Production Research	1	0.93%
International Review of Law and Economics	1	0.93%
Journal of Agricultural and Resource Economics	1	0.93%
Journal of Economic Surveys	1	0.93%
Journal of Macroeconomics	1	0.93%
Journal of the Japanese and International Economies	1	0.93%
Journal of Urban Economics	1	0.93%
Land Economics	1	0.93%
Portuguese Economic Journal	1	0.93%
Singapore Economic Review	1	0.93%
Spanish Economic Review	1	0.93%
Total	**107**	**100.00%**

Finally, it is interesting to analyze the thematic associations of the articles under study, for this we can make use of the citations, *keywords* used and the relations between countries.

In this way, helping us in the bibliographic map created in *VOSviewer* we can represent each theme by a circle that in turn may or may not present a relationship with another theme. The frequency and importance of each *keyword* is related to the size of each circle and the distance between them allows to evaluate the level of association.

Figure 6 shows that the *keywords* used in the articles are related to different themes, represented by the different colors. The keyword *Double Dividend*, used as a filter in this study, is closely related to Environmental Tax *Reform, Environmental Policy, Economic Growth, Employment, Optimal Taxation* and *Carbon Tax*. The other keyword of this study, *Green Tax*, is closely related to *Double Dividends, Sustainable Development, Environmental Fiscal Reform* and *Market-Based Instruments*.

This analysis technique uses the words inserted in the documents to establish and build relationships of conceptual structure and as such can be applied to titles, keywords, abstracts and even to the body of the text.

Finally, in the present work we made the analysis of the references co-cited in the studies that constituted a way of analyzing the connection between two documents based on the frequency with which they are cited together, an additional advantage of data collection in the *SCOPUS* database. (Zupic & Čater, 2015)

As explained earlier, each circle represents an article, the size of this circle indicates the "activity" of the publication of the article. The distance between these indicates the degree of reference co-citations, and the groups of articles are represented by colors.

Figure 6. Keywords used by the authors in scientific articles

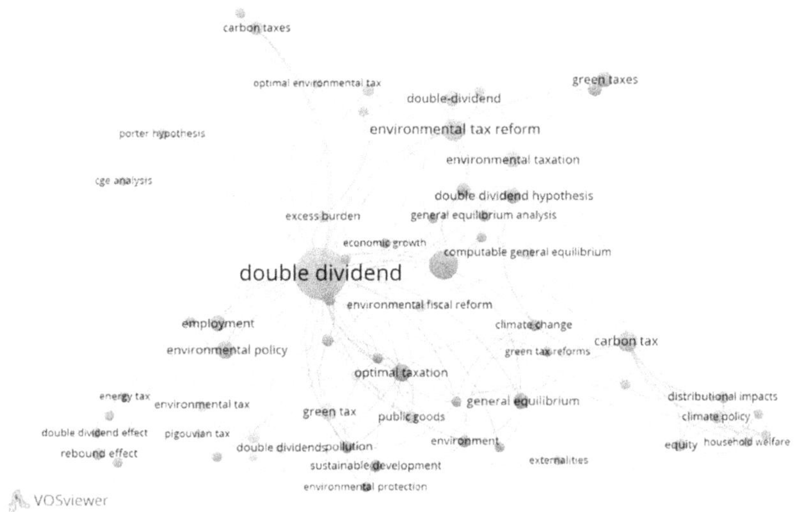

With this it was found that in the total of 3770 citations of references (Figure 7) the most co-cited articles in the bibliographic references are that of *Bovenberg and Mooij* together with *Pearce, Terkla, Parry* and *Tullock*.

Figure 7. Reference co-citations of the articles under study

CONCLUSION

Our study analyzes and reports the evolution of scientific development on the subject of *Green tax* and *double dividend* between 1994 and 2019 in terms of publications available in the *SCOPUS* database. After defining some criteria, except temporal, we obtained 107 articles that allow us to have a comprehensive view of the current state of the literature.

This analysis offers an evolution of the theme providing information on which publications and key authors, as well as the most influential journals, topics, and areas of study most relevant, among others. Thus, serving as a guide for future researchers interested in this theme.

The approach chosen by us to carry out the study also allows us to contribute to the methodologies of bibliometric analysis demonstrating the use of *VOSviewer* as an advanced instrument of these relationship analyses. The specialty of this is the graphical representation of large bibliometric maps so as to provide an easier way to visualize and interpret. (van Eck & Waltman, 2010)

Our results demonstrate that the publication pattern has evolved during the period of analysis (1994-2019), and the theme has recently become more popular and more important. Proof of this is the increase in the number of publications and citations in recent years.

These publications, of articles, are mostly articles affiliated with European countries. However, when we do this analysis by country, we conclude that the country/territory with the most affiliations is the United States of America, which has co-authorship with 14 other countries.

In turn, the most influential authors are *Goulder L.H* and *Bovenberg A.L.* and *van der Ploeg F*.com 506 and 152 citations respectively. *Bovenberg A.L.* He is also the most productive author, having 5 articles published in the sample. Still regarding the type of authorship, we conclude that the tendency is to resort to collaborations in the creation of articles.

Regarding the sources of publication, we identified that the most influential scientific journals are *Environmental and resources* economics and *Energy economics,* where most of the articles are published. However, the *Journal of Cleaner Production* is the one with the highest *CiteScore*, the average number of citations in the last 3 years, being the most influential.

Through the analysis of the *keywords* we could conclude that the focus areas of *green tax* is often related to *Double Dividends, Sustainable Development*, Environmental Fiscal Reform and *Market-Based Instruments*, while the focus areas of *double dividend* are related to *Environmental Tax Reform, Environmental Policy, Economic Growth, Employment, Optimal Taxation* and *Carbon Tax* demonstrating that these are the most influential topics of the subject.

However, because it is a bibliometric, this study presents some limitations derived from its nature, because the projects of a bibliographic nature hardly reach perfection because there is a possibility that relevant authors were not included in the theme. The fact that only the *SCOPUS* database was used, and limits were imposed on the area of study ended up creating a selection of authors by this profile.

It would be useful for the scientific community to add, in future studies, publications from other databases, such as the *Web of Science* or *Google Scholar*, as well as to add to the study the keyword *Environmental Tax Reform* to have a larger sample, and therefore, more representative of the reality of the Green *Tax* and *Double Dividend theme.*

REFERENCES

Baranzini, A., Goldemberg, J., & Speck, S. (2000). A future for carbon taxes. *Ecological Economics*, *32*(3), 395–412. doi:10.1016/S0921-8009(99)00122-6

Borges, F. G. L., & Merlin, L. V. D. C. T. (2018). The (IN) viability of green tax reform in Brazil. *Paths of Law*, *15*(33), 347–375. doi:10.18623/rvd.v15i33.1243

Borrego, A. (2016). Diploma of green taxation - Finally an environmental taxation integrated in Portugal? *Journal of Social and Environmental Management*, *19*. https://www.researchgate.net/publication/312128897%0ADiploma

Bosello, F., & Carraro, C. (2001). Recycling energy taxes: Impacts on a disaggregated labour market. *Energy Economics*, *23*(5), 569–594. doi:10.1016/S0140-9883(00)00083-9

Bovenberg, A. Lans, Mooij, R. A. (1994). Environmental Levies and Distortionary Taxation. *American Economic Review*, *84*(4), 1085–1089. http://search.ebscohost.com/login.aspx?direct=true&db=bth&AN=9410250177&site=eds-live

Bovenberg, A. L., & van der Ploeg, F. (1994). Environmental policy, public finance and the labour market in a second-best world. *Journal of Public Economics*, *55*(3), 349–390. doi:10.1016/0047-2727(93)01398-T

Bovenberg, A. L., & Van der Ploeg, F. (1998). Consequences of environmental tax reform for unemployment and welfare. *Environmental and Resource Economics*, *12*(2), 137–150. doi:10.1023/A:1016040327622

Chaturvedi, A., Saluja, M. S., Banerjee, A., & Arora, R. (2014). Environmental fiscal reforms. *IIMB Management Review*, *26*(3), 193–205. doi:10.1016/j.iimb.2014.06.004

Chiroleu-Assouline, M., & Fodha, M. (2009). Double dividend and distribution of welfare: Advanced results and empirical considerations. *Economie Internationale*, *120*(4), 91–108.

Clinch, J. P., Dunne, L., & Dresner, S. (2006). Environmental and wider implications of political impediments to environmental tax reform. *Energy Policy*, *34*(8), 960–970. doi:10.1016/j.enpol.2004.08.048

Costa, T., Fernández-Llimós, S., & Lopes, M. (2012). *Bibliometrics and the evaluation of scientific production: indicators and tools,* 19–37 NIH.

Ekins, P., & Barker, D. D. T. (2001). Carbon taxes and carbon emissions trading. *Journal of Economic Surveys*, *15*(3), 325–376. http://www.scopus.com/scopus/openurl/link.url?ctx_ver=Z39.88-2004&ctx_enc=info:ofi/enc:UTF-8&svc_val_fmt=info:ofi/fmt:kev:mtx:sch_svc&svc.citedby=yes&rft_id=info:eid/2-s2.0-0034862828&rfr_id=http://search.ebscohost.com&rfr_dat=partnerID:NnvIuKwx&rfr_dat=m. doi:10.1111/1467-6419.00142

European Commission. (2010). *EUROPE 2020 Strategy for smart, sustainable and inclusive growth* (Vol. 0278). ReVision.

Freire-González, J., & Ho, M. S. (2018). Environmental fiscal reform and the double dividend: Evidence from a dynamic general equilibrium model. *Sustainability (Basel)*, *10*(2), 501. doi:10.3390u10020501

Fullerton, D., & Metcalf, G. (1998). Environmental taxes and the doule-dividend hypothesis: Did you really expect something for nothing? *Chicago-Kent Law Review*, *73*(189), 221–256.

Goulder, L. H. (1995). Environmental taxation and the double dividend: A reader's guide. *International Tax and Public Finance*, *2*(2), 157–183. doi:10.1007/BF00877495

Goulder, L. H. (2013). Climate change policy's interactions with the tax system. *Energy Economics*, *40*, S3–S11. doi:10.1016/j.eneco.2013.09.017

Gouveia, E., & Palma, C. (2018). *Impacts of Environmental Taxation on the Portuguese Economy*. NiH.

Hawkins, R. (2000). The use of economic instruments and green taxes to complement an environmental regulatory regime. *Water, Air, and Soil Pollution*, *123*(1–4), 379–394. doi:10.1023/A:1005294917875

Hjøllund, L., & Svendsen, G. T. (2001). Why green taxation? *Energy & Environment*, *12*(1), 29–38. doi:10.1260/0958305011500571

Jackson, T. (2010). The employment and productivity effects of environmental taxation: Additional dividends or added distractions? *Journal of Environmental Planning and Management*, *43*(3), 389–406. doi:10.1080/09640560050010419

Klok, J., Larsen, A., Hansen, K., & Dahl, A. (2006). Ecological Tax Reform in Denmark: History and social acceptability. *Energy Policy*, *34*(8), 905–916. doi:10.1016/j.enpol.2004.08.044

Koskela, E., Schöb, R., & Sinn, D. H. W. (1998). Pollution, Factor Taxation and Unemployment. *International Tax and Public Finance*, *5*(3), 379–396. doi:10.1023/A:1008642512728

Kuralbayeva, K. (2019). Environmental Taxation, Employment and Public Spending in Developing Countries. *Environmental and Resource Economics*, *72*(4), 877–912. doi:10.100710640-018-0230-3

Lai, Y. B. (2018). The Feasibility of the Double-Dividend Hypothesis in a Democratic Economy. *The Scandinavian Journal of Economics*, *120*(1), 211–241. doi:10.1111joe.12213

Liu, A. A. (2013). Tax evasion and optimal environmental taxes. *Journal of Environmental Economics and Management*, *66*(3), 656–670. doi:10.1016/j.jeem.2013.06.004

Maxim, M. R., Zander, K. K., & Patuelli, R. (2019). Green tax reform and employment double dividend in European and non-European countries: A meta-regression assessment. *International Journal of Energy Economics and Policy*, *9*(4), 342–355. doi:10.32479/ijeep.7776

Metcalf, G. (2014). Using the tax system to address competition issues with carbon tax. *National Tax Journal*, *67*(December), 779–806. doi:10.17310/ntj.2014.4.02

Mongeon, P., & Paul-Hus, A. (2016). The journal coverage of Web of Science and Scopus: A comparative analysis. *Scientometrics*, *106*(1), 213–228. doi:10.100711192-015-1765-5

Mota, D., & Palma, C. (2015). *Green tax reforms: an application to the Portuguese economy*. ISCTE.

OECD. (2018). *Financing climate objectives in cities and regions to deliver sustainable and inclusive growth*. OECD.

Pearce, D. (1991). THE ROLE OF CARBON TAXES IN ADJUSTING TO GLOBAL WARMING. *Economic Journal (London)*, *101*(407), 938–948. http://search.ebscohost.com/login.aspx?direct=true&db=bth&AN=4531951&site=eds-live. doi:10.2307/2233865

Peng, J. T., Wang, Y., Zhang, X., He, Y., Taketani, M., Shi, R., & Zhu, X. D. (2019). Economic and welfare influences of an energy excise tax in Jiangsu province of China: A computable general equilibrium approach. *Journal of Cleaner Production*, *211*, 1403–1411. doi:10.1016/j.jclepro.2018.11.267

Pereira, A. M., & Pereira, R. M. (2014). Environmental Fiscal Reform and Fiscal Consolidation: The Quest for the Third Dividend in Portugal. *Public Finance Review*, *42*(2), 222–253. doi:10.1177/1091142113485803

Sandmo, A. (2011). Atmospheric externalities and environmental taxation. *Energy Economics*, *33*(SUPPL. 1), S4–S12. doi:10.1016/j.eneco.2011.07.021

Shi, H., Qiao, Y., Shao, X., & Wang, P. (2019). The effect of pollutant charges on economic and environmental performances: Evidence from Shandong Province in China. *Journal of Cleaner Production*, *232*, 250–256. doi:10.1016/j.jclepro.2019.05.272

Silva, P., & Chaves, C. (2015). *The Implementation of Environmental Tax Reform in Portugal: How Does it Compare with Other European Experiences?* University of Porto.

Terkla, D. (1984). The efficiency value of effluent tax revenues. *Journal of Environmental Economics and Management*, *11*(2), 107–123. doi:10.1016/0095-0696(84)90010-X

UNFCCC. THE PARIS AGREEMENT, L.9/ Rev.1 § (2015). https://doi.org/FCCC/CP/2015/L.9/Rev.1

van Eck, N. J., & Waltman, L. (2010). Software survey: VOSviewer, a computer program for bibliometric mapping. *Scientometrics*, *84*(2), 523–538. doi:10.100711192-009-0146-3 PMID:20585380

Zupic, I., & Čater, T. (2015). Bibliometric Methods in Management and Organization. *Organizational Research Methods*, *18*(3), 429–472. doi:10.1177/1094428114562629

KEY TERMS AND DEFINITIONS

Double Dividend: It refers to the notion that environmental taxes can both reduce pollution (the first dividend) and reduce the overall economic costs associated with the tax system by using the revenue generated to displace other more distortionary taxes that slow economic growth at the same time (the second dividend) (European Environment Agency).

Environmental Tax: Environmentally related taxes are an important instrument for governments to shape relative prices of goods and services. The characteristics of such taxes included in the database (e.g. revenue, tax base, tax rates, exemptions, etc.) are used to construct the environmentally related tax revenues with a breakdown by environmental domain: energy products (including vehicle fuels); motor vehicles and transport services; measured or estimated emissions to air and water, ozone depleting substances, certain non-point sources of water pollution, waste management and noise, as well as management of water, land, soil, forests, biodiversity, wildlife and fish stocks (OECD).

Environmental Tax Reform: Environmental tax reform (ETR) is a reform of the national tax system where there is a shift of the burden of taxation from conventional taxes, for example on labor, to environmentally damaging activities, such as resource use or pollution. The burden of taxes should fall more on 'bads' (such as pollution or natural resource use) than 'goods' (like employment) so that appropriate signals are given to consumers and producers and the tax burdens across the economy are better distributed from a sustainable development perspective (European Environment Agency).

Green Tax: A green tax is defined as a tax whose taxable base is a physical unit (or a substitute for it) that has a specific proven negative impact on the environment. To do this, it is necessary that the tax base of the environmental tax is related to the problem under consideration (the environmental damage of certain emissions or the use of products intricately linked to said damage) and that the structure of tax rates contributes to the environmental damage or to the achievement of pre-set environmental objectives (CIAT – Inter-American Center of Tax Administrations).

Triple Dividend: The triple dividend of resilience (TDR) is an approach that considers avoided losses (first dividend), induced economic or development benefits (second dividend), and additional social and environmental benefits (third dividend) of adaptation actions. The second and third dividends are especially important since they accrue regardless of whether the actual climate risk materializes (World Resources Institute).

ENDNOTES

1. Negative environmental externalities are seen as reductions in the availability of public goods, such as clean air and water, biological diversity, and global climate. Externalities related to economic and business activities lead to a reduction in the supply of these goods, while the reduction of externalities leads to the restoration of goods. (Sandmo, 2011)
2. They are based on taxing the factors of production: Capital and Labor (Pearce, 1991)

Chapter 7
Green Taxation for SMEs' Digital Transformation of Accounting

Kátia Lemos
https://orcid.org/0000-0002-3961-5964
Research Centre on Accounting and Taxation, Institute Polytechnic of Cávado and Ave, Portugal

Ana Dinis
https://orcid.org/0000-0003-3945-5726
Research Centre on Accounting and Taxation, Institute Polytechnic of Cávado and Ave, Portugal

Sara Serra
https://orcid.org/0000-0003-3107-1752
Research Centre on Accounting and Taxation, Institute Polytechnic of Cávado and Ave, Portugal

ABSTRACT

The dematerialization of documents and the resulting contribution to greener and more sustainable practices should be on the agenda of businesses. Companies, especially SMEs, need to be particularly aware of the relevance of digitalization for environmental sustainability, and the role of governments in promoting the digitalization of companies is very important for its success. The present research focuses on the study of tax incentives for the digital transformation of accounting granted to Portuguese micro and SMEs, specifically for the implementation of SAF-T (PT) for accounting, the QR Code®, and the ATCUD. An exploratory data analysis methodology was adopted, considering all Portuguese micro and SMEs that participated in the tax incentive for digital transformation in 2020 and 2021. The authors conclude that the tax incentives studied have not been a practice adopted by Portuguese micro and SMEs, which may reveal a lack of disclosure by the government and knowledge by companies.

DOI: 10.4018/978-1-6684-8592-7.ch007

INTRODUCTION

Countries around the world are working to achieve the goals of the 2030 Agenda to make their economies and growth more sustainable (van Vuuren et al., 2022). Digital transformation is a process that aims to improve organizations, triggering significant changes through the combination of information, computing, communication, and connectivity technologies, and enabling new forms of collaboration between different actors. This represents a huge potential for innovation and performance in organizations, affecting individuals, organizations, and society (Buck et al., 2023). The digital transformation of organizations is therefore inevitable, but it requires organization and planning to be a successful transition (Trenerry et al., 2021). Countries must, therefore, be prepared to face the challenges and changes inherent in a global transition (Elsevier, 2022) to ensure that it results in greater equality and inclusion of citizens, a strengthening of the democratic pillars, an increase in the competitiveness of the economy and the attraction of domestic and foreign investment, as well as the creation of value for the business fabric (European Commission, 2022).

In this chapter, the focus of the study will be on digital transformation, in its concept of the progressive adoption of new information and communication technologies. It should not be forgotten that digitalization plays an important role in sustainable development strategies, allowing companies, especially small and medium-sized enterprises (SMEs) that need to be brought to attention, to better connect with their audiences and achieve better results. In this sense, the authors' intention in this chapter was to study the use of environmental tax incentives for digitalization by Portuguese SMEs.

This chapter aims to contribute to the tax literature by studying the sustainability of SMEs and how governments can encourage this behavior through tax incentives. The main objectives of this chapter are: (i) present digital technologies as a key contributor to the achievement of the Sustainable Development Goals (SDGs); (ii) discuss green taxation as one of the pillars of environmental sustainability; (iii) present the specific case of tax incentives as a fiscal tool for digital accounting transformation; and (iv) present a critical view of the Portuguese tax system through an empirical study based on a sample of Portuguese micro and SMEs that have used the tax incentives for the implementation of the Standard Audit File for Tax Purposes - Portuguese Version (SAF-T (PT)) for accounting, the two-dimensional bar code (QR Code®), and the unique document code of the Portuguese Tax Agency (ATCUD).

This study is particularly relevant for taxpayers, tax authorities, and tax policymakers, from the perspective that the digital era contributes to environmental sustainability. In addition, companies, especially SMEs, need to be particularly aware of the dematerialization of documents and the consequent contribution to more ecological and sustainable methods, not forgetting the positive environmental impact resulting from the reduction of paper waste, with a boost to the reputation of the profession through a more sustainable ecological footprint.

Regarding the research methodology, two research questions were outlined:

- What are the characteristics of Portuguese micro and SMEs that have taken advantage of the tax incentives of digital accounting?
- Does digital transformation, as measured by access to tax incentives, affect the economic and financial performance of Portuguese micro and SMEs?

The chapter is organized as follows: first, a perspective on accounting digital transformation in the context of environmental sustainability is introduced; second, the role of tax incentives in accounting digitalization is presented; third, the empirical study follows, presenting the research methodology, the results obtained and their discussion; finally, implications for future research are presented and a conclusion is drawn.

ACCOUNTING DIGITALIZATION IN THE CONTEXT OF ENVIRONMENTAL SUSTAINABILITY

The digital transformation began in the mid-20th century with electronic data processing, personal computers, communication technologies, and the Internet (Brunetti et al., 2020). At the start of the 1990s, entities began investing more and more in digitalization (Coman et al., 2022), and society is now increasingly digitalized and connected with computers and algorithms (Dufva & Dufva, 2019).

Accounting has yet to be immune to this phenomenon. Digitalization brings many challenges to accounting, such as the use of big data, cloud computing, continuous accounting, artificial intelligence, and blockchain technology. This digitalization will result in significant changes in how accountants work and think, allowing them to use digital and automated solutions for routine tasks. Digitalization will lead to changes in accounting, including generating daily reports, preparing business plans and strategies, implementing online accounting, and outsourcing accounting to distant countries (Gulin et al., 2019).

According to these authors, the routine and repetitive tasks of accountants will soon be automated, and access to information will be rapid and easy. As a result, accountants will perform their services faster and more economically, and they will have more time for their clients, mainly to discuss their current situation and future needs. As a result, the accounting profession may be transformed into that of a consultant, analyst, or accounting engineer.

Digitalization facilitates communication between economic agents, provides real-time access to information (Coman et al., 2022), and enables the reporting and processing of large volumes of transactions, generating the necessary data for analysis (Lutfi et al., 2022).

Although digitalization is considered to contribute to the improvement of accounting and increase the value of its services, it requires investment in digital technologies, often only within reach of large companies due to their high cost. Furthermore, using these technologies also requires the development of new skills and knowledge, especially in engineering, requiring the reformulation of curricula in many educational institutions (Brunetti et al., 2020; Gulin et al., 2019).

Despite its potential, Lutfi et al. (2022) consider that small and medium-sized enterprises (SMEs) need to be faster to digitalize accounting, which is concerning given that its practical use is critical to the business of all industries. Low et al. (2022) also state that the digitalization of accounting in SMEs has yet to reach its full spectrum, although these entities are increasingly adopting digital technologies. Nevertheless, the authors believe that accounting digitalization is crucial to the success of the digital economy. In the same line of thought, Lutfi et al. (2022) argue that governments should take measures to improve and increase the effectiveness of digitalization implementation among SMEs, such as creating policies and laws related to adopting advanced information technologies and strengthening financial support. In addition, professional and political organizations are also encouraged to establish rules and principles regarding adopting accounting digitalization practices, such as accounting rules and reports and data security measures (Pham & Vu, 2022).

Accounting digitalization, in addition to the benefits it brings to organizations that have implemented it, such as reduced labor costs, time savings, provision of useful information, and improvement of professional qualifications (Lutfi et al., 2022), can have implications for environmental sustainability. In fact, according to Acciarini et al. (2022), digitalization and sustainability are interrelated and need to be considered in business models. In this line of thought, Feroz et al. (2021) state that digital technologies offer organizations opportunities to develop new business models focusing on the environment or to incorporate environmentally sustainable practices.

Similarly, Bednarčíková and Repiská (2021) state that digitalization, digital transformation, and technologies are essential for the environment and the economy. The authors even use the term "green digital transformation" to refer to the implementation of digitalization and the integration of green technologies into processes, activities, products, and business models when the company becomes environmentally sustainable. Green technology represents the integration and subsequent use of digital technologies to innovate business processes and is used in the field of sustainable environmental development of the organization.

Corroborating this idea, Truong (2022) also considers that digital transformation has excellent potential to achieve environmental sustainability objectives. According to the author, the advancement of digital technologies, such as artificial intelligence, big data, and blockchain, can help alleviate adverse environmental effects.

Aware of this reality, in 2019, the Von der Leyen Commission announced in its political agenda the European Green Deal, which aimed to combine sustainability with economic growth. This document is based on the belief that new technologies can contribute to sustainability and environmental change and is considered the central promise to solve ecological problems (Čavoški, 2022).

At the 2021 UN COP26 conference on climate change, the contribution of digital technology and information systems to climate change was also discussed, making it clear that responsible digitalization is crucial to mitigate the environmental impact of technology (Dwivedi et al., 2022).

Mondejar et al. (2021) also refer to intelligent technologies to achieve the United Nations Sustainable Development Goals (SDGs) challenges and build an equitable, environmentally sustainable, and healthy society. Intelligent technologies can benefit sustainable food production, access to clean and safe drinking water, the generation and use of green energy, catalyze the transition to sustainable manufacturing practices, improve citizen health by providing digital access to care, especially for underserved communities, and contribute to addressing the planet's threatened biodiversity and climate change challenges. In addition, artificial intelligence technologies can convert passive data into actionable knowledge to improve operation and lead decision-making in water treatment systems, resulting in cost savings, reduced energy consumption, and water quality assurance. Digitalization can also facilitate equality in access to water through the integration of internet-enabled tools and emerging decentralized intelligent technologies in real time.

Sharing the same opinion, David et al. (2022) state that Fourth Industrial Revolution technologies can have an impact on the environment, particularly in the use of water, energy, and food, ensuring the popularization of cleaner production and sustainable resource security, especially Internet of Things (IoT) and big data.

ACCOUNTING DIGITALIZATION TAX INCENTIVES

Sustainability Through Digital Accounting Transformation: The Role of Tax Incentives

As previously presented, the advantages of digitalization have been recognized, particularly in the tax systems by the different stakeholders, namely taxpayers, tax policymakers, legislators, tax authorities, and accountants (Yamen et al., 2022). Especially for accountants, digitalization has brought about very significant changes as the simplification of procedures and the elimination of physical document flows, such as paper printouts and filing (Purcinelli et al., 2020).

Digitalization has affected all accounting and taxation procedures, a transformation that began some time ago with the use of accounting software and the submission of digital tax returns (Coman et al., 2022). This has also forced public services to digitalize themselves (Edelmann & Virkar, 2023). In addition, tax authorities in several countries have begun to provide online platforms for reporting information (Tyutyuryukov & Guseva, 2021). This online channel has become preferred over traditional paper-based reporting (Bassey et al., 2022). Not to mention the tendency toward eliminating paper invoices to increase taxpayer compliance (Bellon et al., 2022). The connection between these measures and the sustainability of the environment is beyond dispute. Not least because, as Zhou et al. (2022) conclude, tax stickiness persists meaningfully in companies, and digital transformation can significantly reduce it.

The process of digitalization of accounting and tax processes was accelerated by the COVID-19 pandemic (George & Schillebeeckx, 2022). On the one hand, delayed by normal resistance to innovation (Diller et al., 2020), this long-awaited revolutionary change in line with the public policies against climate change, was positive, but on the other hand, more situations related to the unregistered economy and tax fraud and evasion emerged. In this sense, the tax authorities had to be more cautious (Ruggiero, 2022), consistent with OECD Base Erosion and Profit Shifting (BEPS) Action 1: "Tax Challenges Arising From Digitalization" (OECD, 2020).

While the 2030 Agenda for Sustainable Development and the SDGs were not yet in place, several countries were also already planning to implement the Standard Audit File for Tax Purposes (SAF-T), which is designed to give auditors access to data in an easily readable format for system data checks and intended for use by companies and their auditors, from multinational corporations (MNCs) to small and medium-sized enterprises (SMEs), and by tax auditors (OECD, 2010).

In this sense, the role of governments in promoting the successful digitalization of businesses, especially SMEs, is essential, as they are particularly reluctant to do so (Rupeika-Apoga et al., 2022). In fact, according to Feng et al. (2022), the government should improve the level of green innovation of enterprises by promoting the construction of digital infrastructure and through tax incentives, which will promote the green development of the regional economy, thus contributing to the achievement of sustainable development goals.

This is where taxation comes in. In a dual approach that seeks to control fraud and tax evasion while promoting environmental sustainability, emerges green taxation on digitalization, through tax incentives.

Tax incentives are increasingly an advantage for economic activity, providing a financial or social benefit and a source of value for the company. They have three characteristics: i) they are an exception to the general tax rules; ii) they pursue relevant social and economic objectives; and iii) they confer an advantage on their recipients. In the context of this study, they are an important instrument of fiscal measurement.

In their study Li et al. (2023), which uses tax incentives as a measure of government support, the authors conclude that both firm digitalization and the level of regional digital industry innovation can promote firm innovation. Now it turns out that employee-driven thinking has emerged as a new approach to explaining innovation. This is why it's important to assess the relationship between companies and their employees. Especially as they become more innovative through digitalization (Opland et al., 2022).

The study of Rapuluchukwu et al. (2017) shows that tax incentives should be targeted to reward firms' outcomes, and shows that firms' productivity depends to a large extent on intra-firm decisions about how to use tax incentives. Moreover, when linked to digital transformation, and as concluded by Wang et al. (2023), there is a positive and robust effect on the tax incentives of digitalization. However, the authors conclude that government subsidies are more efficient in the digitalization of firms because companies´ access to funds before the digital transition is more effective. And it also depends on the different regions of a country. Nevertheless, as Mpofu (2022) argues, tax incentives encourage business investment. In addition, as Chien et al. (2021) point out, green taxation is an effective solution to environmental problems in the country, and green growth has a positive relationship with the environmental condition of the country.

Portugal has not been indifferent to what has been happening in the international context and has followed this trend, establishing tax incentives for digitalization. Let's take a closer look at the digitalization support measures for the digital transformation of accounting and taxation in micro and SMEs.

Exceptional Fiscal Support for the Implementation of SAF-T (PT) for Accounting, QR Code®, and ATCUD in Portugal

Decree-Law 238/2006, of December 20th, embodies a new initiative in Portugal for the dematerialization of acts and procedures by new information and communication technologies and represents another step forward in the challenge of administrative modernization, in line with the guidelines and commitments made by the Portuguese Government.

In this context, adjustments have been made to the Portuguese Tax Codes to improve the technical and legal wording of some legal provisions and, above all, to eliminate and simplify a wide range of ancillary obligations imposed on taxpayers at the level of different taxes.

Nowadays, companies increasingly use electronic data processing systems to record their assets, for invoicing purposes. And these records are subject to verification by the control services within the framework of their powers to control the tax situation of taxpayers.

To facilitate this task, given the diversity of systems, the creation of the standardized file (SAF-T) has been advocated within the OECD, to allow an easy export, at any time, of a predefined set of accounting records, in a readable and common format, independent of the program used, without affecting the internal structure of the program's database or its functionality.

The adoption of this model was intended to provide companies with a tool that would satisfy the requirements for obtaining information from the control services and facilitate its processing, avoiding the need for auditors to specialize in the various systems, simplifying procedures, and encouraging the use of new technologies.

In this sense, the adoption in Portugal of this standardized data export model has become mandatory in a gradual manner, starting with the invoicing and accounting applications. This is evidenced by the ensuing legislation.

In the first moment, Ministerial Order nr. 321-A/2007, of March 26th[1], established the tax audit model file, the so-called Standard Audit File for Tax Purposes - Portuguese Version (SAF-T(PT)), provided for in the national legislation, which covers the information contained in the invoicing and accounting systems.

Over time, the Portuguese government has continued to prioritize the digitalization of procedures. Thus, in 2019, with Decree-Law No. 28/2019 of February 15, the conditions were created for the so-called "paperless invoice", providing the possibility of abandoning the printing of invoices.

Then, conditions were created for the dematerialization of documents, encouraging the adoption of an electronic invoicing system and electronic filing of documents, allowing companies to reduce the cost of tax compliance, stimulating the development and use by companies of new technological tools, incorporating a philosophy of innovation and (de)bureaucratization. To this end, a substantial reform of the rules applicable to the filing of books, records, databases, and accounting support documents has been introduced.

On the other hand, to fight the informal economy, fraud, and tax evasion, mechanisms have been provided to strengthen the control of the operations carried out by taxpayers, by identifying the invoicing programs sold, the establishments where invoicing terminals are installed, and the obligation for invoices issued to contain a two-dimensional bar code (QR Code®) and a unique document code of the Portuguese Tax Agency (ATCUD).

In addition, also in 2019, and by Ordinance No. 31/2019 of January 24, taxpayers are now also required to validate the SAF-T (PT) file on accounting and send it to the Tax and Customs Agency (AT) – this obligation is currently postponed until 2025.

With the State Budget for 2021, with Law 75-B/2020, of December 31st, and within the framework of an economic policy that, in a framework of strong disruptions caused by the global health crisis caused by COVID-19, sought to mitigate the negative impact of the economic and social level and to relaunch the business economic growth in the medium term. Therefore, it decided to grant extraordinary support for the implementation of the SAF-T (PT) file submission in terms of accounting and unique document code, for expenses incurred as of January 1, 2020, and until the end of 2021.

This support could be granted to companies (is only applicable to micro, small, and medium enterprises) that meet the following conditions:

- The implementation of the SAF-T (PT) accounting file has been completed by the end of the 2021 tax period.
- The QR Code® and the ATCUD code must appear on all invoices and other tax-relevant documents of taxpayers as of January 1, 2022.

The government reinforced these extraordinary measures with the State Budget for 2022, with Law No. 12/2022, of June 27, applicable to expenses incurred as of January 1, 2022.

This support could be granted to micro, small, and medium companies that meet the following conditions:

- Implementation of SAF-T (PT) filing concerning accounting completed by the end of the 2023 tax period.
- Implementation of ATCUD to appear on all invoices and other tax-relevant documents of taxpayers as of January 1, 2023.

Note that the requirement by 2021 for companies to include the QR Code® and the unique document code (ATCUD) on invoices has been suspended with this extraordinary support.

The tax incentive granted consisted of:

(a) for expenses related to 2020 and 2021, and for purposes of determining the taxable income of IRC taxpayers and IRS taxpayers with organized accounting
   i) accountable expenses with the acquisition of goods and services directly necessary for the implementation of SAF-T related to accounting can be considered at 120%.
   ii) can be considered in 120% of the accounted expenses of the period regarding the expenses with the implementation of the QR Code® and ATCUD.
   iii) 140% of the accounted expenses, provided that the taxpayer includes the QR Code® in all its invoices and other tax-relevant documents by the end of the first quarter of 2021.
   iv) 130% of the expenses recorded in the period, provided that the taxpayer includes the QR Code® in all its invoices and other tax-relevant documents by the end of the 1st quarter of 2021.

(b) for expenses as of 2022, and to ensure the prior submission of the SAF-T (PT) file related to accounting within the simplified business information (IES) and the unique document code (ATCUD), and for purposes of determining the taxable income of IRC taxpayers and IRS taxpayers with organized accounting
   i) 120% of the expenses recognized in the period relating to the costs of implementing the submission of the SAF-T for accounting purposes may be considered.
   ii) 120% of the expenses recorded in the period relating to the expenses for the implementation of the ATCUD may be considered.

It should be noted that in cases where the expenses relate to assets subject to depreciation, the tax incentives referred to in the previous paragraphs apply to the expenses recorded for depreciation during the useful life of the asset.

EMPIRICAL EVIDENCE IN PORTUGAL

This study aims to analyze how Portuguese micro and SMEs have taken advantage of the tax incentive under study and what developments have taken place between 2020 and 2021.

In addition, it aims to characterize these companies and identify factors associated with the companies that may be related to the tax incentive for digital transformation.

Research Methodology

An exploratory data analysis methodology was adopted, considering all Portuguese micro and SMEs that participated in the tax incentive for digital transformation in 2020 and 2021. The data were collected from the Portuguese Tax Administration portal, considering aggregated values by type of tax incentive in corporate income tax. Data processing was carried out using SPSS software, through a descriptive analysis of the results and a Pearson correlation.

Sample Characterization

According to the Portuguese database PORDATA, the number of micro and SMEs in Portugal was 457,609. However, of these companies, only 373 took advantage of this tax incentive (see Table 1). Moreover, it appears that in 2020 (the first year of implementation of the tax incentive), only 33 companies took advantage of it. In 2021, this number increases to 342 [2 companies benefited in the two years concerned], but it is still an almost insignificant number, considering the population for which the incentive is intended.

This result shows that the number of companies aware of this tax incentive was higher in 2021, which could be explained by the lack of information about this tax incentive among companies and accounting professionals in the first year of its implementation.

Table 1. Number of micro and SMES that took advantage of the tax incentive for digital transformation (per year)

	2020	2021
Number of micro and SMES	33	342
%	8.8	91.7

It seems that the Portuguese micro and SMEs are concerned with the implementation of accounting digitalization, although the number of companies is still too small. These results reaffirm the need for micro and SMEs to be quick in the implementation of digitalization and that governments should take measures to promote this accounting digitalization (Low et al, 2022; Lutfi et al, 2022; Pham & Vu, 2022).

Regarding the characterization of these beneficiary companies, it can be seen through Table 2 that most companies are headquartered in the North of Portugal (37.8%) and in the Lisbon and Tagus Valley region (28.4%), followed by the Portugal Centre (24.9%). A small number of companies are in the south of the country (Alentejo and Algarve) or in the autonomous regions of Madeira and Azores.

Table 2. Geographic location of companies

Region	N	%
Alentejo	12	3.2
Algarve	10	2.7
Autonomous Regions (Madeira and Azores)	11	2.9
Centre of Portugal	93	24.9
Lisbon and Tejo Valley	106	28.4
North of Portugal	141	37.8
Total	373	100.0

Regarding the legal form of the companies, Table 3 shows that 60.3% of the beneficiary companies are private limited companies (similar to the national context in which they represent the majority of companies in Portugal), 26.8% are public limited companies and 12.9% are single-member limited companies.

Table 3. Companies legal form

Legal form	**N**	%
Private limited company	225	60.3
Public limited liability company	100	26.8
Single member limited company	48	12.9
Total	373	100.0

Regarding the sectors of activity, the analysis of Table 4 shows that the sectors with the highest number of companies are Consumer Discretionary (30.3%) and Industrials (29.0%), followed by Basic Materials (16.1%) and Consumer Staples (14.7%).

Table 4. Sector of activity

Sector of activity	**N**	%
Basic Materials	60	16.1
Consumer Discretionary	113	30.3
Consumer Staples	55	14.7
Energy	9	2.4
Financials	1	0.3
Health Care	10	2.7
Industrials	108	29.0
Technology	10	2.7
Telecommunications	4	1.1
Utilities	3	0.8
Total	373	100.0

Presentation and discussion of results

Considering the tax incentives received by Portuguese micro and SME companies in both years of analysis, it can be concluded that these companies benefited from a minimum tax incentive value of 0.59 euros and a maximum value of 7,985.30 euros, or an average of around 214.68 euros (see Table 5).

Although the number of companies that accessed these tax incentives increased in 2021, the average value obtained decreased slightly (from 310.21 euros in 2020 to 204.72 euros in 2021) (see Table 6).

Table 5. Tax incentives for digital transformation amounts

	N	Minimum	Maximum	Mean	Standard Deviation
Tax Incentive	373	0.59	7,985.30	214.6801	700.76

Table 6. Tax incentive for digital transformation amounts (per year)

Year	Mean	N	Standard Deviation
2020	310.21	33	519.067
2021	204.72	342	713.82

To identify some features that may explain the value of the tax incentives allowed for digital transformation, a Pearson correlation was performed between the variable value of the tax incentives granted to Portuguese micro and SME companies and some variables related to the performance measured by the access to the tax incentives of the companies under analysis. The results are presented in Table 7.

Table 7. Pearson correlation results

		Tax incentive for digital transformation
Number of Employees	Pearson correlation	.097
	Sig.	.062
Total assets per employee	Pearson correlation	.123*
	Sig.	**.019**
Working capital per employee	Pearson correlation	.140**
	Sig.	**.007**
Equity per employee	Pearson correlation	.102*
	Sig.	**.050**
Average cost per employee	Pearson correlation	.043
	Sig.	.413
Cost per employee / Operational income (%)	Pearson correlation	-.076
	Sig.	.144
Operational revenues per employee	Pearson correlation	.109*
	Sig.	**.038**
Profit per employee	Pearson correlation	.041
	Sig.	.433
Software programs	Pearson correlation	.082
	Sig.	.114

continues on following page

Table 7. Continued

		Tax incentive for digital transformation
Administrative equipment	Pearson correlation	.096
	Sig.	.063
Intangible assets	Pearson correlation	.065
	Sig.	.213
Tangible fixed assets	Pearson correlation	.056
	Sig.	.277
Income tax	Pearson correlation	.133*
	Sig.	**.010**
Turnover	Pearson correlation	.117*
	Sig.	**.023**
Operational revenues	Pearson correlation	.126*
	Sig.	**.015**
Leverage	Pearson correlation	-.026
	Sig.	.614
Liquidity ratio	Pearson correlation	-.002
	Sig.	.968
Financial profitability	Pearson correlation	-.006
	Sig.	.911
Economic profitability	Pearson correlation	-.025
	Sig.	.630
Equity	Pearson correlation	.092
	Sig.	.077
Total assets	Pearson correlation	.100
	Sig.	.055
Net profit	Pearson correlation	.181**
	Sig.	**.000**
Director gender	Pearson correlation	-.039
	Sig.	.461
* Correlation is significant at the 0.05 level (2 ends).		
** Correlation is significant at the 0.01 level (2 ends).		

According to the results, for a 99% confidence interval, the value of the tax incentive is positively associated with the mean value of the working capital per employee and the mean net profit, revealing that the most profitable companies on average and with the highest average values of working capital per employee are the ones that sought the tax incentive for digital transformation.

For a 95% confidence interval, the value of the tax incentive is associated with the average values of total assets per employee, average equity per employee, average operational revenue per employee, average income tax, average turnover, and average operational revenues.

These results show that the companies that invest the most on average per employee, with higher average equity per employee, with higher average income tax, average turnover, and average operational revenues, have taken the most advantage of the tax benefit to digital transformation. In this sense, these indicators also show that it is the largest and most profitable companies that have benefited from this tax incentive.

Increasing the confidence interval to 90% allows us to identify variables positively associated with the value of the tax incentive, the average operational revenues per employee, the mean value of equity per employee, and the mean value of total assets per employee, confirming the previous conclusions.

In this way, it seems that among Portuguese micro and SMEs, it is the largest and most profitable companies that have been recognized to be pioneers in digital transformation and those that have been most alert to government incentives to promote the digitalization of accounting.

These are probably also the companies that anticipate the digitalization of their information systems since they are also the ones that have the most resources and benefit the most from the anticipation of its implementation, given their volume of transactions. On the other hand, it is also the companies with the highest average results (operational revenue and net profit) that are more concerned with minimizing corporate income tax, trying to take advantage of the existing tax incentives, and are therefore more aware of these tax cuts.

The results obtained are in line with the literature review since there is an association with some variables related to employee ratios, which suggests the existence of a relationship between employee-driven thinking and business innovation (Opland et al., 2022). On the other hand, the results indicate the existence of a relationship between companies' operational revenues and digitalization (Rapuluchukwu et al., 2017) and the influence of tax incentives on digital transformation (Wang et al., 2023).

SOLUTIONS AND RECOMMENDATIONS

The number of Portuguese micro and SMEs that took advantage of the tax incentive for digital transformation is too small, compared to the universe of Portuguese micro and SMEs. This fact may eventually be explained due to the lack of information that the top managers of companies and their accountants have about this tax incentive. In this way, since the tax incentive remains in force in 2023, the authors highly recommend that both the Portuguese Government and professional and business organizations invest in dissemination and training actions, which promote companies' adherence to digital transformation, taking advantage of existing incentives.

FUTURE RESEARCH DIRECTIONS

Since the tax incentive for digitalization was kept in force during the years 2022 and 2023, this leaves room for further investigation of the tax incentive under study, as well as others of a similar nature. The authors intend to extend the period of analysis to verify the evolution and possible changes during the period of eligibility for this tax incentive. The authors also intend to consider other factors that may be related to the value of the tax advantage used by companies and develop a multiple linear regression model that allows the identification of possible determinants.

CONCLUSION

The importance of digitalization is undeniable. The digitalization of accounting brings benefits to organizations that have implemented it, such as reducing labor costs, saving time, providing useful information, and improving professional skills. In addition, digitalization is an important strategy in the search for sustainable competitive advantage for companies, especially SMEs, which are naturally reluctant to embrace it. In this context, this chapter is dedicated to the study of tax incentives for the digital transformation of accounting granted to Portuguese micro and SMEs, specifically for the implementation of SAF-T (PT) for accounting, the QR Code®, and the ATCUD.

The results of this study indicate that the number of micro and SMEs that have taken advantage of the digitalization of accounting is still very low. This may be because companies are still a little reluctant to digitalize, but also to the lack of information from the government about the existing tax incentives.

The average value of the tax incentive under study is only 214.68 Euros, which is in line with the fact that the number of companies benefiting is tiny.

Regarding the characteristics of the companies that benefited from this incentive, it can be concluded that they are the largest companies (with higher values of assets per employee on average and equity on average per employee) and the most profitable (with higher average values of net profit and net profit per employee, working capital per employee on average, average operational revenues and average operational revenues per employee, and average turnover) and, consequently, those with higher values of average income tax.

ACKNOWLEDGMENT

This work is financed by national funds through the FCT - Foundation for Science and Technology, I.P., under project UIDP/04043/2020.

REFERENCES

Acciarini, C., Borelli, F., Capo, F., Cappa, F., & Sarrocco, C. (2022). Can digitalization favour the emergence of innovative and sustainable business models? A qualitative exploration in the automotive sector. *Journal of Strategy and Management*, *15*(3), 335–352. doi:10.1108/JSMA-02-2021-0033

Bassey, E., Mulligan, E., & Ojo, A. (2022). A conceptual framework for digital tax administration - A systematic review. *Government Information Quarterly*, *39*(4), 101754. doi:10.1016/j.giq.2022.101754

Bednarčíková, D., & Repiská, R. (2021). Digital Transformation in the Context of the European Union and the Use of Digital Technologies as a Tool for Business Sustainability. *SHS Web of Conferences*, 115. 10.1051hsconf/202111501001

Bellon, M., Dabla-Norris, E., Khalid, S., & Lima, F. (2022). Digitalization to improve tax compliance: Evidence from VAT e-Invoicing in Peru. *Journal of Public Economics*, *210*, 104661. doi:10.1016/j.jpubeco.2022.104661 PMID:35711616

Brunetti, F., Matt, D., Bonfanti, A., De Longhi, A., Pedrini, G., & Orzes, G. (2020). Digital transformation challenges: Strategies emerging from a multi-stakeholder approach. *The TQM Journal*, *32*(4), 697–724. doi:10.1108/TQM-12-2019-0309

Buck, C., Clarke, J., de Oliveira, R. T., Desouza, K. C., & Maroufkhani, P. (2023). Digital transformation in asset-intensive organisations: The light and the dark side. *Journal of Innovation & Knowledge*, *8*(100335), 100335. Advance online publication. doi:10.1016/j.jik.2023.100335

Čavoški, A. (2022). The European Green Deal and technological determinism. *Environmental Law Review*, *24*(3), 201–213. doi:10.1177/14614529221104558

Chien, F., Ananzeh, M., Mirza, F., Bakar, A., Vu, H. M., & Ngo, T. Q. (2021). The effects of green growth, environmental-related tax, and eco-innovation towards carbon neutrality target in the US economy. *Journal of Environmental Management*, *299*(113633), 113633. Advance online publication. doi:10.1016/j.jenvman.2021.113633 PMID:34492439

Coman, D. M., Ionescu, C. A., Duică, A., Coman, M. D., Uzlau, M. C., Stanescu, S. G., & State, V. (2022). Digitization of Accounting: The Premise of the Paradigm Shift of Role of the Professional Accountant. *Applied Sciences (Basel, Switzerland)*, *12*(7), 1–25. doi:10.3390/app12073359

David, L., Nwulu, N., Aigbavboa, C., & Adepoju, O. (2022). Integrating fourth industrial revolution (4IR) technologies into the water, energy & food nexus for sustainable security: A bibliometric analysis. *Journal of Cleaner Production*, *363*, 132522. doi:10.1016/j.jclepro.2022.132522

Diller, M., Asen, M., & Späth, T. (2020). The effects of personality traits on digital transformation: Evidence from German tax consulting. *International Journal of Accounting Information Systems*, *37*(100455), 100455. Advance online publication. doi:10.1016/j.accinf.2020.100455

Dufva, T., & Dufva, M. (2019). Grasping the future of the digital society. *Futures*, *107*, 17–28. doi:10.1016/j.futures.2018.11.001

Dwivedi, Y., Hughes, L., Kar, A., Baabdullah, A., Grover, P., Abbas, R., Andreini, D., Abumoghli, I., Barlette, Y., Bunker, D., Kruse, L., Constantiou, I., Davison, R., De', R., Dubey, R., Taylor, H., Gupta, B., He, W., Kodama, M., & Wade, M. (2022). Climate change and COP26: Are digital technologies and information management part of the problem or the solution? An editorial reflection and call to action. *International Journal of Information Management*, *63*, 102456. doi:10.1016/j.ijinfomgt.2021.102456

Edelmann, N., & Virkar, S. (2023). The Impact of Sustainability on Co-Creation of Digital Public Services. *Administrative Sciences*, *13*(2), 43. doi:10.3390/admsci13020043

Elsevier. (2022). *Pathways to Net Zero: Global South Research in the Transition to Clean Energy*. Elsevier. https://elsevierfoundation.org/report-pathways-net-zero-global-south-research-transition-clean-energy/

European Commission. (2022). Towards a green, digital and resilient economy: our European Growth Model (COM/2022/83 final). *Official Journal of the European Union*, 1–16. https://eur-lex.europa.eu/legal-content/EN/TXT/?uri=CELEX:52022DC0083

Feng, H., Wang, F., Song, G., & Liu, L. (2022). Digital Transformation on Enterprise Green Innovation: Effect and Transmission Mechanism. *International Journal of Environmental Research and Public Health, 19*(17), 10614. Advance online publication. doi:10.3390/ijerph191710614 PMID:36078329

Feroz, A. K., Zo, H., & Chiravuri, A. (2021). Digital transformation and environmental sustainability: A review and research agenda. *Sustainability, 13*(3), 1–20. doi:10.3390u13031530 PMID:34676112

George, G., & Schillebeeckx, S. J. D. (2022). Digital transformation, sustainability, and purpose in the multinational enterprise. *Journal of World Business, 57*(101326), 101326. Advance online publication. doi:10.1016/j.jwb.2022.101326

GulinD.HladikaM.ValentaI. (2019). Digitalization and the Challenges for the Accounting Profession. *ENTRENOVA 2019 Proceeding, September*, 502–511. doi:10.2139/ssrn.3492237

Li, S., Gao, L., Hanb, C., Guptac, B., Alhalabid, W., & Almakdie, S. (2023). Exploring the effect of digital transformation on Firms' innovation performance. *Journal of Innovation & Knowledge& Knowledge, 8*(100317), 100317. doi:10.1016/j.jik.2023.100317

Low, M., Seah, C., Cham, T., & Teoh, S. (2022). Digitalization adoption for digital economy: An examination of Malaysian small medium-sized enterprises through the technology–organization–environment framework. *Business Process Management Journal, 28*(1), 1473–1494. doi:10.1108/BPMJ-06-2022-0282

Lutfi, A., Alkelani, S., Al-Khasawneh, M., Alshira'h, A., Alshirah, M., Almaiah, M., Alrawad, M., Alsyouf, A., Saad, M., & Ibrahim, N. (2022). Influence of Digital Accounting System Usage on SMEs Performance: The Moderating Effect of COVID-19. *Sustainability (Basel), 14*(22), 15048. doi:10.3390u142215048

Mondejar, M., Avtar, R., Diaz, H., Dubey, R., Esteban, J., Gómez-Morales, A., Hallam, B., Mbungu, N., Okolo, C., Prasad, K., She, Q., & Garcia-Segura, S. (2021). Digitalization to achieve sustainable development goals: Steps towards a Smart Green Planet. *The Science of the Total Environment, 794*(10), 148539. doi:10.1016/j.scitotenv.2021.148539 PMID:34323742

Mpofu, F. Y. (2022). Taxation of the Digital Economy and Direct Digital Service Taxes: Opportunities, Challenges, and Implications for African Countries. *Economies, 10*(9), 219. Advance online publication. doi:10.3390/economies10090219

OECD. (2010). Forum on Tax Administration: *Guidance Standard Audit File. In Guidance for the Standard Audit File – Tax Version 2.0*. OECD. https://www.oecd.org/tax/administration/45045602.pdf

OECD. (2020). *Tax Challenges Arising from Digitalisation – Economic Impact Assessment: Inclusive Framework on BEPS*. OECD/G20 Base Erosion and Profit Shifting Project. OECD Publishing, Paris. https://doi.org/https://doi.org/10.1787/0e3cc2d4-en

Opland, L. E., Pappas, I. O., Engesmo, J., & Jaccheri, L. (2022). Employee-driven digital innovation: A systematic review and a research agenda. *Journal of Business Research, 143*, 255–271. doi:10.1016/j.jbusres.2022.01.038

Pham, Q., & Vu, K. (2022). Digitalization in small and medium enterprise: A parsimonious model of digitalization of accounting information for sustainable innovation ecosystem value generation. *Asia Pacific Journal of Innovation and Entrepreneurship*, *16*(1), 2–37. doi:10.1108/APJIE-02-2022-0013

Purcinelli, L. M., Abreu, R., & Vasconcelos, A. L. (2020). Sustainable and Smart System: Rethinking Accounting and Taxation in Portugal. In D. Crowther & S. Seifi (Eds.), Governance and Sustainability: International Perspectives. Springer International Publishing. doi:10.1007/978-981-15-6370-6_4

Rapuluchukwu, E. U., Belmondo, T. V., & Ibukun, B. (2017). Incentives and firm productivity: Exploring multidimensional fiscal incentives in a developing country. *World Development Perspectives*, *5*, 56–59. doi:10.1016/j.wdp.2017.03.001

Ruggiero, V. (2022). Sustainability and Financial Crime. *International Criminology*, *2*(2), 143–151. doi:10.100743576-022-00052-3

Rupeika-Apoga, R., Bule, L., & Petrovska, K. (2022). Digital Transformation of Small and Medium Enterprises: Aspects of Public Support. *Journal of Risk and Financial Management*, *15*(2), 45. doi:10.3390/jrfm15020045

Trenerry, B., Chng, S., Wang, Y., Suhaila, Z. S., Lim, S. S., Lu, H. Y., & Oh, P. H. (2021). Preparing Workplaces for Digital Transformation: An Integrative Review and Framework of Multi-Level Factors. *Frontiers in Psychology*, *12*(March), 1–24. doi:10.3389/fpsyg.2021.620766 PMID:33833714

Truong, T. (2022). The impact of digital transformation on environmental sustainability. *Advances in Multimedia*, *2022*, 1–12. doi:10.1155/2022/6324325

Tyutyuryukov, V., & Guseva, N. (2021). From remote work to digital nomads: Tax issues and tax opportunities of digital lifestyle. *IFAC-PapersOnLine*, *54*(13), 188–193. doi:10.1016/j.ifacol.2021.10.443

van Vuuren, D. P., Zimm, C., Busch, S., Kriegler, E., Leininger, J., Messner, D., Nakicenovic, N., Rockstrom, J., Riahi, K., Sperling, F., Bosetti, V., Cornell, S., Gaffney, O., Lucas, P. L., Popp, A., Ruhe, C., von Schiller, A., Schmidt, J. O., & Soergel, B. (2022). Defining a sustainable development target space for 2030 and 2050. *One Earth*, *5*(2), 142–156. doi:10.1016/j.oneear.2022.01.003

Wang, S., Li, X., Li, Z., & Ye, Y. (2023, January 1–20). The effects of government support on enterprises' digital transformation: Evidence from China. *MDE. Managerial and Decision Economics*, *44*(5), 2520–2539. doi:10.1002/mde.3831

Yamen, A., Coskun, A., & Mersni, H. (2022). Digitalization and tax evasion: the moderation effect of corruption. *Economic Research-Ekonomska Istrazivanja*, 1–24. . doi:10.1080/1331677X.2022.2142634

Zhou, S., Zhou, P., & Ji, H. (2022). Can digital transformation alleviate corporate tax stickiness: The mediation effect of tax avoidance. *Technological Forecasting and Social Change*, *184*(122028), 122028. doi:10.1016/j.techfore.2022.122028

ADDITIONAL READING

Chen, C. L., Lin, Y. C., Chen, W. H., Chao, C. F., & Pandia, H. (2021). Role of government to enhance digital transformation in small service business. *Sustainability (Basel)*, *13*(3), 1–26. doi:10.3390u13031028

De Widt, D., & Oats, L. (2022). Imagining cooperative tax regulation: Common origins, divergent paths. *Critical Perspectives on Accounting*, 102446. doi:10.1016/j.cpa.2022.102446

Indriastuti, M., & Chariri, A. (2021). The role of green investment and corporate social responsibility investment on sustainable performance. *Cogent Business and Management*, *8*(1), 1960120. Advance online publication. doi:10.1080/23311975.2021.1960120

Kalkuhl, M., Fernandez Milan, B., Schwerhoff, G., Jakob, M., Hahnen, M., & Creutzig, F. (2018). Can land taxes foster sustainable development? An assessment of fiscal, distributional and implementation issues. *Land Use Policy*, *78*(July), 338–352. doi:10.1016/j.landusepol.2018.07.008

Klein, D., Ludwig, C. A., & Spengel, C. (2022). Taxing the Digital Economy: Investor Reaction to the European Commission's Digital Tax Proposals. *National Tax Journal*, *75*(1), 61–92. doi:10.1086/717518

Kraus, S., Durst, S., Ferreira, J. J., Veiga, P., Kailer, N., & Weinmann, A. (2022). Digital transformation in business and management research: An overview of the current status quo. *International Journal of Information Management, 63*(December 2021). doi:10.1016/j.ijinfomgt.2021.102466

Milios, L. (2021). Towards a Circular Economy Taxation Framework: Expectations and Challenges of Implementation. *Circular Economy and Sustainability*, *1*(2), 477–498. doi:10.100743615-020-00002-z

Tchorzewska, K. B., Garcia-Quevedo, J., & Martinez-Ros, E. (2022). The heterogeneous effects of environmental taxation on green technologies. *Research Policy*, *51*(7), 104541. doi:10.1016/j.respol.2022.104541

KEY TERMS AND DEFINITIONS

ATCUD: A unique document code of the Portuguese Tax Agency, which appears on the invoice document, simplifies the control of commercial operations, and allows the unique identification of each tax document issued, regardless of the issuer, the type of document, and the series used.

BEPS: OECD action plan to combat base erosion and profit shifting to low-tax jurisdictions.

Digitization vs. Digitalization: While digitization refers to the act of converting analog information into digital form, digitalization is about transferring existing processes to digital technologies. In other words, digitization is for information, while digitalization is for processes. In this chapter, the authors' research refers to digitalization.

Green Taxation: Environmental or green taxes, including taxes on energy, transport, pollution, and resources. Tax incentives to encourage investments that reduce environmental impact are also considered.

QR Code®: A two-dimensional bar code created in 1994 by the Japanese company Denso-Wave. The QR is derived from Quick Response because the code is designed to be decoded at high speed. The QR code has become popular for mobile tagging applications. Mandatory on all invoices and related tax documents in Portugal as of January 1, 2022.

Standard Audit File for Tax Purposes (SAF-T): File designed to give auditors access to data in an easily readable format for system data audits, and intended for use by companies and their auditors, as well as tax auditors.

Tax incentive: An exception to the general tax rules, that pursues relevant social and economic objectives and confers on its recipients an advantage that may result in a reduction of the tax burden.

ENDNOTE

1 Amended by Ordinance No. 1192/2009, of October 8, by Ordinance No. 160/2013, of April 23, by Ordinance No. 274/2013, of August 21, and by Ordinance No. 302/2016, of December 2.

Chapter 8
Green Taxes in the Solution of Environmental Problems and the Case of Türkiye

Nagihan Özkanca Andıç
https://orcid.org/0000-0003-4028-264X
Pamukkale University, Turkey

ABSTRACT

Green taxes, which have been discussed since the 1970s, are financial instruments used in the fight against environmental pollution. In addition to taxing, with green taxes, the activities thought to be harmful to the environment, it is also aimed to support environmentally friendly activities through economic incentives. Green taxes also address the failure of the free market in taking into account the ecological impacts that occur as a result of production or consumption activities. Countries discuss policy reforms and try to implement them in practice in order to create sustainable development under the umbrella of international associations and conferences. Green tax policies guide the steps to be taken to intervene in environmental changes. Through these policies, governments provide toolsets for the construction of a green economy by ensuring the internalization of externalities and by making guiding effects.

INTRODUCTION

The devastation caused by economic models that take into account the economic growth and that are not sensitive to the environment, has reached the point where it threatens the natural, living diversity and even human health. Together with environmental risks and the acceptance that natural capital is exhaustible, the problem of climate change has been faced and this situation has forced governments to create green policies to meet the needs of their citizens. The increase in consumption as a result of increasing human needs damages the order of nature and an unsustainable eco-system emerges in the long term. For this reason, it is inevitable to develop environmental protection policies through national and international organizations. Green taxes, which have been discussed since the 1970s, are financial instruments used in the fight against environmental pollution. In addition to taxing, with green taxes, the activities thought

DOI: 10.4018/978-1-6684-8592-7.ch008

to be harmful to the environment, it is also aimed to support environmentally friendly activities through economic incentives. Green taxes also address the failure of the free market in taking into account the ecological impacts that occur as a result of production or consumption activities.

Against the backdrop of the 2008 global financial crisis and the global climate crisis, there has been an increased trend towards green carbon policy instruments that do not harm the ecology. Countries discuss policy reforms and try to implement them in practice in order to create sustainable development under the umbrella of international associations and conferences. Creating a development strategy that offers environmentally friendly economic development strategies by protecting natural resources through the concept of green economy and green taxation associated with this concept has become an issue that attracts the attention of almost all governments. Green tax policies guide the steps to be taken to intervene in environmental changes. Through these policies, governments provide toolsets for the construction of a green economy by ensuring the internalization of externalities and by making guiding effects.

In the fight against environmental changes, the greatest power of states is the use of fiscal policy tools such as taxes, public expenditures and incentives. The periods when the use of fiscal policy tools has increased in the historical process have always been periods of crisis. The use of fiscal policy instruments, which gained popularity after the Great Depression of 1929, served as the first aid in correcting the devastation caused by the market economy and wrong economic policies. The use of green fiscal policy tools by states against ecological problems, which constitute one of the biggest problems of our age, will be an effective method in eliminating ecological destruction and protecting biodiversity. The aim of this study is to determine the fiscal policy tools that can be used to establish a sustainable future and to emphasize the necessity of fiscal policies in the perspective of green economy. In this study, the scope of green taxes will be mentioned, and some taxes that can be defined as green tax in Türkiye and some taxes that can be defined as environmentally sensitive will be evaluated also, the answer to the question of what is expected from public policies within the scope of green taxation will be sought.

The Scope of the Environment Tax

Solid wastes produced by people under the name of garbage cause contamination of soil, water and air. The main reasons of solid waste production are the ever-increasing population, unplanned urbanization, unplanned migration, and the inability to control industrialization. In this contex, people have to pay a public price for the solid wastes they produce under the name of garbage. In Article 2 of the Environmental Law No. 2872 environment is defined as "the biological, physical, social, economic and cultural environment in which living things maintain their relationships throughout their lives and interact mutually". Environmental pollution can be defined as any activity that occurs in the environment and may affect the health of living things, plants, and the ecological system negatively.

Environmental taxation is one of the most effective alternative policies followed in order to achieve environmental quality standards. Since Introduction of environmental taxes to the fiscal literature is based on the early work of A.C. Pigou (1920), this type of tax is also called Pigou tax. Environmental taxes are defined by the Organization for Economic Cooperation and Development (OECD) as type of taxes imposed on a physical or similar unit whose base is scientifically proven to have negative effects on the environment (OECD, 2006). Having worked in the field of welfare economics, Arthur Cecil Pigou is known as the economist who first proposed the theoretical principles of environmental tax and taxation as an instrument for internalizing negative environmental externalities. By establishing a relationship between welfare economics and external economy, A.C. Pigou emphasized the necessity of state intervention for welfare increase in the case of insufficient competition market and pioneered the application of externality theory to environmental problems in the modern sense (Luckin, 2000).

In the literature, Environmental taxes are often called "eco-taxes", "green taxes" or "environmental taxes". One of the most effective ways to protect the environment is the financial contribution of companies and/or individuals to the cost of environmental consumption. Thus, in practice, the increase in the costs of individuals and / or companies and the benefit obtained from the decrease in the environmental pollution levels of those concerned are effectively realized through environmental taxes. According to the OECD definition of environmental taxes; environmental tax is a tax based on a physical unit or part of a physical unit that has a specific negative impact on the environment (OECD, 2009).

Environmental taxes, that regulate the interaction between the economy and the environment constitutes one of these measures. Incomes earned from environmental taxes are allocated to the protection of nature and the promotion of sustainable economies based on greener production (Miceikiene et al., 2018).

Environmental taxes are accepted as a reform on the basis of the idea of providing the necessary financial resources to prevent environmental problems. This fact, also called as green tax reform or ecological tax reform, is not just about imposing new taxes from an environmental perspective. It includes the removal of environmentally harmful financial incentives and tax reductions, and the restructuring of existing taxes in an environmentally friendly manner, and beyond that, the implementation of many new market-based financial instruments, taking into account the environmental factor at every stage of economic functioning (OECD, 2001).

Environmental taxes are classified in four categories: energy taxes, transportation taxes, pollution taxes and natural resource taxes. Energy taxes include taxes mostly imposed on energy products used for stationary and transportation purposes. Energy products used for stationary purposes are electricity, coal, natural gas and fuel-oil, while energy products used for transportation are mostly diesel and petroleum. Transportation taxes represent a grouping related to the ownership and use of motor vehicles. Pollution taxes cover emissions measured in air and in water, as well as noise and solid waste, while resource taxes cover taxes on water extraction from springs, some raw materials and forest (Öz & Kutbay, 2016)

Taxes are collected for the financing of public expenditures in the classical sense, in other words, for financial purposes. However, today they are used for different purposes other than financial purposes, as in the case of environmental taxes. Environmental taxes are monetary values collected from real and legal persons unilaterally and unrequitedly by public administrations in order to prevent the destruction of environmental values and ecological balance, to eliminate environmental degradation, to improve the environment and to prevent environmental pollution. However, in practice, together with the mandatory and completely unrequited payments which are about the protection of the environment, some charges and duties are also expressed as environmental tax. The main purpose of environmental taxes is to protect the environment and at the same time to minimize the harmful effects of environmental pollution. In accordance with the polluter pays principle, it is aimed to internalize the environmental costs, namely negative externality, caused by environmental taxes and environmental pollution (Reyhan, 2014).

OBJECTIVES OF ENVIRONMENTAL TAX

Taxes are now used to regulate economic and social life and to intervene in disruptions along with providing financing for public expenditures. In this context, as support can be provided to the society through tax interventions on certain issues for the benefit of the public; it is possible to prevent the issues that may harm the society or to minimize the damages that may occur in this regard (Yegen, 2019).

The United Nations first discussed whether there is a link between the economy and the environment with the Environment Conference held in Stockholm in 1972. In the conference, environmental problems were discussed on the basis of the development levels of countries, and the relationship between economic terms such as development and growth in the national and international arena and the environment began to be investigated (Kaplan, 1991). The primary objective of environmental taxes, which finds application all over the world is to reduce the use or consumption of substances that have a negative externality to the environment and to restain activities that are harmful to the environment (Sollund, 2007).

Also, it is possible to talk about two effects of environmental taxes: incentive and financial effects. Accordingly, the fact that taxes enable individuals and institutions to be more sensitive to the environment reveals the incentive effects, while the fact that they provide income to the public reveals their financial effects. These are taxes that are not created not discourage pollution of the environment, but rather to create financial funds for the efforts of governments to clean the environment. These types of environmental taxes are also named as secondary environmental taxes. Although the main goal of these taxes is to create financial resources, it can be said that they have an incentive effect on large-scale producers and consumers polluting the environment, especially in the long run, due to their costs (Ekeman, 1998).

It is possible to list the characteristics of environmental taxes, which aim to reduce harmful behaviors towards the environment, as follows (Toprak, 2006):

- As environmental taxes increase the cost of environmentally harmful goods, services or activities, they direct producer and consumer units to activities that are not harmful to the environment,
- Environmental taxes also contribute to technological development through directing the producer units to develop new production techniques,
- In environmental taxes, while tax revenues are increased, positive contributions are made to the economy by reducing the tax burden on labor and capital,
- An important contribution of environmental taxes is that they can eliminate the differences between private social benefits and private social costs that individuals incur as a result of externalities caused by pollution.

Internalization of Negative Externalities

Environmental taxes are based on the "polluter pays" principle. According to this principle, negative externalities, by being internalized are included in prices. In this sense, the main purpose of environmental taxes is to reduce the damage to the environment by increasing the cost of environmentally harmful activities. Thus, it is aimed that consumers and producers take into account the damage they cause to the environment due to their economic activities and that they reduce the damage they cause to the environment (Markandya, 2002). Polluters often do not allow their damage to be reflected in the market. For this reason, the whole society, not those who pollute the environment, bears the externalities or external costs. Environmental taxes are the most accepted method of internalizing such external costs.

Environmental taxes are included in the Pigovian taxes, which envisage levying a tax equal to the difference between the marginal private cost and the marginal social cost on firms that emit negative externalities (Burrows, 1979). According to Pigou, an appropriate tax per unit should be imposed on the use of consumption or production goods that emit a negative externality. While determining this amount, tax should be applied by taking into account the difference between the marginal private cost and the marginal social cost, that is, the marginal loss due to the negative externality. Therefore, imposing a tax per unit as much as the amount of marginal loss will help to eliminate market failure. The reverse of this situation is also valid in the application of subsidies in case of positive externalities, and in this case, it is recommended to apply an amount of subsidy equal to the external benefit (Yüksel, 2013). This tax, also known as emission or pollution tax, aims to internalize the costs incurred by firms that generate environmental externalities and impose harm on society. It is envisaged that the government should impose a tax equal to the external costs. By doing so, it would effectively limit the production of activities that cause externalities and consequently reduce the external damages. Since it is not possible to measure external costs accurately, this proposed criterion becomes a necessity in terms of efficiency in resource allocation (Pigou, 1920). However, accurately quantifying external costs can be challenging. It can be complex to determine all aspects of external effects and their impacts on society. Therefore, determining tax rates and mechanisms can be a difficult process. Developing methods to measure and assess external costs as accurately as possible is crucial for an effective emission or pollution tax system.

The essence of Pigouon taxation is based on a simple idea. Accordingly, the external cost that will arise from the decisions of the buyer and the seller as a result of the negative externality, which refers to a situation where the production or consumption of some goods harms someone other than the buyer or seller of that good, is not taken into account. This indicates a market failure. An unregulated free market will often result in an inefficiently high supply of any good with negative externalities. At this point, imposing a tax on the goods that create an externality will be able to fix the externality that will occur. If the applicable tax rate is equal to the marginal external cost, i.e. the total loss to parties other than the buyer and seller as a result of one additional unit of delivery of the good, the external cost to be incurred will be included in the process and thus the buyer will be able to pay the entire marginal social loss. Thus, through the incentive provided by the tax, in the absence of other uncorrected market failures, it will be possible to produce the good effectively in the market (Williams III, 2016). According to this situation identified by Pigou, environmental taxes increase social welfare by reducing environmental pollution. Here the goal is not to reduce the firm's output to zero, but to limit excess production that leads to negative externalities. Therefore, we can say that environmental taxes have the potential to increase social welfare.

Reducing the Tax Burden on Wages/Labor

The main reason for shifting the tax burden from good ones such as income tax on labor and social security premiums and corporate tax on capital to "bad" ones such as emissions is the more equitable distribution of the tax burden (Lutz & Meyer, 2010). Such a tax reform which aims at transfering taxes from 'good' like labor and capital to 'bad' like pollution is called ecological or green tax reform. Thus, while the burden of activities related to the environment will increase, positive welfare gain will be achieved by reducing the tax burden of labor and capital, and activities that reduce welfare will be avoided. On the other hand, at the point of income neutrality, by reducing the burden of th one and increasing that of the other, there will be no change in the total tax burden at the national level, and the budget deficit arising from the reduction of labor taxes will be balanced (Speck, 2007). It is thought that the income earned from environmental taxes can be used to reduce the distorting tax rates on labor.

Providing Income

In the literature, the substitution of income from environmental taxes for income from some taxes such as personal income and corporate taxes, that is, the use of income from environmental taxes to reduce the rates or share of some other taxes is called the double dividend hypothesis. For example, governments have aimed to achieve a "double dividend" with an increase in employment as well as a cleaner environment, while implementing environmental tax reforms in EU member states that shift the tax burden from labor to pollution. (Bovenberg, 1998) The desire to gain double dividend has become a necessity due to the conditions that the countries face. Yet, the reductions in labor taxes in order to increase employment have led to the emergence of an alternative financing need. In this sense, environmental taxes stand out as a good financing alternative.

Besides the environmental goal, green taxes also have a financial goal of increasing public revenues. In many cases, green taxes can reduce environmentally harmful behavior and at the same time they can increase public revenues. However, role sharing may be required between these two objectives, because the environmental objective reduces the consumption and production of the taxed good and this affects public revenues negatively (Cottrell & et al., 2016). Although the main purpose of environmental taxes is to reduce pollution or improve natural resources, it is an undeniable fact that some environmental taxes are an important source of income for governments today. For example, some taxes, such as the fuel tax, were initially considered as environmental taxes, but over time they have come to serve the purpose of generating income. The income earned from environmental taxes can be used to generate income, especially in countries with public deficits. In recent years, the increasing revenue potential of environmental taxes constitutes an important revenue item in the budgets of governments in this sense.

DEVELOPMENT OF ENVIRONMENTAL POLICIES AND SUSTAINABLE DEVELOPMENT IN TÜRKIYE

Until today, economic development has been discussed as one of the leading issues of economic thought in order to meet the unlimited desires of human beings. It has gained different contents in different periods pursuant to the different development processes of societies. The concept is intertwined with concepts that sometimes have similar meanings, such as industrialization, modernization, progress, growth and structural change and has been used instead of them and naturally has undergone a shift in meaning (Yavilioğlu, 2002). Even today, the content of the concept is not completely clear and understandable, it is sometimes used as industrialization, sometimes as growth, sometimes just as development.

Environmental policy, on the other hand, is defined as the whole of the rules, principles and objectives that are created to regulate the relationship between the environment and society, and the measures taken for the protection of the environment. Environmental taxes have been discussed in the literature since the 1970s. It is a financial instrument that is used to solve negative externalities caused by environmental pollution and that almost all countries apply for taxation (Özkanca Andıç & Erasa Akça, 2019).

The concept of sustainable development was first used by the International Union for Conservation of Nature and Natural Resources in the 1980s. The popularity of this concept is based on the "Brundtland Report" prepared by the World Environment and Development Commission (WCED) in 1987 (Özçağ & Hotunluoğlu, 2015). Sustainable development concept also includes environmental, economic and social considerations. Emphasis is placed on economic development within economic, social and environmental integrity. According to the report of the World Commission on Environment and Development in 1987, sustainable development is the ability to meet the needs of the present without compromising the ability of future generations to meet their needs (Pearce et al., 1994).

Sustainable development states the balance of the society in terms of economic, social and environmental aspects. Sustainable development requires a balanced consideration of these three factors in development. The interconnections of economic, social and environmental aspects are also important in ensuring sustainable development. Economic and social aspects are in interaction with issues such as income distribution, poverty reduction, and the solution of unemployment problems. Once these socio-economic conditions of the society become stable, its approach to the environment will also become sustainable. Yet developing societies continue their lives dependent to natural resources. Societies with better socio-economic conditions will have more regular demands from natural capital. The social and environmental aspect envisages the equal use of natural resources as well as equality in income distribution. Future generations should have at least as much natural resources as today. Also, full public participation is required in the management of environmental resources. The economic and environmental aspect envisages the internalization of environmental valuation and the negative effects of economic activities. Undoubtedly, while all economic activities ensure the increase the welfare level of the society, they will also have negative effects on environmental quality. In the implementation of economic activities, while certain amounts of goods are demanded from natural resources, certain amounts of waste materials are released to nature. It is also a requirement of sustainable development that the negativities created by all these activities are clearly revealed. Sustainable development requires the adoption of a set of actions for long-term improvements in human well-being. Economic activities aiming at short-term profits that are harmful to the environment are kept out from sustainable development strategies (Gürlük, 2010).

The environmentally harmful practices attracted attention, especially after the industrial revolution, when countries started to pollute the environment irresponsibly in the cause of economic development. Although this was not taken into account and not noticed at first, the damage to the environment caused by industrialization and output growth have increased day by day. In addition, the first and second world wars also caused significant damage to the environment. Industrialization process that rapidly developed for the recovery of post-war economies triggered environmental pollution. Increasing welfare with industrialization and increase in production have enabled people to live a more modern life and to meet their needs. Economic development, industrial growth and rearmament after the war caused environmental problems to be ignored after the war as well as during the war (Dinç, 2008).

Short-term political priorities such as accelerating economic development, preventing unemployment or controlling inflation have occupied the agenda of economies for many years, local and central governments have had to give priority to production-related variables in their policies because of their economic growth and development concerns. These developments and "interesting parallelism", both in theory and practice, prevented the formation of environmental awareness not only in developing countries but also in industrialized countries for many years, however, in the post-1970s period, especially the United Nations Organization, OECD, EU and other voluntary organizations have increased their efforts to protect the environment. Especially since the 1980s, with the effect of globalization, the conceptual and theoretical content of economic development has begun to change, with the view that the policies followed can reduce the welfare of future generations, thus, by emphasizing the destruction of economic development policies on the natural environment, the traditional short-term understanding of economic management has been replaced by a long-term perspective, and environmental problems have also been brought to the fore.

The effects of economic activities on the environment are inevitable. While the environment ensures the continuity and growth of the economic system with the resources it provides, the ecosystem is negatively affected as a result of production and consumption activities. It is now known that sustainable economic growth affects the welfare level of countries and the living standard of individuals, and environmental factors are also affected by all these works and transactions. It is also known that the increase in poverty and the increase of environmental damage are in direct interaction with each other. Because people damages nature in order to maintain their lives, and in this case, green economy is of great importance in order to sustain economic and environmental development (Kuşat, 2013) In particular, the assumptions and approaches of classical economists that the economy will be automatically balanced by an invisible hand and that natural resources will not run out due to self-renewal, caused central and local administrators to still give priority to economic growth and development even today and caused environmental pollution to be put into the background for many years. However, there is a close link between economic growth and social development. The economic growth of societies with a strong socio-cultural structure is easier and more possible. It is not possible for societies that have social problems due to law and inequality, deal with diseases, suffer from medicine shortage due to poverty, have high death rates from diseases and therefore low participation in the workforce to grow and develop. Therefore, in order to grow quantitatively, the social structure must be healthy. The strategies to be followed in sustainable development can be listed as follows (Adshead et al., 2006);

- Protection of Environment
- A healthy and equitable society
- A sustainable and competitive economy
- Guidance
- Contribution to international sustainable development

In Türkiye, the concepts related to environmental pollution emerged in the 1970s when the European Union faced this problem in the same years. In the Third Five-Year Development Plan, which was first prepared in 1973, environmental problems were addressed and in this plan, it was emphasized that the regulations regarding the protection of the environment should be included in the legislation without hindering rapid economic development policies. Afterwards, in the development plans made until 2000, environmental assessments always remained in the background, the plans were always formed on the basis of the possibility that national development efforts could be damaged by environmental policies, but with the eighth five-year development plan, efforts were made to harmonize the policies implemented and the decisions taken for the solution of environmental problems with European Union norms and international standards. The National Environmental Strategy and Action Program, has an important place in shaping Türkiye's environmental policy. The objectives of this program include preventing pollution, facilitating access to environmental infrastructure and services, encouraging the use of renewable resources, and establishing policies that will enable the environment and economy to remain sustainable at the same time (Serdaroğlu, 2003).

Under the title of "Health services and protection of the environment" in Article 56 of the 1982 Constitution, "Everyone has the right to live in a healthy and balanced environment. It is the duty of the State and citizens to develop the environment, to protect environmental health and to prevent environmental pollution." direct provisions regarding the right to the environment and the protection of the environment are in place. It is seen that the direct regulation at the legal level regarding the environment in Türkiye was made with 1982 Constitution. With the acceptance of the 1982 Constitution, the concept of environmental protection was included in the Constitution for the first time. The point that needs to be emphasized regarding Türkiye's environmental policies is the fact that environmental problems are given a special place in the constitution. Although the 1982 Constitution, which generally sets the basic principles of Turkish Environmental Policies, mentions that the environment should be healthy and balanced, it is seen that how the ideal environment should be or which elements are included in the environment are not mentioned.

We can say that the effects of ecological modernization and sustainable development, which are new environmental policy expressions, have begun to be observed since the 1990s. After the publication of the Brundtland Report and especially in the period following the Rio Conference, sustainable development started to affect the environmental policies in Türkiye as an important environmental policy expression over time, although it was not initially binding. This period has been a period in which the expression of sustainability and sustainable development came to the fore and rose. However, in this process, sustainability and sustainable development were handled in a very limited framework. Significant problems were encountered in the implementation of policy principles, which require the integration of policies that need to be implemented in order to achieve sustainability goals, the participation of stakeholders in decision-making and implementation processes, and the consideration of development from a different perspective. Considering the expressions of the politicians of the period, environmental protection was seen as a prerequisite for development and presented as a kind of ecological modernization project, a project to reduce costs and increase competitiveness, and to be realized for development (Orhan, 2013)

Today, however, market-based (environmental taxes, negotiable licenses, collateral return systems) more cost-reducing economic instruments have begun to direct the preferences of policy makers, rather than commanding and controlling factors. has come to the fore, global interest in the subject has increased and environmental taxes have found an increasingly widespread application area in many countries. However, the necessity of global tax cooperation between countries is especially emphasized in order to prevent unfair tax competition in the global sense, as well as to ensure fair environmental taxation (Özdemir, 2009).

CURRENT GREEN TAX PRACTICES IN TÜRKIYE

The fact that the effects of environmental pollution have been felt intensely, especially in the last thirty years, has been one of the most important reasons that drew the attention of developed countries to environmental policies. Environmental taxes, that are applied to reduce the harmful effects of this pollution, also attract attention with their increasing income potential. The double benefit hypothesis asserts that environmental taxes on pollution will both have a positive impact on the environment and improve the distortions in the system by reducing the deviating taxes that exist in the tax system. According to this, a benefit that increases social welfare can be created by using the revenues obtained with an income-neutral environmental tax application in reducing the deviating taxes that impair economic efficiency (Metcalf, 2003).

In Türkiye, there is a need for the development of mechanisms such as taxation and incentives to protect the environment. At this point, fiscal mechanisms should be diversified, and there is a need for tax policy regulations in areas such as renewable energy incentives and waste management. However, even though the excessive taxation of products like natural gas and oil in Türkiye could be presented as an example of green taxation, the primary objective is more inclined towards generating public revenue rather than emphasizing environmental sensitivity (Vanii&Uçar, 2021).

Compared to the EU and OECD countries, it is seen that the taxes that can be described as environmental taxes are at a very limited level in Türkiye. Environmental Cleaning Tax is mentioned as "the only type of environmental tax that directly serves environmental purposes" In Türkiye (Reyhan, 2014). However, Motor Vehicles Tax levied on land, air and sea vehicles, Special Consumption Tax levied on petroleum and petroleum products (List I) and on motor vehicles (List II) and Value Added Tax levied on natural gas, petroleum and their products are also evaluated in the context of "environment related taxes" or "environmentally beneficial taxes". Since these taxes cause taxation of some products that harm the environment if used or consumed, they are indirectly accepted as environmental tax. It can be said that these taxes may have an indirect effect on preventing pollution spreading to the environment by reducing the consumption or use of these products by taxing the products that emit pollution to the environment However, even though they have such an indirect effect, when the Motor Vehicles Tax, Special Consumption Tax and Value Added Tax applications in our country are examined, it is seen that the financial aim to finance public services comes to the fore more than environmental objectives (Biyan & Gök, 2014)

Environmental Cleaning Tax

The main objective of environmental taxes is to direct the preferences of producers and consumers towards goods and services that do not emit negative externalities to the environment by increasing the price of goods and services that create negative externalities to the environment. In addition, through environmental taxes, companies will start to invest in production techniques that are not harmful to the environment and will also serve environmental technological development. Therefore, the purpose of generating income in environmental taxes will remain in the background. It is noteworthy that the only tax that fully includes environmental regulations in Türkiye is the environmental cleaning tax.

The Environmental Cleaning Tax application, which started in 1993, was amended in 2003 with the Law No. 5035, and the distinction of the Environmental Cleaning Tax for solid waste and the Environmental Cleaning Tax for waste water was abolished. The practice, which started in 2003 and is still valid today, is as follows: The subject of the Environmental Cleanliness Tax consists of "buildings used in residences, workplaces and other forms that are located within the municipal boundaries and adjacent areas and benefit from the environmental cleaning services of the municipalities". The taxpayers are those who use these buildings. The Law has differentiated the Environmental Cleaning Tax application for residences and workplaces. The Environmental Cleaning Tax for the residences is an amount determined per cubic meter based on the water consumption amount of the residences. The tariff is determined separately for metropolitan municipalities and other municipalities. The amount of tax determined in the law and to be charged per cubic meter is increased every year at the rate of re-evaluation.

The Environmental Cleaning Tax to be applied for workplaces and buildings used in other ways is collected according to the tax tariff determined by the law in 7 groups and with 5 degrees. This tax tariff determined in the law is increased every year at the rate of revaluation, as in the tariff for residences. Also, this tariff is applied with an increase of 25 percent in metropolitan municipalities.

It is debatable to what extent the Environmental Cleaning Tax, which is defined as direct environmental tax among the existing practices in Türkiye and which is included in the group of taxes on pollution, can meet this purpose. Since the tax levied on the residences in today's practice is determined on the basis of the amount of water consumption, this part of the tax may have an effect on reducing the water consumption in the residences. On the other hand, it can be said that this tax levied on residences contributes partially to the internalization of negative externalities caused by wastewater. However, the Environmental Cleaning Tax collected from the residences only considers the amount of water consumed and has no relation with the amount of solid waste. The tax collected from workplaces, on the other hand, is collected by associating it with the building groups included in the law, independently of both waste water and solid waste. In this case, internalization of negative externalities are out of the question (Ercan, 2015).

Motor Vehicle Tax

Although the Motor Vehicle Tax is not directly effective in solving the environment problems, it contributes indirectly. Vehicles subject to motor vehicle tax are stated as motorized air and land vehicles that cause carbon emissions with the fuels they consume (Ercan, 2015). MTV entered into force with Law No. 197 in 1963 regarding the taxation of motor vehicles that are one of the wealth elements, is accepted as an indicator of financial power, and has survived to the present day with various changes. The tax is accrued according to the tariffs listed in the MTV Law. Vehicles listed in the tariff are motorized air and land vehicles that emit carbon emissions as a result of the fuel they consume. MTV has functions such as reducing the use of these vehicles that pollute the environment and emitting negative externalities or promoting new technologies that will cause less pollution.

Motor vehicle tax (MTV) is a type of tax that has been applied in many countries. Today, with the increase in the welfare standard worldwide, the use of motor vehicles is increasing and motor vehicle owners impose serious costs in terms of maintenance and repair on roads, parking areas, bridges and underpasses (Ulusoy&Akdemir, 2013).

Motor vehicle tax has a specific tax base. Tariffs are arranged according to the criteria of the type, age, cylinder volume, seating, maximum total weight, engine power and take-off weight of the vehicles. In addition to the implementation of the motor vehicle tax in this way in Türkiye, there are three taxes levied on vehicle ownership in the world, in relation to engine size, measured by cylinder capacity in cubic centimeters (cc3). There is a tax applied if the engine capacity is below a certain cc, or another tax applied if the engine capacity is more than a certain cc, and a one-time special consumption tax on the vehicle price. In addition, a one-time registration fee is charged for imported vehicles, depending on the engine size. Registration fees are repaid if vehicles are subsequently sold, and there is an annual road tax, depending on engine size. Making cars more costly to own through all existing taxes reduces overall demand for new cars. In addition to the motor vehicle tax calculated according to the cylinder volume, imposing a consumption tax on CO2 per kilometer is considered as an alternative option. The advantage of basing motor vehicle taxes on CO2 per kilometer rather than engine size would encourage a wider range of possibilities for improving emissions damages.

Special Consumption Tax

Some tax types are presented as environmental taxation aimed at reducing environmental damage by tax regulators. However, these taxes are mostly applied to increase public revenues rather than environmental purposes. For example, in the special consumption tax on hydrocarbons in relation to the taxation of gasoline and diesel fuels, the consumer has to indirectly pay an amount greater than the value of the product. Therefore, when purchasing gasoline, the consumer pays a price that is the sum of the value of the product, which represents most of the final amount, including the indirect special consumption tax and value added tax (Espadafor, 2018).

In 2002, the fuel tax was abolished and the special consumption tax was put into effect in Türkiye. For this reason, special consumption taxes are also a kind of tax imposed for environmental purposes. Fuel tax is a more direct form of eco-tax because the tax varies based on the amount of fuel consumed and this affects vehicle emissions. At the same time, the fuel tax is easy to enforce, difficult to abstain or avoid, and easily changed. It is known that most OECD countries also collect fuel tax. It is possible to apply a quality tax difference in favor of unleaded gasoline as an incentive to use less environmentally harmful fuels. The part of the SCT that is related to the environment is the A and B charts of the first list. In Chart A; there are petroleum derivatives such as aviation fuel, unleaded super and normal gasoline, natural gas, fuel oil, jet fuel, base oil and diesel, in the chart B; there are benzol, solvent, varnish, ether, thinner and similar products. It is claimed that these products are taxed for the purpose of dissuading the consumer due to the damage they cause to the environment as well as the financial purpose. However, there are many criticisms about the fuel tax in Türkiye today. The high tax burden makes us think that this tax is collected for financial purposes rather than environmental protection and exceeds the limits of optimality.

Plastic Bag Tax

In Türkiye, a fee has been introduced for the use of bags in the context of the polluter pays principle. The first regulation regarding the use of bags was with the Law No. 7153 of the Environment Law and Some Laws, which was put into effect on 29.11.2018. Pursuant to Article 1 of Law No. 7153 Amending the Environmental Law and Some Laws: "Obligatory standards for environmental protection, prevention and elimination of environmental pollution, taxes, fees, contribution fees, promotion of renewable energy sources and clean technologies, recycling contribution, reduction of plastic bags and plastic packaging use, deposit application, emission fee, pollution fee and Market-based mechanisms such as obtaining guarantees for pollution prevention and carbon trading, as well as economic instruments and incentives are used. Administrative and technical procedures and principles regarding these issues are determined by the regulations to be issued by the Ministry".

However, the first concrete regulation regarding the restriction of the use of plastic bags is the "Law on Amending the Environmental Law and Some Laws", which was published in the Official Gazette No. 30621 on 10.12.2018 With this law, it is aimed to sell plastic bags for a fee. In the practice in force in Türkiye, the main objective of the law, which requires at least 25 kuruş per bag, is to reduce the use of plastic bags and to replace them with cloth or paper bags. 15 cents of the minimum 25 cents fee to be applied to plastic bags consists of plastic bag tax and 10 cents consists of 18% VAT and cost. Plastic bags that are subject to a fee have a barcode and are shown as pieces on the sales document.

Although the main purpose of the law, which was planned to enter into force in January 2018, but started to be implemented on 1 January 2019 with a delay of one year, was not to generate income, it also caused some reactions with the start of its implementation. The reaction of the public is based on the fact that it is an unnecessary expense to charge the bags that they have been accustomed to using for years free of charge and as much as they want (Özçelik, 2019). It is thought that it is not enough to collect a fee on the bag, and it is thought that training programs should be organized for public awareness and the damage caused by such wastes to the environment should be explained. Thus, it is thought that the use of unnecessary bags will be limited even if it is not completely finished, and the consumption of not only bags but also disposable plastic tools that harm nature will decrease significantly.

EXPECTATIONS FROM PUBLIC POLICY AND ITS PLACE IN GREEN TAXATION

In the prevention of environmental pollution, environmental taxes are used as a policy instrument that is frequently referred in public policies. However, environmental taxes are not seen as a sufficient instrument for reducing environmental pollution alone. Environmental taxes provide pricing of costs rather than reducing environmentally harmful activities. For this reason, there is a need for countries to take more responsibility and making new public policies that prevent or reduce activities harmful to the environment. The structure of incentives that can be applied to businesses and individuals through tax policies has a fundamental effect on both consumption and investment decisions. On the other hand, how governments spend their tax revenues not only on recurring costs but also on investments that support public infrastructure or technology development plays a critical role in economic development and the transition to a green economy. For this reason, the transition to a green economy should be based on a broader and more robust practice, such as more systematic taxation of fossil fuel-based energy and other natural resources (UNEP, 2010).

The main idea behind the use of environmental tools, particularly environmental taxes, to reduce environmental damage is the recognition of negative environmental externalities in unregulated economies. Negative externalities refer to costs that one economic unit imposes on another without considering them in production or consumption decisions. When the costs associated with pollution or resource use are not internalized, it leads to overproduction or overconsumption. This inefficiency in the market results in social costs. Additionally, goods and services like clean air and clean water are considered public goods, and property rights may be uncertain. Individuals often use these resources without considering the impact of their actions on other economic agents and future generations (OECD, 2001). By implementing taxes, the production costs of companies intentionally increase. Consequently, companies may either reduce their production or adopt alternatives that have a lesser impact on the environment. The foundation of the green taxation approach lies in the polluter pays principle, where economic units that pollute the environment and create negative externalities bear the costs through taxation. This approach aims to discourage such behavior and internalize environmental costs.

Tax incentives are considered as another type of important tax policy instrument applied in ensuring the green economic order. It is possible to direct the production and consumption preferences of individuals and companies through use of tax incentives. Tax incentives are waivers providef by government for tax revenues to support less polluting types of consumption and investment to achieve environmental policy goals and thus address positive externalities. For example, people can be encouraged to use electric vehicles by using tax deductions or exemptions. The basic logic here is not to ensure that economic actors avoid environmentally destructive practices by increasing costs, but to encourage people to use practices that help the environment. Tax deductions reduce high capital costs and enable the private sector to shift investment opportunities to climate-friendly investments. With appropriate policies, local innovations and technologies that can bring advantages both regionally and globally can be supported.

In most of the European countries, there are tax applications such as carbon tax, which is directly within the scope of environmental taxes. The most widely used tax type is energy taxes collected from goods and services such as natural gas, oil and electricity. On the other hand, many countries have tax incentives for green equipment, investment and electric vehicles. In Türkiye, there is only environmental cleaning tax application which is included in the scope of direct environmental taxes. Although the primary purpose of applications such as Motor Vehicles Tax and Special Consumption Tax is not to protect the environment, they can be evaluated within the scope of environmental tax as they indirectly affect the environment. The absence of a tax application such as carbon tax in Türkiye appears as a lack in the perspective of sustainable development.

In this regard, the following recommendations can be made for Türkiye to further advance its environmental sustainability efforts:

Promoting the Use of Renewable Energy Sources: Türkiye can invest in renewable energy sources such as solar, wind, hydroelectric, and biomass. Regulations and incentives can be implemented to encourage the utilization of these resources. Emphasis can be placed on high-potential areas such as solar energy and wind energy.

Enhancing Energy Efficiency: Türkiye can further focus on energy efficiency. It can promote energy-efficient building standards to increase energy efficiency in buildings. Energy efficiency programs and incentives can be implemented in the industrial sector to reduce energy consumption.

Emphasizing Waste Management: Türkiye can adopt more comprehensive and effective waste management policies. Recycling can be encouraged, and waste sorting facilities and recycling infrastructure can be developed. Widely implementing wastewater treatment plants and proper disposal of environmentally harmful waste are crucial.

Developing Public Transportation: Public transportation systems in Türkiye can be improved and expanded. Investments can be increased to promote the use of public transport through measures such as price policies and the expansion of transportation networks. Encouraging the transition to electric public transport vehicles can also reduce environmental impacts.

Awareness and Education: Increasing awareness and education on environmental issues is important. Campaigns can be organized to encourage environmentally friendly behavior, environmental education can be integrated into schools, and information on environmental impacts can be provided to the public.

These suggestions can aid Türkiye in advancing its environmental sustainability efforts. However, a more detailed analysis considering factors such as feasibility, cost, and other considerations is necessary to determine the emphasis on each recommendation.

The use of revenues from green taxation is another important issue for the purposes of the public sector. In practice, the effect of green taxes on reducing pollution and generating revenue for the government are discussed. Instead, while determining the area of use of the revenues obtained from green taxes, its use should be evaluated not only in terms of environmental purposes but also in terms of economic efficiency, in line with environmental purposes, or more generally, in order to combat climate change and ensure sustainable development. On the other hand, the aim of increasing revenues should be abandoned and the said revenues should be used both for environmental purposes and for financing the reduction of taxes, which are considered as a deflector that creates a loss of efficiency in the economy. In this way, the loss of efficiency in the economy will be reduced and the income obtained will be used more efficiently.

Public expenditures such as environmental protection spending, investment spending, and subsidies are effective fiscal policy tools that can be used to support the green economy. Governments can support sustainable development by making investment expenditures in areas such as sustainable energy and transportation where the private sector may not invest or fall short. Additionally, without directly making investments, governments can direct the private sector towards green economic activities through subsidies. Subsidies for environmental research and development projects, initiatives, and resource utilization will be influential in directing economic actors towards these areas. In addition to these measures, tariff exemptions, direct regulations, emissions trading, and similar systems designed by the government are other policy tools that can be used to create an environmentally compatible economic system. Considering all of these factors, private economic actors in the market tend to be reluctant to engage in environmentally positive activities due to the costs involved and their focus on personal interests. In this regard, the fiscal policy tools of governments play a critical role in directing actors within the economy towards a sustainable future. It is crucial for governments to consider not only their own financial structures but also the economic and social structures in which they operate when using fiscal policy tools. Otherwise, the existing system may lead to the emergence of new crises. Therefore, in the perspective of the green economy, it is important to emphasize the need to increase the scope and amounts of public expenditures and shape tax systems with a more environmentally positive outlook (Ağcakaya&Kaya, 2022).

CONCLUSION

While trying to achieve economic growth and development, the aspect of technology that makes life easier is taken into account, whereas the damage it causes to the environment is often overlooked. The unproportional movement of the increasing population with natural resources causes a rapid decrease in natural resources in the world. For example, the use of fossil fuels and carbon dioxide emissions from the use of motor vehicles is one of the main causes of environmental pollution. The sensitiveness of the people is not enough in minimizing these causes. In addition to individual efforts, public administrations should also take some measures and support this effort through green taxes.

One of the most fundamental duties of the state is undoubtedly to establish a sustainable life cycle and a sustainable welfare standard. For this reason, all economic activities carried out by the state should be based on the protection of the ecological system and natural resources. Therefore, providing economic development without harming the nature should be its main goal in all its public activities. At this point, while trying to provide economic development, it should support sustainable development by prioritizing investments in green economy. However, the strategies and policies to be implemented in order to achieve green growth vary according to the conditions and development indicators of each country.

In many developing countries such as Türkiye, which follow the economic development process and do not include environmental protection among their primary objectives, environmental policy should be reviewed before the tax system. The environmental policy should also be designed to adopt the aim of protecting the environment and preventing environmental problems as the primary objective. In such countries, most of the environmental taxes are collected for the purpose of generating income for the general budget. Therefore, the high income potential does not make sense unless it is used for environmental purposes. In addition, in countries like Türkiye, where there is a poverty problem with income distribution problems, subsistence problems are more prominent instead of environmental awareness. For a poor society and individual, it is more important that consumption be cheap rather than harming the environment. Again, the lack of education that comes with poverty is an important deficiency in terms of the establishment of sustainable development awareness in societies. Sustainable development is only possible with well-educated individuals and societies who are conscious of the subject. Both awareness and support of individuals who manage the countries on sustainable development, and conciousness of the households that perform the consumption process in the development process are very important.

When the application area of green taxes in Türkiye is examined, it is seen that there are four taxes that can be evaluated in the green tax category. It is seen that the taxes applied are insufficient in terms of environmental protection policies. It can be said that these taxes are far from a real environmental protection policy, but are taxes collected for financial purposes. For example, when the Environmental Cleaning Tax is examined, it is seen that the aim here is to transfer resources to the municipalities. When the Special Consumption Tax is examined, it is seen that there is no tax on chemical wastes, and that a product with a high carbon emission rate such as Jet Fuel is not taxed. When the current MTV is examined, it is seen that it is far from being a green tax. In practice, as the age of the vehicle increases, the amount of tax to be paid decreases. As the engine cylinder volume increases, the tax payable also increases. However, as the age of the vehicle increases, its technology becomes obsolete and carbon emission production increases. If Türkiye adopts a progressive taxation system based on Carbon Emissions, like EU countries, it will be possible to act more harmoniously with the aim of green taxation.

Türkiye can make strategic investments in various areas to achieve its environmental sustainability goals using tax revenues generated from environmental sources. Examples of these strategic areas include the development and promotion of renewable energy sources, sustainable building designs, and energy efficiency. Increasing clean energy production, raising environmental awareness through education, and preserving natural resources can also be prioritized. It is crucial to collaborate with internal and external stakeholders and establish effective mechanisms for resource allocation to ensure the efficient use of funds. However, it can be said that integrating environmental taxes into the tax system in societies like Türkiye negatively affects low-income people. For this reason, it may be preferable to use the tax revenues obtained from green taxes to eliminate the negative effects on income distribution. Transfer expenditures to low-income people can be increased through green taxes, tax reductions that will affect low-income families in particular can be put into effect. There is no doubt that the transfer expenditure policy shaped in line with this purpose will have a positive contribution on the regressive effects. However, it should be taken into account that tax reforms based on environmental taxs, which include tax reductions in existing taxes without changing the total tax revenue, is also likely to have an additional positive effect in the long run, by influencing real wages and employment, in addition to its short-term positive effect on income distribution. In this framework, if the changes that alter the distribution of the tax burden in favor of low-income people also increase social welfare, the objectives of efficiency and justice in the distribution of resources will be achieved together.

DECLARATION

The author declares that the manuscript follows ethical standards and there are no potential conflicts of interest with respect to the research, authorship and/or publication of this chapter.

DISCLAIMER

ACKNOWLEDGMENT

The author extends sincere gratitude to

- The Editor and the International Editorial Advisory Board (IEAB) of this book, who initially desk reviewed, arranged a rigorous double/triple-blind review process, and conducted a thorough, minute, and critical final review before accepting the chapter for publication.
- All anonymous reviewers who provided very constructive feedback for thorough revision, improvement, extension, and fine-tuning of the chapter.
- All colleagues, assistants and well-wishers who assisted to complete this task.

REFERENCES

Adshead, F., Thorpe, A., & Rutter, J. (2006). Sustainable development and public health. *Public Health, 11*(12), 1102–1105. doi:10.1016/j.puhe.2006.10.003 PMID:17097119

Ağcakaya, S. & Kaya, I. (2022). Sürdürülebilir kalkinma ve yeşil ekonomi perspektifinden yeşil maliye politikalari uygulamalari, *Çukurova Üniversitesi Sosyal Bilimler Enstitüsü Dergisi, 2*, 512-525.

Biyan, Ö., & Gök, M. (2014). Çevre politikaları kapsamında Avrupa Birliği ve Türkiye'de çevre vergilerinin uygulanışı: Karşılaştırmalı bir analiz. *Hitit Üniversitesi Sosyal Bilimler Enstitüsü Dergisi, 2*, 281–310.

Bovenberg, A. L. (1998). Environmental taxes and the double dividend. *Empirica, 25*(1), 15–35. doi:10.1023/A:1006836720884

Burrows, P. (1979). Pigovian taxes, polluter subsidies, regulation, and the size of a polluting industry. *The Canadian Journal of Economics. Revue Canadienne d'Economique, 12*(3), 494–501. doi:10.2307/134738

Cottrell, J., Kai, S., Matthias, R., & Alexander, M. (2016). Environmental tax reform in developing, emerging and transition economies, The German Development Institute, 1-126.

Dinç, G. (2008). *Avrupa insan hakları sözleşmesi'ne göre çevre ve insan.* Türkiye Barolar Birliği Yayınları.

Ekeman, E. (1998). *Avrupa Birliği ve Türkiye'nin çevre politikalarının karşılaştırmalı incelemesi.* İktisadi Kalkınma Vakfı Yayını.

Ercan, E. (2015). Türkiye'de çevre vergileri mi çevre vergisi mi? *TBB Dergisi, 119*, 209–226.

Gürlük, S. (2010). Sürdürülebilir kalkınma gelişmekte olan ülkelerde uygulanabilir mi? *Eskişehir Osmangazi Üniversitesi İktisadi ve İdari Bilimler Fakültesi Dergisi, 5*, 85–99.

Kaplan, A. (1991). Çevre sorunları ve ekonomi. *Ankara Üniversitesi Siyasal Bilgiler Fakültesi Dergisi, 1*, 279–287.

Kuşat, N. (2013). Yeşil sürdürülebilirlik için yeşil ekonomi: Avantaj ve dezavantajları- Türkiye incelemesi. *Journal of Yasar University, 29*, 4896–4916.

Luckin, D. (2000). Environmental taxation and red-green politics. *Capital and Class*, *24*(3), 161–189. doi:10.1177/030981680007200108

Lutz, C., & Bern, M. (2010). Environmental tax reform in the European Union: Import on CO2 emissions and the economy. *Z. Energiewirstseh*, *34*(1), 1–10. doi:10.100712398-010-0009-x

Markandya, A. (2002). Environmental implacations of non-environmental policies. Handbook of Environmental Economics, 3, Stockholm.

Metcalf, E. G. (2003). Environmental levies and distortionary taxation: Pigou, taxation and pollution. *Journal of Public Economics*, *87*(2), 313–322. doi:10.1016/S0047-2727(01)00116-5

Miceikiene, A., Rauluskeviciene, J., Streimikine, D., & Ciuleviciene, V. (2018). Assessment of the effect of environmental taxes on environmental protection [Journal of Economics]. *Ekonomicky Casopis*, *66*, 286–308.

OECD. (2001). *Environmentally related taxes: issues and strategies Policy Brief. OECD*. www.oecd.org/pcd/2674642.pdf

OECD (2006). Subsidy Reform and Sustainable, . doi:10.1787/9789264019379-en

OECD. (2009). *Environmentally related taxes in the OECD countries*. OECD. https://www.oecd.org/dataoecd/27/23/36966499.pdf

Orhan, G. (2013). Türkiye'de çevre politikaları: Değişen söylemler değişmeyen öncelikler. *Memleket Siyaset Yönetim*, *19*, 1–24.

Öz, E., & Kutbay, H. (2016). Ekolojik vergileme: Seçilmiş bazı dünya ülkeleri ile türkiye verilerinin karşılaştırılması. *Eskişehir Osmangazi Üniversitesi İktisadi ve İdari Bilimler Fakültesi Dergisi, Vol, 11*, 247–271.

Özçağ, M., & Hotunluoğlu, H. (2015). Kalkınma anlayışında yeni bir boyut: Yeşil ekonomi. *CBÜ Sosyal Bilimler Dergisi*, *13*, 304–324.

Özdemir, B. (2009). Küresel kirlenme sürdürülebilir ekonomik büyüme ve çevre vergileri. *Maliye Dergisi*, *156*, 1–36.

Özkanca Andıç, N., & Erasa Akça, İ. (2019). Küresel kamusal bir mal olan çevrenin korunmasına yönelik vergileme yeşil vergiler ve Türkiye uygulaması. *Journal of Social and Humanıtıes Sciences Research*, *6*, 3447–3453.

Pearce, D. W., Atkinson, G. D., & Dubourg, W. R. (1994). The economics of sustainable development. *Annual Review of Energy and the Environment*, *19*(1), 454–474. doi:10.1146/annurev.eg.19.110194.002325

Pigou, A. C. (1920). *The Economics of Welfare* (1st ed.). Macmillan London.

Reyhan, A. S. (2014). Çevre ekonomisinde çevre vergileri uygulamaları. *Hitit Üniversitesi Sosyal Bilimler Enstitüsü Dergisi*, *7*, 110–120.

Serdaroğlu, S. (2003). *Avrupa bütünleşmesine katkıda bulunan bir faktör olarak çevre politikası, dünden bugüne AB*. Boyut Yayınları.

Sollund, S. (2007). *Environment taxes. United Nations Group of Experts on Domestic Resource Mobilisation*, Rome.

Speck, S. (2007). Overview of environmental tax reforms in EU member states, COMERTFinal Report to the European Commission, DG Research and DG Taxation and Customs Union. EU.

Toprak, D. (2006). Sürdürülebilir kalkınma çerçevesinde çevre politikaları ve mali araçlar. *Süleyman Demirel Üniversitesi Sosyal Bilimler Enstitüsü Dergisi*, *4*, 46–69.

Ulusoy, A., & Akdemir, T. (2013). Yerel yönetimlerin finansman sorunlarının çözümünde motorlu taşıtlar vergisi önerisi. *Sosyoekonomi*, *19*, 87–116.

UNEP. (2010). *Green economy: driving a green economy through public finance and fiscal policy reform*. UNEP. https://www.greengrowthknowledge.org/sites/default/files/downloads/resource//Driving_a_GE_through_public_finance_and_fi scal_policy_reform_UNEP.pdf

Vanii, S., & Uçar, A. (2021). Çevre ve yeşil vergiler üzerine Türkiye özelinde değerlendirmeler, *International Journal of Social* [IJSPFR]. *Political and Financial Researches*, *1*, 68–81.

Williams, R. C., III. (2016). *Environmental taxation*. NBER Working Paper Series 22303. https://www.nber.org/papers/w22303/

Yavilioğlu, C. (2002). Kalkınmanın anlambilimsel tarihi ve kavramsal kökenleri, *C. Ü. İktisadi ve İdari Bilimler Dergisi*, *3*, 59-77.

Yegen, B. (2019). Toplum sağlığına vergisel bir müdahale: Obezite vergisi. *Fiscaoeconomia*, *2*, 146–158.

Yüksel, C. (2013). *Kamu ekonomisi ve çevre sorunları*. Ekin Kitabevi Yayınları.

KEY TERMS AND DEFINITIONS

Environmental Cleaning Tax: The Environmental Cleaning Tax, which was established in 1993 in Turkey, is in the nature of a fee in the theoretical sense, and the subject of the tax is a local administration tax that is collected from the residential, workplace and other buildings that benefit from the solid waste collection services of the municipalities, according to the fixed tariff stipulated.

Green Tax: If a tax actually encourages the protection of natural resources and the environment or prevents harm, that tax can be expressed as a "Green Tax".

Pigouvian Tax: Pigou tax, introduced by Arthur Cecil Pigou, is the taxation of negative externalities to ensure efficiency in resource allocation.

Pollution Pay Principle: The "polluter pays" principle is a simple common-sense principle: The polluter – and this may be the polluting actors or activity – must pay the price to right the wrong. This may require cleaning the contaminated area or reimbursing the health costs of affected people.

Public Policies: Public policy is concerned with how binding decisions are made in a society, how social issues come to the agenda of decision-makers, how policy choices are made, what role actors play in the policy-making process, and how the success or failure of policy outcomes is evaluated.

Sustainable Development: The concept of sustainable development was first defined in the Brundtland Report prepared by the World Commission on Environment and Development in 1987 as "development that meets the needs of the present without compromising the ability of future generations to meet their needs".

Tax Burden: The tax burden, which can be defined as the ratio of taxes paid to income in general, is an indicator of the obligation and effort of the society to meet public expenditures.

Chapter 9
Legal, Financial and Fiscal Measures to Support the Collection Solutions of Biowaste in Portugal

Fátima David
https://orcid.org/0000-0001-7075-5159
CICF-IPCA, UDI-IPG, Polytechnic of Guarda, Portugal

Pedro Melo Rodrigues
https://orcid.org/0000-0002-7790-3473
CIQUP-IMS-DGAOT-FCUP, UDI-IPG, Polytechnic of Guarda, Portugal

Elisabete Soares
https://orcid.org/0000-0003-0488-2306
UDI-IPG, Polytechnic of Guarda, Portugal

Tânia Cristina Azevedo
https://orcid.org/0000-0003-4242-4576
State University of Feira de Santana (UEFS), Brazil

ABSTRACT

The Portuguese Decree-Law No. 102-D/2020, of December 10 (PCM, 2020b), dictates the implementation of biowaste selective collection networks or the separation and recycling of biowaste at the source until December 31, 2023. This chapter aims to analyze the financial support provided to municipalities (local authorities) for implementing biowaste collection solutions, such as on public roads or through door-to-door collection, and to citizens for adopting biowaste separation and recycling practices through domestic or community composting. The authors also explore the application of the polluter-pays principle, which requires citizens to pay environmental taxes and fees to support sustainable development. Regarding this focus, the chapter highlights the legal, financial, and fiscal measures used to support biowaste collection and recycling efforts in Portugal, providing insight into the steps needed to achieve a sustainable environment while addressing the economic and social needs of the population.

DOI: 10.4018/978-1-6684-8592-7.ch009

INTRODUCTION

Climate change is an undeniable reality already affecting all regions of the Earth with profound impacts on the environment and socio-economic activities. Human activity is the primary cause, leading to biodiversity loss and a general increase in pollution, releasing large amounts of greenhouse gases (GHGs) into the atmosphere, such as carbon dioxide (CO_2), methane (CH_4), and nitrous oxide (N_2O), which contribute to climate change (EPA, 2023). Thus, addressing climate change requires a multifaceted approach (Hawken, 2017; Fazey et al., 2018; IPCC, 2019) that involves reducing emissions, protecting and restoring ecosystems, and transitioning towards more sustainable economic models.

In response to these issues, the European Commission presented the "*European Green Deal*", COM(2019) 640 final, of December 11. This agreement aims to shape the European Union's (EU) growth policy for the coming years, to achieve carbon neutrality by 2050 through the "*European Climate Law*", which ensures that all EU policies contribute to achieving net-zero GHG emissions, mainly by reducing emissions, investing in green technologies, and protecting the natural environment. The cost of carbon is internalized into the economy by incorporating it into the final price of the products (Weitzman, 2015; Baranzini et al., 2017; Trinks et al., 2022).

Similarly, the "*Circular Economy Action Plan*", approved by COM(2020) 98 final of March 3, aims to promote the development of a more sustainable production and consumption system to pursue carbon neutrality and biodiversity conservation. More than half of the emissions produced worldwide, over 90% of biodiversity loss, and hydrological stress are related to the extraction and processing of raw natural resources. In this context, waste prevention and recycling are among the top priorities, with ambitious recycling targets and an emphasis on avoiding waste production, seeking that in the production processes, secondary raw materials can replace the extractive raw materials, particularly in sectors such as textiles, construction, electronics, and plastics.

Like other Western European countries, such as Spain, France and Italy, Portugal aims to increase the sorting and recycling of solid waste to reduce the amount of waste sent to incineration or landfill. As of 2021, biowaste collection in mainland Portugal accounted for 37.42% of total municipal waste collected, with a significant proportion (45.48%) in undifferentiated waste and a smaller proportion (7.86%) in the selective collection (APA, 2022a). In contrast, the EU has set a target of at least 65% recycling of solid waste for households (Dijkgraaf & Gradus, 2016). By separating biowaste, Portugal can remove the main component responsible for GHG emissions and leach from landfills, maximizing its potential for high-quality recycling and increasing the use of high-quality secondary raw materials, thus enabling a more circular economy (Rodrigues et al., 2022).

This chapter aims to examine the measures taken by the Portuguese Government to ensure that the country complies with the mandate to implement selective biowaste collection networks or the separation and recycling of biowaste at the source by December 31, 2023, as approved by the Decree-Law No. 102-D/2020, of December 10. As a result, biowaste that is selectively collected cannot be incinerated or deposited in landfills. The amount of biodegradable municipal waste that enters the aerobic or anaerobic treatment can be counted as recycled if it produces compost or other output with a similar quantity of recycled material and only if it is separately collected or separated at the source (Rodrigues et al., 2022). Furthermore, if the resulting material from the treatment is used in soil, it can only be considered recycled if it benefits agriculture or the environment. Through this examination, the chapter sheds light on the actions taken by the Portuguese Government to promote sustainable waste management and achieve a circular economy.

Therefore, the established methodology is qualitative and descriptive (Maxwell, 1992; Sandelowski, 2000; Mack et al., 2005; Sandelowski, 2010; Nassaji, 2015; Manjunatha, 2019; Elliott & Timulak, 2021; Cropley, 2022) and includes a review of the legal, financial, and fiscal measures that govern the selective collection or separation and recycling of biowaste. In this sense, the chapter presents the legal measures that ensure that biowaste recycling contributes to carbon neutrality in Portugal by 2050. Additionally, it outlines governmental financial mechanisms that allow, on the one hand, municipalities (local authorities) to implement biowaste selective collection solutions (on public roads or door-to-door) and, on the other hand, citizens to adopt separation and recycling at the source (whether through domestic or community composting). The chapter also addresses fiscal measures, such as environmental taxes and fees (charges paid by citizens), which follow the polluter-pays principle and ensure a sustainable environment. Citizens, institutions, and businesses (such as hotels and the catering industry) typically finance municipal solid waste (MSW) management through fees and taxes (World Bank, 2018). Finally, the authors present conclusions based on their reflections.

LEGAL MEASURES

The Constitution of the Portuguese Republic (CPR) states in paragraph 1 of article 66 that "*everyone shall possess the right to a healthy and ecologically balanced human living environment and the duty to defend it*". In order to ensure this fundamental right of all citizens within an overall framework of sustainable development, paragraph 2 of the same article requires to Portuguese State to "*prevent and control pollution and its effects and the harmful forms of erosion;* (…) *promote the rational use of natural resources, safeguarding their ability to renew themselves and maintain ecological stability, with respect for the principle of inter-generational solidarity; promote, in cooperation with local authorities, the environmental quality of rural settlements and urban life* (…)*; promote the integration of environmental objectives into the various policies of a sectoral nature; promote environmental education and respect for environmental values;* [and] *ensuring that fiscal policy renders development compatible with the protection of the environment and the quality of life*". Thus, the Constitution seeks to preserve the environment for future generations, foster a culture of environmental responsibility, and ensure that economic progress is not achieved at the expense of the natural world.

Likewise, other Western European countries have references to environmental concerns in their constitutions. For example, the Spanish Constitution, adopted in 1978, recognizes the right to environmental protection and emphasizes the significance of environmental conservation and sustainable development. Article 45 of the Constitution states that everyone has the right to enjoy an environment suitable for personal development and the duty to preserve it. Similarly, the French Constitution, promulgated in 1958, includes provisions that address environmental aspects and the protection of the environment. The Preamble of the French Constitution emphasizes the importance of preserving the environment for present and future generations. It states that the nation must protect the common heritage, including nature, the environment, and historical and cultural heritage. The Constitutional Council has recognized environmental protection as a fundamental and essential principle. Therefore, the Constitution implicitly guarantees the right to a healthy environment. The precautionary principle in French legislation and enshrined in the Constitution states that authorities must take preventive measures to avoid environmental or human health harm in the presence of scientific uncertainty regarding a risk.

Furthermore, the Italian Constitution, enacted in 1947, recognizes the fundamental right to a healthy environment. Article 9 of the Fundamental Principles of the Constitution states that the Republic must protect the natural environment and cultural heritage. Additionally, Alinea S of article 117, Title V – Regions, Provinces – Municipalities, specifies that the State has exclusive legislative powers to protect the environment, the ecosystem, and cultural heritage.

Achieving environmental sustainability requires a multilevel governance approach (McGuinn et al., 2017; Wagner et al., 2021) involving different actors and institutions. Di Foggia & Beccarello (2023) argue that, at the national level, the central State establishes laws and targets to guide environmental policy, as is the case of Law No. 19/2014, of April 14, which approved the Portuguese "*Environmental Framework Law*", establishing the national environmental policy to optimize and guarantee the permanent use of natural resources, both qualitatively and quantitatively, as well as the sustainable development. Thus, the environmental policy aims to manage, preserve, and improve the natural components of the environment, such as air, water and sea, biodiversity, soil and subsoil, and the landscape and recognizes and values the importance of natural resources and ecosystem, goods and services, in order to prevent environmental dysfunctions and minimize adverse effects on human health and the environment. The environmental policy also addresses the components linked to human behavior, namely climate change, waste, noise, and chemical products. Similarly, outer European countries have established legal frameworks to address environmental concerns. In Spain, the Environmental Protection Law (Law No. 26/2007, of October 23) sets the legal framework for environmental protection and sustainable development. This law covers various aspects such as environmental conservation, pollution prevention, waste management, and natural resource management. It also establishes the responsibilities of public authorities, individuals, and businesses in safeguarding the environment. France has the Environmental Code (Decree No. 2007-397, of March 22), consolidating the country's environmental legislation. The Environmental Code covers many environmental topics, including pollution control, waste management, nature conservation, water resources, and energy efficiency. It provides the legal framework for environmental protection, with its provisions being implemented through regulations and decrees. Similarly, Italy also has an Environmental Code (Legislative Decree No 152/2006, of April 3), which governs environmental matters in the country with the same goals mentioned for Spain and France. It addresses various environmental concerns and regulations.

For that, Di Foggia & Beccarello (2023) also emphasize that, at the regional level, Regions develop plans to achieve these targets to reflect their specific circumstances and challenges. Meanwhile, Regulators oversee environmental, technical, and economic issues, and Municipalities assume local waste collection organization responsibility. At the individual level, Citizens play a crucial role in implementing environmental sustainability policies, particularly regarding separating and disposing of waste, including biowaste. By involving all these stakeholders, a comprehensive and coordinated approach to environmental management can be achieved, helping to ensure a cleaner and healthier planet for future generations.

In line with the EU environmental directives, the Portuguese State has implemented Decree-Law No. 102-D/2020 of December 10, which requires the establishment of networks for the selective collection of biowaste or the implementation of separation and recycling of biowaste at-source until December 31, 2023. This legislation transposed, among others, Directive (EU) 2018/851 of the European Parliament and the Council of May 30, which aims to promote a circular economy and reduce the amount of waste in landfills. The circular economy has become one of the mainstream concepts to address environmental issues (Afum et al., 2022) and contributes globally to mitigate the negative externalities created by companies and the environmental burden on society (Harris et al., 2021; Baah et al., 2021; Pfister & Mathys, 2022). For a truly circular economy, adopting additional measures to promote sustainable production and consumption is essential, focusing on the entire product life cycle in a way that preserves resources and closes the loop. By mandating the separation and recycling of biowaste, the Portuguese Government aims to reduce environmental pollution, promote the sustainable use of resources, and improve the ecosystem's overall health. To this end, the "*Landfill Waste Legal Framework*", approved by Decree-Law No. 183/2009 of August 10, has been revoked. Specifically, the goal is to ensure that, by 2030, landfills will no longer accept waste suitable for recycling or other forms of recovery, such as MSW. The reuse and recycling targets were 55%, 60% and 65% for 2025, 2030, and 2035, respectively. In addition, Decree-Law No. 102-D/2020 of December 10 requires the diversion of biowaste from landfills and the recovery of previously landfilled waste through landfill mining, as well as the methodology for calculating recycling rates has been updated. The target for landfill waste has been reduced to only inert materials or waste that cannot be recovered, with a maximum limit of 10% until 2035. These changes align with the "*waste management hierarchy*" and represent a clear and ambitious vision for sustainable MSW management in Portugal.

Thus, the "*National Waste Management Plan*" (Portuguese acronym: PNGR), approved by Decree-Law No. 178/2006 of September 5, has been revised. The current legislation introduces several programmatic aspects, such as the definition of a new waste planning and management structure and the densification of the content of the "*National Waste Plans*". Furthermore, it provides updated targets, including those for municipal waste production, hazardous waste reduction and food waste reduction (mainly to ensure the separate collection of biowaste), providing the general outlines of the new requirements for extended producer liability regimes. It also adjusts the structure and scope of the waste management fee (Portuguese acronym: TGR) to penalize less noble treatment operations in the waste hierarchy.

Currently, the national waste policy is supported, at the macro level, by the PNGR 2030, approved by Order No. 4242/2020 of April 7, and subsequently published through Council of Ministers Resolution No. 31/2023 of March 24. This plan serves as a macro-level planning instrument for waste management, defining strategic guidelines and ensuring coherence with other plans such as the "*Strategic Plan for Municipal Waste*" (Portuguese acronym: PERSU 2030) and the "*Strategic Plan for Non-Municipal Waste*" (Portuguese acronym: PERNU 2030), both published by Council of Ministers Resolution No. 30/2023, of March 24. Table 1 shows the strategic objectives, indicators, and targets of the PNGR 2030.

In this context, Table 2 sets out the strategic targets of PERSU 2030, underlying the PNGR 2030 and focusing on implementing the waste hierarchy, emphasizing prevention. In the coming years, is expected a reversal of the trend of increasing waste production, namely through measures that encourage the reuse, recycling, and/or the extension of the lifespan of products. Regarding the production of waste that cannot be avoided, a significant increase in the quantities collected selectively is foreseen, to increase the quality of waste recovered, an essential condition for obtaining products with a higher added value, a key factor for the transition to a circular economy with a high level of efficiency in the use of the resources.

Table 3 exhibits the monitoring of the Portuguese Operational Objectives (Portuguese acronym: OP), defined in the PERNU 2030, which contribute to the global challenge of reducing non-municipal waste production, as well as the environmental impacts resulting from its management. These OPs are:

- OP1 - Prevent waste generation in terms of quantity and hazardousness;
- OP2 - Reduce the environmental impact of waste management;

- OP3 - Raise awareness, train, and disseminate, at academic and organizational levels, on waste prevention and management; and
- OP4 - Increase investment capacity and expenditure on R&D+I (Research and Development + Innovation) aimed at prevention and waste management.

Table 4 shows the targets for each of the strategic indicators to be monitored in fulfilling the PERNU 2030, translating the national guidelines for the prevention and management of non-municipal waste, contributing to the implementation of the principles laid down in the different strategic initiatives and legislation at national, Community and international level for the protection of the environment.

Table 1. PNGR 2030's indicators and targets per strategic objectives

Strategic Target	Indicator	Reference Value (2018)	Target 2023	Target 2027	Target 2030
Strategic Objective 1 – Prevent waste production in terms of quantity and hazardousness					
Reduce waste generation	Waste generation (value in 2018 = index 100)	100	96.8	89.6	85.6
Reduce the generation of hazardous waste compared to total waste generation	Hazardous waste generation/Waste generation (%)	7.0	5.1	4.8	4.4
Strategic Objective 2 – Promote resource efficiency, contributing to a circular economy					
Decouple economic growth from material consumption	Gross domestic product/Domestic consumption of materials (k€/t)	1.18	1.25	1.51	1.68
Decouple economic growth from waste generation	Waste generation/Gross domestic product (t/k€)	0.080	0.0736	0.063	0.059
Increase the availability of waste for the economy	Recovery (non-energy)/Waste generation (%)	65.0	73.1	77.4	80.5
Strategic Objective 3 – Reduce negative environmental impacts through an integrated and sustainable waste management					
Reduce the amount of waste sent for disposal	Waste sent for disposal	100	85.0	56.5	41.5
Reduce GHG emissions from the waste sector	Mt $CO_{2eq.}$ emissions from the waste sector	6.50	5.28	4.86	4.55

Source: Adapted from Council of Ministers Resolution No. 31/2023. Legend: K€ - 1000 Euros; t - ton.

Table 2. PERSU 2030's strategic targets

Strategic Target	Target 2020	Target 2025	Target 2030	Target 2035
Municipal waste prepared for reuse or recycled (%)	50	55	60	65
Municipal biodegradable waste landfilled (%)	35			
Separate collection for recycling (Kg/inhabitants year)	47			
Reduction of municipal waste landfilled (%)				10
Packaging waste recycling (%)		65	70	
Separate collection of single-use plastic beverage bottles (%)		77	90 (year 2029)	
Reduction of food waste generation (indicative) (%)		30	50	

Source: Adapted from Decree-Law No. 102-D/2020 and Council of Ministers Resolution No. 30/2023.

Table 3. PERNU 2030's strategic targets and indicators per operational objectives

Objectives	Strategic Target	Indicator
OP1, OP2, OP3	Reduce non-municipal waste generation	Non-municipal waste generation
	Decrease in quality and representativeness of the generation of non-municipal hazardous waste	Non-municipal hazardous waste generation
OP2, OP3	Decouple economic growth from material consumption	Gross domestic product/Domestic consumption of materials
	Decrease the ratio between waste generation and gross domestic product	Non-municipal waste generation/Gross domestic product
	Reintroduction of waste into the economy	Recovery (non-energy) of non-municipal waste/ Non-municipal waste generation
	Maintain the growing trend in the representativeness of recovery operations	Recovery (including energy) of non-municipal waste/ Non-municipal waste generation
OP2, OP3	Decrease the amount of non-municipal waste sent to disposal operations	Non-municipal waste sent to disposal
OP4	Increase total expenditure and respective representativeness of R&D investments with environmental socio-economic objectives	Expenditure on R&D with environmental objectives/Total expenditure on R&D

Source: Adapted from APA (2022b).

Table 4. PERNU 2030's targets per indicator

Indicator	Reference Value (2019)	Target 2023	Target 2027	Target 2030
Non-municipal waste generation (t)	11 427 435	10 210 698	9 611 428	9 320 010
Non-municipal hazardous waste generation (t)	1 066 055	812 555	677 130	599 261
Gross domestic product/Domestic consumption of materials (k€/t)	1.181	1.253	1.507	1.666
Non-municipal waste generation/Gross domestic product (t/k€)	0.056	0.049	0.043	0.040
Non-municipal waste sent to disposal (t)	1 751 501	1 272 452	975 592	811 087
Recovery (non-energy) of non-municipal waste/ Non-municipal waste generation (%)	82.7	85.9	88.5	90.2
Recovery (including energy) of non-municipal waste/ Non-municipal waste generation (%)	84.7	87.5	89.8	91.3
Expenditure on R&D with environmental objectives/Total expenditure on R&D (%)	6.4	9.1	13.1	17.1

Source: Adapted from APA (2022b). Legend: t – ton; K€ - 1000 Euros.

Additionally, Decree-Law No. 102-D/2020, of December 10, also contains a very ample set of changes to the "*Management of Specific Waste Streams Framework*", provided for in Decree-Law No. 152-D/2017, of December 11, seeking to regulate the procedures and responsibilities of industries and economic sectors, particularly concerning the reuse of packaging, used oils, used tires, electrical and electronic equipment, batteries and accumulators, and end-of-life vehicles.

Moreover, the national circular economy strategy, approved by Council of Ministers Resolution No. 190-A/2017 of December 11, defends an economy that actively promotes the efficient use and productivity of resources through products, processes and business models that prioritize dematerialization, reuse, recycling, and material recovery. The aim is to maximize materials' economic value and utility for as long as possible, in cycles powered by renewable sources, with benefits for both suppliers and users. By prioritizing the circular economy, the strategy also provides various environmental benefits. For instance, it reduces the need to extract and import raw materials, reducing the carbon footprint associated with these activities and helps to minimize waste production and associated emissions, making it a sustainable and responsible way to operate.

The national strategy is a forward-thinking approach to economic growth that aligns it with sustainability principles. The existence of waste and biowaste collection systems are linked to sustainable development, whether environmental, social, or economic, to improve living conditions and preserve the surrounding environment in the short and, above all, in the medium and long term. They must respond to the needs of the present without compromising the ability of future generations to grow.

According to the National Energy and Climate Plan (Portuguese acronym: PNEC) for 2021-2030, approved by the Council of Ministers Resolution No. 53/2020 of July 10, the recycling of biowaste may positively contribute to other objectives, such as limiting GHG emissions, building a strategy based on renewable energy sources on the way to a carbon-neutral economy by 2050 (electrification of the economy and consumption, promotion of electric vehicles and introduction of renewable gases), as well as diversifying energy sources through the development of endogenous renewable energy resources and promoting the interconnection of energy markets and systems.

Therefore, in Portugal, the responsibility for the selective collection and recycling of biowaste at source lies with the municipalities and lower administrative entities, entailing working closely with higher administrative entities responsible for treating and recovering this type of waste. This means that all municipalities are required to plan and implement separate biowaste collection schemes, either through door-to-door collection or by providing citizens with different biowaste collection points. Alternatively, they can promote the home or community composting units to encourage citizens to recycle their biowaste directly at the source. This approach aims to prevent biowaste from ending up in the public collection system, thereby reducing the waste that requires treatment and minimizing its environmental impact. By prioritizing selective collection and recycling at the source, Portugal is taking an essential step towards achieving a more sustainable and responsible municipal waste management system.

Similarly, Spain has transposed the Waste Framework Directive through the Law on Waste and Contaminated Soils (Law No. 7/2022, of April 8). This law establishes waste management principles, including waste prevention, recycling, and reducing landfilling. It sets out the legal framework for waste management planning, waste classification, and the responsibilities of waste producers, managers, and public authorities. The Waste Framework Directive in France has also been transposed through the Environmental Code. The Environmental Code contains provisions on waste management, including waste prevention, recycling, and reducing landfilling. It establishes requirements for waste collection, treatment, and disposal and outlines the responsibilities of waste producers, operators, and public authorities. Italy has implemented the Waste Framework Directive through the Consolidated Environmental Law. This legislation encompasses waste management provisions, including waste prevention, recycling, and landfill reduction. It sets out requirements for waste collection, treatment, and disposal and outlines the roles and responsibilities of waste producers, operators, and public authorities.

GOVERNMENTAL FINANCIAL MECHANISMS

The option to pursue a development model that assumes unlimited natural resources and the ability of nature to handle the waste generated by human activities has caused severe environmental problems, including the pollution of soils, waters, and air, and also contributing to climate change. The escalation of economic activities and their impact on the depletion of natural resources and the decreasing environment's capacity to handle waste has become a significant concern at the highest levels of policymaking and within society. Consequently, integrating environmental policies has become vital for achieving sustainable and well-balanced economic and social development.

Similarly, like other countries, the Portuguese Government's policy promotes several strategic pillars in the environmental field, including:

- Territory planning through biodiversity conservation;
- Decarbonization, with a focus on climate action, energy efficiency, and sustainable mobility; and
- Circular economy, emphasizing dematerialization, efficient resource utilization, and recycling.

Within the framework of a circular economy, resources, including materials and energy, are managed to maintain their value for as long as possible, thereby reducing the reliance on raw materials and minimizing waste generation. This approach leads to cost savings and mobilizes capital that can be redirected towards wages, job creation, and research and development (R&D). Portugal is actively developing its national action plan for the circular economy and making significant investments in this area, mainly through establishing the Environment Fund (EF).

According to Resor (1997), EFs in each country play a crucial role in implementing national environmental action plans and agendas. From a financial standpoint, EFs are considered long-term funding sources for environmental protection and sustainable development. They serve as effective instruments for managing funds, mobilizing additional resources, and promoting cost-effective approaches. From an environmental perspective, EFs support national environmental strategies, strengthen the capacity of local environmental organizations, and provide opportunities for public-private partnerships and decentralized decision-making (Meyer, 1997). A similar situation can be observed in other European countries. In Spain, there is the *Fondo para la Protección del Medio Ambiente y la Acción por el Clima - Fondo Clima* (in English, Environmental and Climate Change Fund). It is a financial instrument managed by the Ministry for Ecological Transition and Demographic Challenge. The fund supports projects and investments focused on environmental protection, climate change mitigation, and sustainability. In France, there is the *Agence de la transition écologique - ADEME* (in English, Environment and Energy Management Agency), a public institution dedicated to supporting ecological transition and sustainable development. ADEME provides financial assistance, grants, and funding programs to individuals, businesses, and local authorities for environmental projects and investments. While in Italy, the Ministry of Ecological Transition oversees various funding mechanisms and institutions to support environmental investments. For example, the Ministry collaborates with the Italian Institute for Environmental Protection and Research to provide technical and financial support for environmental initiatives. These initiatives aim to promote environmental protection, sustainability, and the transition to a greener economy.

In line with this, the Portuguese EF, established by Decree-Law No. 42-A/2016 of August 12, aims to support environmental policies that align with sustainable development objectives and also contribute to the fulfilment of national and international commitments related to climate change, water resources, waste management, nature conservation, biodiversity, and the transition to a circular economy. The EF provides financial support to institutions, public or private, activities or projects that address the following objectives:

- Climate change mitigation, adaptation, and cooperation;
- Carbon sequestration;
- Participation and utilization of the carbon market to meet international targets;
- Efficient use of water, protection of water resources, and sustainability of water services;
- Prevention and remediation of environmental damage;
- Compliance with national and Community targets and objectives for municipal waste management;
- Transition to a circular economy;
- Protection and conservation of nature and biodiversity;
- Environmental awareness and education; and
- Environmental R&D.

The Portuguese EF has approved several programs and funding initiatives in the framework of waste and the circular economy since 2018. Table 5 illustrates one of these programs, called "*Circular Economy in Municipalities (JUNTAr+)*", which was published in Notice No. 6519/2019 of April 9. This program aims to support and promote local solutions within the circular economy, highlighting their significant economic, social, and environmental benefits. Specifically, it caters for the needs of municipalities. In 2018 and 2019, the funded projects under this program encompassed a diverse range of areas, including establishing local composting networks, reusing school manuals, toys, and childcare products, as well as implementing community kitchens. Additionally, in 2021, following the COVID-19 pandemic, the program shifted its focus to the construction sector, particularly materials and components derived from construction or demolition/rehabilitation works, textiles, post-consumer furniture, and electrical and electronic equipment with potential for reuse.

Table 5. Circular economy in municipalities (JUNTAr+)

	2018	2019	2021
Funding value (€)	1 000 000	1 000 000	500 000
Number of candidatures	110	99	48
Number of projects funded	49	44	19
Number of projects non-funded	61	55	29

Source: Adapted from FA (2019; 2021).

Another program funded by the Portuguese EF is the "*Accelerator: Creating and Rethinking Circular Businesses - CIRCULAr: Startups*" outlined in Notice No. 7659/2019, of May 3. The program, with a maximum allocation of €150,000, aims to support an entity that provides a circular economy acceleration program for startups and the business community, with a specific focus on:

- Startups that aim to develop and expand business opportunities identified within the circular economy; and
- Existing companies that seek to develop or scale up business opportunities within the circular economy or adapt their products, services, or business models to align with circular economy principles.

Also, in 2019, the Portuguese EF provided funding for the program titled "*Support for incentive system for the return of non-reusable plastic beverage packaging*", published in Notice No. 12599/2019, on August 7. The program received a maximum allocation of €1,665,000 to support implementing a pilot project to test such a system. In this context, "*non-reusable plastic beverage packaging*" refers to single-use beverage packaging made of polyethylene terephthalate (PET) plastic, with a capacity ranging from 0.1 to 2 litres inclusive, and it includes all categories of beverages such as water, juices, soft drinks, and alcoholic beverages, excluding dairy drinks. The funding's scope is restricted to facilitate the operationalization of the pilot project and focus learning efforts on PET plastic, the predominant material. This consideration takes into account the future EU target of achieving a separate collection of beverage bottles under 3 litres, including their caps and closures: 77% by 2025 and 90% by 2029. Additionally, companies bound by recycling agreements have committed to achieving a 90% take-back rate by 2025.

In 2020, the Portuguese EF provided further funding through Order No. 7262/2020 of July 17. This funding created the "*Support Program for the Elaboration of Municipal Studies for the Development of Biowaste Collection Systems*" with a total allocation of €1,400,000. This funding applies to mainland Portugal, and the eligible beneficiaries include municipalities, managing entities responsible for municipal waste management systems, and managing entities of inter-municipal waste management systems delegated with this responsibility by the municipalities. The incentive aims to support the preparation of a diagnosis leading to formulating an action and investment plan for implementing biowaste selective collection. This involves the recovery of biowaste through establishing a selective biowaste collection network or source separation and recycling via domestic or community composting. These efforts align with the strategies already defined or that will still be defined by the municipal waste management systems.

Subsequently, through Notice nº 8247/2021, of May 4, the Portuguese EF approved the program "*Biobairros - da terra à terra*" with a maximum allocation of €500,000. The program provides 85% co-financing, up to a maximum value of €31,250 per application. It aims to incentivize municipalities to implement more effective separation and recycling solutions at the source, reducing waste disposal in landfills and simultaneously encouraging the circular economy. The proposed solutions should focus on achieving source separation objectives, utilizing biowaste produced by the individual producer (domestic composting) and establishing a local network for biowaste reception and distribution of compost (community composting).

In 2021, the Portuguese EF also funded the program "*Packaging reuse systems in ready-to-eat and take-away or home delivery*", approved under Notice No. 19975/2021, October 22. The program received a maximum allocation of €150,000 with 85% co-financing, up to a maximum value of €75,000 per application. Its purpose is to support the creation and implementation of innovative systems that promote the availability of reusable packaging in take-away and/or delivery systems. The general objective is to reduce the use of disposable packaging in preparing and distributing meals, which often generates a significant volume of waste. The program aims to encourage the adoption of reusable alternatives that meet food hygiene and consumer safety criteria while raising consumer awareness about the negative impact of excessive disposable packaging. Its specific objectives are to provide cost-effective solutions that allow restaurants and catering establishments to replace disposable packaging with reusable alternatives for ready-to-eat, take-away and/or home delivery services, thereby promoting these activities' environmental, social, and economic sustainability and offering consumers a choice.

In 2022, the Portuguese EF supported the program "*RecolhaBio - Support for the implementation of projects for the selective collection of biowaste*" (FA, 2023a). This initiative contributes to achieving carbon neutrality by 2050 and supports the PNEC. Furthermore, it plays a central role in the future national circular bioeconomy strategy while creating job opportunities. In this sense, the program's objective is to promote and support implementing projects for the selective collection of biowaste in Portuguese municipalities. These projects aim to increase the quantity and quality of waste recycling, focusing on biowaste. By financing these projects through a return of 30% of the increase in the TGR to the municipalities, the program aims to promote the reduction of waste disposal in landfills (FA, 2023b).

This year, 2023, the Portuguese EF supported the program "*Sê-lo Verde 2023*", as published in Notice No. ° 5557-B/2023, of March 15. The program has a maximum allocation of €600,000 with 70% co-financing per application. It aims to encourage the adoption of innovative and environmentally friendly practices that positively impact significant events. It provides financing for green measures to be implemented at these events, with the following overarching objectives:

- Encouraging the adoption of environmental criteria that contribute to reducing impacts and promoting the efficient use of material and energy resources;
- Promoting the adoption of innovative approaches, such as the integration of renewable technologies, the promotion of the collaborative economy, and the implementation of ecological design; and
- Contributing to the education and environmental awareness of the various stakeholders involved, including event promoters, brands, municipalities, spectators, and nearby local businesses.

Through this program, the Portuguese EF aims to drive sustainability and foster a more environmentally conscious approach to organizing major events. By encouraging the implementation of green practices and embracing new technologies, the program seeks to create a positive environmental, social, and economic impact while raising awareness about the importance of environmental responsibility among all stakeholders.

In summary, Table 6 presents an overview of the programs funded by the Portuguese EF in the framework of waste management and the circular economy, covering the period from 2018 to 2023. It provides an overview of the funded programs. It highlights the Portuguese Government's commitment to addressing environmental challenges, promoting recycling and waste reduction, and driving the transition towards a more sustainable society and a profitable and well-suited circular economy.

Given the above, the waste policy in Portugal has undergone significant developments to align with the environmental directives of the EU. The primary objective is to achieve sustainable management of the materials, aiming to protect, preserve and improve the quality of the environment while promoting the principles of the circular economy. In order to accomplish this, municipalities must implement effective biowaste collection solutions, which can include initiatives such as collection on public roads or through door-to-door services, or other innovative technical solutions. Furthermore, active participation from citizens is essential, as they should embrace biowaste separation and recycling practices, whether through domestic or community composting initiatives. By adopting these measures, Portugal can make

Table 6. Programs financed by the Portuguese Environment Fund from 2018 to 2023

Programs	2018	2019	2020	2021	2022	2023
Circular economy in municipalities (JUNTAr+)	*	*		*		
Accelerator: creating and rethinking circular businesses - CIRCULAr: Startups		*				
Support for incentive system for the return of non-reusable plastic beverage packaging		*				
Support program for the elaboration of municipal studies for the development of biowaste collection systems			*			
'Biobairros - da terra à terra'				*		
Packaging reuse systems in ready-to-eat and take-away or home delivery				*		
'RecolhaBio' - Support for the implementation of selective collection projects for biowaste					*	
'Sê-lo Verde 2023'						*

significant strides towards a more sustainable and efficient waste management system while contributing to environmental preservation and promoting circular economy principles.

FISCAL MEASURES

Tax law holds a significant role in addressing negative environmental externalities, whereby tax law instruments are used to prevent, reduce, and repair environmental damage (Teles, 2015; Liu & Ge, 2023). Through environmental taxes, which are payments imposed by the Government without any corresponding benefit, and through fees, which are mandatory payments primarily levied by municipalities for waste collection and treatment, the Government promotes the responsible use of natural resources. These measures align with the polluter-pays principle, directly incorporating the costs associated with services and environmental damage into the prices of goods, services, and activities and contributing to integrating environmental policies within economic policies. Borrego (2016) recalls that it is a means of holding accountable the cause of environmental damage through acts committed in the environment. This principle has become the best known among nations and served as the basis of environmental legislation.

Consequently, the Government ensures that the true environmental costs are accounted for, incentivizing businesses and individuals to embrace sustainable practices and fostering a more environmentally conscious and economically sustainable society. This context requires Government planning that prioritizes public policy instruments, including environmental sustainability. Furthermore, the alignment between tax law and environmental objectives is crucial for promoting a greener and more sustainable future.

As emphasized by Freire-González et al. (2022), policymakers should impose taxes on landfill and incineration to promote circularity in the economy and alleviate environmental burdens. Studies conducted by Panzone et al. (2021) and Mottershead et al. (2021) support the idea that landfill taxation plays a vital role in transforming waste disposal practices at both household and local authority levels. Additionally, Briguglio (2021) advocates for the implementation of waste taxes as a means to discourage household disposal of biodegradable waste and encourage the adoption of composting alternatives.

In 2014, the Portuguese Government approved the Law No. 82-D/2014 of December 31, known as the Green Taxation Law (GTL), which resulted from the European and international commitments undertaken by the State in the scope of the environmental tax reform efforts. This law reformed the country's environmental tax system and promoted the following measures: incidence on plastic bags; TGR; and incentive for purchasing hybrid or electric vehicles by reducing the tax on old vehicles (Castilho, 2016; Mota, 2016). The GTL represents a significant milestone in pursuing sustainable development and reflects Portugal's commitment to aligning its fiscal policies with environmental goals. This law has the potential to generate a triple dividend, as it manages to protect the environment and reduce national energy dependence, promote employment and economic growth, contribute to budgetary balance, and reduce external imbalances (Pereira & Pereira, 2014; Rodríguez et al., 2019; Maxim, 2020). Before its implementation, environmental taxation in Portugal was fragmented, consisting of unplanned regulations. Also, the green taxation framework in Spain is established under the Environmental Tax Law (in Spanish, *Ley de Impuestos sobre el Medio Ambiente*). This law aims to promote environmental protection and sustainable development through fiscal instruments. It introduces taxes and charges on activities with negative environmental impacts, such as energy consumption, air pollution, waste generation, and water usage. The revenue generated from these taxes is intended to finance environmental programs, projects, and initiatives. Various green taxation measures in France have been implemented as part of the country's environmental policy. One significant piece of legislation is the General Tax on Polluting Activities (in French, *Taxe Générale sur les Activités Polluantes - TGAP*). The TGAP imposes taxes on activities that generate pollution or contribute to environmental degradation, including waste disposal, water pollution, and air emissions. The revenue generated from the TGAP is also used to fund environmental programs and projects. In Italy, green taxation is primarily regulated under the Environmental Tax Law (in Italian, *Legge Tributaria Ambientale*). This law includes provisions for taxes, duties, and charges on activities that have environmental implications. It covers energy consumption, waste management, water usage, and air pollution. The revenue generated from these environmental taxes is allocated to support environmental protection, conservation, and sustainable development initiatives.

According to Borrego (2016), the GTL had two primary objectives. Firstly, it aimed to promote environmentally responsible behavior among taxpayers, encouraging adopting practices that minimize the need for 'green' taxation. Secondly, it aimed to generate tax revenue through environmental taxation. Following the fiscal neutrality principle underpinning this reform, the focus is on taxing consumption and pollution rather than production and income.

The GTL involves various measures, including the landfill tax and the revision of the TGR, which aims to reduce the amount of waste incinerated or deposited in landfills, recognizing the crucial role of the waste sector in the economy, the public health, and the overall well-being of the population. Marques et al. (2018) explain that in Portugal, the waste sector consists of three different markets: the primary market, which encompasses collection and street cleaning services (the so-called "retail" service); the secondary market, which encompasses waste disposal services in landfills or other treatment facilities and waste transportation (the so-called "wholesale" services); and the tertiary market, which involves recycling, reuse, and other waste-related purposes.

It is important to note that implementing the GTL does not necessarily have to occur all at once; it can be done gradually. There are several advantages to a gradual implementation approach. Firstly, it minimizes the costs associated with adapting to new policies, reducing political opposition and lowering short-term costs. Additionally, gradually implementing reforms helps to mitigate the potential economic downturns compared to rapid wide-range reforms. Gradualism is a valuable tool in addressing the initial efficiency costs associated with policy implementation (Santos, 2021).

For example, the TGR, which was approved by Decree-Law No. 178/2006 of September 5, revised by Law No. 82-D/2014 of December 31, and recently updated by Decree-Law No. 102-D/2020 of December 10, has experienced a gradual and significant increase over the years, especially since 2015. Table 7 shows the evolution of TGR values per ton of waste from 2015 to 2025. According to Ordinance No. 278/2015 of September 11, waste service management entities are responsible for paying the TGR, which is subsequently passed to the customers through the tariffs added to their water bills.

Table 7. Evolution of the TGR value from 2015 to 2025

	2015	**2016**	**2017**	**2018**	**2019**	**2020**	**2021**	**2022**	**2023**	**2024**	**2025**
TGR value (€/t)	5.50	6.60	7.70	8.80	9.90	11.00	22.00	22.00	25.00	30.00	35.00

Source: Adapted from Law No. 82-D/2014 and Decree-Law No. 102-D/2020. Legend: € - Euros; t - ton.

The TGR, which is applied to municipal waste management systems and waste treatment facilities, is determined annually based on factors such as the quantity of waste generated and its final destination, as illustrated in Table 8.

Table 8. Percentage of TGR payable by type of operation

Operation	**2020**	**2021**	**2022**	**After 2022**
Landfill disposal – elimination (%)	100	100	100	100
Incineration – elimination (%)	70	85	85	85
Energy recovery – valorization (%)	25	25	20	20

Source: Adapted from Law No. 82-D/2014 and Decree-Law No. 102-D/2020.

Also, it was necessary to establish a comprehensive legal framework that allows for exemptions from applying the TGR, in exceptional cases. These exceptions are specific and time-limited, ensuring that they do not compromise the achievement of environmental objectives or have implications for shaping behaviors, such as:

- Mandatory incineration of certain Hospital Waste, which necessitates proper disposal; and
- Landfill disposal of Construction and Demolition Waste containing asbestos requires proper landfill disposal due to its specific characteristics.

Furthermore, the GTL introduced a fee for lightweight plastic bags to promote more sustainable behavior, among consumers, producers, and traders. This fee, amounting to 8 cents per bag, was implemented on January 1, 2015, and is registered on invoices. The responsibility of paying this fee falls on the end consumer, and the economic agents must include this contribution in the final price. However, certain plastic bags are exempt from this fee, such as those that are in contact with foodstuffs (including ice), bags exported by the taxable person, bags shipped or transported to another EU Member State by the taxable person or a third party on their behalf, and those used for donations to social solidarity institutions.

Additionally, a fee for single-use packaging made of plastic or multi-material with plastics came into effect on July 1, 2022, while the fee for single-use packaging made of aluminum or multi-material with aluminum took effect on January 1, 2023. These fees, amounting to 30 cents per packaging unit, were regulated by Law No. 75-B/2020 of December 31. Part of the revenue generated from these fees is allocated to support the Portuguese Nature Conservation Fund and finance various municipal projects that contribute to preserving and enhancing local environments.

The implementation of these fiscal measures is expected to have a significant impact on reducing plastic usage and encouraging the adoption of alternative forms of bags and packaging. As a result of the price increase for the end consumers, a shift towards more sustainable options is anticipated. In line with these objectives, the Portuguese EF has launched the "*Packaging reuse systems in ready-to-eat and take-away or home delivery*" program. This program aims to support developing and implementing innovative systems that promote reusable packaging in ready-to-eat, take-out, and/or home delivery services. As discussed earlier in this chapter, the program aims to promote sustainable practices and reduce the environmental impact associated with packaging waste.

Drawing from the methodology outlined by Craizer et al. (2020), this point addressed several key issues regarding charging systems. These included proposals to improve waste management practices and ensure their effectiveness, forcing more responsible usage. Suggestions embraced increasing the TGR for landfill and incineration to ensure comprehensive coverage of treatment costs, raising the value of the TGR in cases of non-compliance with MSW management targets, and reevaluating charges paid by citizens to increase their overall contribution gradually. However, to promote fairness, it is recommended to eliminate the current practice of linking charges to water bills because the citizens pay for waste collection, regardless of how much or how little waste they generate.

Effectively, the authors defend the implementation of a system that would enable the adoption of PAYT (pay-as-you-throw) programs, which have the potential to incentivize waste sorting and penalize those who neglect to do so. This approach would enable the recognition and reward of citizens who actively participate in waste sorting efforts while discouraging those who continue disregarding proper waste management practices. Thus, with PAYT, the municipalities would charge the citizens a fee for each bag of waste they generate or the weight of their waste. Under PAYT programs, citizens pay only for what they throw away.

Consistent with previous findings, the authors also support the suggestions presented by Vence & López Pérez (2021), prioritizing taxes on non-renewable resources while reducing or eliminating taxes on renewable resources. By considering these measures, policymakers can establish more effective waste management systems and promote a shift towards more responsible resource management. This contributes to establishing a greener and more sustainable future where waste is minimized, resources are utilized efficiently, and environmental impacts are mitigated.

CONCLUSION

The national solid waste policy in Portugal, aligned with the EU framework, focuses on implementing source separation and composting schemes at the municipal level and establishing central composting or anaerobic digestion facilities. In this regard, local authorities, as the primary stakeholders responsible for implementing biowaste policies, must intensify their effort to effectively organize and carry out biowaste management activities to protect the environment and reduce management costs for the public's benefit. Simultaneously, it is essential to develop programs that raise awareness among citizens and businesses about the importance of recycling, waste reduction, and changes in collection systems.

The implemented solutions should ensure that source separation encourages biowaste recycling by the producer (domestic composting) and/or provides a network of composters at the source to receive the biowaste and then distribute the resulting compost (community composting). However, conducting a study to examine the factors that influence waste sorting behaviors is crucial for evaluating municipal waste management schemes and determining additional policies that can be implemented effectively (Saldivia-Gonzatti et al., 2022).

Through these efforts, Portugal is moving towards more sustainable waste management practices while promoting a greener and more environmentally conscious society. In this sense, the legal and fiscal measures aim to reduce the amount of waste sent to landfills and reduce GHG in the atmosphere, improving people's quality of life. Despite recent decreases in total municipal waste generation and direct deposition in landfills, the prevalence of undifferentiated collection remains a major challenge. Addressing this challenge will require substantial efforts, especially considering the time constraints imposed by the "European Green Deal" and the "Circular Economy Action Plan" at the European level, as well as the goals set by the PNEC in Portugal to aims to achieve a carbon-neutral and more sustainable and circular society.

On the one hand, Portugal has made significant strides towards achieving a circular economy by implementing a national action plan funded by the governmental EF. The EF aims to promote a more sustainable and resource-efficient economy, where the principles of reduction, reuse, and recycling are deeply integrated into several programs and funding initiatives, involving collaboration between key stakeholders, including policymakers and civil society. By adopting a holistic approach and involving all stakeholders, Portugal is taking concrete steps towards achieving its circular economy objectives and positioning itself as a contributor to global sustainable development.

On the other hand, taxation is a crucial instrument in transitioning to a carbon-neutral society to promote cleaner technologies and remove perverse incentives for fossil fuels or investments that increase GHG emissions. Through the GTL, such as the fee for lightweight plastic bags and single-use packaging, as well as the revision of the TGR, Portugal aims to encourage the users and the polluters to reduce their environmental impact by minimizing the number of pollutants generated and disposed into the environment. It is worth noting, as Carattini et al. (2018) emphasised, environmental taxes could be more widespread if people had the opportunity to experience their functioning and correct their beliefs.

In conclusion, by analyzing the Portuguese Government's measures to support solid wastes, particularly biowaste management and recycling, this chapter provides insights into the efforts required to create a sustainable environment while also meeting the economic and social needs of the population. Thus, the principles of sustainable development, polluter-pays, and international cooperation are important in the relationship between tax law and environmental protection.

REFERENCES

Afum, R.; Agyabeng-Mensah, Y.; Baah, C.; Agyapong, G.; Armas, A. & Al Farooque, O. (2022). Prioritizing zero-waste performance and green differentiation advantage through the Prism of circular principles adoption: A mediated approach. *Journal of Cleaner Production*, *361*(August 10), 132182. doi:10.1016/j.jclepro.2022.132182

APA - Agência Portuguesa do Ambiente. (2022a). *Relatório Anual de Resíduos Urbanos 2021*. APA. https://apambiente.pt/sites/default/files/_Residuos/Producao_Gest%C3%A3o_Residuos/Dados%20RU/RARU_2021.pdf [Accessed in February 2023].

APA - Agência Portuguesa do Ambiente. (2022b). *PERNU 2030: Plano Estratégico para os Resíduos Não Urbanos*. APA. https://participa.pt/contents/consultationdocument/PERNU%202030.pdf

Baah, C., Afum, E., Agyabeng-Mensah, Y., & Agyeman, D. O. (2021). Stakeholder influence on adoption of circular economy principles: Measuring implications for satisfaction and green legitimacy. *Circular Economy and Sustainability*, *2*(1), 91–111. doi:10.100743615-021-00093-2

Baranzini, A., van den Bergh, J. C. J. M., Carattini, S., Howarth, R. B., Padilla, E., & Roca, J. (2017). Carbon pricing in climate policy: Seven reasons, complementary instruments, and political economy considerations. *Wiley Interdisciplinary Reviews: Climate Change*, *8*(4), e462. doi:10.1002/wcc.462

Borrego, A. (2016). Diploma of green taxation - Finally an environmental taxation integrated in Portugal? *Revista de Gestão Social e Ambiental*, *10*(3), 51–68. doi:10.24857/rgsa.v10i3.1163

Briguglio, M. (2021). Taxing household waste: Intended and unintended consequences. *Journal of Cleaner Production*, *304*, 127034. doi:10.1016/j.jclepro.2021.127034

Carattini, S.; Baranzini, A. & Lalive, R. (2018). Is Taxing Waste a Waste of Time? Evidence from a Supreme Court Decision. *Ecological Economics*, *148*, 131.151. doi:10.1016/j.ecolecon.2018.02.001

Castilho, A. F. A. N. (2016). A extrafiscalidade no Brasil. *Revista Jurídica Luso Brasileira*, *2*(4), 135-159. https://www.cidp.pt/revistas/rjlb/2016/4/2016_04_0135_0159.pdf

Craizer, R., Frantzis, I., & Bacolas, N. C. (2020). *Economic instruments to improve waste management in Greece – Including a pre-feasibility study on a deposit refund system, Final Report*. Brussels: European Commission.

Cropley, A. J. (2022). *Qualitative research methods: A practice-oriented introduction*. Editura Intaglio., doi:10.13140/RG.2.1.3095.6888/1

Di Foggia, G., & Beccarello, M. (2023). Designing circular economy-compliant municipal solid waste management charging schemes. *Utilities Policy*, *81*, 101506. doi:10.1016/j.jup.2023.101506

Dijkgraaf, E., & Gradus, R. (2016). An EU recycling target: What does the Dutch evidence tell us? *Environmental and Resource Economics*, *68*(3), 501–526. doi:10.100710640-016-0027-1

Elliott, R., & Timulak, L. (2021). Why a generic approach to descriptive-interpretive qualitative research? In R. Elliott & L. Timulak (Eds.), *Essentials of descriptive-interpretive qualitative research: A generic approach* (pp. 3–14). American Psychological Association. doi:10.1037/0000224-001

EPA - U.S. Environmental Protection Agency. (2023). *Overview of Greenhouse Gases*. EPA. https://www.epa.gov/ghgemissions/overview-greenhouse-gases

FA - Fundo Ambiental. (2019). *Relatório Final - Economia Circular em Freguesias (Programa JUNTAr+)*. FA. https://www.fundoambiental.pt/ficheiros/relatorio-final-juntar-mais-pdf.aspx .

FA - Fundo Ambiental. (2021). *Relatório Final - Economia Circular em Freguesias (Programa JUNTAr+)*. FA. https://www.fundoambiental.pt/ficheiros/relatorio-final-aviso-71822021-juntar--pdf.aspx

FA - Fundo Ambiental. (2023a). *RecolhaBio - Apoio à implementação de projetos de recolha seletiva de biorresíduos*. FA. https://www.fundoambiental.pt/avisos-anteriores/apoios-2022/residuos-e-economia-circular/recolhabio-apoio-a-implementacao-de-projetos-de-recolha-seletiva-de-biorresiduos.aspx [Accessed in March 2023].

FA - Fundo Ambiental. (2023b). *O que é o programa Recolhabio?* FA. https://www.fundoambiental.pt/recolhabio-apoio-a-implementacao-de-projetos-de-recolha-seletiva-de-biorresiduos/ficheiros/faqs-recolhabio-pdf.aspx [Accessed in March 2023].

Fazey, I., Schäpke, N., Caniglia, G., Patterson, J., Hultman, J., van Mierlo, B., Säwe, F., Wiek, A., Wittmayer, J., Aldunce, P., Al Waer, H., Battacharya, N., Bradbury, H., Carmen, E., Colvin, J., Cvitanovic, C., D'Souza, M., Gopel, M., Goldstein, B., & Wyborn, C. (2018). Ten essentials for action-oriented and second order energy transitions, transformations and climate change research. *Energy Research & Social Science*, *40*, 54–70. doi:10.1016/j.erss.2017.11.026

Freire-González, J.; Martinez-Sanchez, V. & Puig-Ventosa, I. (2022). Tools for a circular economy: Assessing waste taxation in a CGE multi-pollutant framework. *Waste Management*, 139 (February 15), 50-59. doi:10.1016/j.wasman.2021.12.016

Harris, S., Martin, M., & Diener, D. (2021). Circularity for circularity's sake? Scoping review of assessment methods for environmental performance in the circular economy. *Sustainable Production and Consumption*, *26*, 172–186. doi:10.1016/j.spc.2020.09.018

Hawken, P. (2017). *Drawdown: The Most Comprehensive Plan Ever Proposed to Roll Back Global Warming*. Penguin USA.

IPCC - Intergovernmental Panel on Climate Change. (2019). *Global Warming of 1.5°C*. IPCC. https://www.ipcc.ch/site/assets/uploads/sites/2/2022/06/SR15_Full_Report_HR.pdf

Liu, B., & Ge, J. (2023). The optimal choice of environmental tax revenue usage: Incentives for cleaner production or end-of-pipe treatment? *Journal of Environmental Management*, *329*, 117106. doi:10.1016/j.jenvman.2022.117106 PMID:36566734

Mack, N., Woodsong, C., Macqueen, K. M., Guest, G., & Namey, E. (2005). *Qualitative Research Methods: A Data Collector's Field Guide*. Family Health International.

Manjunatha, N. (2019). Descriptive Research. *Journal of Emerging Technologies and Innovative Research*, *6*(6), 863–867. https://www.jetir.org/papers/JETIR1908597.pdf

Marques, R. C., Simões, P., & Pinto, F. S. (2018). Tariff regulation in the waste sector: An unavoidable future. *Waste Management (New York, N.Y.)*, *78*, 292–300. doi:10.1016/j.wasman.2018.05.028 PMID:32559914

Maxim, M. R. (2020). Environmental fiscal reform and the possibility of triple dividend in European and non-European countries: Evidence from a meta-regression analysis. *Environmental Economics and Policy Studies*, *22*(4), 633–656. doi:10.100710018-020-00273-8

Maxwell, J. A. (1992). Understanding and validity in qualitative research. *Harvard Educational Review*, *62*(3), 279–300. doi:10.17763/haer.62.3.8323320856251826

McGuinn, J., Oulès, L., Bradley, H., & McNeill, A. (2017). *Effective multilevel environmental governance for a better implementation of EU environment legislation: Study*. Europa. https://cor.europa.eu/en/engage/studies/Documents/Environmental-governance.pdf

Meyer, C. A. (1997). Public-Nonprofit Partnerships and North-South Green Finance. *Journal of Environment & Development*, *6*(2), 123–146. https://www.jstor.org/stable/44319261. doi:10.1177/107049659700600203

Mota, P. M. F. (2016). *Avaliação do desempenho das políticas de Ambiente em Portugal*. UNL. https://run.unl.pt/bitstream/10362/18561/1/Mota_2016.pdf

Mottershead, D., Watkins, E., Gore, T., Andersen, M. S., Pedersen, A. B., Whiteoak, K., Bruyn, S., Vergeer, R., Chowdhury, T., Hummelen, S. V., & Chewpreecha, U. (2021). *Green taxation and other economic instruments – Internalizing environmental costs to make the polluter pay*. European Commission.

Nassaji, H. (2015). Qualitative and descriptive research: Data type versus data analysis. *Language Teaching Research*, *19*(2), 129–132. doi:10.1177/1362168815572747

Panzone, L., Ulph, A., Areal, F., & Grippo, V. (2021). A ridge regression approach to estimate the relationship between landfill taxation and waste collection and disposal in England. *Waste Management (New York, N.Y.)*, *129*, 95–110. doi:10.1016/j.wasman.2021.04.054 PMID:34051583

Pereira, A. M., & Pereira, R. M. (2014). Environmental fiscal reform and fiscal consolidation: The quest for the third dividend in Portugal. *Public Finance Review*, *42*(2), 222–253. doi:10.1177/1091142113485803

Pfister, N., & Mathys, N. A. (2022). Waste taxes at work: Evidence from the canton of Vaud in Switzerland. *Ecological Economics*, *193*, 107314. doi:10.1016/j.ecolecon.2021.107314

Resor, J. P. (1997). Debt For Nature Swaps: A Decade of Experience and New Directions for the Future. *Unasylva*, *48*(188), 15–22.

Rodrigues, P., David, F., Soares, E., Monteiro, E., Melo, N., Gregório, J., & Rodrigues, R. (2022). Economic Aspects of a Collection Solution of Biowaste in an Association of Municipalities in Portugal. In: IWWG–International Waste Working Group (Eds). *Proceedings of the 9th International Symposium on Energy from Biomass and Waste*. Padova (Italy): Cisa Publisher.

Rodríguez, M., Robaina, M., & Teotónio, C. (2019). Sectoral effects of a Green Tax Reform in Portugal. *Renewable & Sustainable Energy Reviews*, *104*, 408–418. doi:10.1016/j.rser.2019.01.016

Saldivia-Gonzatti, L. I., Jannes, G., & Barreal, J. (2022). Factors influencing the rate of sorted solid waste collection: An empirical analysis towards local management in Catalonia (NE Spain). *Cities (London, England)*, *131*, 104038. doi:10.1016/j.cities.2022.104038

Sandelowski, M. (2000). Whatever Happened to Qualitative Description? *Research in Nursing & Health*, *23*(4), 334–340. doi:10.1002/1098-240X(200008)23:4<334::AID-NUR9>3.0.CO;2-G PMID:10940958

Sandelowski, M. (2010). What's in a name? Qualitative description revisited. *Research in Nursing & Health*, *33*(1), 77–84. doi:10.1002/nur.20362 PMID:20014004

Santos, M. L. (2021). *O setor dos resíduos em Portugal: o impacto da Reforma da Fiscalidade Verde e da regulação sobre os consumidores finais*. Bit Stream. https://repositorio-aberto.up.pt/bitstream/10216/136556/2/502062.pdf

Teles, N. S. (2015). *A Fiscalidade Verde: Do Princípio da Equivalência à Neutralidade Fiscal*. Bit Stream. https://repositorio-aberto.up.pt/bitstream/10216/83754/2/37256.pdf .

Trinks, A., Mulder, M., & Scholtens, B. (2022). External carbon costs and internal carbon pricing. *Renewable & Sustainable Energy Reviews*, *168*(October), 112780. doi:10.1016/j.rser.2022.112780

Vence, X., & López Pérez, S. J. (2021). Taxation for a Circular Economy: New Instruments, Reforms, and Architectural Changes in the Fiscal System. *Sustainability (Basel)*, *13*(8), 4581. doi:10.3390u13084581

Wagner, P. M., Torney, D., & Ylä-Anttila, T. (2021). Governing a multilevel and cross-sectoral climate policy implementation network. *Environmental Policy and Governance*, *31*(5), 417–431. doi:10.1002/eet.1942

Weitzman, M. L. (2015). Internalizing the Climate Externality: Can a Uniform Price Commitment Help? *Economics of Energy & Environmental Policy*, *4*(2), 37–50. doi:10.5547/2160-5890.4.2.mwei

World Bank. (2018). *Municipal Solid Waste Management: A Roadmap for Reform for Policy Makers*. World Bank Group.

KEY TERMS AND DEFINITIONS

Anaerobic Decomposition: This is a natural process that takes place in environments absent of oxygen, such as waterlogged soils or anaerobic organic waste treatment systems. During biochemical reactions, anaerobic microorganisms, including bacteria and archaea, break down complex organic matter into simpler compounds, such as methane, carbon dioxide, hydrogen sulfide, and organic acids. Anaerobic decomposition is an essential ecological process that helps to break down organic matter and recycle nutrients, but it can also produce harmful byproducts such as methane and hydrogen sulfide.

Aerobic Decomposition: This is a natural process that occurs in the presence of oxygen and is facilitated by aerobic microorganisms, including bacteria and fungi. During this process, organic matter is broken down into simpler compounds such as water, and carbon dioxide, and nutrients such as nitrogen and phosphorus. This decomposition process converts organic matter into energy and biomass. The byproducts of this process are beneficial to the environment as they can enrich the soil and facilitate plant growth.

Biowaste: Refers to any biodegradable waste material that can be decomposed through either anaerobic or aerobic processes. This type of waste can include food and kitchen waste from homes, restaurants, canteens, catering, and retail units, as well as similar waste from food processing plants. Biowaste can also include yard waste such as leaves, grass, and branches from gardens and parks. Biowaste is typically managed through composting, anaerobic digestion, or other similar processes, which can produce useful end-products such as fertilizer, biogas, or soil amendment.

Circular Economy: An innovative production and consumption model that promotes the principles of sharing, renting, reusing, repairing, renovating, and recycling existing materials and products. Its main goal is to prolong the life cycle of products, optimizing resource utilization, reducing extraction, promoting reuse, and enhancing overall efficiency. By adopting this approach, we can foster sustainability and minimize waste generation, leading to a more environmentally conscious and economically viable system.

Green Taxation: A policy tool used by governments to promote sustainable behavior by imposing taxes and fees on activities that pollute and degrade the environment. The goal of green taxation is to incentivize individuals and organizations to reduce their environmental impact by penalizing those who engage in unsustainable practices and rewarding those who adopt more sustainable behaviors. Green taxes and fees can take various forms, such as taxes on carbon emissions, fees for the disposal of hazardous waste, or taxes on plastic bags. The revenue generated from these taxes and fees can be used to environment fund programs and incentives.

Polluter-Pays Principle: An environmental policy principle that holds that those who cause pollution or environmental damage should bear the costs associated with preventing, controlling, and remedying that pollution or damage. Under this principle, polluters are held financially responsible for the environmental harm they cause, and the cost of measures taken to prevent and mitigate such harm. This principle serves as a powerful incentive for polluters to avoid environmental damage and encourages them to adopt sustainable practices. By holding polluters accountable for the costs of their actions, the polluter-pays principle helps to shift the burden of environmental damage from society to those who are responsible for causing it.

Selective Collection of Biowaste: This is a waste management practice that involves the separation and collection of organic and biological waste, such as food waste and garden and park waste, at the source of generation. This waste is then treated through various processes, such as composting or anaerobic digestion, to produce organic compost that can be used as a soil amendment in agriculture and/or energetic valorization. The selective collection of biowaste reduces the amount of waste sent to landfills, which helps to reduce greenhouse gas emissions and promote sustainable waste management practices, enhancing the circular economy models.

Separation and Recycling at the Source of Biowaste: Refers to the practice of separating organic waste, such as food scraps and yard waste, from other types of waste at the point of generation. This can be done through measures such as domestic or community composting, local recycling solutions, or selective collection for transport to recycling facilities, such as composting and anaerobic digestion facilities. The purpose of this practice is to prevent biowaste from being mixed with other waste streams, particularly the organic fraction of unsorted waste, and to promote sustainable waste management practices.

Sustainable Environment: The responsible management and use of natural resources in a way that ensures their preservation and promotes ecological balance for the benefit of both current and future generations. It is a critical component of sustainable development, that requires a multifaceted approach to balance economic, social, and environmental concerns and to emphasize the importance of reducing waste and pollution, promoting renewable energy sources, conserving biodiversity, and minimizing the negative impact of human activities on the planet.

Waste: Any material or substance that is left over from human and animal activities or from productive processes. This can include organic matter, such as food waste and agricultural residues, as well as domestic waste, industrial effluents, and gases released in industrial processes or by engines. Waste is often classified according to its source or characteristics, such as hazardous waste, municipal solid waste, or electronic waste.

Chapter 10
Sustainable Taxi Fleet:
An Approach to the Contribution of Green Taxation

Ana Arromba Dinis
https://orcid.org/0000-0003-3945-5726
Research Centre on Accounting and Taxation, Polytechnic Institute of Cávado and Ave, Portugal

André Carvalho
https://orcid.org/0000-0002-7896-0947
Research Centre on Accounting and Taxation, Polytechnic Institute of Cávado and Ave, Portugal

Sara Luís Dias
Polytechnic Institute of Cávado and Ave, Portugal

ABSTRACT

It is a well-known fact that the automotive sector plays a key role in a country's socio-economic development, due to the large contribution of vehicles to greenhouse gas emissions (GHG). Alternatives such as car-sharing, public transport, cycling, the transition to electric vehicles, or walking can help reduce emissions. Therefore, it becomes a sector where environmental regulation should be demanded. This research aims to contribute to the tax literature by extending previous studies on corporate taxation by evaluating green tax incentives as an effective environmental tax measure. Considering taxis as an important part of urban public transport systems, the authors set out to investigate Portuguese taxi companies that have taken advantage of tax incentives for less environmentally polluting vehicles. The authors conclude that the income tax level, the economic profit, and the number of employees, are determining factors in these firms 'access to the tax incentives under study.

DOI: 10.4018/978-1-6684-8592-7.ch010

INTRODUCTION

Fast economic development is unequivocally associated with excessive energy consumption, especially in emerging economies, being, as a result, a high risk of environmental damage (Ekins & Zenghelis, 2021).

In this context, environmental sustainability must be the basis for current government action. Sustainable development strategies focusing on energy emissions and reducing greenhouse gas emissions (GHG) to achieve climate neutrality are seen as key policies for a sustainable climate agenda in the European Union (EU), as stated by Panarello & Gatto (2023) and Akdag & Yıldırım (2020). And in the Organization for Economic Co-operation and Development (OECD) countries under the OECD International Program for Action on Climate (IPAC), which support country progress towards net-zero greenhouse gas (GHG) emissions and a more resilient economy by 2050.

It is noted that the transport sector is widely recognized globally as a significant and growing source of air pollution (Williams & Blyth, 2023). It represents an excessive source of pollution, currently producing around a quarter of Europe's GHG emissions, and 33% of total urban GHG emissions in major cities are caused by transport (OECD, 2020). Thus, each country must start aligning its public policies to reduce GHG emissions caused by transport and achieve sustainable carbon neutrality (Kazancoglu et al., 2021; Xu et al., 2020).

This is where environmental taxation comes in. Green taxation, as it is also referred to, includes taxes on energy, transport, and pollution, and can generate, if correctly applied, a significant environmental impact on a particular country (Mosquera- Valderrama & Balharová, 2021; Oliveira & Mendes, 2017).

At this time, the relationship between environmental policy and tax policy is seen by governments as an opportunity to adjust the tax system to a more energy-efficient economy in the use of resources, making it innovative, with low carbon emissions, and more sustainable (Delgado et al., 2022; Han & Lee, 2022).

Taxis [in this study, the authors consider taxi companies to be those engaged in the rental of passenger transport] are a significant part of urban public transportation systems in addition to their advantages over other types of mobility, also considering their comfort and commodity (Qiu et al., 2022). Therefore, the replacement of taxis powered by conventional fuels with electric/hybrid vehicles is essential. To this extent, and in the case of the taxi sector, the Portuguese tax system provides tax incentives, namely for vehicles that are equipped with hybrid engines, plug-in hybrid engines, or that use exclusively natural gas or electric power. The authors set out to investigate Portuguese taxi companies that have accessed tax incentives for less environmentally polluting vehicles.

This chapter intends to contribute to the tax literature by extending previous studies on corporate taxation evaluating green tax incentives as an effective tax environmental measure. The main objectives of this chapter are: (i) to highlight the importance of green taxation for environmental protection; (ii) to bring to the climate change discussion the role of tax incentives for the transport sector; (iii) to study the Portuguese taxi companies that have accessed tax incentives for less-polluting cars.

The authors consider this study relevant, namely in the current context where road transport is responsible for the largest share of GHG emissions around the world. Furthermore, the taxi sector needs special attention due to the importance it represents for cities to become more sustainable. Portugal registered 24,684 taxi drivers in 2016, spread across 9,997 companies. By 2021 the number of taxis decreased to 21,902, also decreasing the number of companies operating in this sector to 9,262 (data according to the Portuguese Statistical Yearbooks of Mobility and Transport). These data are a cause for concern, as one would expect the trend to be the other way around. With this research, and using the Portuguese case study, the authors hope to contribute to the discussion of tax policies to encourage the adoption of more sustainable behaviors, namely from the firm sector (and taxi companies in particular).

As far as research methodology is concerned and to achieve the objectives initially defined, three research questions were outlined:

*Which are the cities where the taxi companies that have accessed the tax incentives for the most sustainable vehicles are headquartered?

- What are the characteristics of the taxi companies that have taken advantage of the tax incentives for less polluting cars in Portugal?

*Are taxi companies that take advantage of tax incentives for less polluting vehicles more profit-efficient?

The structure of this chapter is as follows. Firstly, some preliminary considerations on gas emissions in the international context will follow, presenting an overview of the problem and the urban sustainability agenda. Secondly, the contributions of green tax incentives will be addressed. Thirdly, the tax incentives for taxi fleet sustainability will be analyzed as a tax policy to support climate change in Portugal. Fourthly, the authors analyze the data and discuss the results obtained. Fifthly, some outlines for future research are presented. Finally, in the sixth and last place, the main conclusions will be presented.

BACKGROUND

An overview of the problem of gas emissions in the international context

As Table 1 shows, among the 10 European countries that contributed most to the emission of GHG in 2020, Germany stands out by holding 22.1% of ton carbon dioxide equivalent (tCO2eq) of European emissions in 2020, decreasing from 1990 (-3.5 percent points tCO2eq).

Table 1. Top ten European countries that contributed most to the emission of GHG, from 1990 to 2020 (%)

Country\Year	1990	2000	2010	2020	Variation 1990_2020
Germany	25.6	23.3	22.4	22.1	-3,5
France	11.2	12.4	12.2	11.9	+0.7
Italy	10.7	12.5	12.4	11.6	+0.9
Poland	9.8	8.9	9.9	11.4	+1.6
Spain	6.0	8.7	8.6	8.3	+2.3
Netherlands	4.5	4.9	5.1	5.0	+0.5
Czech Republic	4.1	3.4	3.4	3.4	-0.7
Romania	5.2	3.1	2.9	3.3	-1.9
Belgium	3.0	3.4	3.2	3.2	+0.2
Greece	2.1	2.8	2.8	2.3	+0.2

Source: Data extracted on 01 Jan 2023, from Pordata. https://www.pordata.pt//

The Czech Republic and Romania are countries with a decrease compared with the proportion of Europe's emissions of 0.7 percent points and 1.9 percent points tCO2eq, respectively. Spain and Poland show an increase in the weight of GHG emissions in Europe, with the high leverage of the contribution of their emissions in 2.3 percent points and 1.6 percent points tCO2eq, respectively.

Table 2. Net greenhouse gas emissions by selected European countries, from 1990 to 2020

Country\Year	1990	2000	2010	2020	Variation 1990_2020
Portugal	100	139.4	119.3	98.6	-1.4
Germany	100	84.2	76.6	59.2	-40.8
France	100	101.9	94.7	72.5	-27.5
Italy	100	107.8	100.5	73.5	-26.5
Poland	100	83.4	87	79.2	-20.8
Spain	100	134.9	125.8	95.4	-4.6
Netherlands	100	101.3	98.8	76	-24
Czech Republic	100	75.9	70.9	57	-43
Romania	100	55.7	49.3	43.9	-56.1
Belgium	100	103.2	92.6	74	-26
Greece	100	121.8	114.3	71.9	-28.1
EU 27 (2020)	100	92.4	87.3	68.4	-31.6

Source: Data extracted on 01 Jan 2023, from Pordata. https://www.pordata.pt//

However, as shown in Table 2, Germany, France, Italy, Poland, Netherlands, Czech Republic, Romania, Belgium, and Greece have a net emission reduction of GHG greater than 20 tCO2eq. Compared to the EU-27 average GHG reduction (31.6 tCO2eq), only three countries reduced GHG, namely Germany, the Czech Republic, and Romania. Comparing Portugal with the top 10 countries with the largest weight of net emission of GHG in Europe, in 2020, it is apparent that from 1990 to 2020, Portugal was the country with the lowest net emission reduction of GHG, with a reduction of 1.4 tCO2eq.

As highlighted by the European Environment Agency (2022), road transport was responsible for 71.7% of GHG in 2019. As shown in Table 3, it should be also noted that in road transport GHG has increased over the last three decades, registering an 11.24% increase between 1990 and 2020 (in 2019 the rate was 27.86%, and despite the decline in 2020, caused by the covid-19 pandemic, the expectation is that this percentage will increase again).

These data are in line with the fact that cars are among the biggest environmental poisoners, with high percentages of GHG emissions (Lozhkin, Lozhkina, & Dobromirov, 2018).

As in other countries, in Portugal the automotive sector is an extremely important sector for the future of the Portuguese economy, due to its ability to create skilled jobs, the boosting of new industry sectors, increase exports, and also because it is an important motivation for research and development in the country (Novas et al., 2022; Dieguez et al., 2020). Nevertheless, there is an alarmingly increasing positive tendency towards the acquisition of diesel-powered vehicles, which account for 65.18% of the total number of vehicles in circulation in Portugal in 2021, as Table 4 displays.

Table 3. Greenhouse gas emissions from transport in the EU, change in emission level from 1990 (index1990=100)

Year\ Variation 1990_2020 (%)	International aviation	International maritime transport	Domestic Aviation	Road transport	Railways	Domestic navigation
1990	100	100	100	100	100	100
1995	+21.99	+0.37	+9.65	+8.96	-23.02	-5.87
2000	+57.39	+25.30	+45.49	+20.51	-36.65	-13.45
2005	+77.75	+49.86	+41.48	+28.51	-51.17	-10.31
2010	+85.96	+46.43	+32.47	+24.43	-57.71	-16.51
2015	+101.35	+26.44	+9.15	+22.13	-64.90	-35.15
2020	+3.51	+19.36	-33.21	+11.24	-71.75	-30.44

Source: Data extracted on 21 Mar 2023, from EEA, https://www.eea.europa.eu/

Table 4. Cars in circulation in Portugal: total by fuel type (2010-2021)

Year\Fuel Type	Total Cars	Diesel Cars	Petrol Cars	Liquefied Petroleum Gas Cars	Biodiesel, Plug-in hybrid electric, non-plug-in hybrid electric, and Other Cars
2010	6 182 051	57.37%	41.85%	0.63%	0.14%
2011	6 181 188	58.56%	40.66%	0.62%	0.16%
2012	5 556 041	60.01%	39.11%	0.68%	0.20%
2013	5 624 964	60.48%	38.57%	0.74%	0.21%
2014	6 095 506	61.46%	37.51%	0.78%	0.24%
2015	6 083 694	62.76%	36.11%	0.80%	0.32%
2016	6 208 350	63.85%	34.92%	0.81%	0.42%
2017	6 447 241	64.74%	33.84%	0.82%	0.60%
2018	6 705 331	65.03%	33.25%	0.85%	0.87%
2019	7 027 599	65.47%	32.46%	0.83%	1.24%
2020	7 021 112	65.33%	32.09%	0.85%	1.74%
2021	7 090 889	65.18%	31.46%	0.88%	2.48%
Variation 2010_2021	+ 908 838	+7.81	-10.39	+0.25	+2.34

Source: Source: Data extracted on 01 Jan 2023, from Pordata. https://www.pordata.pt//

In contrast, petrol-powered cars have consequently reduced their number over the years, representing around 32% of the total number of vehicles in circulation in 2021. In the remaining cases, there is a slight increase in liquefied petroleum gas-powered vehicles (LPG), representing 0.88%, and, more positively, the vehicles considered less polluting for the environment (biodiesel, pure electric, plug-in hybrid, non-plug-in hybrid) have registered significant increases since 2010, registering in 2021 an increase of 2.34 percentage points, nonetheless represent only around 2.5% of the total number of vehicles in circulation.

Public Transportation in the Urban Sustainability Agenda: The Case of the Taxi Sector

While it is a well-known fact that the automotive sector plays a key role in a country's socio-economic development, due to the strong contribution of vehicles to GHG emissions, alternatives such as car-sharing, public transport, cycling, the transition to electric vehicles, or walking can help reduce emissions. It, therefore, becomes a sector where environmental regulation should demand. For example, regulating the adoption of electric vehicles can contribute to the mitigation of GHG, reducing global climate change, fuel dependency, air pollution, and the associated impacts of the transport sector on human health, which provides environmental, economic, and social benefits compared to traditional vehicles (Cihat Onat, 2022; Xu et al., 2020). Additionally, the transition to electric solutions is a requirement of the EU and the OECD for countries to become climate neutral by 2050 (Tang et al., 2021; Antón, 2022), targeting a sustainable energy system to respond to climate change mitigation (Zhao & Baker, 2022) and achieve the European Green Deal and also take care of the International Program for Action on Climate (IPAC). However, the problem that appears in the automotive industry is a slow energy-sustainable transition from traditional vehicle solutions to more sustainable solutions.

In this regard, it is important to highlight the contribution of public transport modes and car-sharing, seen as crucial to mitigate pollution, fossil energy consumption, traffic congestion, and city noise (Calise et al., 2019; Kang & Lee, 2019).

Table 5. Passenger transport: road passenger transport by buses and coaches (passenger kilometers, millions)

Country\Year	2000	2005	2010	2015	2020	2021	Variation 2000_2021
Australia	16 997	17 842	19 541	21 301	16 812	12 973	-4 024
Belgium	13 270	17 540	17 385	15 170	..	..	..
Denmark	7 418	7 169	6 853	6 682	..	..	..
Finland	7 700	7 540	7 540	7 540	6 800	6 000	-1 700
France	49 501	50 056	54 152	57 824	37 578	40 957	-8 544
Germany	69 000	82 489	78 092	81 771*	45 731*	..	..
Greece	6 000*	6 226*	..	..	..	..	..
Italy	93 550	100 954	102 219	102 508	58 634	80 307	-13 243
Japan	81 582	84 266	77 750	71 443	25 593	..	..
Korea	..*	58 213*	114 582	109 260	61 049	..	..
Netherlands	15 453*	16 034*	..	..	..	..	..
Spain	50 278	53 176	50 902	46 389	18 929	..	..
Sweden	9 618*	9 727	9 922	10 439	9 530	9 332	-286
United Kingdom	46 500	42 464	44 723	39 367	..	..	..
United States	258 973	238 170	469 790	553 732	..	..	..

.. missing data

*Estimated value/Break/Provisional value

Source: Data extracted on 23 Mar 2023 16:23 UTC (GMT) from OECD.Stat. https://stats.oecd.org/

The data in Table 5 refer to road passenger transport by bus and coach for the countries selected for analysis. The data cover the period from 2000 to 2021.

Table 5 exhibits that in Italy the number of passenger kilometers by bus and coach has increased significantly since 2000, and it can be inferred that this circulation has been affected by the covid_19 period, in 2020 and 2021 with the circulation bans (having decreased). The same is true for France and Australia. The upward trend is also evident in most of the selected countries.

In addition, Table 6 shows data on road passenger transport by passenger car for the countries selected for analysis, from 2000 to 2021.

From the data in Table 6, Italy, the Netherlands, and Sweden stand out as countries that have significantly reduced the number of passenger cars on the road. And this trend already occurred before the covid_19 period (years 2020 and 2021), which is quite relevant to consider. On the contrary, France shows an increasing trend, which is worrying.

Due to their characteristics, and associated with the daily distances traveled in cities, taxis lead to increased demand for public transport in urban areas, and this can generate a greater impact on the reduction of GHG emissions and atmospheric pollutants per vehicle (Hao et al., 2020; Hassouna & Assad, 2020; Kang & Lee, 2019).

Table 7 exhibits the number of companies licensed to provide taxi services in Portugal and the number of taxi drivers, for the period 2016 to 2021.

Table 6. Passenger transport: road passenger transport by passenger car (passenger kilometers, millions)

Country\ Year	2000	2005	2010	2015	2020	2021	Variation 2000_2021
Australia	239 809	262 063	262 229	280 144	265 627	269 476	29 667
Belgium	106 130	109 420	109 388	107 070	..	..	..
Denmark	56 776	58 348	59 759	60 862	..	..	..
Finland	55 700	61 910	64 745	66 295	64 100	63 600	7 900
France	697 553	717 234	709 789	793 393	680 478	728 190	30 637
Germany	831 300	856 900	884 800	927 000*	809 200*	..	..
Greece	34 458*	36 258	..	..	..	..	..
Italy	713 931*	677 014 *	698 390	676 350	488 299	518 275	-195 656
Japan	869 669*	848 739	799 201	808 492	743 847	..	..
Korea	..	142 566	264 281	268 784	278 488	..	..
Netherlands	141 100	148 800	144 200	139 500	105 250	115 750	-25 350
Spain	279 985*	337 797	341 629*	317 553	265 538	..	..
Sweden	103 655*	107 983	108 013	111 953	90 135	89 452	-14 203
United Kingdom	638 566	666 714	651 061	682 018	..	..	..
United States	4 094 907*	4 344 110	5 520 043	5 839 310	4 555 240	..	..

.. missing data

*Estimated value/Break/Provisional value

Source: Data extracted on 23 Mar 2023 15:42 UTC (GMT) from OECD.Stat. https://stats.oecd.org/

Table 7. Companies licensed to provide taxi services and taxi drivers in Portugal (2016-2021)

Taxi Info/Year	2016	2017	2018	2019	2020	2021	Variation 2016_2021 (%)
Taxi drivers	24,684	28,982	25,699	26,384	26,984	21,902	-11.27
Taxi companies	9,997	10,059	10,058	9,909	9,685	9,262	-7.35

Source: Portuguese Statistical Yearbook of Mobility and Transport, Institute for Mobility and Transport, I.P. https://www.imt-ip.pt/

In a trend that was expected to be the opposite, the number of taxi drivers and taxi companies in Portugal decreased in 2021 (11.27% and 7.35%, respectively). This is worrying data, as it was expected that this number would increase, given the importance of taxis in climate action, as it did in 2017. It is likely to reflect the covid-19 period, as the trend before was upward.

The provision of taxi services requires the possession of a license for each vehicle in use. Table 8 shows their distribution by region in Portugal in the year 2017 (the last year with data available for analysis).

As can be seen from Table 8, the Lisboa region concentrates 34% of the taxi licenses in Portugal, followed by the North and Center regions, which represent 27% and 20% of the licenses, respectively. These data are justified since these regions are also the most populated in Portugal[1].

Table 8. Geographical distribution of licensed taxis by region in Portugal (2017)

Region	% of licensed taxis (2017)
Alentejo	4.58
Algarve	4.00
Lisboa	**34.00**
Center	20.00
North	27.00
Madeira	6.00
Açores	4.00

Source: Adapted from Portuguese Mobility and Transport Authority. Data extracted on 5 Apr 2023. https://www.amt-autoridade.pt/

TAX INCENTIVES IN PORTUGAL FOR THE USE OF VEHICLES WITH REDUCED ENVIRONMENTAL IMPACTS

The Contributions of Green Tax Incentives Addressing Climate Change

When it comes to the contribution of green taxation, tax incentives are emerging as a climate policy instrument. And especially when dealing with purchasing vehicles. Reducing vehicle taxes through environmental taxation has a high potential to reduce GHG (Fridstrøm & Østli, 2017).

It is well known that car traffic is a problem that urgently needs to be solved in the context of sustainable cities (Cai et al., 2019). Therefore, within the scope of this study, countries are required to implement measures to reduce GHG emissions from cars in cities.

For this reason, several authors have used the specific experiences of their countries in their studies to justify and highlight the relevance of government environmental protection policies. For example, the

Norwegian experience presented by Yan & Eskeland (2018) shows that the country's vehicle registration tax, which is linked to the CO2 intensity of vehicles, significantly shifts consumers toward lower-emission vehicles. In Iceland, Shafiei et al. (2018) have shown how tax incentives can more or less encourage the purchase of electric vehicles. Similarly, van Eck et al. (2019) have shown the importance of tax incentives for CO2 abatement in the case of The Netherlands. Urrutia-Mosquera & Fábrega (2021) concluded that Chilean consumers are always waiting for the most attractive tax incentive to purchase more sustainable vehicles. In the same vein, Deyganto (2022) points out that tax incentives have promoted corporate sustainability in Ethiopia, even in times of crisis. Therefore, Qi et al. (2023) conducted their study to show that for Chinese companies, fiscal policy under environmental legislation with tax reduction through investment tax incentives can lead them to reduce GHG.

Yang et al. (2022) also point out that sustainable companies achieve higher levels of economic profitability better gross value added (Harzendorf et al., 2022), higher net profits (Litan et al., 2023) through stronger turnover (Sanoran, 2023) and operational results (Magon et al., 2018), which is reflected in companies' performance (Alonso-Martinez et al., 2021) and companies' profit margin (Zhang & Xie, 2022). As a result, sustainable companies are more tax compliant and pay more income tax (Saka et al., 2020). At the same time, they have higher levels of financial sustainability, with lower debt ratios (He & Liu, 2023), higher solvency ratios (Aziz et al., 2021), and high equity ratios (Gonçalves et al., 2022).

From this point of view, in addition to maximizing tax revenues, taxes can stimulate economic development. This can be done through the provision of tax incentives and/or preferential tax regimes. In this respect, previous studies have shown that firms respond more effectively to tax incentives than to subsidies (Sarmento, 2022; Wu et al., 2023)

Green tax incentives are a helpful tool not only because of their direct effect on reducing the tax burden but also because they address the emerging concern for environmental protection given the impact of pollution on environmental sustainability. In particular, companies choosing to acquire new environmentally friendly assets through the use of tax incentives (Pástor et al., 2021), will reduce pollution and improve corporate social responsibility (CSR) performance (Tang & Wang, 2022; Kong & Zhu, 2022; Zhang, 2021). However, for Badulescu et al. (2018), the younger a company is, the less likely it is to engage in CSR. Nevertheless, it is important to highlight how public environmental policies can also promote employment (Harzendorf et al., 2022).

Investment tax incentives and income tax deductions are widely used as fiscal policies to promote corporate sustainability and have an important impact on corporate sustainable investment (Shafi et al., 2023). However, other environmental tax measures, such as reduced rates and exemptions, are also important to help companies become more sustainable and tend to invest more in fixed assets (Rahman, 2022), particularly in transport equipment (Barcellos de Paula & Marins, 2018). Therefore, Li et al. (2022) emphasize the importance of the design of environmental tax policies, otherwise, the effects obtained will be contrary to what governments plan.

To encourage firms to purchase sustainable vehicles (Yan, 2018), such as all-electric vehicles (BEVs), hybrid vehicles (HVs), and plug-in hybrid vehicles (PHVs), governments should address their public policies, especially their tax policies, by subsidizing these purchases and exempting consumers from taxation (Ji et al., 2022).

A case in point is the experience of Portugal, which will be presented in the following sections.

The Green Taxation Reform in Portugal: Short Background Note

Portugal, as a member state of the EU, follows the European Regulations and instructions related to green taxation and the measures to address the effects of climate change and to promote sustainability and the usage of resources with more efficiency and less impact on the environment.

The armed conflict between Russia and Ukraine provoked a huge impact on the energy markets, especially regarding natural gas prices, which increase significantly. The green taxes and their role in the economy will become more important and they can be used not only to follow environmental purposes and to reduce pollution but also to respond to the economic instability in the petrol and gas market.

The Green Taxation reform in the Portuguese legal system was introduced by the Law No. 82-D/2014, of December 31(known as the Green Taxation Law), which creates a set of new environmental tax rules, following the EU legislation and international commitments. On the one hand, it aims to promote more sustainable behaviors in the production and consumption and usage more efficient of natural resources and, on the other hand, penalize polluting behaviors and activities. As mentioned in the preamble of the Law, it introduced some changes in different fields such as energy and emissions sectors, transports, water, waste, territorial planning, forests, and biodiversity, a new tax regime for plastic bags and an incentive regime for the scrapping of end-of-life vehicles.

It is also important to refer to the revenue generated by some of these taxes. According to data revealed by the Portuguese statistical office (2022), in 2021, the value of taxes with environmental relevance is around 5 billion Euros, corresponding to 6.6% of the revenue from taxes and social contributions (6.3% more than in 2020). Green taxation also becomes an additional source of revenue, which is not insignificant.

At this moment, as described by Fernandes & Moreira (2022), the tax policy options in Portugal can be divided into three schemes of green taxation: adapting indirect taxes (namely Value Added Tax (VAT)) to target renewable energy; increasing parafiscal taxation in different sectors of activity and multiply the incentives for renewables (for example, the Incentive System for the Support of Renewable Hydrogen Production and Other Renewable Gases, approved by Order No. 98-A/2022 on February 18, 2022).

Regarding the categories, the authors can distinguish four relevant fields in green taxation: transports (taxes related to the acquisition and utilization of some kind of transport modes), energy (taxes on energy products), pollution (taxes related to polluting emissions) and resources (related with the extraction and usage of natural resources).

Tax Incentives for Taxi Fleet Sustainability Under the Portuguese Vehicle Tax Code

The green taxes with more relevance in Portugal are related to vehicles: Tax on oil products (included in the Excise Tax Code (CIEC)), vehicle tax (ISV), and Single Circulation Tax (IUC).

Concerning ISV, paid when the vehicle acquires Portuguese matriculation and calculated by the CO2 emissions, there is a special regime for taxis and other vehicles used for rent in article no 53 of the Portuguese Vehicle Code.

When these cars have no more than four years of usage, counted as of the granting of the first registration and respective documents, and do not have CO emission levels (index 2) NEDC higher than 160 g/km or CO emission levels (index 2) WLTP higher than 184 g/km (confirmed by the respective conformity certificate) are exempted in 70 percent. of correspondent tax. If these cars are equipped with engines prepared for the exclusive use of natural gas or electric energy in their propulsion system, or with hybrid engines, prepared for the use of electric or solar energy or gasoline or diesel in their propulsion system, they are completely exempted from this tax.

The recognition of these exemptions depends on a request addressed to the Directorate General of Customs and Excise, before or concomitant to the presentation of the Vehicle Customs Declaration (DAV). This request must be done with a copy of the license to exercise the activity of passengers' transportation, a taxi license, and proof of the corporate form of the taxpayer.

In other cases, taxis can also benefit from an exemption of 40 percent in this tax when the vehicles must have an NEDC CO (index 2) emission level of up to 120 g/km or a WLTP CO (index 2) emission level of up to 138 g/km or, in the case of motorhomes, an NEDC CO (index 2) emission level of up to 165 g/km or a WLTP CO (index 2) emission level of up to 190 g/km (these emission levels must be confirmed by a certificate of compliance). The beneficiary companies of this partial exemption must be licensed exclusively for the activity of car rental. Vehicles rented or leased for more than 3 months to the same person or entity, during 12 consecutive months, or to persons or entities legally or economically linked to the beneficiary entity cannot benefit from this exemption.

Under article 5.º of the ISV code, taxis are also exempt from this tax for vehicles of category B having an NEDC CO2 emission level of up to 180 g/km or a WLTP CO2 emission level of up to 205 g/km and vehicles of category A.

As the authors could verify, as part of a broader tax policy mix, the Portuguese Vehicle Tax Code (VTC) contains a set of tax incentives that work mostly through rate reduction and tax exemption, promoting through lower taxation, the acquisition of vehicles with low fuel consumption and lower GHG emission, and consequently the change of sustainable environmental awareness associated with this type of vehicle, and of the firm sector.

The authors believe that the intention of the Portuguese tax legislator was focused on this mode of transport, which daily produces high emissions of GHG around the world since taxis as public transport consequently become less polluting, and increasingly sustainable, providing economic, social, and environmental benefits to cities that embrace this type of policy (following Guyader et al., 2021).

Study on the Portuguese Taxi Companies That Have Accessed Tax Incentives for Less Polluting Cars

The authors used public information on companies that have received tax incentives for less polluting cars (in vehicle tax – information provided by the Portuguese Tax Authority). The authors adopted the case study of vehicle tax incentives whose extra tax objective is the protection of the environment, through less polluting cars (the starting point being the tax incentives granted from two thousand and fifteen to two thousand and twenty-one).

From the data obtained for the period under study (2015-2021), the authors verified that the attribution of the tax incentive for less polluting cars to taxi companies has generated a tax expenditure of almost 935 thousand euros– which represents 0.5% of the total amount granted to companies in Portugal during this period (note that no amount was allocated to taxi companies in the "Tax incentive for the scrapping of end-of-life vehicles"). Data also shows that the number of taxi companies that took advantage of the tax incentive for less polluting vehicles was 237, with most taxi companies (190) focusing on the benefit of "Exemption for taxis and rental vehicles with engines designed to run exclusively on natural gas, electricity or hybrids", to which they were granted almost 98% of the total benefits in the period (547,310.59 €).

Table 9. Tax incentives for taxi companies (2015-2021)

Tax incentives for less polluting cars (in vehicle tax)	End-of-life vehicle scrapping tax incentive	Deduction for a negative environmental component in the cylinder capacity component	Tax reduction for cars with hybrid engines	Tax reduction for cars that use only natural gas*	Tax reduction for cars with plug-in hybrid engines	Exemption for taxis and rental vehicles with engines designed to run exclusively on natural gas, electricity, or hybrids	Total
Number of taxi companies that took advantage of tax incentives	0	120	168	1	17	190	237
Total amount of tax incentives granted to taxi companies (€)	0	3,489.48	312,770.4	1,172.02	70,167.09	547,310.59	934,909.58
2015/2021 Total amount of tax incentives granted (€)	224,405.94	1,835,635.19	72,545,963.23	51,568.81	111,742,886.74	558,520.07	186,958,979.98
Tax incentive granted to taxi companies (%)	0	0.19	0.43	2.27	0.06	97.99	0.50

* The previous law also considered vehicles powered by liquefied petroleum gas (LPG), but this was withdrawn by Law 2/2020, of March 31.

Source: Data extracted on 27 Dec 2022 11:05 UTC (GMT) from https://www.portaldasfinancas.gov.pt/

Research Methodology and Results

This section presents the research methodology, which includes research hypotheses, data collection, sample characterization, and results. The authors used version 28 of the software Statistical Package for Social Science for Windows (SPSS).

Following the previous literature review, the sustainability data sources used in this study are listed in Table 10.

According to the literature review, several determinants lead firms to take advantage of environmental tax incentives. The authors conducted the study based on taxi companies.

Research Hypotheses

Hypothesis One: The most sustainable taxi companies have higher operational results.

Hypothesis Two: Taxi companies with access to environmental tax benefits will have higher levels of investment in assets.

Hypothesis Three: Taxi companies that take advantage of environmental tax benefits have better economic-financial metrics.

Table 10. Data sources

Variables	Studies
Tax incentives for less polluting cars	Urrutia-Mosquera & Fábrega, 2021
Assets	Pástor et al., 2021
Debt	He & Liu, 2023
Economic profitability	Yang et al., 2022
Number of employees	Ruiz-Pérez et al., 2021
Equity ratio	Gonçalves et al., 2022
Firm age	Badulescu et al., 2018
Gross value-added	Harzendorf et al., 2022
Income tax	Saka et al., 2020
Net profit	Litan et al., 2023
Operational result	Magon et al., 2018
Performance	Alonso-Martinez et al., 2021
Profit margin	Zhang & Xie, 2022
Solvency ratio	Aziz et al., 2021
Transport equipment	Barcellos de Paula & Marins, 2018
Turnover	Sanoran, 2023

Hypothesis Four: The firm age of taxi companies determines the use of environmental tax benefits for less polluting cars.

Hypothesis Five: Taxi companies that hire more workers are the ones that benefit the most from tax incentives for less polluting cars.

This study is based on a data set provided by the Portuguese Tax Administration and SABI database (Iberian Balance Sheet Analysis System), which has allowed the authors to create a database of all the Portuguese taxi companies that have taken advantage of the environmental tax benefits for less polluting cars.

Sample and Descriptive Statistics

The results show that most of the benefiting taxi companies are concentrated in the Lisboa and Vale do Tejo region (59.5%). However, taxi companies in the North region received a higher average of tax incentives for less polluting cars (4,472.53 euros). Thus, although only 16.9% of the taxi companies that took advantage of the tax incentives under study are concentrated in the municipality of Porto (as opposed to almost 49% in Lisboa), they had access to an average of 4,735.13 euros in tax incentives for less polluting cars. It is also noteworthy that the taxi companies that had the most access to the tax incentive on average (5,553.85 euros) are in the municipality of Vila Nova de Gaia, although they represent only 3.4% of the beneficiary companies (compared to almost 13% of the taxi companies in the municipality of Lisboa). In addition, although most taxi companies are limited companies (73.4%), it is the single-member limited taxi companies that benefited more on average from the tax incentives for less polluting cars (of 4,067.80 euros). In short, it is interesting to note that although the percentage of taxi companies benefiting is higher in Lisboa and Vale do Tejo region, the average number of incentives granted is higher in the North of Portugal. Table 8 shows that in 2017, most taxi licenses allocated in Portugal were in the Lisboa region, while the second largest was in the North region. This seems to indicate that companies in the North region are less aware of this type of tax advantage.

Table 11. Detail of taxi companies that benefited from tax incentives for less polluting cars (in vehicle tax)

Data on taxi companies	Percentage	Average vehicle tax incentives for less polluting cars (€)
Region:		
Lisboa and Vale do Tejo	**59.50**	3,693.86
North	30.80	**4,472.53**
Centre	3.80	3,680.81
Alentejo	3.00	4,370.74
Algarve	3.00	3,408.20
District:		
Lisboa	**48.90**	3,720.57
Porto	16.90	**4,735.13**
Braga	9.30	3,952.91
Setúbal	7.20	3,466.17
Évora	3.00	4,370.74
Municipality:		
Lisboa	**12.70**	4,203.47
Sintra	11.00	3,434.15
Amadora	5.10	3,280.63
Braga	5.10	4,070.41
Cascais	4.60	4,049.72
Loures	4.20	3,632.58
Odivelas	3.80	4,359.96
Vila Nova de Gaia	3.40	**5,553.85**
Gondomar	3.00	3,467.40
Legal Form:		
Private limited company	**73.40**	3,812.80
Single member limited company	26.20	**4,067.80**

Source: Output SPSS Statistics; Data extracted on 27 Dec 2022 11:05 UTC (GMT) from https://www.portaldasfinancas.gov.pt

Results: Bivariate Analysis

A bivariate analysis was conducted to study the relationship between the average tax incentives for less polluting cars (in vehicle tax) granted to taxi companies (Group A: a database of 237 taxi companies) and key variables related to the characteristics of the companies. The results are presented in Table 12.

From the results obtained, it is possible to underline that the highest significant and positive correlation relates to companies´ average income tax in the period from 2015 to 2021 ($R=0.552$, $p<0.001$). The same for the average total assets ($R=0.207$, $p=0.001$), average turnover ($R=0.202$, $p=0.002$), average operational result ($R=0.201$, $p=0.002$), average transport equipment ($R=0.188$, $p=0.004$), average gross value-added ($R=0.186$, $p=0.004$), and average net profit ($R=0.132$, $p=0.044$). This shows that as the average tax incentives for less polluting cars increase for taxi companies, the companies' income tax, total assets, turnover, operational result, transport equipment, gross value added, and net profit also increase on average. The same applies to the economic-financial indicators. As the average tax incentives for less polluting cars increase for taxi companies, the companies' economic profitability ($R=0.177$, $p=0.006$), performance ($R=0.168$, $p=0.010$), solvency ratio ($R=0.151$, $p=0.020$), and equity ratio ($R=0.151$, $p=0.020$) increase on average. In addition, the average number of employees ($R=0.169$, $p=0.009$) increases as the average tax incentives for less polluting cars that companies have had access to increase.

Table 12. Pearson correlations

Variables	Tax incentives for less polluting cars (taxb) Group A: Taxi companies with tax incentives	
	(R)	ρ
Operational result *opres*	**0.201****	0.002
Net profit netprof	**0.132***	0.044
Total assets *tasset*	**0.207****	0.001
Economic profitability *econprf*	**0.177****	0.006
Number of employees *emplnb*	**0.169****	0.009
Turnover *turn*	**0.202****	0.002
Performance *perf*	**0.168****	0.010
Gross value-added *gva*	**0.186****	0.004
Solvency ratio *solv*	**0.151***	0.020
Transport equipment *transeq*	**0.188****	0.004
Income tax *inctax*	**0.552****	<0.001
Equity ratio *eqrat*	**0.151***	0.020
Firm age *fage*	-0.027	0.682
Profit margin *profmg*	0.066	0.314
Debt *debt*	-0.125	0.054

Source: Output SPSS Statistics. ***Notes****: Pearson correlations (ρ) are significant at **the 0.01 level (2-tailed) and at *the 0.05 level (2-tailed); (R) = Correlation coefficient;* ***taxb****=average tax incentives for less polluting cars from 2015 to 2021;* ***opres****=average operational result from 2015 to 2021;* ***netprof****= average net profit from 2015 to 2021;* ***tasset****=average of total assets from 2015 to 2021;* ***econprf****=average economic profitability from 2015 to 2021;* ***emplnb****=average number of employees from 2015 to 2021;* ***turn****=average turnover from 2015 to 2021;* ***perf****=average operational performance from 2015 to 2021;* ***gva****=average gross value added from 2015 to 2021;* ***solv****=average percentage of solvency ratio from 2015 to 2021;* ***transeq****=average transport equipment from 2015 to 2021;* ***inctax****= average income tax from 2015 to 2021;* ***eqrat*** *= average equity ratio from 2015 to 2021;* **fage***=firm age; ;* ***profmg****=average average profit margin from 2015 to 2021;* ***debt****=average debt from 2015 to 2021*

The study found no significant correlation between tax incentives and firm age (R=-0.027, p=0.682), suggesting that younger and older taxi companies alike can benefit from tax incentives for less polluting cars.

Results: Regression Model Specification

A detailed database of the tax incentives granted to Portuguese taxi companies has been developed to compare companies that have taken advantage of the tax incentives with others in the same sector that have not. Since it is impossible to analyze all the companies in the taxi sector and using the SABI database (Iberian Balance Sheet Analysis System), the companies with the activity code "Occasional Transportation in Passenger Cars" were selected, namely those that presented profit margin percentage, operating income values in at least one of the periods from 2015 to 2021, as well as all the taxi companies with a positive trend in net profit.

Comparing taxi companies that benefited from tax incentives for less polluting cars (Group A: a database of 237 taxi companies) with others in the same sector that did not (Group B: database of 6,773 taxi companies, including Group A companies), the ANOVA results show that there are differences between companies that did and did not take advantage of the tax benefits under investigation, so the authors applied the regression model to Group A only.

Multiple linear regression was used to verify whether the access to environmental tax benefits for less polluting cars can be predicted by taxi companies´ economic and financial features, i.e., the authors applied a model to explain which characteristics of taxi companies are significant in inducing access to environmental tax incentives under study. The dependent variable is *taxb* [average tax benefits for less polluting cars]. The independent variables are the variables in Table 12. $\mathbf{u}_i$ is the error term, relative to unknown factors. The mathematical specification of the regression model is:

$$taxb_i = b_0 + b_1 opres_{1i} + b_2 netprof_{2i} + b_3 tasset_{3i} + b_4 econprf_{4i} + b_5 emplnb_{5i} + b_6 turn_{6i} + b_7 perf_{7i} + b_8 gva_{8i} + b_9 solv_{9i} + b_{10} transeq_{10i} + b_{11} inctax_{11i} + b_{12} eqrat_{12i} + b_{13} \mathrm{fage}_{13i} + b_{14} profmg_{14i} + b_{15} debt_{15i} + u_i$$

After checking the consistency of the model's assumptions, namely multicollinearity, homoscedasticity, and independence of errors, the model was estimated with a 5% significance level. The method used was the Ordinary Least Squares (OLS), stepwise selection method.

Table 13 reports the multiple regressions result, where the marginal effect of the independent variables can be measured. For Model 1 (Group A) results are significant: [$F(4,226)=35.857$; $p<0.001$; $R^2=0.388$].

Several aspects can be pointed out regarding the results of the regression. First, following the bivariate analysis carried out previously, it was found that the average income tax is the most important determinant of Group A taxi companies' admission to tax incentives for less polluting cars (*t*-test equal to 11.080), being the correlation significant and positive (66.2%). Second, the significant coefficient of the average number of employees from 2015 to 2021 is also positive for Model 1 (38.5%). Third, related to average economic profitability, it turns out the significant coefficient indicates that the correlation (15.3%) within Group A taxi companies' admission to tax incentives for less polluting cars is positive (*t*-test equal to 2.829). Four, the results predicted something unexpected, namely the significant negative correlation (39.2%) with the average value of gross value added (*t*-test equal to -4.455).

Table 13. Regression results: model specifications

Variables	Model 1 Group A		
	Standardized coefficients (β)	*t*	**Sig.**
(Constant)		22.611	<0.001**
Income tax *(inctax)*	0.662	11.080	<0.001**
Economic profitability *(econprf)*	0.153	2.829	0.005**
Number of employees *(emplnb)*	0.385	4.838	<0.001**
Gross value added *(gva)*	-0.392	-4.455	<0.001**
N		237	
R^2		0.388	
Adjusted R^2		0.377	

* Significant at 5%

** Significant at 1%

Source: Output SPSS Statistics. **Notes**: Dependent Variable: **taxb**=average tax incentives for less polluting cars from 2015 to 2021; **Predictors:** (Constant), **inctax**= average income tax from 2015 to 2021; **econprf**=average economic profitability from 2015 to 2021; **emplnb**=average number of employees from 2015 to 2021; **gva**=average gross value added from 2015 to 2021

FINDINGS DISCUSSION

For Group A taxi companies, the results show that the average level of income tax, the average number of employees, and the average economic profitability can be determinants of firms' access to the tax incentives for less polluting cars, demonstrating firms' commitment to sustainability. This is in line with previous studies that conclude that more sustainable companies are also more profitable (Yang et al., 2022), more tax compliant (Saka et al., 2020), and by being more socially responsible also promote further employment opportunities (Ruiz-Pérez et al., 2021). Although the literature review confirms that more sustainable companies have higher gross value added (Harzendorf et al., 2022), it is important to remember that the gross value added of non-financial companies has declined, marked by the strong impact of the Covid-19 pandemic. This has affected the results found, which cover the years 2020 and 2021.

Furthermore, there is no evidence of a significant relationship between the use of tax incentives for less polluting cars and the age of the firm. This is consistent with what has been discussed in the literature about "newer" firms being less committed to social responsibility (Badulescu et al., 2018). In summary, hypotheses one, three, and five are supported, while hypotheses two and four are not supported by the results.

These findings have implications for policymakers and business leaders, highlighting the need for companies to prioritize sustainability and social responsibility to access tax incentives and achieve long-term success.

FUTURE RESEARCH DIRECTIONS

The authors believe that there is a need for more understanding among taxi companies in Portugal about the tax incentives available for more sustainable vehicles. The fact that only 237 out of Portuguese 9,828 taxi companies took advantage of these incentives during the period 2015 to 2021 is a clear indication of a lack of consciousness on this issue. This provides an opportunity for future research to investigate the reasons for this lack of awareness, and to identify strategies to uptake incentives among taxi companies, since, as the chapter demonstrates, sustainable businesses are more profitable.

Furthermore, the concentration of benefiting taxi companies in the Lisboa and Vale do Tejo region and the higher average of incentives granted in the North region suggest that there may be regional differences in awareness and access to these tax incentives. The authors believe that a discussion should be launched to explore these regional differences and the potential role of regional policies and initiatives in promoting sustainable practices in the transport sector, including the level of wealth of the regions.

In terms of future and emerging trends, tax incentives for less polluting cars will likely continue to gain momentum as governments and policymakers seek to address climate change and promote sustainability. This could result in more companies adopting sustainable practices to remain competitive in the marketplace. In addition, advances in technology and the increasing availability of electric and hybrid vehicles may make it easier for companies to switch to less polluting cars. A further study of Portuguese companies in general, for comparison between sectors of activity, will be fundamental. Carrying out the study with other European countries could also be a future line of research.

CONCLUSION

This chapter highlights the importance of tax policies focusing on GHG. The results show that the average income tax, average number of employees, and average economic profitability were important factors for taxi companies' access to tax incentives. The results suggest that companies that prioritize sustainability tend to be more profitable, tax-compliant, and job promoters.

The findings of the study can help the business owners of taxi companies to understand business sustainable practices and tax policy effects. Furthermore, the authors consider that these results have a significant implication for managers and other stakeholders, showing that taxi companies that take advantage of environmental tax benefits have better economic-financial metrics.

Given these results, it is recommended that authorities promote awareness campaigns to encourage more taxi companies to invest in sustainable vehicles. This can include providing information to companies on the benefits of using eco-friendly vehicles, including lower operating costs, and reduced environmental impact through tax incentives.

This study extends the literature on the topic of environmental challenges using tax incentives as an issue of a company's policy outcome and public concern emphasizing the importance of considering the impact of tax incentives on companies' behavior. Overall, this chapter provides a valuable contribution to support governments to evaluate the efficiency of tax incentives.

The study provides evidence that tax incentives can play a significant role in promoting more sustainable behaviors in the transport sector, particularly in the taxi sector, emphasizing the need for more widespread adoption of tax policies that support sustainable development and environmental protection.

ACKNOWLEDGMENT

This work is financed by national funds through the FCT - Foundation for Science and Technology, I.P., under project UIDP/04043/2020.

REFERENCES

Akdag, S., & Yıldırım, H. (2020). Toward a sustainable mitigation approach of energy efficiency to greenhouse gas emissions in the European countries. *Heliyon*, *6*(3), e03396. doi:10.1016/j.heliyon.2020.e03396 PMID:32154402

Alonso-Martinez, D., De Marchi, V., & Di Maria, E. (2021). The sustainability performances of sustainable business models. *Journal of Cleaner Production*, *323*(129145), 129145. doi:10.1016/j.jclepro.2021.129145

Antón, Á. (2022). EU energy taxation framework revision in the context of the green deal. *International Visual Culture Review, 9*(Monográfico), 1–13. doi:10.37467/revvisual.v9.3670

Aziz, S., Rahman, M., Hussain, D., & Nguyen, D. K. (2021). Does corporate environmentalism affect corporate insolvency risk? The role of market power and competitive intensity. *Ecological Economics*, *189*(July), 107182. doi:10.1016/j.ecolecon.2021.107182

Badulescu, A., Badulescu, D., Saveanu, T., & Hatos, R. (2018). The relationship between firm size and age, and its social responsibility actions-Focus on a developing country (Romania). *Sustainability (Basel)*, *10*(3), 805. doi:10.3390u10030805

Barcellos de Paula, L., & Marins, F. A. S. (2018). Algorithms applied in decision-making for sustainable transport. *Journal of Cleaner Production*, *176*, 1133–1143. doi:10.1016/j.jclepro.2017.11.216

Cai, H., Wang, X., Adriaens, P., & Xu, M. (2019). Environmental benefits of taxi ride sharing in Beijing. *Energy*, *174*, 503–508. doi:10.1016/j.energy.2019.02.166

Calise, F., Cappiello, F. L., Cartenì, A., Dentice d'Accadia, M., & Vicidomini, M. (2019). A novel paradigm for a sustainable mobility based on electric vehicles, photovoltaic panels and electric energy storage systems: Case studies for Naples and Salerno (Italy). *Renewable & Sustainable Energy Reviews*, *111*, 97–114. doi:10.1016/j.rser.2019.05.022

Cihat Onat, N. (2022). How to compare sustainability impacts of alternative fuel Vehicles? *Transportation Research Part D, Transport and Environment*, *102*, 103129. doi:10.1016/j.trd.2021.103129

Delgado, F. J., Freire-gonzález, J., & Presno, M. J. (2022). Environmental taxation in the European Union : Are there common trends? *Economic Analysis and Policy*, *73*, 670–682. doi:10.1016/j.eap.2021.12.019

Deyganto, K. O. (2022). The effect of tax incentives practices on the sustainability of micro, small and medium enterprises in Ethiopia during the outbreak of corona virus pandemic. *Journal of Innovation and Entrepreneurship*, *11*(1), 8. doi:10.118613731-022-00194-8 PMID:35106268

Dieguez, T., Ferreira, L. P., Silva, F. J. G., & Tjahjono, B. (2020). Open innovation and sustainable development through industry-academia collaboration: A case study of automotive sector. *Procedia Manufacturing*, *51*, 1773–1778. doi:10.1016/j.promfg.2020.10.246

Ekins, P., & Zenghelis, · Dimitri. (2021). *The costs and benefits of environmental sustainability.* Springer. doi:10.1007/s11625-021-00910-5

European Environment Agency. (2022). *Decarbonising road transport: the role of vehicles, fuels and transport demand.* EEA. doi:10.2800/68902

Fernandes, F. V., & Moreira, T. M. (2022). *The evolution of green taxation in Portugal.* International Law Office. https://www.internationaltaxreview.com/

Fridstrøm, L., & Østli, V. (2017). The vehicle purchase tax as a climate policy instrument. *Transportation Research Part A, Policy and Practice*, *96*, 168–189. doi:10.1016/j.tra.2016.12.011

Gonçalves, T. C., Dias, J., & Barros, V. (2022). Sustainability Performance and the Cost of Capital. *International Journal of Financial Studies*, *10*(3), 1–32. doi:10.3390/ijfs10030063

Guyader, H., Friman, M., & Olsson, L. E. (2021). Shared Mobility: Evolving Practices for Sustainability. *Sustainability (Basel)*, *13*(21), 12148. doi:10.3390u132112148

Han, H., & Lee, T. (2022). Varieties of Green Stimulus Policies: Comparative Analysis of the Green Growth and Green New Deal Policies in South Korea. *Journal of Environment & Development*, *107049652211393*. doi:10.1177/10704965221139303

Hao, X., Wang, H., Lin, Z., & Ouyang, M. (2020). Seasonal effects on electric vehicle energy consumption and driving range: A case study on personal, taxi, and ridesharing vehicles. *Journal of Cleaner Production*, *249*(119403), 119403. doi:10.1016/j.jclepro.2019.119403

Harzendorf, F., Wulf, C., Haase, M., Baumann, M., Ersoy, H., & Zapp, P. (2022). Domestic value added as an indicator for sustainability assessment: A case study on alternative drivetrains in the passenger car sector. *Clean Technologies and Environmental Policy*, *24*(10), 3145–3169. doi:10.100710098-022-02402-1

Hassouna, F. M. A., & Assad, M. (2020). Towards a sustainable public transportation: Replacing the conventional taxis by a hybrid taxi fleet in the west bank, Palestine. *International Journal of Environmental Research and Public Health*, *17*(23), 1–11. doi:10.3390/ijerph17238940 PMID:33271931

He, Y., & Liu, R. (2023). The impact of the level of green finance development on corporate debt financing capacity. *Finance Research Letters, 52*(November 2022), 1–8. doi:10.1016/j.frl.2022.103552

Ji, Q., Wang, C., & Fan, Y. (2022). Environmental and welfare effects of vehicle purchase tax: Evidence from China. *Energy Economics*, *115*, 106377. doi:10.1016/j.eneco.2022.106377

Kang, S.-C., & Lee, H. (2019). Economic appraisal of implementing electric vehicle taxis in Seoul. *Research in Transportation Economics*, *73*, 45–52. doi:10.1016/j.retrec.2018.11.007

Kazancoglu, Y., Ozbiltekin-Pala, M., & Ozkan-Ozen, Y. D. (2021). Prediction and evaluation of greenhouse gas emissions for sustainable road transport within Europe. *Sustainable Cities and Society*, *70*, 102924. doi:10.1016/j.scs.2021.102924

Kong, D., & Zhu, L. (2022). Governments' Fiscal Squeeze and Firms' Pollution Emissions. *Environmental and Resource Economics*, *81*(4), 833–866. doi:10.100710640-022-00656-3

Li, Y., Mao, J., Chen, S., & Yang, D. (2022). Tax-reducing incentive and corporate green performance: What we learn from China. *Renewable Energy*, *199*(April), 791–802. doi:10.1016/j.renene.2022.07.128

Litan, H., Rong, K., Wu, Y., Xie, D., Zhang, H., & Zhao, D. (2023). Profit versus Sustainability in Bike-share. *Sustainable Cities and Society*, *93*(March), 104512. doi:10.1016/j.scs.2023.104512

Magon, R. B., Thomé, A. M. T., Ferrer, A. L. C., & Scavarda, L. F. (2018). Sustainability and performance in operations management research. *Journal of Cleaner Production*, *190*, 104–117. doi:10.1016/j.jclepro.2018.04.140

Mosquera Valderrama, I., & Balharová, M. (2021). *Tax Incentives in Developing Countries: A Case Study—Singapore and Philippines*. Springer. doi:10.1007/978-3-030-64857-2_7

Novas, N., Garcia Salvador, R. M., Portillo, F., Robalo, I., Alcayde, A., Fernández-Ros, M., & Gázquez, J. A. (2022). Global Perspectives on and Research Challenges for Electric Vehicles. *Vehicles*, *4*(4), 1246–1276. doi:10.3390/vehicles4040066

Oliveira, D., & Mendes, I. (2017). Green taxation on competitiveness: The effect of the ISP tax on the Portuguese retail sector of road fuels. *International Journal of Energy Economics and Policy*, *7*(1), 178–184.

Panarello, D., & Gatto, A. (2023). Decarbonising Europe – EU citizens' perception of renewable energy transition amidst the European Green Deal. *Energy Policy*, *172*, 113272. doi:10.1016/j.enpol.2022.113272

Pástor, Ľ., Stambaugh, R. F., & Taylor, L. A. (2021). Sustainable investing in equilibrium. *Journal of Financial Economics*, *142*(2), 550–571. doi:10.1016/j.jfineco.2020.12.011

Qi, Y., Zhang, J., & Chen, J. (2023). Tax incentives, environmental regulation and firms' emission reduction strategies: Evidence from China. *Journal of Environmental Economics and Management*, *117*(November 2021). doi:10.1016/j.jeem.2022.102750

Qiu, J., Huang, K., & Hawkins, J. (2022). The taxi sharing practices: Matching, routing and pricing methods. *Multimodal Transportation*, *1*(1), 100003. doi:10.1016/j.multra.2022.100003

Rahman, M. M. (2022). The effect of taxation on sustainable development goals: Evidence from emerging countries. *Heliyon*, *8*(9), e10512. doi:10.1016/j.heliyon.2022.e10512 PMID:36105472

Ruiz-Pérez, F., Lleo, A., & Ormazabal, M. (2021). Employee sustainable behaviors and their relationship with Corporate Sustainability: A Delphi study. *Journal of Cleaner Production*, *329*, 129742. Advance online publication. doi:10.1016/j.jclepro.2021.129742

Saka, C., Oshika, T., & Jimichi, M. (2020). Does tax avoidance diminish firms' sustainability? *Journal Global Policy and Governance*, *8*(2), 95–114. doi.org/10.14666/2194-7759-8-2-005

Sanoran, K. (2023). Corporate sustainability and sustainable growth: The role of industry sensitivity. *Finance Research Letters*, *53*(103596), 103596. doi:10.1016/j.frl.2022.103596

Sarmento, J. M. (2022). Preface. In R. Marques & S. Reis (Eds.), *Código Fiscal do Investimento Anotado e Comentado* (pp. 7–10). Almedina.

Shafi, M., Ramos-Meza, C. S., Jain, V., Salman, A., Kamal, M., Shabbir, M. S., & Rehman, M. (2023). The dynamic relationship between green tax incentives and environmental protection. *Environmental Science and Pollution Research International, 30*(12), 32184–32192. doi:10.100711356-023-25482-y PMID:36723845

Shafiei, E., Davidsdottir, B., Fazeli, R., Leaver, J., Stefansson, H., & Asgeirsson, E. I. (2018). Macroeconomic effects of fiscal incentives to promote electric vehicles in Iceland: Implications for government and consumer costs. *Energy Policy, 114*(May 2017), 431–443. doi:10.1016/j.enpol.2017.12.034

Tang, M., & Wang, Y. (2022). Tax incentives and corporate social responsibility: The role of cash savings from accelerated depreciation policy. *Economic Modelling, 116*, 106040. doi:10.1016/j.econmod.2022.106040

Tang, Y., Cockerill, T. T., Pimm, A. J., & Yuan, X. (2021). Reducing the life cycle environmental impact of electric vehicles through emissions-responsive charging. *iScience, 24*(12), 103499. doi:10.1016/j.isci.2021.103499 PMID:34927031

Urrutia-Mosquera, J., & Fábrega, J. (2021). Impact of fiscal incentives in the consumption of low emission vehicles. *Case Studies on Transport Policy, 9*(3), 1151–1159. doi:10.1016/j.cstp.2021.06.003

van Eck, G., de Jong, G., Wesseling, B., & van Meerkerk, J. (2019). Simulating the impact of tax incentives using a type choice model for lease cars. *Case Studies on Transport Policy, 7*(4), 814–822. doi:10.1016/j.cstp.2019.07.014

Williams, I. D., & Blyth, M. (2023). Autogeddon or autoheaven: Environmental and social effects of the automotive industry from launch to present. *The Science of the Total Environment, 159987*, 159987. doi:10.1016/j.scitotenv.2022.159987 PMID:36372167

Wu, W., Hu, Y., & Wu, Q. (2023). Subsidies and tax incentives - Does it make a difference on TFP? Evidences from China's photovoltaic and wind listed companies. *Renewable Energy, 208*(June 2022), 645–656. doi:10.1016/j.renene.2023.03.080

Xu, L., Yilmaz, H. Ü., Wang, Z., Poganietz, W. R., & Jochem, P. (2020). Greenhouse gas emissions of electric vehicles in Europe considering different charging strategies. *Transportation Research Part D, Transport and Environment, 87*, 102534. doi:10.1016/j.trd.2020.102534

Yan, S. (2018). The economic and environmental impacts of tax incentives for battery electric vehicles in Europe. *Energy Policy, 123*, 53–63. doi:10.1016/j.enpol.2018.08.032

Yan, S., & Eskeland, G. S. (2018). Greening the vehicle fleet: Norway's CO2-Differentiated registration tax. *Journal of Environmental Economics and Management, 91*, 247–262. doi:10.1016/j.jeem.2018.08.018

Yang, L., Zhou, Y., Meng, B., Li, H., Zhan, J., Xiong, H., Zhao, H., Cong, W., Wang, X., Zhang, W., Lakshmanan, P., Deng, Y., Shi, X., Chen, X., & Zhang, F. (2022). Reconciling productivity, profitability and sustainability of small-holder sugarcane farms : A combined life cycle and data envelopment analysis. *Agricultural Systems, 199*, 103392. Advance online publication. doi:10.1016/j.agsy.2022.103392

Zhang, D. (2021). How environmental performance affects firms' access to credit: Evidence from EU countries. *Journal of Cleaner Production, 315*, 128294. doi:10.1016/j.jclepro.2021.128294

Zhang, D., & Xie, Y. (2022). Customer environmental concerns and profit margin: Evidence from manufacturing firms. *Journal of Economics and Business, 120*(November 2021), 106057. doi:10.1016/j.jeconbus.2022.106057

Zhao, G., & Baker, J. (2022). Effects on environmental impacts of introducing electric vehicle batteries as storage - A case study of the United Kingdom. *Energy Strategy Reviews, 40*, 100819. doi:10.1016/j.esr.2022.100819

ADDITIONAL READING

Carattini, S., Baranzini, A., Thalmann, P., Varone, F., & Vöhringer, F. (2017). Green Taxes in a Post-Paris World: Are Millions of Nays Inevitable? *Environmental and Resource Economics, 68*(1), 97–128. doi:10.100710640-017-0133-8

Dinis, A. A. (2022). *A Tributação Autónoma em Portugal - Uma Abordagem Jurisprudencial*. Almedina.

Eroğlu, E. (2019). An Analysis of Green Taxation in Turkey for Sustainable Growth. In S.-B. Tsai, C.-H. Shen, H. Song, & B. Niu (Eds.), *Green Finance for Sustainable Global Growth* (pp. 270–296). IGI Global.

Filipiak, B. Z., & Wyszkowska, D. (2020). Impact of green taxes on the public financial system: An example of european union countries. In M. Ziolo (Ed.), *Social, economic, and environmental impacts between sustainable financial systems and financial markets* (pp. 96–119). IGI Global. doi:10.4018/978-1-7998-1033-9.ch006

László, C. (2021). The Green Tax Revolution. *Inter Economics, 56*(5), 284–287. doi:10.100710272-021-1000-y PMID:34629506

Lobo, C. (2019). *Finanças e Fiscalidade do Ambiente e da Energia*. Almedina.

Monbiot, G. (2009). *Heat: How to Stop the Planet From Burning*. South End Press.

Pereira, A. (2020). Benefícios Fiscais à Eficiência Energética. In S. Vasques (Ed.), *Fiscalidade da Energia* (pp. 16–30). Almedina.

van der Ploeg, F. (2023). Fiscal Costs of Climate Policies: Role of Tax, Political, and Behavioural Distortions. *De Economist, 0123456789*(2), 119–137. Advance online publication. doi:10.100710645-023-09419-x

KEY TERMS AND DEFINITIONS

All-Electric Vehicles: Vehicles that have an electric motor instead of an internal combustion engine, vehicles also referred to as battery electric vehicles (BEVs).

Climate-Neutral Impact: By 2050, to reach the point of net zero emissions, countries must dramatically reduce their greenhouse gas emissions and find ways to offset remaining and unavoidable emissions.

Environmental Sustainability: The ability to maintain the ecological balance of our planet's natural environment and conserve natural resources to support the well-being of present and future generations.

Green Taxation: Environmental or green taxes include taxes on energy, transport, pollution, and resources. It is seen by governments as an opportunity to adapt the tax system to a more energy-efficient economy in the use of resources, making it innovative, low-carbon, and more sustainable. It can take the form of tax incentives.

Greenhouse Gas Emissions (GHG): A term to describe a gas that absorbs and emits radiant energy within the thermal infrared range, causing the greenhouse effect. Carbon dioxide (CO2), methane (CH4), nitrous oxide (N2O), perfluorocarbons (PFCs), hydrofluorocarbons (HFCs), sulfur hexafluoride (SF6), and nitrogen trifluoride (NF3) are the most important greenhouse gases.

Hybrid Electric-Petroleum Vehicles: Vehicles powered by an internal combustion engine and one or more electric motors, which use energy stored in batteries.

Sustainable City: A city that has been able to introduce greenery into the urban environment to reduce CO2 emissions and improve air quality, has promoted renewable energy to conserve and protect natural resources, and has successfully implemented sustainable mobility and the use of public transportation.

Tax Incentives: Concessions in the tax codes whose extra-fiscal purpose overlaps tax revenues and represent a deliberate loss of government budget revenues.

Taxi: A motor vehicle licensed to transport passengers in return for payment of a fare and typically fitted with a taximeter.

ENDNOTES

1 In this sense, see: OECD (2022), *OECD Regions and Cities at a Glance 2022*, OECD Publishing, Paris, https://doi.org/https://doi.org/10.1787/14108660-en.

Chapter 11
Taxation Policies as an Environmental Protection Instrument:
The Portuguese Case

Sara Luís Dias
Research Centre on Accounting and Taxation, Polytechnic Institute of Cávado and Ave, Portugal

João Sérgio Ribeiro
School of Law of University of Minho, Portugal & The Research Centre for Justice and Governance (JusGov), Portugal

ABSTRACT

The "green taxation," which usually includes taxes on energy, transport, pollution, and resources, is being used to achieve environmental and sustainability purposes. The European Union is promoting this strategy, creating different provisions related to the usage of taxes to change environmental behaviors. Portugal, as a member-estate of European Union, follows this path establishing a policy where taxes are being used for these purposes. However, "green taxation" is also a significant form of revenue and, currently, it is difficult to define the criteria for the creation of these taxes, as they are not considering the ability to pay (one of the main principles in the creation of taxes) affecting the equitable distribution of the tax burden. The authors can conclude that, with the masking of "taxes required for environmental protection," these taxes are being used as an additional source of revenue, distorting the principles underlying the creation and collection of taxes.

INTRODUCTION

Currently, one of the most important concerns of our States is controlling the environmental damage provoked by the evolution of economical and industrial activities. In particular, the European Union is committed to becoming the first climate neutral continent in 2050, by reducing polluting emissions,

DOI: 10.4018/978-1-6684-8592-7.ch011

fighting the energy poverty, reducing external energy dependency, and improving the general wellbeing (*European Green Deal provisions*). Therefore, a lot of strategies and measures are being considered to promote the desired environmental protection and the sustainable development.

In this context, we are verifying that most of the countries are using taxation to achieve these environmental and sustainability purposes. We refer to the already known "green taxation", which usually includes taxes on energy, transport, pollution, and resources.

However, we must not forget that the main purpose of taxes is to generate revenue providing the States with the financial means to fulfil the collective needs of their citizens. When taxes are used to achieve other purposes such as those mentioned above (related to environmental protection and sustainability), we are entering the field of corrective taxes or pigouvian taxes which allows the State to intervene actively in the economic and social domain. Taxes are a very efficient and useful mechanism of dealing with socially undesirable activities, such as the environmental pollution, water and soil contamination, CO2 emissions, use of polluting energies, among others.

The purpose of this study is to analyze the benefits and risks of adopting tax policies based on these corrective purposes, understanding the main strategies adopted, the constitutional and legal framework of these taxes, their limitations and the use of revenues generated through the collection of these taxes. This seems to be relevant specially when a lot of countries, as Belgium, Denmark, Ireland, Italy, Lithuania, Poland, Portugal, Romania and Slovakia, are trying to adopt green tax measures but with little or no detail on specific policies (Gore, Urios & Karamperi, 2022).

In fact, the factors underlying the creation, determination and collection of these taxes are always difficult to define. Which are the most effective taxes? Taxes levied on the consumption of goods and services, on the manufacturing and production of assets, on the exercise of some economical activities? How can Governments evaluate the damages and the impact of these prejudicial actions and the consequent role of these types of taxes? If the idea is controlling and discourage these risky and threatening environmental behaviors, taxes must be determined on this basis. Does a real green tax reform must use just the ecological elements as the basis and main factor for the definition of the general tax base of a country? Is the main objective of the green tax reform the replacement of ordinary taxation by environmental taxation?

It seems relevant trying to determine the way of implementing this type of taxes. It is important to reflect on the impact of the replacement of taxes and contributions on income and applied progressively, which consider the ability to pay of the individual by taxes that affect all equally (based, for example, on the polluter pays principle), regardless of the ability to contribute. It can generate equality and social justice problems, reducing the redistribution function of tax systems (Borges & Merlin, 2018). It cannot work in countries with huge economic and social inequalities as it can seriously affect the general principle of an equitable distribution of the tax burden.

Taking these ideas into consideration, and referring, in particular, to the Portuguese case, the authors consider this study of particular relevance and interest, hoping to contribute to the debate on tax policies and to establish a framework for green taxes in which fundamental tax principles and the coherence of the tax system as whole are duly observed.

Considering this context, the specific objectives of this chapter are:

- Understand the main principles associated with the use of "green taxes";
- Verify the revenue generated in Portugal with the collection of "green taxes";

- Analyze the criteria used for the creation e collection of these taxes;
- Discuss the problems generated by this taxation and its impact for the contributors.

This research will be based on the literature review, making reference to some published studies and authors opinions who study this matter, and the legal system will be described and analyzed. Trying to achieve the mentioned objectives it will also be analyzed the evolution of green taxes in Portugal (case study) and the conclusions will be achieved.

BACKGROUND

The Green Taxation as an EU Priority: Short Background Note

Currently, the sustainable development and climate protection are priority objectives for the European Union's internal and external policies and, for this reason, a lot of strategies and measures are trying to be adopted to promote these goals.

The *2030 Agenda for Sustainable Development*, adopted by the international community in September 2015, is ambitious and approach different perspectives of the sustainable development. It includes seventeen specific objectives: The sustainable development Goals (SDGs) with a large content about ending poverty, promote the equality, reducing pollution, protect the environment, reducing external energy dependency and improving the general wellbeing and sustainable behaviors.

The concept of "sustainable development" is defined in the Report of the World Commission on Environment and Development: Our Common Future as «a process of change in which the exploitation of resources, the direction of investments, the orientation of technological development, and institutional change are made consistent with future as well as present needs». In article 37 of the Charter of Fundamental Rights of the European Union, this concept of "sustainable development" is presented as an objective for the States: «a high level of environmental protection and the improvement of the quality of the environment must be integrated into the policies of the Union and ensured in accordance with the principle of sustainable development».

To achieve this "sustainable development", we verify a growing concern of States to create and develop environmental protection mechanisms: direct method, in which the State institutes an environmental policy, approving mandatory rules and the consequent sanctions; or the indirect method, in which the State limits itself to creating incentives of a positive or negative nature in order to change contaminating behaviors (Oliveira, 2004).

In 2019, the European Parliament declared the "climate and environmental emergency", asking the Commission to adopt urgent measures. It was presented the "European Green Pact", a roadmap for the European Union to achieve climate neutrality by 2050, stablishing a set of legislative proposals with the aim of making the EU's climate, energy, transport and taxation policies suitable for achieving a reduction of at least 55% of greenhouse gas emissions by 2030 compared to 1990 levels.

A lot of strategies and measures are being considered to promote the desired environmental protection and the sustainable development. We can, for example, highlight the EU taxonomy, approved by the Regulation (EU) 2020/852 of the European Parliament and of the Council of 18 June 2020. It is a classification system, including a list of environmentally sustainable economic activities, its definitions and setting the conditions that an economic activity must meet to be classified as environmentally sustainable: "European Green Deal".

In this context, most of the countries are also using taxation to achieve these environmental and sustainability purposes. We refer to the already known "green taxation", which usually includes taxes on energy, transport, pollution, and resources.

In the European Union, we can find some provisions related with these specific goals. We can, for instance, mention the Regulation (EU) 2021/119 of the European Parliament and of the Council of 30 June 2021, establishing the framework for achieving climate neutrality and amending Regulations (EC) No 401/2009 and (EU) 2018/1999 ("European Climate Law") and the recent proposal for a council directive related with «restructuring the Union framework for the taxation of energy products and electricity (recast)». This proposal highlights that «the taxation of energy products and electricity plays an important role in the area of climate and energy policy».

According to the European Commission's vision, «green taxation can also help promote sustainable growth, support intergenerational fairness and maintain tax revenue levels for EU Member States while allowing them to cut other, more distortive, taxes such as those on labour. Finally, and since the EU Green Deal is also the EU's growth strategy, green taxation can help reboot the EU's economy as we recover from the COVID-19 pandemic» (https://taxation-customs.ec.europa.eu/green-taxation-0_en). The idea seems to achieve more than corrective purposes, related to environment and sustainability. These taxes can also represent a significant means of collecting revenue for the States. Based on this idea, the "green taxation" can be interpreted as an additional form of taxation and additional revenue source, taking the focus off the corrective goals. Striking a balance in the interplay of these interests can be difficult and it can generate situations that end up punishing taxpayers and failing to mitigate the serious environmental problems affecting the planet.

Given the current resource scarcity scenario and the effects of the armed conflict between Russia and Ukraine, which caused a huge impact in the energy markets, especially regarding the increasing of natural gas prices, the green taxes and its role in the economy will become more important and they can be used not only to follow environmental purposes and to reduce the pollution and preserve the natural resources, but also to respond to the economic instability on the petrol and gas market.

TAX POLICIES IN PORTUGAL AND GREEN TAXATION: GENERAL OVERVIEW

As a member-state of European Union, Portugal implements the European Regulations and measures related with green taxation and there is an increasing concern with the adoption of policies directed towards the environmental protection, sustainable growth and to promote sustainability and the usage of resources with more efficiency and less impact in the environment.

The environment is, in Portuguese legal system, an essential and indispensable legal resource. It has been recognized in article 66 of the Constitution of the Portuguese Republic (CRP) and it also appears as a fundamental task of the State (article 9, e) of CRP). Therefore, the State must orientate the economical and social policies to the protection and promotion of a health and preserved environment. The Law no 19/2014 of April 14 defines the grounds of environmental policy. We can also find the "Climate Framework Law" (Law no. 98/2021 of December 31) which reinforce the objectives, principles, and obligations for the different levels of Governance for climate action through public policies and establishes new provisions of climate policy.

The protection of the environment and the promotion of a sustainable development are now in the order of the day, both for economic and social public policies but also for the citizens, who now give more value to these matters, showing more concern about the pollution and its effects in climate, natural ambient and resources. The economic and social development can no longer be faced from a reductionist economistic perspective but as a sustainable and harmonious development model, which integrates ecological concerns and more sustainable policies (Oliveira, 2004).

Following the EU legislation and international commitments, Portugal approved the Law No. 82-D/2014, of December 31 ("Green Taxation Law"), which modified some environmental tax rules in the energy and emissions sectors, transports, water, waste, territorial planning, forests, and biodiversity, also introducing a special tax regime for plastic bags and an incentive regime for the scrapping of end-of-life vehicles.

Green taxes are rules to steer the economy, focusing on the consumption of certain goods, increasing their price and thus seeking to reduce their use, taking into account the high *negative externalities* associated with this consumption acts (Sanches, 2007). A *negative externality* occurs when someone adopts some behavior or practice some act that impacts negatively on the welfare of a third party (Borges & Merlin, 2018). The utilization of public policies related with the creation of some tax instruments to measure, control and reduce negative environmental externalities results from a change of thinking which consider the environmental problems not only in the physical aspect but also as a problem that can affect human lives and activities (Borges & Merlin, 2018).

Therefore, regarding the environment protection, we can find two specific objectives for the green taxes: The idea is using these taxes to promote more sustainable behaviors in production and consuming and a usage more efficient of the natural resources and, on the other hand, to penalize polluting behaviors and activities.

According to Vasques (2011), the link between Tax law and Environmental law has been strengthened, based on the "polluter pays principle" and on the idea that it is possible and desirable to use economic mechanisms to control environmental effects, forcing taxpayers to bear the costs that their personal choices and behaviors imply for society at the environmental level.

The tax policy options in Portugal can be divided in three schemes of green taxation (Moreira & Fernandes, 2022): adapting indirect taxes (namely Value Added Tax (VAT)) to target renewable energy; increasing parafiscal taxation in different sectors of activity and multiply the incentives for renewables (for example, the Incentive System for the Support of Renewable Hydrogen Production and Other Renewable Gases, approved by Order No. 98-A/2022 on February 18).

Regarding the categories of taxes with relevance to environment and sustainable protection, we can distinguish four relevant fields in green taxation: transports (taxes related with the ownership and utilization of some kind of transport modes, including circulation and importation of vehicles), energy (taxes on energy products as gasoline, fuel oil, diesel, natural gas and electricity), pollution (taxes related with the polluting emissions into air or water) and resources (related with the extraction and usage of natural resources such as oil and gas, forest, water, forest, wildlife, among other, as these activities consume natural resources).

Specifically, we can highlight Law No. 22-A/2007 of 29 June which introduced the legal regime of ISV (vehicle tax) and IUC (single circulation tax). ISV applies the principle of equivalence, seeking to burden taxpayers for the costs they cause in the areas of the environment, road infrastructure and road accidents (article 1 of ISV Code). It is calculated taking into consideration the levels of CO2 emissions and CC of the vehicles. IUC also applies the principle of equivalence, taxing for the environmental and circulation costs produced by taxpayers. It depends on the vehicle's registration year.

The Authors will now verify the relevance of green taxes existing in Portugal, trying to understand which type of green taxes are more relevant and its evolution during the last years.

Table 1. Relevance of green taxes existing in Portugal in 2021 (%)

Green taxes (type)	Relevance (%)		
	2019	2020	2021
Single circulation tax	12,9%	14,3%	13,8%
Vehicle tax	13,8%	9,4%	8,6%
Resource taxes	0,4%	0,4%	0,5%
Pollution taxes	0,3%	0,4%	0,5%
Petroleum and energy taxes	67,5%	69,2%	70,7%
Other energy taxes	5,1%	6,3%	5,9%

Source: Adapted from National Institute of Statistics. Data extracted on 23 Apr 2023. https://www.ine.pt/

Analyzing these data, we can verify that the Petroleum and energy taxes are the most relevant tax with environmental purposes, increasing over the three years (from 67,5% in 2019 to 69,2% in 2020 and 70,7% in 2021). On the other hand, the other energy taxes, which include gas emission allowances, decreased their relevance in 2021 (from 6,3% in 2020 to 5,9 in 2021).

The most relevant reduction is seen on vehicle tax which has decreased over these years, specially between 2019 (13,8%) to 2020 (9,4%), continuing the reduction in 2021 (8,6%). We can verify a difference of -5,2% in this tax revenue between 2019 and 2021. According to the National Institute of Statistics' report of 2021, this decrease is not related to changes in the taxpayers' environmental behavior (there has been an increase in the number of registered vehicles) but rather to the changes introduced by Tax Authority in the calculation of this tax. According to the Court of Justice of European Union's jurisprudence, namely the last decision in case no. C 169/20 of September 2nd of 2021 (which analyze article no. 11 of Portuguese Vehicle Tax Code), the subjecting of used vehicles imported from other Member-States to a higher tax burden than the residual tax contained in similar used vehicles traded on the domestic market (by disregarding the devaluation of the environmental component) is incompatible with European Union law, violating the provisions of article 110 of Treaty on the functioning of the European Union. Therefore, the devaluation of the environmental component must also be considered in the calculation of the value applicable to second-hand vehicles put into circulation in Portuguese territory and purchased in another Member-State (Press release no 146/21 issued by the Court of Justice of European Union available at https://curia.europa.eu/jcms/upload/docs/application/pdf/2021-09/cp210146pt.pdf).

On the same path, we find several other Portuguese Courts' decisions (as example, see Arbitral Court´s decisions in cases 498/2021-T of June 24th of 2022 and 271/2021-T of June 12th of 2021, available at https://caad.org.pt/tributario/decisoes).

We can also conclude that the taxes related to the purchase and usage of vehicles (single circulation tax, vehicle tax and petroleum and energy tax) represented more than 90% of "green taxation" in Portugal (94,2% in 2019, 92,9% in 2020 and 93,1% in 2021). The increase in the other green taxes (resource and pollution taxes) is not significant (just 0,1% between 2019 and 2020).

REVENUE GENERATED BY "GREEN TAXATION" IN PORTUGAL: AN OVERVIEW OF THE RELEVANCE OF THIS TAXATION

It is also relevant to refer the revenue generated by some of these taxes. According to data revealed by the Portuguese statistical office (2022), in 2021 the value of taxes with environmental relevance is around 5 billion Euros, corresponding to 6.6% of the total revenue from taxes and social contributions (more 6.3% than 2020).

According to Borges & Merlin (2018), the implementation of a green tax reform allows the multiplication of income´s sources by the displacement of the tax burden on income for other facts related with environmental behaviors and the replacement of ordinary taxes for environmental taxes, whose revenues would serve both the purposes of environmental policies and the maintenance of tax revenues, guaranteeing the social activity of the State.

In fact, the authors believe that, considering the Portuguese´s case, the green taxes are not a substitutive way of tax revenue but a real additional source of revenue, which is not insignificant and occupies an important position in the financing of the state.

The Authors will now verify some data related with the revenue generated by green taxes in Portugal, analyzing its evolution during the last years.

Table 2. Evolution of the revenue related to green taxes in Portugal (between 2010 and 2021)

Year	Revenue (in billions of Euros)
2010	4.361,60
2011	4.084,00
2012	3.681,60
2013	3.757,70
2014	3.932,10
2015	4.344,50
2016	4.819,30
2017	5.051,90
2018	5.271,90
2019	5.418,10
2020	4.728,80
2021	5.024,60

Source: Adapted from National Institute of Statistics. Data extracted on 23 Apr 2023. https://www.ine.pt/

In 2021, the revenue of green taxes exceeds 5 billion Euros (5.024,60), increasing from the previous year (4.728,80) which reflects the recovery in the fuel consumption and in the usage of vehicles. In general, after 2012, we can assist to an annual increase of revenue. The reduction verified in 2020 (5.418,10 in 2019 to 4.728,80 in 2020) confirms the significant impact of the effects of the pandemic COVID-19 on tax revenues associated with the purchase and use of motor vehicles cars.

According to the National Institute of Statistics' report of 2021, the revenue of 2021 corresponds to 6.6% of the total revenue from taxes and social contributions and, in 2020, it represents 6.7%. According to this report, the relevance of these taxes in total tax revenue (including social contributions) was higher in Portugal (6.7%) compared to the European Union average (5.6%).

We believe this difference is not due to a greater concern of the Portuguese State with environmental and sustainability issues, but rather to the growing importance that these taxes represent for the State's revenue.

Value added tax (VAT) was excluded from these data because it is a neutral tax (it can be deducted by the companies) and it is does not change the price of goods and services as a relevant green tax does. Also, the special taxes on consumption as tobacco and alcohol are also excluded because these taxes do not have a special relevance in the negative impact of the environment.

As Saldanha Sanches (2007) has already advanced, the creation of green taxes also arises with the objective of ensuring the generation of revenues, as an alternative to other sources of income (other taxes and contributions levied on other legal facts).

THE EFFICIENCY OF GREEN TAXES AND ITS PURPOSES: AN ADDITIONAL REVENUE SOURCE?

One of the most complex issues in the area of green taxation is how to strike a balance between the corrective purpose and the revenue related dimension. Truly green taxes would be entirely corrective taxes, that is, they would have their mission accomplished when no more revenue is generated which would mean that the behaviours to be combated would have ceased. In practice, there are very few strictly corrective taxes because the revenue aspect, in a context of increasingly pressing financial needs, is not neglected. We are therefore trapped by the circumstance that taxes which supposedly aim at combating negative externalities, perversely we would say, depend, to a large extent, on their collection in order to meet growing financial needs.

This hard to keep balance can, in a way, discredit and make corrective taxation such as green taxes more difficult to accept by citizens who see the fight against negative environmental externalities and related taxes, mere justifications for increasing the tax burden imposed on them. The merely corrective effect, in fact, can't stand alone, given the chronic lack of revenue affecting the states. On the other hand, we stress, the large revenues they produce generate discomfort among citizens who look at this type of taxation mostly as a burden that falls on them.

THE (DIFFICULT) DETERMINATION OF GREEN TAXES

The Determination Criteria of This Taxation

It is very hard to establish a link between the environmental damage and the corresponding tax. In fact, the negative externality raises problems not only in terms of its observation, but also in terms of its quantification and connection to the behaviour of a given subject. This is because, in many situations, the negative externality is just an eventuality. If we refer, as an example, to the emission of polluting fumes, objectively there is a verifiable pollution because the emission can be perceived. However, when it comes to the verification of actual damage and its quantification, everything becomes more complicated.

The assessment of the damage will depend on various factors, namely the natural conditions of the site: if there are winds, if it is a valley or mountainous place, if there are other polluting industries, if it is a wooded area or not, and other conditions that, even without engaging in technical debates, the common sense tells that they influence the degree of pollution or environmental damage. What we have just said will obviously greatly complicate the calculation of the measure of damage and, therefore, the amount of green tax to be paid. As a result, it is not easy for policy-makers, given the scarcity of information about the real value of externalities, to determine the measure of corrective taxes. The value of those must necessarily be symbolic, the authors believe. Hence, even if, in theoretical terms, the calculation can refer to the marginal loss of deterioration of the environment, translating the marginal value into the sum of the losses suffered by all the people affected by externalities, in practical terms, the loss or damage is almost impossible to estimate.

The solution ultimately lies in the application of presumptive methods or symbolic values, precisely because information about the losses suffered by millions of people is impossible, not only to assess, but also to process.

The circumstances described justify the frequent use of the volume of emissions, an objective factor, rather than the damage which, on top of being merely eventual, not verifiable, can only be determined with great difficulty; making it difficult to apply the taxes targeting those environmental damages, as happens in the so-called green taxes (Fullerton & Wolverton, 2003).

The already identified difficulties are maximized by the fact that most types of negative environmental externalities, are transnational. If we take as an example the emissions and greenhouse gases that lead to global warming, all countries are simultaneously polluters and victims of this pollution.

THE RISK OF VIOLATION OF FUNDAMENTAL TAX PRINCIPLES BASED ON EQUITABLE DISTRIBUTION OF TAXES BURDEN

Corrective taxation such as green taxation has a number of social costs that are felt in the economic dimension, but also in the area of rights, freedoms and guarantees.

In general, the so-called corrective taxes allow taxpayers to pass them on, which ends up essentially burdening consumers and, among them, the most vulnerable. Regardless of the repercussions, it should be noted that corrective taxation in the environmental area is strongly regressive.

Energy costs, for example, weigh more heavily on low-income citizens. It should also be noted that many of the most environmentally damaging goods are consumed by people who are worse off economically. We see them using: older vehicles and, therefore, more polluting, less energy-efficient housing, firewood for heating, among other examples.

It can be said that, in the long run, a healthier environment will ultimately benefit the most disadvantaged.

There is, in fact, evidence that environmental deterioration, climate change and other negative effects mainly affect people in a state of greatest vulnerability. There are, however, in the short and medium run some fringes of the population that require special attention.

In the area of rights, freedoms and guarantees, problems also arise. We highlight the somewhat perverse character of corrective taxation, since through the payment of taxes, the harmful behaviours that it seeks to prevent seem to be legitimized. As a consequence, environmental taxes often act as a license to pollute or engage in harmful activities, contributing to some extent to legitimization of such activities; with the aggravating factor, already mentioned, of being often passed on to consumers.

FUTURE PERSPECTIVES FOR GREEN TAXATION AND ITS CHALLENGES

The Assessment of the Negative Externality

With regard to the issues identified, namely the assessment of the negative externality, there are, in fact, problems that are very difficult to overcome, which justifies that, in an effort to mitigate them, we resort both to presumptions and to the central role that is given to objective indicators, such as the volume of emissions.

Even though there is no magic solution to the problem, authors are, however, convinced that the best way to deal with this difficulty is to see corrective taxation not as a way of covering costs, always uncertain and hard to measure, but a manner of preventing the verification of damages that, in the long run, will be reflected in people's health and lives.

If corrective taxes are to truly fulfil their potential, they must not be determined solely by economic considerations, that is, by taking into account the value of negative externalities. This determination should be made by reference to social and ethical considerations. Economic efficiency should therefore be only an argument and not the only or predominant one. Also because what in the short term is a loss can, in the long term, bring gains. Therefore, that and other positive consequences also have a monetary value that is often misjudged.

Such difficulties, as stated earlier, stem from the fact that many of the externalities have a transnational impact. However, this circumstance demands that they are tackled at the international level as well. One of the possible solutions could be an increase in carbon taxation (IMF/OECD (2021), COM (2021) and Falcão, 2019) by, *inter alia*, creating a truly European tax on it, which would be administered by the European Union itself; with the support of the tax authorities of the Member-States, as is already the case with the Common Customs Tariff.

Another possible response could be the use of customs duties – even if difficulties may arise (Falcão, 2019) – to penalize the importation of products that imply, in their production, practices that translate into negative externalities, such as pollution or depletion of resources.

TACKLING SOCIAL COSTS

Within the social costs, as it results clearly from the analysis made above, there is some diversity, which naturally impacts the type of response. With regard to the social costs with economic impact, we assume that taxation in general, and, therefore, also corrective taxation that targets externalities such as environmental degradation, affects the most vulnerable in a more serious way.

To start with, there is an immediate loss of well-being, without prejudice to the fact that in the medium and long term the changes promoted by taxation may favour, in particular, these taxable persons who always end up being the most affected by the consequences of negative externalities, as happens, for example, with climate change.

In the short term, however, given the strongly regressive character of corrective taxation (Schöb, 2003), it will mainly affect the most disadvantaged to the extent that they are the ones who normally consume less environmental friendly products (because they are often less expensive (In developing countries people resort to wood to heat their dwellings)), but also because the most powerful taxpayers that may be affect by environmental taxation manage to pass on tax costs to those who are more vulnerable.

One possible solution for the problem we have just put forward could be to use the revenue generated by corrective taxes to cut other taxes, reducing the burden on taxpayers. That is, corrective taxes, such as environmental would improve the quality of the environment and life in general and simultaneously reduce the distortions of other existing taxes.

This advantage coupled with the protection of the environment justifies the fact that many authors associate this type of taxation with the idea of a double dividend (Schöb, 2003). There are, however, some differing views. This position was challenged by several authors that assert that the double dividend only exists if the excessive tax burden of the tax system as a whole is reduced (Goulder, 1995). Some scholars go further and state that environmental taxes rather than correcting distortions maximize them, even if the revenue is used to lower taxes that cause these distortions (Bovenberg & Mooij, 1994).

The weakness in the use of corrective tax revenues to lower other taxes lies in the fact that some taxable persons with less income are normally exempt from income taxation, which, in practice, would mean that they would not see their tax burden relieved.

For those taxpayers and even, we would say, for taxpayers in general, it would be preferable to benefit from a state service, financed directly with the revenues of corrective taxation, in order to be compensated for the damage and perils suffered, which at the end of the day, are the *rationale* for this type of taxation. In fact, many of these damages result from the use by some groups of the resources that belong to everyone, without proper compensation ever being made to those affected. The fact that the revenues from these taxes flow, generally, into the general budget of the State, causes several problems. This happens because even if there is some degree of compensation by using the added revenue generated by corrective taxes to lower other taxes, thus alleviating the tax burden, it does not solve all the issues. Despite the apparent similarity, it would not be an unconditional basic income, but a payment that would be based strictly on the actual or potential damage to citizens.

The fact that revenue would be distributed to all would, on the one hand, lessen the impact of corrective taxation (the fact that the most part of the revenue will be allocated to citizens is a good argument to refute the critique that environmental taxes are regressive) on the most vulnerable people; on the other hand, it would allow not only these people, but everyone in general, to adjust their consumption habits, opting for more environmentally friendly products, in order to prevent loosing, through corrective taxation, what they received as compensation for the actual and potential damages (which are at the basis of corrective taxation). The recipients of the stipend may not be subject to corrective taxes. It is more likely that those corrective taxes will be imposed on companies that pollute. It may happen, however, that if those taxes are imposed on individual persons those individuals may end up in a situation where they are simultaneously recipients of the stipend which would be paid to every individual, but also direct payers if they incur in negative behaviours or indirectly if the corrective tax is passed onto them by polluting companies. In this situation they would feel encouraged to refrain from negative environmental behaviours triggering corrective taxation or from consuming products sold by companies that are subject to corrective taxation. By adopting this behaviour, they would avoid that the stipend received by them would become a mere illusion, to the extent that it will be funded directly or indirectly by them.

The potential to encourage both producers and consumers to adjust their behaviour would be great. There would also be the advantage of breaking the monopoly the State has on the use tax revenues, which would have, in addition to a moderating effect on the use of these revenues by the state, also the benefit of freeing citizens from part of the paternalism of the State, allowing them to decide on the use to be made of a portion of public revenues. It would also make possible, to a modest extent, to mitigate the risk of a less successful use of public money. The recent pandemic has shown that the approach based on direct payments to citizens and businesses, injecting money, is feasible and can have an interesting effect on the economy. As matter of fact, many of the benefits and compensation which were provided during that unfortunate period share, in material terms, the nature of the benefits we are advocating.

Another complex problem, within the social costs, is the fact that corrective taxation can, in a perverse way, launder such negative behaviours by legitimizing them whenever the tax they trigger is paid.

The solution of attributing the revenue from these taxes to the citizens in general, because they are the aggrieved ones, can have a moralizing effect. Consequently, the State would not gain directly from this taxation. That is, it would not generate revenue at the expense of ethically objectionable behaviours, but would limit his involvement to the task of compensating citizens for the improper use of resources that belong to everyone, or that have an impact on everyone.

The stigma arising from the fact that certain taxpayers bear corrective taxation may have the positive effect of encouraging them to change their behaviour so that they stop being subject (or are to a lesser extent) to such taxes. Moreover, citizens themselves, through their choices, which are increasingly motivated by social and environmental concerns, will play an important role by preferring products that are more socially acceptable and not harmful for the environment.

PROMOTING A BALANCE BETWEEN THE CORRECTIVE AND REVENUE DIMENSION OF GREEN TAXES

The type of taxation to which authors refer, although characterized essentially by the corrective dimension, has also a non-negligible revenue dimension which, in the current context, competes with the first. Dealing with this difficult relationship implies that one determines, at the outset, whether revenue generation is something desirable in the area of corrective taxation and should be sought and kept or, on the contrary, it should be seen as a side effect to be borne on a temporary basis until the objectives underlying corrective taxation are achieved.

In connection with the dilemma we have just put forward arises the question of determining what destination should be given to the revenue resulting from the taxation under analysis. There are, of course, several possible positions regarding the interplay between the corrective and the revenue aspects.

At one extreme, we have the approach that conceives corrective taxes as restrictive taxes that aim at extinguishing behaviours considered negative and that, therefore, reach their goal when they produce absolutely no revenue. This position proves to be radical, since it requires a too swift adjustment with very strong impact on the economic activity and the lives of the taxpayers directly affected, enhancing the difficulties already felt by the most vulnerable. In addition, it implies that significant revenue that could be used to deal with the damages inflicted on the environment is relinquished, increasing, as a consequence, public spending.

Another conceivable approach could be sustained by a slight raise of corrective taxes, in order to reinforce their impact on behaviour modelling, but not to the point of turning them into strictly corrective taxes that do not seek to generate any revenue. This approach would convey an important message, given that if the corrective taxation is too easy to accommodate on the part of those who are subject to it, in addition to discrediting it, it also reinforces the idea that combating the negative externality is only an excuse to ensure a stable and continuous level of revenue. The revenues produced could be used, in the first place, to mitigate the damage caused by externalities, by covering the increased expenses, for example, with ecosystem restoration. The revenue could also be used for awareness-raising campaigns and to support the reconversion of activities that produce the negative externalities (Zannoni, 2018 and Kogels, 2019). This approach is usually identified, in international literature, as earmarking, that is, the allocation of revenues for purposes consistent with combating the harmful effects of the various negative externalities and also preventing them.

The solution that authors have just advocated seems to them much more correct than the one in which the revenues of corrective taxation are used to meet the various expenditures, diluting them in the budget, without any earmarking. On the one hand, by using corrective taxes to deal with financial problems in a certain way, it does not represent a stimulus to an efficient management of public financial resources; on the other, it sends the wrong message to taxpayers, since environmental taxation, or any other expression of corrective taxation, is seen only as a way to increase revenues and, therefore, taxation in general, which certainly has political costs for those who implement such measures (not to mention the obvious social costs). The yellow vests protests in France are a good example.

Transferring directly to citizens, part of the revenues from corrective taxation, in the terms authors described earlier, whilst still allowing some earmarking (to cover the additional expenses necessary to respond to the damages arising from negative externalities), would also be a very interesting alternative with several advantages. First, it would be a way of mitigating the particular impact that this form of taxation has on the most vulnerable. Then, it would be a way to encourage good behaviours, to the extent that the beneficiaries of this revenue would try not to lose it through a less environmentally adjusted consumption or behaviour (object of corrective taxation). It would also be a promising way of mitigating the paternalism of the state with regard to the management and application of public revenues, often done inefficiently, allowing citizens to participate in this management in the full use of their autonomy.

Implementing some corrective taxes, with this design, at the level of the European Union, would certainly have also many various merits. It would prevent a distortion of competition in the internal market, since by doing so at European level would eliminate any disparities that might exist in terms of corrective taxation. It would also have the potential to bring citizens closer to the European Union, because when tax matters are at stake citizens are more prone to participate. Furthermore, in order to respect the centuries-old slogan that there is no taxation without participation, it would be necessary to make decision-making procedures in tax matters more democratic, possibly strengthening the involvement of the European Parliament. Finally, the fact that the European Union would pay citizens directly a stipend, even if modest, within the framework of the suggested distribution of revenue, would make citizens more willing to have another approach and receptivity towards the European Union, providing it the cohesion it has been seeking.

FUTURE RESEARCH DIRECTIONS

The authors believe that there is a need for more understanding about the effectiveness of those green taxes in Portugal. Study the impact of this taxation in the modification of some environmental behaviors to understand if they are useful for that purpose. This provides an opportunity for future research to investigate the evolution of this taxation and predict the future impact, and to identify strategies to make a better use of tax policies with impact in the environment and sustainability.

At the same time, the authors believe it is also important to develop the study of the impact of these green taxes in people's lives in terms of ability to contribute, understanding if the current green taxes take into consideration the polluter pay principle, achieving the equality and proportionality in taxation.

The authors think is important to understand if the allocation of tax benefits could be more effective (or equally effective) than the creation of additional taxes, realizing if the revenue generated with these green taxes are an essential condition to keep this taxation.

These studies can be significant to the evolution of the tax policies related with the environment and sustainability purposes.

CONCLUSION

This chapter highlights the growing importance that environmental and sustainability issues have assumed in tax policy-making, reflected in the creation and management of green taxes, including taxes on energy, transport, pollution, and resources.

As a result of commitments made by several European Union countries, numerous measures have been adopted that aim to promote sustainability and environmental protection and discourage the continuity of polluting behaviors. Most of the countries are using taxation to achieve these environmental and sustainability purposes.

As verified, the revenue generated by these taxes are significant in Portugal, where the revenue is, on average, higher than in the rest of States of European Union and this fact is important to understand the purposes of this taxation and its effectiveness.

By way of conclusion, we would say that the use of green taxation, despite its long history, has gained a new life. Especially due to the growing climate emergency. The use of this powerful form of taxation requires, despite everything, moderation and the search for a difficult balance between corrective and revenue generating purposes.

To achieve this objective, we believe that the assignment of part of the revenue obtained to the mitigation and prevention of damages should be privileged, to the detriment of its use, in a diluted form, in the general budget, for purposes other than corrective. If that approach is not adopted, counterproductive political and social tensions will definitely arise. That sort of outcome is mostly certain if we try to solve financial problems that need urgent resolution, through green taxation.

Finally, it seems to us that the EU framework may be ideal for testing cross-border taxation. In this context, however, priority should be given (for all the advantages we pointed out) to the transfer to citizens of part of the revenue obtained from the misuse of resources that belong to everyone by a restricted group of individuals or companies, and which until now has never generated the necessary compensation.

REFERENCES

Borges, F. G. L. & Merlin, L. V. C. T. (2018, December). The (in) viability of green tax reform in Brazil. *Veredas do Direito, 15*(33). doi:10.18623/rvd.v15i33.1243

Bovenberg, A. L., & Mooij, R. A. (1994). Environmental Levies and Distortionary Taxation. *The American Economic Review*, *84*, 1085–1089.

Dias, J. E. O. F. (2007). *Direito Constitucional e Administrativo do Ambiente*. Almedina.

Falcão, T. (2015). Is Carbon Tax Compatible with the WTO Treaties? Intertax, 43(10), 571-799. doi:10.54648/TAXI2015056

Falcão, T. (2019). Taxing Carbon Emissions from International Shipping. Intertax, 47(10), 832-844. doi:10.54648/TAXI2019085

Fullerton, D., & Wolverton, A. (2003, August), The Two-Part Instrument in a Second-Best World. *NCEE Working Paper Series*. (Working Paper # 03-04).

Gore, T., Urios, J. & Karamperi, M. (2022, February). Green and fair taxation in the EU - Green EU tax systems tend to also be more progressive. *Institute for European Environmental Policy*.

Goulder, L. H. (1995). Environmental Taxation and the 'Double Dividend'. *A Reader's Guide*. *International Tax and Public Finance*, *2*(2), 157–184. doi:10.1007/BF00877495

IMF/OECD (2021, April), Tax Policy and Climate Change: IMF/OECD Report for the G20 Finance Ministers and Central Bank Governors. *Communication from the Commission to the European Parliament and the Council - Business Taxation for the 21st Century*. OECD.

Instituto Nacional de Estatística. (2021, October). *Impostos e taxas com relevância ambiental 2020*. INE. www.ine.pt

Instituto Nacional de Estatística. (2022, October). *Impostos e taxas com relevância ambiental 2021*. INE. www.ine.pt

Jr., R. A. (2020). Direito Internacional do Ambiente: a Responsabilidade no Antropoceno. *In Revista Jurídica Portucalense*. (No. 28). doi:10.34625/issn.2183-2705(28)2020.ic-05

Kogels, H. (2019). *The Energy Transition and Fiscal Policy* (Vol. 1).

Moreira, T. M., & Fernandes, F. V. (2022, April). The evolution of green taxation in Portugal. *International Tax Review*. https://www.internationaltaxreview.com/article/2a6abcbtm7tszx6frm680/the-evolution-of-green-taxation-in-portugal

Nabais, J. C. (2005). *Por um Estado Fiscal Suportável Estudos de Direito Fiscal*. Almedina.

Nabais, J. C. (2021). Direito fiscal e tutela do ambiente em Portugal. *Fiscalidade: Revista de Direito e Gestão Fiscal*, (21). Business & Economics School. https://www.isg.pt/wp-content/uploads/2021/02/21_1_CASALTA.23-11-2004.pdf

Oliveira, M. O. B. (2004). Os impostos ambientais – alguns aspectos da sua temática. *Fiscalidade: Revista de Direito e Gestão Fiscal* (no 19/20). Business & Economics School. https://www.isg.pt/wp-content/uploads/2021/02/19_20_3_Impostos-ambientais-Revista-Fiscalidade.pdf

Sanches, J. L. Saldanha (2007). Manual de Direito Fiscal. (3.º ed.). Almedina.

Schöb, R. (2003, August). *The Double Dividend Hypothesis of Environmental taxes: A Survey*. (CESIFO Working Paper - NCEE Working Paper Series).

Vasques, S. (2011). *Manual de Direito Fiscal*. Almedina.

Vítor, J. (2014). *Os Desafios Ambientais do Sistema Fiscal*. Faculdade de Direito da Universidade Católica. https://repositorio.ucp.pt/bitstream/10400.14/16439/1/Os%20 desafios%20ambientais%20do%20sistema%20fiscal.pdf

Zannoni, D. (2018). Fiscal Aid to Promote Clean Transport. *EC Tax Review*.

ADDITIONAL READING

Condesso, F. R., & Condesso, R. A. A. (2022). *Direito do Urbanismo e do Ambiente*. Almedina.

László, C. (2021). The Green Tax Revolution. *Inter Economics*, *56*(5), 284–287. doi:10.100710272-021-1000-y PMID:34629506

Pereira, P. R. (2020). *Princípios do Direito Fiscal Internacional*. Almedina.

KEY TERMS AND DEFINITIONS

Environmental Behaviors: The human acts and attitudes that can produce effects or impact on natural environment.

Equally: An equal and identical treatment, based on the principles of generality and uniformity, for situations with the same characteristics.

Green Taxation: Environmental or green taxes include taxes on energy, transport, pollution, and resources or the assignment of tax incentives. This taxation is related with environmental policy goals to promote more sustainable and greener habits and develops an efficient use of resources.

Sustainability: Ability to maintain a practice of using natural resources in a responsible way in private lives but also in the exercise of economic activities to keep the balance of our planet's natural environment and protect natural resources to support the well-being of present and future generations.

Tax Burden: The weight of taxes paid by individuals and organizations taking into consideration the gross domestic product. It is related with citizens' responsibility to pay taxes. The burden of taxation should be shared in accordance with taxpayers' respective ability to pay (the ability to pay' principle).

Tax Policies: The adoption of public strategies related with taxation in order to become taxes more effective to achieve some social, economic and political objectives.

Tax Principles: The constitutional principles applied to taxation in order to allow and limit the power of the State to create and charge taxes, such as principle of legality (no taxation without representation, rule of law), equity, neutrality, equality, legal certainty.

Tax Revenue: The income generated with the collection of taxes applied on profits, goods, services, transactions, property to be used by the States to support social expenses.

Chapter 12
The Contribution of the Environmental Tax on the Reduction of Carbon Dioxide Emissions:
An Exploratory Study in Portugal and Spain

Maria Goreti Dâmaso
Polytechnic Institute of Santarém, Portugal

Nuno Carlos Leitão
https://orcid.org/0000-0002-7602-2615
Polytechnic Institute of Santarém, Portugal & CEFAGE, University of Évora, Portugal & CEsA, Lisbon University, Portugal

ABSTRACT

This chapter aims to analyze the environmental tax's contribution to reducing carbon dioxide emissions in Portugal and Spain. Moreover, this study analyses the impact of the use of renewable energies and the gross domestic product (GDP) on carbon dioxide emissions. The data are collected from the Organization for Economic Cooperation and Development (OECD) and Word Bank indicators from 1990 to 2019. The methodology is the time series (two- stage least squares; fully modified OLS; DOLS- dynamic OLS canonical cointegration regression, and granger causality). The unit-roots test demonstrates that the variables used in this investigation are integrated with the first differences. The results demonstrate the importance of the environmental tax and the use of renewable energies in reducing carbon dioxide emissions, both in Portugal and Spain. The results also show the positive effect of economic growth on Spain's increase in carbon dioxide emissions.

DOI: 10.4018/978-1-6684-8592-7.ch012

INTRODUCTION

This chapter aims to analyse the contribution of Environmental Taxes to reducing carbon dioxide emissions in Portugal and Spain. In addition, this study analyses the impact of the use of renewable energies and the Gross Domestic Product (GDP) on carbon dioxide emissions.

The increased global importance of reducing carbon dioxide (CO_2) emissions, reinforced by the 2030 Agenda for Sustainable Development of the Organization for Economic Cooperation and Development (OECD), could influence a country's climate policies (Engberg-Pedersen & Zwart, 2018). The Sustainable Development Goals adopted by the United Nations reinforce the importance of working together for a better world (United Nations, 2023).

Usually, the global objectives for climate policies are environmental effectiveness, economic efficiency, and equity (Vona, 2021). In this context, climate policies must consider environmental fiscal policies.

According to empirical studies, two types of environmental fiscal policies can be assumed. The penalizing policies, in which the tax penalizes what negatively affects the environment. Furthermore, the policies promote what benefits the environment through benefits and incentives (Bontems & Bourgeon, 2005; Qi et al., 2023; Shafi; et al., 2023; Wolde-Rufael & Mulat-Weldemeskel, 2021; Yu et al., 2019). This chapter will be focus in the first type, tax penalization.

Several countries have implemented environmental taxes to promote sustainable attitudes, mechanisms, and systems. This implementation of green taxes was carried out fundamentally through the so-called ecological tax reforms. In 2014, the green tax reform in Portugal encouraged and boosted concerns about environmental issues. Some studies investigated the impact of this reform (Borrego, 2016; Pereira et al., 2016; Pereira & Pereira, 2023; Rodríguez et al., 2019; Sousa, 2021). In Spain, environmental concerns and the corresponding reforms have also increased over time (Böhringer et al., 2019; Labandeira et al., 2019). These environmental tax reforms are carried out to obtain a triple dividend. The first dividend is the reduction of carbon dioxide emissions. The second dividend is the gross domestic product (GDP) increase. Moreover, the third dividend is the highest employment (Maxim, 2020). Therefore, the authors consider it essential to analyze whether green taxation impacts environmental indicators (the first dividend). Research on this topic has increased recently (e.g., Bashir et al., 2021). Some studies relate environmental taxation to carbon dioxide emissions in just one country (Sarigül & Topcu, 2021). Other regional studies, such as the OECD or European Countries or comparisons to different countries (Depren et al., 2023; Ghazouani et al., 2020; He et al., 2019; Rafique et al., 2022).

The study of the impacts of environmental taxes on environmental issues gained importance in academic terms with the Kyoto Protocol. From the various empirical studies in the literature, econometric models apply both to time series (Esen & Dündar, 2021; Sarigül & Topcu, 2021; Sun et al., 2020; Ulucak et al., 2020) and panel data (Bashir et al., 2021; Rafique et al., 2022).

This research is compared Portugal and Spain, using time series from 1990 to 2019 as an econometric strategy, following the contributions of previous studies by Ghazouani et al. (2020), Islam et al. (2022) and Telatar & Birinci (2022). Based on the empirical research, is tested the effect of environmental taxes on carbon dioxide emissions, such as the studies by Sarigül & Topcu (2021), Ghazouani (2020), and Hájek et al. (2019). The studies usually demonstrate a negative e significant correlation between environmental taxes and emissions. In addition, the impact of renewable energies on carbon dioxide emissions is tested. The literature shows a negative correlation between cleaner energies and climate change. However, as demonstrated by the works of Fuinhas et al. (2021), Balsalobre-Lorente et al. (2021) and Leitão et al. (2022), the use of renewable energies makes it possible to improve the environment. In this context, based on the principle of sustainable development, the hypothesis associated with the environmental Kuznets curve was formulated, where it is defended as arguments that, in an initial phase, per capita income is positively correlated with carbon dioxide emissions and the extent to which countries earn awareness of environmental issues economic growth is negatively correlated with carbon dioxide emissions (Alola & Ozturk, 2021; Le & Ozturk, 2020; Leitão, 2021; Leitão et al., 2022).

This hypothesis is supported by the Kyoto Protocol conferences (1997), the Paris Agreement (2015), and the 2030 Agenda for Sustainable Development.

The chapter begins with the introduction, where it is proposed to present the issues under study, followed by the literature review, which observes and explains the state-of-the-art (literature review), resorting to recent empirical studies. Then appears the methodology and data, where the hypotheses to be tested are presented, as well as the formulation of the econometric model. Next, in the empirical results, it discusses the results found. Finally, in conclusion, the main findings of this research are presented and the implications for fiscal, economic, and social policy.

The following section presents the literature review that shows the links between the various inter-dependence relationships under study.

BACKGROUND

This literature review focuses on more recent and relevant studies that analyze the relationship between environmental tax and using cleaning or renewable energy to reduce CO_2 emissions. In addition, the relationship between economic growth and CO_2 emissions is also considered in this study.

The empirical studies presented below seek to assess the relationship between fiscal policy and carbon dioxide emissions, the effect of using renewable energies on polluting emissions and the relationship between economic growth and carbon dioxide emissions. From the consulted references, it is observed that researchers use two different types of methodologies: time series and panel data.

In general, empirical studies conclude that fiscal policy and the use of renewable energies improve environmental efficiency.

In this context, Sarigül & Topcu (2021) studied the effect of environmental taxes on CO_2 emissions in Turkey. They found that ecological taxes have a negative impact on CO_2 emissions. In the same country (Turkey), Telatar & Birinci (2022) analyze the environmental tax effect on CO_2 emissions and the Ecological Footprint (from 1994 to 2019). The authors conclude the environmental tax doesn't have long-run effects on CO_2 emissions and the ecological footprint. Different results in the same country are found by Esen & Dündar (2021) in the study of the impact of energy taxes on the carbon footprint (from 1994 to 2017). They found a negative impact of energy taxes on the carbon footprint in the long term.

A comparative study of European countries that adopted and not-adopted carbon taxation policies was made by Ghazouani et al. (2020). They consider the implementation of carbon taxation policies to promote the reduction of CO_2 emissions.

The study by Hájek et al. (2019) focused on the efficiency of a carbon tax on the energy industries in the following countries: Denmark, Ireland, Finland, Sweden and Slovenia. The results of his study suggest that the carbon tax is environmentally effective in reducing greenhouse gas production in the energy industries.

Sharif et al. (2023) analyzes the impact of environmental tax, green technology, and green energy on CO_2 emissions (from 1995 to 2020) in Denmark, Finland, Iceland, Norway and Sweden (Nordic countries).

The results of this study show that environmental tax, green technology, and green energy negatively impact CO_2 emissions. On the other hand, Depren et al. (2023) studied the impact of an environmental tax on ecological quality in the same countries, and the results were heterogeneous in the five countries.

In the OECD countries, Chiu & Zhang (2023) analyze the moderating impact of financial development on renewable energy in the CO_2 emissions (from 1990 to 2015). They found the use of renewable energy may decrease CO_2 emissions.

He et al. (2019) analyzes if environmental tax policy can help to reduce pollution emissions through a panel data model in 35 OECD countries (from 1994 to 2016) and 31 provinces of China (from 2004 to 2016). They conclude that there is a relationship between environmental tax and the reduction of pollution emissions.

The research on the effect of the use of electric vehicles on greenhouse gas emissions in 39 European Union countries is considered by Fuinhas et al. (2021). The results of this study show that economic growth positively impacts greenhouse gas emissions, but the promotion of electric vehicles mitigates that impact.

Chu & Le (2022) analyze the impact of economic policy uncertainty, economic complexity, renewable energy and energy intensity in the CO_2 emissions and ecological footprint for the G7 experience.

The results in the long-term show the utilization of renewable energy and economic policy uncertainty reduces environmental degradation. But the energy intensity has the opposite impact.

Along the same line, Doğan et al. (2022) analyze the relationship between environmental taxes and CO2 emissions. The empirical results of this study "suggest that environmental taxes effectively reduce emissions for the G7 countries and confirm that the marginal effects of the environmental tax on traditional energy consumption" (Doğan et al., 2022, p. 644).

In the study of Balsalobre-Lorente et al. (2021), the sample countries are France, Germany, Italy, Spain, and the United Kingdom, normally called EU-5 countries. They analyze the impact of renewable energy, economic growth, international tourism, and technological innovation on CO_2 emissions (from 1990 to 2015), considering a panel FMOLS and panel causality. The results demonstrate that renewable energy negatively impacts CO_2 emissions, revealing an improvement in environmental policy. Then, economic growth could restore the positive impact of the CO_2 emissions, i.e., the improvements in the process of growth are at the same time increases in the environmental concerns. Similar results are found by Alola & Ozturk (2021) in the United States (from 1984 to 2017). This study suggests sustainable economic growth in which this evolution is coordinated with environmental policies.

Shayanmehr et al. (2023) analyze the impact of environmental tax and renewable energy on the ecological footprint. In this study, 27 countries constitute the sample, and the authors found that environmental tax and renewable energy negatively impact the ecological footprint. They also suggest improving economic development sustainability, applying the environmental tax and the incentives to use renewable energy.

Islam et al. (2022) analyze the impact of green taxation on CO_2 emission, focusing on the industrial sector (from 2011 to 2020). The countries in the analysis are Australia, Canada, France, Germany, Japan, Korea, Switzerland, the United Kingdom, the United States, China, and India. The results demonstrate the green tax's impact on decreasing CO_2 emissions. Furthermore, in the opinion of these authors, the green tax promotes improvements in eco-friendly industries.

In the other particular sector, Passenger Vehicles, Meireles et al. (2021) analyze the impact of Environmental Taxes in Reducing CO_2 emissions in the Mediterranean countries (from 2008 to 2018). The countries examined are Bulgaria, Croatia, Cyprus, France, Greece, Italy, Malta, Portugal, Romania, Slovenia, and Spain. The authors used a panel cointegration and ARDL model as an econometric methodology. They found that the environmental tax is essential in decreasing CO_2 emissions. In this line, the transportation sector contributes to CO_2 emissions, and tax legislation could reduce this impact (Runst & Höhle, 2022).

Leitão et al. (2023) analyze the impact of economic growth, population, foreign direct investment, and renewable energy on CO_2 emissions (from 1990 to 2018) in the following's countries: Poland, the Czech Republic, Slovakia, and Hungary. The research applied a panel ARDL (autoregressive distributed lag model) to test the environmental Kuznets curve (EKC) and a panel Fully modified OLS (FMOLS) to analyze the economic growth determinants. The results demonstrated that EKC is valid in Visegrad countries, except in Hungary. Additionally, the authors found a negative correlation between foreign direct investment and CO_2 emissions, showing that the result is according to halo pollution emissions. Using FMOLS, the econometric results to explain the economic growth demonstrate that renewable, non-renewable, and foreign direct investment promote economic growth. However, the urban population decreased the economic growth in this group of countries.

Le & Ozturk (2020) study the relationship between globalization, financial development, institutional quality, government expenditure, energy consumption, economic growth, and CO_2 emissions (from 1990 to 2014) in Emerging markets and Developing Economies. This study finds that income per capita and squared income per capita are impacted positively and negatively by carbon dioxide emissions. According to the environment Kuznets curve (EKC) hypotheses, these results show that long-term economic growth is directly associated with sustainable practices. Moreover, the authors also found that the variables of globalization, financial development, energy consumption and energy consumption promote climate change in this group of countries.

Leitão (2021) study evaluated the relationship concerning trade intensity, energy consumption, income per capita, and CO_2 emissions in Portugal. The empirical analysis used an economic strategy time series (ARDL-autoregressive distributed lag model, quantile regressions, and cointegration models). Considering the Long run effects of the ARDL model, the econometric results show that the CO_2 emissions in Portugal decreased during the analysis period. Besides, the variables of trade intensity and economic growth contribute to environmental improvements. Nevertheless, energy consumption is directly impacted by carbon dioxide emissions. This research confirms the same tendency with cointegration models (FMOLS- Fully- modified OLS, DOLS-Dynamic OLS and CCR- Canonical Cointegration Regression).

In the same country, Portugal, Kirikkaleli et al. (2023) analyze the impact of energy consumption, renewable energy, and economic growth on CO_2 emissions from 1990 to 2019. The results of the long-term estimations reveal a positive impact of energy consumption and economic growth on CO2 emissions. The use of renewable energy has a positive effect on environmental quality. Similar results are found by Adebayo et al. (2022) in the same country, Portugal (from 1980 and 2019). The authors suggest that "policymakers in Portugal should stimulate investment in renewable energy sources, establish restrictive laws, and enhance energy innovation" (Adebayo et al., 2022, p. 23887).

The impact of economic growth on a decrease in environmental quality was analyzed by Moutinho et al. (2017) in 13 sectors in Portugal and Spain (from 1975 to 2012). The results demonstrate that impact but in different ways in the two countries. Portugal presents an N-shape relationship, and Spain presents mixed evidence. These two countries, Portugal, and Spain, which share a common history, will be analyzed in this chapter.

A study of the seven emerging economies it's made by Wolde-Rufael & Mulat-Weldemeskel (2021). The authors studied the impact of environmental tax and environmental stringency policies on CO_2 emissions. The countries in this research are the Czech Republic, Greece, Hungary, Korea, Poland, South Africa, and Turkey (from 1994 to 2015). The results show the environmental tax contributes to the reduction of CO_2 emissions. The same authors analyze the impact of ecological taxation and renewable energy on CO_2 emissions in Latin America and the Caribbean (from 1994 to 2018) (Wolde-Rufael & Mulat-Weldemeskel, 2022). The results demonstrate the importance of the environmental tax and renewable energy in the decrease of CO_2 emissions. But this importance is more efficient in high-emission countries.

Kahia et al. (2023) studied the impact of environmental innovation and renewable energy on CO_2 emissions in Saudi Arabia (from 1990 to 2018). The empirical study of Kahia et al. (2023) considered an ARDL model, and they found that using renewable energy improves CO_2 emissions in the long run. Nevertheless, economic growth and foreign direct investment accelerate climate change. These results are according to the pollution haven hypotheses theory and EKC.

Considering the conclusions of a theoretical study conducted by Shahzad (2020), the implementation of environmental tax and legislation, the use of renewable energy and the implementation of green technologies are essential to environmental quality and the sustainability of the countries.

Lastly, most studies demonstrate the importance of implementing environmental tax and using renewable energy to decrease CO_2 emissions. Moreover, economic growth could have implications for the increase of CO_2 emissions. But when this economic growth is sustainable, the rise of CO_2 emissions is not proportional.

Although our study does not assess the relationship between environmental standards and regulations, there has been more recent interest from the academic community in studying the impact of environmental taxes on renewable energies. There seems to be some duality in some econometric results, as Bashir et al. (2022) reported. However, the development of that study finds a negative association between environmental taxes and renewable energy, in which the ecological standards do not facilitate the consumption of renewable energy.

Next, we will present the methodology to be applied to this empirical study, the variables under investigation and the respective hypotheses to be tested in the econometric models.

METHODOLOGY AND DATA

The contribution of environmental taxes in reducing climate change via carbon dioxide (CO_2) emissions is evaluated in this research. The empirical study collects data for Portugal and Spain, comparing the two Iberian countries, considering a recent period from 1990 to 2019. In addition, the impact of renewable energies and economic growth on carbon dioxide emissions are also considered in this investigation. In terms of econometric models, time series were used, namely the TSLS- Two – Stage least squares estimator and cointegration regression, namely FMOLS- Fully- modified OLS, DOLS-Dynamic OLS, CCR- Canonical Cointegration Regression, and Granger Causality. At first, the variables under study are analyzed to determine if there is stationary in the levels or if the variables are integrated into the first differences. Subsequently, the econometric models are applied.

Considering the literature review presented in the previous section, we present below a set of formulated hypotheses:

Hypothesis One: Does environmental tax reduce carbon dioxide emissions?

The first hypothesis was constructed based on the assumption that the environmental tax contributes to reducing carbon dioxide emissions and is consequently associated with decarbonization.

The studies by Sarigül & Topcu (2021), Ghazouani (2020), and Hájek et al. (2019) support our hypothesis, demonstrating that there is a negative correlation between environmental taxes and pollution emissions. Also, a recent study by Giménez & Marco (2020) demonstrates that the environmental tax reduces carbon dioxide emissions.

To test this hypothesis, the following variables were used:

ETPT-Portuguese environmental tax in the percentage of GDP from OECD.

ETSP-Spanish environmental tax in the percentage of GDP from OECD.

Hypothesis Two: Does the use of renewable decreasing carbon dioxide emissions?

The second hypothesis seeks to test the impact of renewable energies on carbon dioxide emissions. This determinant in the equation aims to show the extent to which renewable energies allow reducing carbon dioxide emissions, thus improving air quality and global warming.

Several empirical studies such as Fuinhas et al. (2021), Balsalobre-Lorente et al. (2021), Leitão et al. (2023), and Chu and Le (2022) show that renewable energies are negatively correlated with carbon dioxide emissions, allowing to reduce climate change.

RENPT-Portuguese renewable energy in the total final energy consumption percentage from World Bank Indicators.

RENSP-Spanish renewable energy in the total final energy consumption percentage from World Bank Indicators.

Hypothesis Three: To what extent do economic growth and economic activities increase carbon dioxide emissions?

As a rule, energy and environmental economics studies reveal a positive association between economic growth and carbon dioxide emissions. This determinant was introduced into the equation to understand the extent to which economic growth stimulates an increase in carbon dioxide emissions, using some of the assumptions of the Kuznets environmental curve.

In this context, the studies by Leitão et al. (2023), Le & Ozturk (2020), Alola & Ozturk (2021), and Leitão (2021) prove that economic growth is positively correlated with pollution emissions.

GDPPT- Portuguese gross domestic product per capita in current US dollars from World Bank Indicators.

GDPSP- Spanish gross domestic product per capita in current US dollars from World Bank Indicators.

After presenting the variables and hypotheses under study, is formulated the following equations:

$$CO_2PT= \beta_0+\beta_1ETPT+\beta_2RENPT+\beta_3GDPPT+ \mu it \quad (1)$$

Where the dependent variable is Portuguese carbon dioxide emissions (CO_2), and μit represents the random residual term or error. Then, it is expected that $\beta_1 ETPT<0$; $\beta_2 RENPT<0$; $\beta_3 GDPPT>0$.

In the second equation, is tested the empirical results for the Spanish experience:

$$CO_2SP= \beta_0+\beta_1ETSP+\beta_2RENSP+\beta_3GDPSP+ \mu it \quad (2)$$

Thus, $\beta_1 ETSP<0$; $\beta_2 RENSP<0$; $\beta_3 GDPSP>0$.

Next, is presented a summary table with a description of the variables to be used and the respective expectations, considering the previously formulated hypotheses and the sources where the data were collected.

Table 1. The description of variables used in this research

Variables	Description	Expected Sign	Source
Dependent variable			
CO_2PT CO_2SP	Portuguese carbon dioxide emissions per capita Spanish carbon dioxide emissions per capita		World Bank
Explanatory variables			
ETPT	Portuguese environmental tax (% of GDP)	[-]	OECD
ETSP	Spanish environmental tax (% of GDP)	[-]	OECD
RENPT	Portuguese renewable energy in the total final energy consumption percentage	[-]	World Bank
RENSP	Spanish renewable energy in the total final energy consumption percentage	[-]	World Bank
GDPPT	Portuguese gross domestic product per capita (current US dollars)	[+]	World Bank
GDPPSP	Spanish gross domestic product per capita (current US dollars)	[+]	World Bank

Source: Composition prepared by the authors.

EMPIRICAL RESULTS

This section considers the econometric results that aim to evaluate the impact of environmental taxes, renewable energies, and economic growth on carbon dioxide emissions for the case study of Portugal and Spain. Then, the correlations between the variables under investigation for the two equations are considered. Finally, we discuss independently and separately for Portugal and Spain the effects of environmental taxes, renewable energies and economic growth on carbon dioxide emissions using the Two-Stage Least Squares (TSLS) estimator and the FMOLS- Fully-modified OLS, DOLS- Dynamic OLS, CCR- Canonical Cointegration Regression, and Granger Causality. As mentioned in the methodology section, data were collected from the World Bank and the OECD (see Table 1). We selected a recent period, 1990 to 2019, for each country under analysis. In each estimator, it is also possible to visualize the number of observations for our sample.

General Statistics

Next, the properties of the variables used in this research are presented. First, we consider unit root tests to see if the variables under study have unit root problems or are stationary in levels or first differences. Subsequently, the correlations between the variables for the Portuguese and Spanish economies are considered.

Table 2 below shows the unit roots test using the Augmented Dickey-Fuler (ADF) criterion. The results demonstrate that the variables Portuguese carbon dioxide emissions (CO_2PT), Portuguese environmental taxes (ETPT), Portuguese renewable energies (RENPT) and Portuguese per capita income (GDPPT) are stationary in the first differences. Similarly, it can be concluded that Spanish carbon dioxide emissions (CO_2SP), Spanish environmental taxes (ETSP), Spanish renewable energies (RENSP), and Spanish per capita income (GDPSP) are the first differences.

Table 2. Unit root test considering augmented Dickey-Fuller (ADF) test statistic

Variables	Level	First Differences
CO_2 PT	-1.584 (0.477)	-4.888*** (0.000)
ETPT	-1.591 (0.472)	-5.078*** (0.000)
RENPT	0.038 (0.95)	-6.651*** (0.000)
GDPPT	-2.869* (0.061)	-5.083*** (0.000)
CO_2SP	-0.943 (0.758)	-4.011*** (0.005)
ETSP	-1.429 (0.552)	-4.676*** (0.001)
RENSP	-0.942 (0.755)	-3.944*** (0.008)
GDPSP	-2.366 (0.159)	-5.081*** (0.000)

Source: Statistics were obtained through the EViews 12 software.
Notes: ***, and * represent statistically significant at 1%, 10% levels.

Table 3 and Table 4 show the correlations for equations 1 and 2. Both equations positively correlate environmental taxes and per capita income with carbon dioxide emissions. Furthermore, Portuguese, and Spanish renewable energies are negatively correlated with polluting emissions.

Table 3. Correlations between the variables used for Portuguese model

Variables	CO_2 PT	ETPT	RENPT	GDPPT
CO_2 PT	1.000			
ETPT	0.379	1.000		
RENPT	-0.663	-0.888	1.000	
GDPPT	0.277	0.683	-0.627	1.000

Source: Statistics were obtained through the EViews 12 software.

Table 4. Correlations between the variables used for Spanish model

Variables	CO_2 SP	ETSP	RENSP	GDPSP
CO_2 SP	1.000			
ETSP	0.098	1.000		
RENSP	-0.493	-0.692	1.000	
GDPSP	0.504	0.731	-0.653	1.000

Source: Statistics were obtained through the EViews 12 software.

The Empirical Results for Portugal and Discussion

This section will present the econometric results using the two-stage least squares estimator (TSLS) and the cointegration models (FMOLS, DOLS, and CCR). Using different estimators aims to observe if there are similarities or differences between the coefficients obtained.

Table 5 shows the econometric results considering two-stage least squares (TSLS) for the Portuguese economy. Environmental tax (ETPT) and renewable energy (RENPT) coefficients are statistically significant at a 1% level. These results validate hypotheses 1 and 2 formulated. According to the Durbin-Watson statistics, the model doesn't present the problem of serial correlation. The variable of environmental tax (ETPT) aims to decrease pollution emissions. The empirical studies of Sarigül & Topcu (2021), Ghazouani (2020), and Hájek et al. (2019) also found a negative link between environmental tax (ETPT) and carbon dioxide emissions. The result found demonstrates that fiscal policy contributes to environmental efficiency.

Table 5. The contribution of an environmental tax on Portuguese CO_2 emissions with two- stage least squares (TSLS)

Variables	**TSLS** **Coef.**
ETPT	-1.509*** (0.007)
RENPT	-0.086*** (0.000)
GDPPT	-0.003 (0.944)
C	10.625*** (0.000)
Adj. R^2	0.593
SE of regression	0.388
Number of Instruments	5
Mean dep var	5.328
S.D dep var	0.607
Sum squ. resi.	2.705
Durbin- Watson stat	1.385
Observations	22

Source: The econometric model was obtained through the EViews 12 software.
Note: *** represents statistically significant at 1% level.

Moreover, renewable energy (RENPT) negatively affects carbon dioxide emissions, showing that the use of cleaner energy promotes the improvement of environmental and air quality. Fuinhas et al. (2021), Balsalobre-Lorente et al. (2021), Leitão et al. (2023), and Chu and Le (2022) also found a negative association between renewable energy and carbon dioxide emissions.

The econometric results for the Portuguese economy through the cointegration estimators (FMOLS-Fully modified OLS, DOLS-Dynamic OLS and CCR- Canonical Cointegration Regression model) are presented table 6. In general, we observe that the results are identical to each other and those previously shown through the TSLS estimator (Two- Stage least Squares).

Thus, the hypothesis of the variable of environmental tax (ETPT) contributing to the reduction of climate change is validated, and this has statistical significance at 1% for the estimator FMOLS and DOLS-Dynamic OLS and 5% for the CCR- Canonical Cointegration Regression model. This result is in line with the Kyoto Protocol.

Regarding renewable energy (RENPT), note that it promotes environmental quality. This result is according to the Kyoto Protocol conferences (1997), the Paris Agreement (2015), and the 2030 Agenda for Sustainable Development. The studies by Kahia et al. (2023) and Chiu & Zhang (2023) also find a negative relationship between renewable energies and carbon dioxide emissions The result validates the hypothesis, showing that renewable energy improves the environment and reduces climate change.

Table 6. The contribution of an environmental tax on Portuguese CO_2 emissions with cointegration models

Variables	FMOLS Coef.	DOLS Coef.	CCR
ETPT	-1.952*** (0.000)	-4.214*** (0.002)	-2.066** (0.029)
RENPT	-0.099*** (0.000)	-0.171*** (0.000)	-0.103*** (0.000)
GDPPT	0.013 (0.85)	0.145 (0.223)	0.016 (0.762)
C	12.032*** (0.000)	18.920*** (0.000)	12.379*** (0.000)
Adj. R^2	0.59	0.81	0.57
SE of regression	0.389	0.273	0.397
Long-run var	0.152	0.052	0.152
Mean dep var	5.36	5.401	5.369
S.D dep var	0.607	0.621	0.607
Sum squ. resi.	2.573	0.446	2.687
Observations	21	19	21

Source: The econometric models were obtained through the EViews 12 software.
Notes: ***represents statistically significant at 1% level.

Table 7 assesses the interdependence relationships between the variables used for the Portuguese economy. According to our results, we observe there are only unidirectional relationships, namely between the Portuguese environmental tax (ETPT) and carbon dioxide emissions (CO_2PT); the same is true between renewable energies (RENPT) and carbon dioxide emissions. carbon (CO_2PT).

Finally, there is also a unidirectional relationship between the Portuguese environmental tax (ETPT) and economic growth (GDPPT) and Portuguese economic growth (GDPPT) and renewable energies (RENPT).

The Empirical Results for Spain and Discussion

As in the previous section, we used different techniques (the two-stage least squares estimator and cointegration models) to observe whether there are similarities or differences between the results obtained.

The results for the Spanish model are shown in Table 8 using the Two-Stage Least Squares (TSLS) estimator. It is observed that the variables renewable energy (RENSP) and income per capita (GDPSP) are statistically significant at 1%, and the variable of environmental tax (ETSP) is statistically significant at 5%.

The Spanish variable of environmental tax (ETSP) demonstrates a decrease in carbon dioxide emissions and agrees with the studies by Sarigül & Topcu (2021), Ghazouani (2020), and Hájek et al. (2019).

Table 7. Granger causality between the variables for Portugal

Null Hypothesis:	Obs	F-Statistic	Prob.
ETPT does not Granger Cause CO_2PT	25	4.840**	0.038
CO_2PT does not Granger Cause ETPT		0.531	0.474
RENPT does not Granger Cause CO_2PT	25	3.643*	0.069
CO_2PT does not Granger Cause RENPT		2.696	0.115
GDPPT does not Granger Cause CO_2PT	29	3.584*	0.069
CO_2PT does not Granger Cause GDPPT		0.228	0.637
RENPT does not Granger Cause ETPT	21	1.820	0.194
ETPT does not Granger Cause RENPT		5.216**	0.034
GDPPT does not Granger Cause ETPT	25	0.194	0.664
ETPT does not Granger Cause GDPPT		3.901*	0.071
GDPPT does not Granger Cause RENPT	25	3.183*	0.088
RENPT does not Granger Cause GDPPT		1.354	0.257

Source: Statistics were obtained through the EViews 12 software.
Notes: **, * represents statistically significant at 5%, and 10% level.

Table 8. The contribution of an environmental tax on Spanish CO_2 emissions with two- stage least squares (TSLS)

Variables	TSLS Coef.
ETSP	-3.899*** (0.002)
RENSP	-0.062** (0.012)
GDPSP	0.276*** (0.005)
C	14.117*** (0.000)
Adj. R^2	0.59
SE of regression	0.643
Number of Instruments	5
Mean dep var	6.57
S.D dep var	0.928
Sum squ. resi.	7.446
Durbin- Watson stat	0.803
Observations	22

Source: The econometric model was obtained through the EViews 12 software.
Notes: *** represents statistically significant at 1% level.

Spanish renewable energy (RENSP) finds the sign advanced by the literature (Kahia et al. 2023; Chiu & Zhang 2023; Leitão et al. 2023, and Chu & Le 2022) demonstrates that the use of more efficient and cleaner technologies helps to reduce carbon dioxide emissions.

Regarding economic growth (GDPSP), it is observed that there is a positive correlation with carbon dioxide emissions. This result demonstrates that economic activities stimulate climate change. Furthermore, the studies by Leitão et al. (2023), Le & Ozturk (2020), Alola & Ozturk (2021), and Leitão (2021) also find a positive association between per capita income and carbon dioxide emissions.

The results for the cointegration estimators (FMOLS- Fully modified OLS, DOLS-Dynamic OLS and CCR- Canonical Cointegration Regression) are shown in Table 9. As can be seen, the best results are those provided by the FMOLS estimator.

Table 9. The contribution of an environmental tax on Spanish CO_2 emissions with cointegration models

Variables	FMOLS Coef.	DOLS Coef.	CCR
ETSP	-4.407*** (0.006)	-1.714 (0.686)	-4.538** (0.015)
RENSP	-0.052* (0.088)	0.036 (0.745)	-0.045 (0.244)
GDPSP	0.374*** (0.004)	0.695** (0.038)	0.430*** (0.007)
C	14.733*** (0.001)	7.074 (0.436)	14.779*** (0.000)
Adj. R^2	0.47	0.70	0.41
SE of regression	0.678	0.506	0.718
Long-run var	0.656	0.409	0.657
Mean dep var	6.614	6.701	6.614
S.D dep var	0.933	0.935	0.933
Sum squ. resi.	7.829	1.542	8.754
Observations	21	19	21

Source: The econometric models were obtained through the EViews 12 software.
Notes: ***, and * represent statistically significant at 1%, 10% levels.

The Spanish variable of environmental tax (ETSP) has statistical significance at 1% and meets the signal expected by recent empirical studies (Ghazouani et al., 2020; Hájek et al., 2019; Murat Telatar & Birinci, 2022; Sarigül & Topcu, 2021).

Spanish renewable energy (RENSP) has statistical significance at 1%, and one more proves to contribute to improving the environment.

Thus, Spanish economic growth (GDPSP) continues demonstrating that it accelerates carbon dioxide emissions, showing statistical significance at 1%. Once again, this result is supported by several previous studies such as Le & Ozturk (2020), Alola & Ozturk (2021), and Leitão (2021).

The following table complements the analysis carried out earlier. The following table complements the analysis carried out earlier. The causal relationships between the variables used for the Spanish economy are considered as in the previous section. As can be seen, there is a bidirectional relationship between renewable energies (RENSP) and carbon dioxide emissions (CO_2SP).

Table 10. Granger causality between the variables for Spain

Null Hypothesis:	Obs	F-Statistic	Prob.
ETSP does not Granger Cause CO_2SP	25	17.081***	0.000
CO_2SP does not Granger Cause ETSP		2.13923	0.157
GDPSP does not Granger Cause CO_2SP	29	6.531**	0.016
CO_2SP does not Granger Cause GDPSP		0.488	0.491
RENSP does not Granger Cause CO_2SP	25	4.149*	0.054
CO_2SP does not Granger Cause RENSP		5.222**	0.032
GDPSP does not Granger Cause ETSP	25	4.464**	0.046
ETSP does not Granger Cause GDPSP		16.469***	0.000
RENSP does not Granger Cause ETSP	21	0.157	0.696
ETSP does not Granger Cause RENSP		36.908***	0.000
RENSP does not Granger Cause GDPSP	25	0.059	0.810
GDPSP does not Granger Cause RENSP		0.753	0.395

Source: Statistics were obtained through the EViews 12 software.
Notes: ***, **, and * represent statistically significant at 1%, 5%, 10% levels.

Furthermore, a bidirectional association is also observed between economic growth (GDPSP) and the Spanish environmental tax (ETSP). Again, there is a unidirectional relationship between the Spanish environmental tax (ETSP) and carbon dioxide emissions (CO_2SP), Spanish economic growth (GDPSP) and carbon dioxide emissions (CO_2SP) and Spanish environmental tax (ETSP) and Spanish renewable energies (RENSP).

FUTURE RESEARCH DIRECTIONS

For future studies, we propose the analysis of several countries. For example, an analysis of OECD and non-OECD countries could verify the impact of the sustainable development indicators of the 2030 Agenda. Besides, it seems interesting to us an analysis of panel data applied to the EU 27 countries, introducing new control variables such as democratization and corruption. Using panel data in future works will allow other strategies, such as second-generation unit root and cross-section dependence tests and other more robust econometric models with the quantile moment panel or estimating the respective decision's fixed and random effects through the Hausman test. Furthermore, using the dynamic panel model through the ARDL model (autoregressive distributed lag) or generalized moments (GMM) would be interesting.

Another interesting area for future work is the impact of environmental standards on the consumption of renewable energies. Several studies like ours find a negative relationship between environmental taxes and carbon dioxide emissions. However, as analyses by Bashir et al. (2022), the relationship between ecological standards and renewable energy consumption seems to obey a paradigm shift, although it needs to be more evident, as the mentioned study proves, that environmental standards promote the consumption of renewable energy.

CONCLUSION

This Chapter analyzes the impact of the environmental tax on CO_2 emissions in Portugal and Spain using time series from 1990 to 2019. It also analyzes the effects of renewable energy and economic growth on carbon dioxide emissions.

The literature review demonstrated the importance of tax environment and the use of renewables energies to reduce CO_2 emissions (Chu & Le, 2022; Depren et al., 2023; Ghazouani et al., 2020; He et al., 2019; Rafique et al., 2022; Shayanmehr et al., 2023).

In the previous studies, economic growth contributes to the increase in CO_2 emissions (Alola & Ozturk, 2021; Balsalobre-Lorente et al., 2021; Fuinhas et al.,2021 Le & Ozturk, 2020; Leitão, 2021; Leitão et al., 2022).

The data are collected from the OECD and Word Bank indicators from 1990 to 2019 and analyzed through the unit root tests and the estimations using the Two-Stage Least Squares (TSLS) estimator, and cointegration models such as fully modified ordinary least squares (FMOLS), Dynamic OLS (DOLS), Canonical Cointegration Regression model (CCR), and Granger causality.

Considering the unit root tests, it was observed that the variables used in this study are stationary in the first differences.

The results show the importance of the environmental tax and the use of renewable energies in decreasing carbon dioxide emissions, both in Portugal and Spain. Besides, the results of this investigation confirm the finds of previous studies (Chu & Le, 2022; Depren et al., 2023; Ghazouani et al., 2020; He et al., 2019; Rafique et al., 2022; Shayanmehr et al., 2023). According to these results, the authors suggest governments can use environmental taxes to reduce CO_2 emissions. Additionally, governments can use the benefits and incentives to promote the use of renewable energy. The efficient coordination of these two mechanisms should promote more sustainable attitudes and improve the environment.

The results also show the positive effect of economic growth on the increase of carbon dioxide emissions in Spain. These results are consistent with the preceding literature. The results of Sarigül & Topcu (2021), Shahzad (2020), and He et al. (2019) support our results.

In addition, the Granger causality methodology was applied to each of the economies. It was observed that the relationship between the Spanish economy is almost always bidirectional, while in Portugal, it is always unidirectional.

In this context, it is observed that in the Spanish economy, there are bidirectional relationships between renewable energies (RENSP) and carbon dioxide emissions (CO_2SP), economic growth (GDPSP) and the Spanish environmental tax (ETSP).

In summary, considering the results of this investigation, it is possible to state that fiscal policy, via the environmental tax, contributes to greater ecological efficiency in the Portuguese and Spanish economies, which improves the environment and reduces the effects of pollution in general, namely, the greenhouse effects, contributing sustainably to both economies.

Furthermore, as referenced by recent empirical studies, using renewable energy makes it possible to reduce carbon dioxide emissions. However, economic growth is associated with an increase in polluting emissions.

All scientific investigations have limitations. In this study, the fact the analysis is made in only two countries is one of them.

REFERENCES

Adebayo, T. S., Oladipupo, S. D., Adeshola, I., & Rjoub, H. (2022). Wavelet analysis of impact of renewable energy consumption and technological innovation on CO_2 emissions: Evidence from Portugal. *Environmental Science and Pollution Research International, 29*(16), 23887–23904. doi:10.100711356-021-17708-8 PMID:34817815

Alola, A. A., & Ozturk, I. (2021). Mirroring risk to investment within the EKC hypothesis in the United States. *Journal of Environmental Management, 293*, 112890. doi:10.1016/j.jenvman.2021.112890 PMID:34082348

Balsalobre-Lorente, D., Driha, O. M., Leitão, N. C., & Murshed, M. (2021). The carbon dioxide neutralizing effect of energy innovation on international tourism in EU-5 countries under the prism of the EKC hypothesis. *Journal of Environmental Management, 298*, 113513. doi:10.1016/j.jenvman.2021.113513 PMID:34403918

Bashir, M. F., Ma, B., Bashir, M. A., Radulescu, M., & Shahzad, U. (2022). Investigating the role of environmental taxes and regulations for renewable energy consumption: Evidence from developed economies. *Ekonomska Istrazivanja, 35*(1), 1262–1284. doi:10.1080/1331677X.2021.1962383

Bashir, M. F., Ma, B., Bilal, Komal, B., & Bashir, M. A. (2021). Analysis of environmental taxes publications: A bibliometric and systematic literature review. *Environmental Science and Pollution Research International, 28*(16), 20700–20716. doi:10.100711356-020-12123-x PMID:33405155

Böhringer, C., Garcia-Muros, X., & González-Eguino, M. (2019). *Greener and fairer: a progressive environmental tax reform for Spain*. https://www.econstor.eu/handle/10419/191538

Bontems, P., & Bourgeon, J. M. (2005). Optimal environmental taxation and enforcement policy. *European Economic Review, 49*(2), 409–435. doi:10.1016/S0014-2921(03)00036-9

Borrego, A. (2016). Diploma of green taxation - Finally an environmental taxation integrated in Portugal? *Revista de Gestão Social e Ambiental, 10*(3), 51–69. https://go.gale.com/ps/i.do?id=GALE%7CA554040581&sid=googleScholar&v=2.1&it=r&linkaccess=abs&issn=1981982X&p=IFME&sw=w

Chiu, Y., & Zhang, W. (2023). Moderating Effect of Financial Development on the Relationship between Renewable Energy and Carbon Emissions. *Energies, 16*(3), 1467. https://www.mdpi.com/1996-1073/16/3/1467. doi:10.3390/en16031467

Chu, L. K., & Le, N. T. M. (2022). Environmental quality and the role of economic policy uncertainty, economic complexity, renewable energy, and energy intensity: The case of G7 countries. *Environmental Science and Pollution Research International, 29*(2), 2866–2882. doi:10.100711356-021-15666-9 PMID:34382169

Depren, Ö., Kartal, M. T., Ayhan, F., & Kılıç Depren, S. (2023). Heterogeneous impact of environmental taxes on environmental quality: Tax domain based evidence from the nordic countries by nonparametric quantile approaches. *Journal of Environmental Management, 329*, 117031. doi:10.1016/j.jenvman.2022.117031 PMID:36528942

Doğan, B., Chu, L. K., Ghosh, S., Diep Truong, H. H., & Balsalobre-Lorente, D. (2022). How environmental taxes and carbon emissions are related in the G7 economies? *Renewable Energy*, *187*, 645–656. doi:10.1016/j.renene.2022.01.077

Engberg-Pedersen, P., & Zwart, R. (2018). *The 2030 Agenda and Development Co-operation Results*. OECD. https://www.oecd-ilibrary.org/content/paper/2f391534-en

Esen, Ö., & Dündar, M. (2021). Do energy taxes reduce the carbon footprint? Evidence from Turkey. *Journal of Emerging Economies and Policy*, *6*(2), 179–186. https://dergipark.org.tr/joeep

Fuinhas, J. A., Koengkan, M., Leitão, N. C., Nwani, C., Uzuner, G., Dehdar, F., Relva, S., & Peyerl, D. (2021). Effect of Battery Electric Vehicles on Greenhouse Gas Emissions in 29 European Union Countries. *Sustainability, 13*(24), 13611. doi:10.3390/su132413611

Ghazouani, A., Xia, W., Ben Jebli, M., & Shahzad, U. (2020). Exploring the Role of Carbon Taxation Policies on CO_2 Emissions: Contextual Evidence from Tax Implementation and Non-Implementation European Countries. *Sustainability (Basel)*, *12*(8680), 8680. doi:10.3390u12208680

Giménez, J. V., & Marco, A. Z. (2020). A Dynamic Spatial Panel of Subnational GHG Emissions: Environmental Effectiveness of Emissions Taxes in Spanish Regions. *Sustainability (Basel)*, *12*(7), 2872. doi:10.3390u12072872

Hájek, M., Zimmermannová, J., Helman, K., & Rozenský, L. (2019). Analysis of carbon tax efficiency in energy industries of selected EU countries. *Energy Policy*, *134*, 110955. doi:10.1016/j.enpol.2019.110955

He, P., Ning, J., Yu, Z., Xiong, H., Shen, H., & Jin, H. (2019). Can Environmental Tax Policy Really Help to Reduce Pollutant Emissions? An Empirical Study of a Panel ARDL Model Based on OECD Countries and China. *Sustainability, 11*(16), 4384. doi:10.3390/su11164384

Islam, M., & Rahman, N., SarzulAlom Sodesh, M., & Saha, M. (2022). Green Taxation Impact On Industrial Sectors Which Reduces Carbon Emission In A Country. *Journal of Business and Management*, *24*(12), 30–46. doi:10.9790/487X-2412063046

Kahia, M., Jarraya, B., Kahouli, B., & Omri, A. (2023). Do Environmental Innovation and Green Energy Matter for Environmental Sustainability? Evidence from Saudi Arabia (1990–2018). *Energies*, *16*(3), 1376. doi:10.3390/en16031376

Kirikkaleli, D., Awosusi, A. A., Adebayo, T. S., & Otrakçı, C. (2023). Enhancing environmental quality in Portugal: Can CO2 intensity of GDP and renewable energy consumption be the solution? *Environmental Science and Pollution Research International*, *30*(18), 53796–53806. doi:10.100711356-023-26191-2 PMID:36867330

Labandeira, X., Labeaga, J. M., & López-Otero, X. (2019). New Green Tax Reforms: Ex-Ante Assessments for Spain. *Sustainability (Basel)*, *11*(20), 5640. doi:10.3390u11205640

Le, H. P., & Ozturk, I. (2020). The impacts of globalization, financial development, government expenditures, and institutional quality on CO_2 emissions in the presence of environmental Kuznets curve. *Environmental Science and Pollution Research International*, *27*(18), 22680–22697. doi:10.100711356-020-08812-2 PMID:32323231

Leitão, N. C. (2021). Testing the Role of Trade on Carbon Dioxide Emissions in Portugal. *Economies, 9*(1), 22. doi:10.3390/economies9010022

Leitão, N. C., Dos Santos Parente, C. C., Balsalobre-Lorente, D., & Cantos Cantos, J. M. (2023). Revisiting the effects of energy, population, foreign direct investment, and economic growth in Visegrad countries under the EKC scheme. *Environmental Science and Pollution Research International, 30*(6), 15102–15114. doi:10.100711356-022-23188-1 PMID:36168012

Leitão, N. C., Koengkan, M., & Fuinhas, J. A. (2022). The Role of Intra-Industry Trade, Foreign Direct Investment, and Renewable Energy on Portuguese Carbon Dioxide Emissions. *Sustainability, 14*(22), 15131. doi:10.3390/su142215131

Maxim, M. R. (2020). Environmental fiscal reform and the possibility of triple dividend in European and non-European countries: Evidence from a meta-regression analysis. *Environmental Economics and Policy Studies, 22*(4), 633–656. doi:10.100710018-020-00273-8

Meireles, M., Robaina, M., Magueta, D., Han, H., & Meegoda, J. N. (2021). The Effectiveness of Environmental Taxes in Reducing CO2 Emissions in Passenger Vehicles: The Case of Mediterranean Countries. *International Journal of Environmental Research and Public Health, 18*(10), 5442. doi:10.3390/ijerph18105442 PMID:34069643

Moutinho, V., Varum, C., & Madaleno, M. (2017). How economic growth affects emissions? An investigation of the environmental Kuznets curve in Portuguese and Spanish economic activity sectors. *Energy Policy, 106*, 326–344. doi:10.1016/j.enpol.2017.03.069

Murat Telatar, O., & Birinci, N. (2022). The effects of environmental tax on Ecological Footprint and Carbon dioxide emissions: A nonlinear cointegration analysis on Turkey. *Environmental Science and Pollution Research International, 29*(29), 44335–44347. doi:10.100711356-022-18740-y PMID:35129752

Pereira, A. M., & Pereira, R. M. (2023). Energy Taxation Reform with an Environmental Focus in Portugal. *Energies, 16*(3), 1232. doi:10.3390/en16031232

Pereira, A. M., Pereira, R. M., & Rodrigues, P. G. (2016). A new carbon tax in Portugal: A missed opportunity to achieve the triple dividend? *Energy Policy, 93*, 110–118. doi:10.1016/j.enpol.2016.03.002

Qi, Y., Zhang, J., & Chen, J. (2023). Tax incentives, environmental regulation and firms' emission reduction strategies: Evidence from China. *Journal of Environmental Economics and Management, 117*, 102750. doi:10.1016/j.jeem.2022.102750

Rafique, M. Z., Fareed, Z., Ferraz, D., Ikram, M., & Huang, S. (2022). Exploring the heterogenous impacts of environmental taxes on environmental footprints: An empirical assessment from developed economies. *Energy, 238*, 121753. doi:10.1016/j.energy.2021.121753

Rodríguez, M., Robaina, M., & Teotónio, C. (2019). Sectoral effects of a Green Tax Reform in Portugal. *Renewable & Sustainable Energy Reviews, 104*, 408–418. doi:10.1016/j.rser.2019.01.016

Runst, P., & Höhle, D. (2022). The German eco tax and its impact on CO_2 emissions. *Energy Policy, 160*, 112655. doi:10.1016/j.enpol.2021.112655

Sarigül, S. S., & Topcu, B. A. (2021). The Impact of Environmental Taxes on Carbon Dioxide Emissions in Turkey. *International Journal of Business and Economic Studies*, *3*(1), 43–54. 10.08.2021

Shafi, M., Ramos-Meza, C. S., Jain, V., Salman, A., Kamal, M., Shabbir, M. S., & Rehman, M. U. (2023). The dynamic relationship between green tax incentives and environmental protection. *Environmental Science and Pollution Research International*, *30*(12), 32184–32192. doi:10.100711356-023-25482-y PMID:36723845

Shahzad, U. (2020). Environmental taxes, energy consumption, and environmental quality: Theoretical survey with policy implications. *Environmental Science and Pollution Research, 27*(20), 24848–24862. doi:10.1007/s11356-020-08349-4

Sharif, A., Kartal, M. T., Bekun, F. V., Pata, U. K., Foon, C. L., & Kılıç Depren, S. (2023). Role of green technology, environmental taxes, and green energy towards sustainable environment: Insights from sovereign Nordic countries by CS-ARDL approach. *Gondwana Research*, *117*, 194–206. doi:10.1016/j.gr.2023.01.009

Shayanmehr, S., Radmehr, R., Ali, E. B., Ofori, E. K., Adebayo, T. S., & Gyamfi, B. A. (2023). *How do environmental tax and renewable energy contribute to ecological sustainability? New evidence from top renewable energy countries*. Taylor & Francis. doi:10.1080/13504509.2023.2186961

Sousa, S. (2021). *Environmental Taxation in Portugal: A Contribution to Sustainability*. Springer. doi:10.1007/978-3-030-63149-9_23

Sun, Y., Mao, X., Liu, G., Yin, X., & Zhao, Y. (2020). Modelling the effects of energy taxes on ecological footprint transfers in China's foreign trade. *Ecological Modelling*, *431*, 109200. doi:10.1016/j.ecolmodel.2020.109200

Telatar, O. M., & Birinci, N. (2022). The effects of environmental tax on Ecological Footprint and Carbon dioxide emissions: A nonlinear cointegration analysis on Turkey. *Environmental Science and Pollution Research International*, *29*(29), 44335–44347. doi:10.100711356-022-18740-y PMID:35129752

Ulucak, R., Danish, & Kassouri, Y. (2020). An assessment of the environmental sustainability corridor: Investigating the non-linear effects of environmental taxation on CO_2 emissions. *Sustainable Development (Bradford)*, *28*(4), 1010–1018. doi:10.1002d.2057

United Nations. (2015). *The Paris Agreement*. UNFCCC. https://unfccc.int/process-and-meetings/the-paris-agreement

United Nations. (2023). *Transforming our world: the 2030 Agenda for Sustainable Development*. UN. https://sdgs.un.org/2030agenda

United Nations Treaty Collection. (1997). *Kyoto Protocol to the United Nations Framework Convention on Climate Change*. Kyoto Protocol to the United Nations Framework Convention on Climate Change. https://treaties.un.org/Pages/ViewDetails.aspx?src=IND&mtdsg_no=XXVII-7-a&chapter=27&clang=_en

Vona, F. (2021). *Managing the distributional effects of environmental and climate policies: The narrow path for a triple dividend*. OECD. https://www.oecd-ilibrary.org/environment/managing-the-distributional-effects-of-environmental-and-climate-policies_361126bd-en

Wolde-Rufael, Y., & Mulat-Weldemeskel, E. (2021). Do environmental taxes and environmental stringency policies reduce CO_2 emissions? Evidence from 7 emerging economies. *Environmental Science and Pollution Research International*, *28*(18), 22392–22408. doi:10.100711356-020-11475-8 PMID:33417133

Wolde-Rufael, Y., & Mulat-Weldemeskel, E. (2022). The moderating role of environmental tax and renewable energy in CO_2 emissions in Latin America and Caribbean countries: Evidence from method of moments quantile regression. *Environmental Challenges*, *6*, 100412. doi:10.1016/j.envc.2021.100412

Yu, M., Cruz, J. M., & Li, D. M. (2019). The sustainable supply chain network competition with environmental tax policies. *International Journal of Production Economics*, *217*, 218–231. doi:10.1016/j.ijpe.2018.08.005

ADDITIONAL READING

Balsalobre-Lorente, D., Leitão, N. C., & Bekun, F. V. (2021). Fresh Validation of the Low Carbon Development Hypothesis under the EKC Scheme in Portugal, Italy, Greece and Spain. *Energies*, *14*(1), 250. doi:10.3390/en14010250

Khatiwada, D., Vasudevan, R. A., & Santos, B. H. (2022). Decarbonization of natural gas systems in the EU–costs, barriers, and constraints of hydrogen production with a case study in Portugal. *Renewable & Sustainable Energy Reviews*, *168*, 112775. doi:10.1016/j.rser.2022.112775

Leitão, N. C. (2021). The Effects of Corruption, Renewable Energy, Trade and CO_2 Emissions. *Economies*, *9*(2), 62. doi:10.3390/economies9020062

Leitão, N. C., & Balsalobre-Lorente, D. (2020). *The Causality between Energy Consumption, Urban Population, Carbon Dioxide Emissions, and Economic Growth. Econometrics of Green Energy Handbook*. Springer International Publishing. doi:10.1007/978-3-030-46847-7_5

Leitão, N. C., & Balsalobre-Lorente, D. (2021). *The Effects of Tourism Arrivals, Renewable Energy on Carbon Dioxide Emissions. Sustainable Tourism and Macro- Economy in Era of Globalization Handbook. Publisher*. Springer International Publishing. doi:10.1007/978-3-030-59675-0_4

Leitão, N. C., Balsalobre-Lorente, D., & Cantos-Cantos, J. M. (2021). The Impact of Renewable Energy and Economic Complexity on Carbon Emissions in BRICS Countries under the EKC Scheme. *Energies*, *14*(16), 4908. doi:10.3390/en14164908

Liu, H., Zafar, M. W., Sinha, A., & Khan, I. (2023). The path to sustainable environment: Do environmental taxes and governance matter? *Sustainable Development (Bradford)*, sd.2505. Advance online publication. doi:10.1002d.2505

Wang, W. H., Espinosa, V. I., & Huerta de Soto, J. (2022). A Free-Market Environmentalist Enquiry on Spain's Energy Transition along with Its Recent Increasing Electricity Prices. *International Journal of Environmental Research and Public Health*, *19*(15), 9493. doi:10.3390/ijerph19159493 PMID:35954848

KEY TERMS AND DEFINITIONS

Carbon Dioxide Emissions: This polluting effect is responsible for gases that stimulate greenhouse effects. Community directives (EU) and the various international conferences on the environment and quality of life seek to promote reducing greenhouse effects.

Climate change: Associated with practices that do not respect sustainability principles. As a rule, measures are based on negative environmental and biodiversity impacts but are often understood as global warming.

Economic growth: The increase of the economic goods and services that a society produces and consumes in one period compared with a previous period.

Environmental Taxes: Include taxes on energy, transport, pollution, and resources.

GDP: Gross domestic product, being a measure that is used to measure economic growth. According to macroeconomics manuals, this measure is part of national accounting and can be determined from the production, expenditure, and income perspective.

GDP Per Capita: According to macroeconomic theories, this measure allows for assessing the income distribution of a given country or economy. In a way, it provides for determining the country's degree of development since there are other ways to measure the degree of development of an economy.

Renewable Energy: This type of energy is associated with sustainability issues. As a rule, EU community directives usually list five types: solar, biomass, hydrogen, wind, and oceans.

Chapter 13
The Portuguese Environmental Tax Burden: A Systematic Approach

Gisela Pinto Oliveira
Universidade de Aveiro, Portugal

Cecília M. R. Carmo
https://orcid.org/0000-0002-1883-5490
Universidade de Aveiro, Portugal

Sérgio Ravara Cruz
Universidade de Aveiro, Portugal

ABSTRACT

Environmental taxation is an important tool in the hands of member states to combat against climate change and promote environmental sustainability. This chapter describes in detail the taxes that are part of the Portuguese tax system and that have the social function of protecting the environment. For this purpose, a background of the Portuguese Green Tax Reform is provided and each of the environmental taxes considered to be the main burdens of the Portuguese tax system are explained. This study aims to contribute to providing academics and practitioners with more information on environmental taxation and to motivate regulators to develop more articulated green taxation strategies in the future.

INTRODUCTION

Human beings are largely responsible for climate change and global warming. To control carbon dioxide emissions, consumers should be discouraged from excessive consumption of products, especially those that have negative effects on the environment. Action on climate change should be preventive and severe, because it is better to make a mistake that leads to a small climate change, than to deal with the consequences of not doing anything about the problem (Hassler et al., 2021).

DOI: 10.4018/978-1-6684-8592-7.ch013

It is therefore the responsibility of the State, not individuals or private companies, to take the lead in the fight against climate change. This chapter discusses some of the fiscal measures adopted by the Portuguese State to protect the environment, highlighting the social function of taxes. To this end, we will summarize the Portuguese path towards a more sustainable future and explain the implementation of the Green Tax Reform in Portugal, taking into account that it is a member of the European Union (EU) and has to follow the guidelines established by it.

Portugal ranks very high among the more ambitious countries when it comes to the topic of renewable energy targets, even though environment-related taxes reached over 5 billion euros of the Nacional Gross Domestic Product (GDP) in 2021. Environmental taxes represent 6.6 percent of the total revenue that comes from taxes and social contributions, and about 2.3 percent of the national GDP, according to the Portuguese National Statistics Institute (INE, 2022).

When it comes to climate change, around 33 percent of Europeans believe that their national state is not doing enough to tackle on this battle, according to the 2021 report released by the European Commission (2021). Portugal is no different, ranking second among the most dissatisfied countries, with 85 percent of the respondents believing that their national state is not doing enough to achieve a more sustainable future. In addition, the Portuguese State admits that the targets set by the European Commission are not ambitious enough (European Commission, 2021; INE, 2022).

The effects of fiscal measures, whether they have explicit environmental objectives, depends on whether they encourage or discourage certain behaviors (Avison, 2022). Empirical evidence showed that the impact of the environmental taxes on the environment and on consumer choices are difficult to measure.

Esen et al. (2021) examined the impact of environmental taxes on environmental performance for fifteen EU countries over the period 1995 to 2016. The authors analyzed the overall ecological balance and its main components. The results of the study showed that revenue from environmental taxes as a share of GDP significantly reduces ecological deficits after a certain threshold has been exceeded. Thus, well-designed environmental taxes at an optimal level have the potential to reduce environmental problems if they are not accompanied by policies such as tax exemptions, refunds or tax allowances that limit their impact. Aydin and Esen (2018) studied the impact of environmental taxes on carbon dioxide emissions for the EU member states. The research, which was conducted from 1995 to 2013, reveals an asymmetric relationship between environmental taxes and carbon dioxide emissions. The result shows that after a certain level, the effect of environmental taxes on carbon dioxide emissions changes from significantly positive to significantly negative. To investigate the link between green growth and carbon neutrality targets, Chien et al. (2021) examined the role of environmental taxes and green energy in the United States of America between 1970 and 2015. The findings showed a significant and negative impact of green growth and environmental taxes in determining the carbon dioxide (CO_2) emissions. It was also found that green growth, environmental taxes, and renewable energy played an important role in reducing pollution.

The impact of the Portuguese Green Tax Reform has been studied by Santos (2021), who concluded that it has had a positive evolution between 2012 and 2019, making it possible to pursue the objectives of the strategic plans and optimize the efficiency of waste management. This study did not find any significant shifts in the waste collection, as the reduction in waste was not significant. However, the tax revenue generated by the new Green Tax Reform shows an exponential increase after the year of its adoption. Also in Portugal, Rebelo (2016) analyzed the impact of two environmental measures: the tax on plastic bags and the ban on the availability of free shopping bags, regardless of their material. The results of an online survey of 692 individuals showed a high level of support for both measures, demonstrating their effectiveness in reducing the use of lightweight plastic bags and shopping bags, as well as raising environmental awareness among the participants. There was also an increase in reuse habits and a greater willingness to pay more environmental taxes and reduce living standards for the sake of the environment.

The present study describes in detail the taxes that are part of the Portuguese tax system and that seek to have positive impacts on the environment, here referred to as environmental taxes. To this end, the next section provides a background of the Portuguese Green Tax Reform, mentioning the polluter pays principle and the environmental tax reform in the European Union. In the following section, each of the environmental taxes considered to be the main burdens of the Portuguese tax system is explained in detail – the tax on petroleum products, the single circulation tax, the motor vehicle tax, the energy tax, the pollution tax, the resource tax, the autonomous taxation of motor vehicles, and the tax on single use plastics. The study concludes with some reflections on the environmental protection approach of the Portuguese tax system.

BACKGROUND

The Polluter Pays Principle

The Organization for Economic Co-operation and Development (OECD) first introduced the polluter pays principle in 1972. The polluter pays principle states that whoever is found guilty of polluting the environment must bear the cost of implementing the pollution prevention and the control measures put in place by public authorities to ensure that the environment is in an acceptable state (European Court of Auditors, 2021; Pinto-Bazurco, 2022).

According to the polluter pays principle, polluters are incentivized to reduce and avoid causing environmental damage, and, in addition, they are held responsible for all the damage they cause to their surrounding environment (Luppi et al., 2012). In addition, it takes the burden of covering the costs of pollution away from taxpayers and place it on to the person or persons actually responsible for the damaging act, forcing polluters to internalize the negative environmental externalities they cause in their production costs. Passing on the costs of pollution on to the polluting behavior places the costs only to on those who are responsible for that behavior and increases the price of the polluting good or service to include those costs. As customers prefer lower prices, the increase will act as an incentive for producers to market less polluting products (European Union, 2012; European Court of Auditors, 2021).

Environmental Tax Reform in the EU

Environmental taxes were introduced in Europe in the early 1990s and have become one of the most widely used environmental policy instruments across the European Union (EU) (European Environment Agency, 2005).

Every two years, EU countries report to the European Commission on their progress in promoting and usage of renewable energy in line with the 2020 targets. The European Commission draws up and sets guidelines for its member States, addressing the main concerns of each country. The results were very positive and by the end of 2020 the EU had fully met its targets. But there is still a long way to go, and the European Commission wants to be more ambitious and accelerate the deployment of renewables to contribute to and achieve the goal of reducing greenhouse gas emissions and becoming a climate neutral continent by the end of 2050 (European Commission, 2021).

The environmental tax reform was proposed by the European Environment Agency (EEA), in 2005. The EEA was set up to support sustainable development and to help countries achieve significant and measurable environmental improvements. The EEA provides policymakers and the public with reliable and comparable information on the environment. Its areas of activity include, among other, the following: air and water quality, waste management, noise emissions and chemical substances.

With this reform, the EU has been able to shift the tax burden from goods, such as capital and labor, to polluting factors. Therefore, applying environmental taxation as a tool to correct negative externalities, reduces general tax distortions and improves social welfare, as stated by Castiglione et al. (2014) and Liu (2013) in their studies on the impact of the Green Tax Reform.

The main characteristic of taxes is its one-sidedness, in the sense that the collection of revenue ensures the provision of goods and services for the society, whether they be education, health, safety, or others. There is no universal definition of environmental taxes. The EU describes environmental taxes as taxes "whose tax base is physical unit, or a proxy of it, of something that has a proven, specific negative impact on the environment" (European Union, 2011, article 2). The tax base can be estimated according to the amount of emissions of a given pollutant substance. Due to the nature of the tax base, it can be a quite difficult to measure the impact accurately, and because of this obstacle, the tax base turns out to be an estimate of the damage.

Furthermore, environmental taxes can be defined according to their main objective. On the one hand, there are the "real" environmental taxes, whose sole purpose is to reduce a harmful behavior that has been proven to damage the planet. On the other hand, there are the "fake" environmental taxes, whose main purpose is to collect revenue, even if it is to further redistribute the costs associated with the ecological damage caused, disguised behind a sustainable façade (Novais, 2022).

Environmental taxes work just like the sugar tax, as they are designed to change behavior. The enforcement of these taxes is a success, even though the revenue from these types of taxes is bound to decline over time, as behavior changes and consumption of these products declines (Avison, 2022).

However, environmental taxation also has drawbacks, such as preserving the right to pollute, increasing fiscal pressure, reducing competitiveness and consumption, and even providing incentives for rent-seeking activities (Eisenack et al., 2012; Castiglione et al., 2014).

The Portuguese Green Tax Reform

The Constitution of the Portuguese Republic (CRP, 2005) established the right to environmental protection: "everybody has the right to a humane, healthy and ecologically sound environment and the duty to defend it" [own translation, article 66, n.º 1]. It was also in the Constitution that the task of protection was assigned to the Portuguese State, as stated in article 9, to "protect and enhance the cultural heritance to its people, defend the nature and environment, preserve natural resources and ensuring a correct territorial planning" [own translation].

The national States have a range of tools at their disposal to contribute to environmental sustainability, such as regulations, information programmes, innovation policies, environmental subsidies and taxes (Sousa, 2021).

Over the years, environmental law has evolved and merged not only with Constitutional law but also with tax law, not only in Portugal but throughout the world. The tax law comes as a tool to enforce environmentally friendly policies, solving some of the main problems and improving the performance of the country (Oliveira, 2016; Novais, 2022). In 2014, the Portuguese State approved a new law, Law n.º 82-D/2014, of 31st of December. This Law is better known as the Green Tax Law, because it brings changes to the Portuguese reality, adding a new set of environmental policies in line with the EU guidelines for a more sustainable world. In addition, this new law has triggered the adoption of several international commitments.

The approval of the law, known as the Green Tax Reform, was followed by several legislative amendments, at numerous extents. From changes in the income consumption tax, to variations in the energy taxes and even modifications at the parafiscal level (Moreira & Fernandes, 2022).

The Green Tax Reform introduced by the Portuguese State does not aim to solve every single environmental problem and concern. The core objective is clear: to contribute to a healthier environment and a more sustainable world, and the Green Tax Reform intends to do this with the support of other policies, that are directly and indirectly related to the environment (Moreira & Fernandes, 2022).

The assumptions for the Green Tax Reform are to achieve the triple dividend, which is based on three assumptions. The first assumption is to achieve fiscal neutrality, which means that the net increase in revenues from the new environmental taxes created by the law must be used to reduce the revenue that derives from other types of taxes, especially the income taxes. The second assumption, which translates the main focus of the reform, is to protect the environment and dependence on foreign energy, to promote growth and empowerment. The third assumption is to contribute to fiscal responsibility and reduce external imbalances (Law n.º 82-D/2014; MAOTE, 2015).

On the one hand, the Green Tax Reform aimed to reduce the tax burden and redistribute the new revenues generated by the reform. On the other hand, it aimed to promote behavioral change among individuals, society and businesses, encouraging more sustainable conduct while promoting economic growth (Abreu, 2011; MAOTE, 2015).

THE PORTUGUESE ENVIRONMENTAL TAXES

Overview

According to the ESA (European System of Accounts), there are three main categories of taxes (Eurostat, 2010):

- Taxes on production and imports;
- Current taxes on income, wealth, etc.; and
- Capital taxes.

As far as environmental taxes are concerned, they can be classified as follows (INE, 2022):

- **Energy taxes:** this group includes taxes on energy products, such as petrol, diesel, fuel, oil, natural gas, coal and electricity. In addition, the revenues from greenhouse gas emission permits are recorded as an allowance to the national accounts;
- **Transport taxes:** this group mainly includes taxes that are related to the ownership and the use of motor vehicles. Taxes on other means of transport, such as planes, trains, and ships, are also

included in this category, as far as they fall within the general definition of environmental taxes. Transport taxes may be a recurring tax, such as an annual road tax, or a single tax, such as an import or sales tax;

- **Pollution taxes:** this category of taxes includes taxes on measured or estimated emissions to air and water, management of soil waste and noise;
- **Resource taxes:** finally, this group of taxes includes taxes related to the extraction or to the use of natural resources, such as oil, gas, water, wild flora, and fauna, etc., since these activities deplete natural resources.

The Portuguese tax system includes several taxes with environmental relevance (INE, 2022): the tax on petroleum and energy products (TPP), the single circulation tax (SCT), the motor vehicle tax, the energy tax, the tax on resources and the tax on pollution, which includes the greenhouse gas emission permits. Through the observation of Figure 1, we can see the significant weight of each of the environmentally related taxes in the year 2021. It is undeniable the importance of the TPP in Portugal with 70.6 percent of the collective burden of all environmental taxes. In the second place is the revenue derived from the SCT, with 13.8 percent of the total revenue form environmentally related taxes. The motor vehicle tax accounts for 8.5 percent of revenues in 2021. Finally, with little significance in the total revenue, we have the energy tax, the pollution tax, and the resource tax, with about 5.6 percent, 0.75 percent, and 0.75 percent, respectively (INE, 2022).

Figure 1 shows the importance of the revenues from the acquisition and use of motor vehicles. On the one hand, the TPP, the SCT and the motor vehicle tax represent almost 93 percent of the total environmental tax revenue. On the other hand, we can also see the lack of importance of pollution and resource taxes, which represent only 1 percent of the tax burden in Portugal. Sousa (2021) had already documented this situation in 2017.

Figure 1. Environmental taxes in 2021
Source: Based on INE (2022)

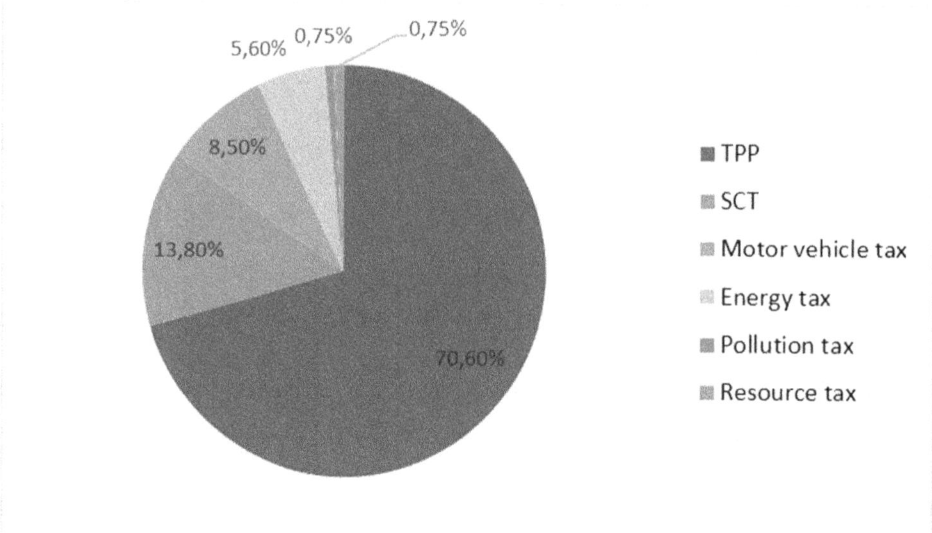

Each of the taxes presented in Figure 1 will be analyzed individually in the next section, from those that account for the largest share of the environmental tax burden to those that have little impact on the environmental tax revenues.

The Tax on Petroleum Products

The TPP is levied on a wide range of goods, not just fuel. According to the Special Excise Tax Code (CIEC, 2023), in its article 88, are included in this tax petrol products and energy, any other product that is used, placed for sale or consumption as a fuel, other hydrocarbons and electricity. The revenues generated by the application of the TPP will be partially transferred to the Permanent Forest Fund, in order to contribute to the protection and recovery the forest area (Fusco, 2016).

The largest contribution to the TPP tax revenue comes from fuel prices: diesel and petrol. This is an integrated tax that is included in fuel prices. The TPP is special because it integrates in its tax burden the component of the Special Excise Tax, and the Value Added Tax, since the Special Excise Tax is levied through the Value Added Tax.

In its origins, the TPP, just like other excise taxes, was designed to change a harmful behavior by making taxpayers pay for the costs they cause to the environment and public health through the emission of air pollutants (Amorim, 2018).

The taxes provided for in the Special Excise Tax Code (CIEC, 2023) are intended to act as conditioners of consumption, and if we examine the behavior of consumers after an increase in the tax burden, we see a significant decrease in the purchase of tobacco and alcohol. However, in the case of fuel, the increase in the tax may lead to a slight decrease in consumption, but this is not the only factor determining this change (Fusco, 2016; INE, 2022).

In 2021, as mentioned above, the TPP represented 70 percent of the environmental tax burden. However, as of March 2022, in order to combat the spiraling inflation in fuel prices, the Portuguese State declared, as an exceptional measure, it would review the TPP on a weekly basis.

Single Circulation Tax

The SCT is levied on every single vehicle with an active number plate in Portugal. This tax is paid once a year in order to compensate the environmental damage and the road deterioration caused by the use of the vehicle. The revenue generated by the SCT goes to the municipality where the vehicle is registered, but also to the State in its environmental component. However, the SCT does not only apply to road vehicles, as ships and aircraft are also covered by the Single Circulation Tax Code (CIUC, 2023).

Two components are taken into account when determining the amount of the SCT: the cylinder capacity of the vehicle and the carbon dioxide emissions. The amounts are usually updated every year (CIUC, 2023).

The STC tax is consistent with the polluter pays principle, so the tax is higher for vehicles that are considered more polluting and lower for those that are less polluting (Ochôa, 2014; Amorim, 2018). Electric cars, for example, are exempt from the SCT.

The Motor Vehicle Tax

As previously mentioned, the motor vehicle tax makes taxpayers pay for their behavior. It is paid every time a new number plate is issued in Portugal, and the revenue is used mainly to pay for environmental damage (CISV, 2023; Amorim, 2018). The motor vehicle tax considers the cylinder capacity of the vehicle and an environmental component, which is determined as a fraction of the carbon dioxide emissions (CISV, 2023).

As in the case of the SCT, the motor vehicle tax is in line with the polluter pays principle, as vehicles with more powerful engines have a higher cost due to their greater environmental impact. As in the case of the STC, the electric vehicles are exempted from the motor vehicle tax, because their emissions are not harmful to the environment (CISV, 2023; Fusco, 2016).

Energy Tax, Pollution Tax, and Resource Tax

Energy tax, pollution tax and resource tax represent a significant part of the final price paid by the consumer and can have a significant impact on the consumer's final decision. These taxes are a subcategory of the Special Excise Tax Code (CIEC, 2023) and they revert to the Portuguese State.

The energy tax covers electricity and natural gas. It is reflected in different rates of energy taxation, with significant differences between household and industrial use, between different energy sources such as diesel and other fuels, and with rates that bear little relation to the energy content or externalities such as carbon dioxide emissions or air pollution.

The pollution tax is levied on the amount of pollution emitted by a given company, which creates an incentive to find ways of reducing emissions in order to reduce its fixed costs (Cardoso, 2015).

The resource tax is included in the water bill and is intended to compensate the entity responsible for providing the service to the consumer. The water resource tax consists of two sub-taxes, the first for water and the second is for the basic sanitation (Amorim, 2018).

Corporate Income Tax

In this section we address autonomous taxation and the deductibility of vehicle depreciation about light passenger vehicles.

Autonomous taxation is an additional form of taxation in Portugal, provided for in the Corporate Income Tax Code (CIRC, 2023) and in the Personal Income Tax Code (CIRS, 2023). With this type of taxation, the legislator's main objective is to penalize and prevent certain abuses by taxpayers in the pursuit of fiscal justice (Sousa & Silva, 2017).

This tax is indirect in nature and completely independent ("autonomous") of income, since the taxpayer is subject to autonomous taxation whether or not he makes a profit at the end of the economic period (Mesquita, 2014; Palma, 2012).

The enlargement of the autonomous taxation base and the respective nominal rates had the clear purpose of obtaining more tax revenue but also considered the sustainability and environment component (Nabais, 2015).

The expenses with motor vehicles are the main target of the autonomous taxation in the Corporate Income Tax Code, since most companies have motor vehicles in their assets, and in most cases, it is difficult for the legislator to understand who benefits from the use of these vehicles – the company or the employee (Sousa & Silva, 2017). As a rule, expenses with these vehicles are deductible, autonomous taxation acts as a penalty. For example are subjected to autonomous taxation depreciation, rental, leasing, insurance, maintenance, repairs, fuel, and taxes. Rates depending on the type of vehicle and the acquisition cost and are regardless the year of acquisition (Amorim, 2018; CIRC, 2023).

At the time we are writing this chapter, the autonomous taxation rates for vehicles by type, established by article 88 of Corporate Income Tax Code (CIRC, 2023), are those presented in Table 1. It should be noted, however, that the autonomous taxation rates are updated very frequently.

Table 1. Autonomous taxation rates by acquisition cost and by type of vehicle

Acquisition cost (euros)	Type of vehicle		
	Hybrid Plug-in [(1)] Compressed Natural Gas	**Exclusively Electric**	**Other**
Less than 27.500	2,5%	-	10%
[27.500; 35.000[	7,5%	-	27,5%
Equal to or greater than 35.000	15%	-	35%
Greater than 62.500	-	10%	35%
[(1)] That has a minimum electric autonomy of 50 km and official emissions of less than 50gCO_2/km.			

Source: Based on Corporate Income Tax Code (CIRC, 2023)

As can be seen in Table 1, the autonomous taxation rates increase from hybrid plug-in vehicles or natural gas vehicles to vehicles using "other" fuels. Furthermore, electric vehicles are not subject to any autonomous taxation if their purchase cost does not exceed 62,500 euros.

The type of fuel or energy used by the motor vehicle to operate its engine is now a determining factor in the application of autonomous taxation rates, and companies must take this information into account when deciding on which vehicles are best for their business (Palma, 2012; Amorim, 2018).

In the context of sustainability, autonomous taxation does not aim to penalize bad behavior, but to stimulate the search for more environmentally friendly options, while at the same time generating revenue for the State (Mesquita, 2014).

The autonomous taxation is not the end of the penalization for motor vehicles. The Portuguese State has set limits on the deductibility of depreciation for vehicles, as stated in article 34 of Corporate Income Tax Code (CIRC, 2023). The limits have been set by Order n.º 467/2010 and by the Law n.º 82-D/2014 (with the Green Tax Law), as presented in Table 2.

Table 2 shows the depreciation limit per type of motor vehicle, taking into account the year of acquisition of the vehicle. Note that before 2015, only electric vehicles had a depreciation limit above 25,000 euros, but now the depreciation limits are different for each type of motor vehicle.

It should be noted that the types of vehicles are currently grouped differently for the purpose of defining autonomous taxation rates (Table 1) and depreciation limits (Table 2).

Table 2. Depreciation limits by type of vehicles and acquisition year

Type of Vehicle	Depreciation limits (euros)	
	Vehicles acquired between 2012 and 2014	Vehicles acquired from 2015
Exclusively Electric	50,000	62,500
Hybrid Plug-in	25,000	50,000
Compressed Natural Gas	25,000	37,500
Other	25,000	25,000

Source: Based on Order n.° 467/2010 and Law n.° 82-D/2014.

Single Use Plastics

In 2015, another tax was introduced, the light plastic bag tax, as part of the Green Tax Reform implemented by Law n.° 82-D/2014. This aim of this measure was to promote more sustainable habits and reinforce the responsibility of citizens and companies. The main objective was to significantly reduce the annual consumption of plastic bags, from 466 bags per capita in 2014, to only 50 and 35 bags per capita in 2015 and 2016.

The final consumer should be charged 10 cents (0.10 euro) for each light plastic bag. The revenue from single-use plastic bags is shared among many entities, and only 75% of the revenue goes to the Portuguese State.

Luís et al. (2020), in their study on the impact of the plastic bag taxation in Portugal, found a 94 percent decrease in plastic bags consumption between 2014 and 2016. In addition, the authors found that citizens are open to extend the plastic bag taxation on to other types of plastic bags and products.

In recent years, there have been other legislative changes aimed at reducing the use of plastics and other single-use materials. The Decree-Law n.° 102-D/2020 prohibited the availability of any type of bag at the checkout, regardless of the material. The Decree-Law n.° 78/2021 introduced a fee for single-use plastics, for packaging made of plastic or multi-material with plastic, such as bottles and takeaway trays, and, at the end of 2023, this fee will also include packaging made of aluminium or multi-material with aluminium.

Green Tax Benefits

As shown in Table 1, electric vehicles with a purchase price until 62,500 euros are not taxed autonomously. In the case of plug-in hybrid cars that have a minimum range of 50 km in electric mode and official emissions below 50 gCO_2/km, autonomous taxation rates are lower compared to vehicles powered by other fuel (CIRC, 2023). Still at the corporate income tax, as we showed in Table 2, less polluting vehicles are more likely to have their depreciation expenses accepted for tax purposes (Order n.° 467/2010; Law n.° 82-D/2014; CIRC, 2023).

In the case of Value Added Tax, both for electric cars for companies and for plug-in hybrids, there is a full deduction of the tax, involving expenses related to acquisition, provided that the acquisition cost respects the limits expressed in Table 2. The cars that use compressed natural gas or the liquefied petroleum gas can deduct 50% of tax, when then acquisition cost agree with those limits. Other important situation is the type of fuel, that can be possible deduction of the tax in 50% (most situations), 100% (energy) and 0% to light passenger vehicles gasoline powered (article 21 of Value Added Tax Code) (CIVA, 2023).

For companies, the exclusively electric cars are exempt from paying the motor vehicle tax. Hybrid vehicles also have several benefits. On the one hand, the cars with a minimum range of 50 km in electric mode and official emissions below 50 gCO_2/km benefit from a 75% reduction. On the other hand, the reduction is 40% if these vehicles have a minimum range of 50 km in electric mode and official emissions below 50 gCO_2/km (CISV, 2023).

The Single Circulation Tax Code (CIUC, 2023) establishes exemption for exclusively electric cars.

The Tax Benefits Statute (EBF, 2023) provides for tax benefits for urban buildings dedicated exclusively to the production of energy from renewable sources. In fact, land located in an urban agglomeration and not intended for construction is subject to a 50% reduction in the municipal property tax rate, for 5 years (article 44-A).

Other environmental benefits are provided for in article 44-B of the Tax Benefits Statute (EBF, 2023). Municipalities may, by deliberation of the municipal assembly, set a reduction of up to 25% of the municipal property tax rate, applicable to energy-efficient urban buildings. In addition, Municipalities, through the deliberation of the municipal assembly, may establish a reduction, up to 50% of the municipal property tax rate, to apply to rustic buildings integrated in classified areas that provide services of ecosystem that cannot be appropriated by the market. The duration of this benefit is 5 years. For this purpose, these buildings must be recognized as such by the Institute for the Conservation of Nature and Forests, I.P..

CONCLUSION

In this chapter, we will look at some of the most burdensome environmental taxes in Portugal, from the tax on petroleum and energy products, to the single circulation tax, the vehicle tax, the energy tax, the resource and pollution tax, the autonomous taxation of vehicles and depreciation limits, and even the tax on plastic bags.

From the analysis carried out, we conclude that the Portuguese State lacks uniqueness when it comes to environmental taxes and ends up taxing the same type of behavior with different taxes. This is particularly evident when it comes to the taxation of light passenger cars, which are taxed in the context of corporate income tax (autonomous taxation and deductible depreciation) and value added tax (choice of car type and fuel to full or partial deduction of the tax). It was also found that the contribution of cars to the Portuguese environmental tax burden is very significant. Nevertheless, the Tax Benefits Statute provides for green tax benefits related to urban buildings.

Little is known about the impact of green taxation on the environment and on the choices made by firms and individuals in their investment and consumption decisions. The difficulty of measuring the impact of a such complex puzzle of taxes, benefits, exemptions, and refunds certainly contributes to the lack of research in this area.

The main contribution of this study is to provide academics and practitioners with more information on environmental taxation, allowing them to choose the best strategy for promoting environmental sustainability through the application of efficient environmental taxation. As far as regulators are concerned, we hope that more articulated green taxation strategies can be promoted in the future.

REFERENCES

Abreu, L. C. (2011). *Fiscalidade e protecção do ambiente*. In G. Teixeira & A. F. Cunha (Eds.), *I Congresso de Direito Fiscal* (pp. 233-245). Vida Económica.

Amorim, J. D. C. (2018). Os incentivos fiscais em matérias de fiscalidade verde. *RCEJ Rebules, 30*, 303–318. https://parc.ipp.pt/index.php/rebules/article/view/3165/1085

Avison, Z. (2022). *A green tax road map for the future*. Green Alliance. https://green-alliance.org.uk/wp-content/uploads/2022/11/A-green-tax-roadmap-for-the-future.pdf

Aydin, C., & Esen, Ö. (2018). Reducing CO_2 emissions in the EU member states: Do environmental taxes work? *Journal of Environmental Planning and Management, 61*(13), 2396–2420. doi:10.1080/09640568.2017.1395731

Cardoso, A. C. (2015). *O imposto ambiental como instrumento de desenvolvimento sustentável*. [Master's thesis, FDUP]. https://repositorio-aberto.up.pt/bitstream/10216/81181/2/37025.pdf

Castiglione, C., Infante, D., Minervini, M. T., & Smirnova, J. (2014). Environmental taxation in Europe: What does it depend on? *Cogent Economics & Finance, 2*(1), 967362. doi:10.1080/23322039.2014.967362

Chien, F., Ananzeh, M., Mirza, F., Bakar, A., Vu, H. M., & Ngo, T. Q. (2021). The effects of green growth, environmental-related tax, and eco-innovation towards carbon neutrality target in the US economy. *Journal of Environmental Management, 299*(1), 113633. doi:10.1016/j.jenvman.2021.113633 PMID:34492439

CIEC. Special Excise Tax Code. (2023). CIEC. https://www.pgdlisboa.pt/leis/lei_mostra_articulado.php?nid=1598&tabela=leis&so_miolo=

CIRC. Corporate Income Tax Code. (2023). CIRC. https://info.portaldasfinancas.gov.pt/pt/informacao_fiscal/codigos_tributarios/Cod_download/Documents/CIRC.pdf

CIRS. Personal Income Tax Code. (2023). CIRS. https://info.portaldasfinancas.gov.pt/pt/informacao_fiscal/codigos_tributarios/Cod_download/Documents/CIRS.pdf

CISV. Motor Vehicle Tax Code. (2023). CISV. https://www.pgdlisboa.pt/leis/lei_mostra_articulado.php?nid=976&tabela=leis

CIUC. Single Circulation Tax Code. (2023). CIUC. https://www.pgdlisboa.pt/leis/lei_mostra_articulado.php?nid=976&tabela=leis

CIVA. Value Added Tax Code. (2023). CIVA. https://info.portaldasfinancas.gov.pt/pt/informacao_fiscal/codigos_tributarios/Cod_download/Documents/CIVA.pdf

CRP. Constitution of the Portuguese Republic. (2005). CRP. https://www.pgdlisboa.pt/leis/lei_mostra_articulado.php?tabela=leis&nid=565&pagina=1&ficha=1

Decree-Law n.º 102-D/2020, 10th December, General waste management regime. https://www.pgdlisboa.pt/leis/lei_mostra_articulado.php?nid=3398&tabela=leis&so_miolo=S

Decree-Law n.o 78/2021, 24th September, Transposes the Directive (EU) 2019/904 on the reduction of the impact of certain plastic products on the environment. https://files.dre.pt/1s/2021/09/18700/0000300027.pdf

Tax Benefits Statute. (2023) EBF. https://info.portaldasfinancas.gov.pt/pt/informacao_fiscal/codigos_tributarios/Cod_download/Documents/EBF.pdf

Eisenack, K., Edenhofer, O., & Kalkuhl, M. (2012). Resource rents: The effects of energy taxes and quantity instruments for climate protection. *Energy Policy*, *48*, 159–166. doi:10.1016/j.enpol.2012.05.001

Esen, Ö., Yıldırım, D. Ç., & Yıldırım, S. (2021). Pollute less or tax more? Asymmetries in the EU environmental taxes – Ecological balance nexus. *Environmental Impact Assessment Review*, *91*, 106662. Advance online publication. doi:10.1016/j.eiar.2021.106662

European Commission. (2021). *Special Eurobarometer 513: Climate change*. EC. https://europa.eu/eurobarometer/surveys/detail/2273

European Court of Auditors. (2021). *Special report 12/2021: The polluter pays principle: Inconsistent application across EU environmental policies and actions*. ECA. https://www.eca.europa.eu/en/publications?did=58811

European Environment Agency. (2005). *Market-based instruments for environmental policy in Europe*. EEA. https://www.eea.europa.eu/publications/technical_report_2005_8

European Union. (2011). *Regulation (EU) n.º 691/2011 of the European Parliament and of the Council of 6 July 2011 on European environmental economic accounts*. EU. https://eur-lex.europa.eu/legal-content/EN/TXT/?uri=celex:32011R0691

European Union. (2012). *Consolidated version of the treaty on the functioning of the European Union*. EU. https://eur-lex.europa.eu/LexUriServ/LexUriServ.do?uri=CELEX:12012E/TXT:en:PDF

Eurostat. (2010). *European System of Accounts – ESA 2010*. Eurostat.https://ec.europa.eu/eurostat/web/products-manuals-and-guidelines/-/KS-02-13-269

Fusco, G. (2016). *Impostos especiais sobre o consumo em Portugal: A extrafiscalidade e a saúde*. [ISEG, Master's thesis]. http://hdl.handle.net/10400.26/14263

Hassler, J., Krusell, P., & Olovsson, C. (2021). Suboptimal climate policy. *Journal of the European Economic Association*, *19*(6), 2895–2928. doi:10.1093/jeea/jvab048

INE (Portuguese National Statistics Institute). (2022). *Environmentally related taxes reached 5 billion euro – 2021*. INE. https://www.ine.pt/xportal/xmain?xpid=INE&xpgid=ine_destaques&DESTAQUESdest_boui=414630553&DESTAQUESmodo=2&xlang=en

Law n.º 82-D/2014, 31st December, Green Tax Law. https://www.pgdlisboa.pt/leis/lei_mostra_articulado.php?nid=2253&tabela=leis&so_miolo=

Liu, A. A. (2013). Tax evasion and optimal environmental taxes. *Journal of Environmental Economics and Management*, *66*(3), 656–670. doi:10.1016/j.jeem.2013.06.004

Luís, S., Roseta-Palma, C., Matos, M., Lima, M. L., & Sousa, C. (2020). Psychosocial and economic impacts of a charge in lightweight plastic carrier bags in Portugal: Keep calm and carry on? *Resources, Conservation and Recycling*, *161*, 104962. doi:10.1016/j.resconrec.2020.104962

Luppi, B., Parisi, F., & Rajagopalan, S. (2012). The rise and fall of the polluter-pays principle in developing countries. *International Review of Law and Economics*, *32*(1), 135–144. doi:10.1016/j.irle.2011.10.002

MAOTE. Ministry of the Environment, Spatial Planning and Energy. (2015). *Reforma fiscalidade verde*. MAOTE. https://www.crescimentoverde.gov.pt/wp-content/uploads/2014/10/ReformaFiscalidade-Verde_GreenTaxReform_emagazine.pdf

Mesquita, M. (2014). *A tributação autónoma no CIRC: A sua (in)coerência* [Master's thesis, UCP]. https://repositorio.ucp.pt/handle/10400.14/16231

Moreira, T., & Fernandes, F. (2022). *Evolution of green taxation in Portugal*. International Law Office. https://www.vda.pt/xms/files/05_Publicacoes/2022/Artigos/Evolution_of_green_taxation_in_Portugal.pdf

Nabais, J. (2015). *Direito Fiscal*. Almedina.

Novais, A. (2022). *Os impostos ambientais enquanto instrumento de proteção ambiental*. [Master's thesis, ISCAL]. http://hdl.handle.net/10400.21/15310

Ochôa, J. (2014). Breve reflexão sobre a incidência subjetiva do imposto único de circulação. *Cadernos de Justiça Tributária, 6*(Outubro-Dezembro), 3-15. https://www.tfra.pt/wp-content/uploads/2020/08/Breve-reflexao-sobre-a-incidencia-sujectiva-do-Imposto-Unico-de-Circulacao-1.pdf

Oliveira, V. (2016). *O impacte das tributações autónomas nas microentidades*. [Master's thesis, UA]. http://hdl.handle.net/10773/16772

Order n.º 467/2010, Definition of the acquisition cost or revaluation value of passenger or mixed-use vehicles. https://www.pgdlisboa.pt/leis/lei_mostra_articulado.php?artigo_id=2253A0024&nid=2253&tabela=leis&pagina=1&ficha=1&nversao=

Palma, C. (2012). As tributações autónomas vistas pelo Tribunal Constitucional. Comentário ao acórdão do Tribunal Constitucional n.o 310/2012, de 20 de junho de 2012. *Revista de Finanças Públicas e Direito Fiscal*, *5*(2), 241–255.

Pinto-Bazurco, J. F. (2022). How to enforce the polluter-pays principle. *International Institute for Sustainable Development Earth Negotiations Bulletin*, 1–8. IISD. https://www.iisd.org/system/files/2022-02/still-one-earth-polluter-pays-principle.pdf

Rebelo, A. R. B. (2016). *Taxa sobre sacos de plástico leves e proibição da disponibilização gratuita de sacos de caixa: perceções e impacto nos hábitos de consumo dos portugueses*. [Master's thesis, ISCTE]. https://www.iscte-iul.pt/tese/13754

Santos, M. L. (2021). *O setor dos resíduos em Portugal: o impacto da reforma da fiscalidade verde e da regulação sobre os consumidores finais* [Master's thesis, FEP]. https://hdl.handle.net/10216/136556

Sousa, A. S. N., & Silva, A. F. J. (2017). As tributações autónomas no sistema fiscal Português. *XVI Congresso Internacional de Contabilidade e Auditoria*. OCC. https://www.occ.pt/pt/dtrab/trabalhos/listacc/trabalhoscc.php

Sousa, S. (2021). Environmental taxation in Portugal: A contribution to sustainability. In M. H. Bilgin, H. Danis, E. Demir, & S. Vale (Eds.), *Eurasian Economic Perspectives* (pp. 369–382). Springer. doi:10.1007/978-3-030-63149-9_23

KEY TERMS AND DEFINITIONS

Autonomous Taxation: This taxation is levied on certain costs borne by companies which, by their very nature, may have a more ambiguous link with the realization of taxable income or the maintenance of the source of production. Autonomous taxation is intended to discourage certain types of excessive expenditure and their negative impact on the generation of tax income.

Environmental Sustainability: The ability to maintain the ecological balance of the planet by conserving natural resources and protecting global ecosystems that support the health and well-being of present and future generations.

Environmental Taxation (or Green Taxation): Consists of using taxes and fees (negative incentives), or through tax benefits (positive incentives), to encourage taxpayers to adopt more environmentally friendly production and consumption patterns. Environmental or green taxes include taxes on energy, transport, pollution, and resources. It is seen by the State as an opportunity to adapt the tax system to a more resource-efficient economy, making it innovative, low-carbon, and more sustainable. It can take the form of tax incentives.

Green Tax Reform: The main objective is to contribute to a healthier environment and a more sustainable world, and this reform aims to achieve this with the support of other policies, that are directly or indirectly related to the environment. The assumptions for the Green Tax Reform are to achieve the triple dividend: 1) to achieve fiscal neutrality, which means that the net increase in the revenue that derives from the new environmental taxes created by the law must be used to reduce the revenue that derives from other types of taxes, particularly income taxes; 2) to protect the environment and dependence on foreign energy, to promote growth and empowerment; and 3) to contribute to fiscal responsibility and reduce external imbalances.

Green Tax Benefits: A set of situations, defined by the legislator, that allow taxpayers to benefit from certain environmentally friendly goods in their activity, by paying no tax or a lower amount of tax.

Hybrid Electric-Petroleum Vehicles: Vehicles powered by an internal combustion engine and one or more electric motors, which use energy stored in batteries.

Polluter Pays Principle: The polluter pays principle states that those responsible for polluting the environment should bear the costs of implementing the pollution prevention and the control measures put in place by public authorities to ensure an acceptable state of the environment.

Social Function of Taxes: The social function of taxes is extended to the environmental area, so that the tax can also be used as an indirect form of environmental protection, to avoid economic, social, and environmental risks, and to safeguard resources for future generations.

Chapter 14
The Role of Sustainability Reporting in Corporate Tax Transparency

Sónia Monteiro
Research Centre on Accounting and Taxation, Polytechnic Institute of Cávado and Ave, Portugal

Beatriz Aibar Guzmán
Santiago de Compostela University, Spain

Isabel-María García Sánchez
https://orcid.org/0000-0003-4711-8631
Universidad de Salamanca, Spain

ABSTRACT

Tax transparency is often associated with standardized reporting. Among the goals of the 2030 agenda, goal 12.6 encourages organizations to adopt sustainable practices and integrate sustainability information into their reporting cycle. The Global Reporting Initiative (GRI) published the GRI 207- Tax (2019), aiming to increase transparency and accountability in corporate tax practices, by providing disclosure guidance for companies to disclose information about their tax strategy, governance, and risk management. As it is a recent standard, research on GRI-207 is still scarce. Thus, this chapter aims to contribute to the debate on the link between corporate reporting and fiscal responsibility by providing an overview on the background and structure of the GRI 207, which could serve as a basis for future research in the field of tax disclosure, as well as contributing to the policy debate. This chapter could have political and practical implications, as transparency in the disclosure of tax-related issues in corporate reporting could mitigate the reputational tax risk of companies.

DOI: 10.4018/978-1-6684-8592-7.ch014

INTRODUCTION

Tax, fair share, and tax avoidance issues have been increasingly debated over the past few years. Companies perceived as corporate taxpayers may employ aggressive tax avoidance measures by exploiting loopholes in regulations. Tax avoidance is deemed socially irresponsible when it involves schemes, strategies, or transactions without economic substance and can only be justified by the desire to minimize tax liabilities (De la Cuesta-González & Pardo, 2019). In some cases, companies try to reduce their tax liabilities by entering the field of tax evasion through illegal and intentional actions in order to minimize the amount of taxes to be paid (Alm & Torgler, 2011).

These behaviors, along with the associated corruption, represent a major obstacle to the political, social and economic development of countries (Branco & Delgado, 2014), so national and international regulatory bodies have developed several initiatives to reduce and/or eradicate them. However, corporate social responsibility (CSR) can also play a key role in this regard, by reinforcing information transparency.

The need for transparency and social responsibility are the cornerstone topics within this global debate (Albuquerque & Neves, 2021). Paying taxes is seen as an element of corporate social responsibility (CRS). It is an important way for firms to positively engage with society (Zeng, 2018). Paying taxes without engaging in aggressive tax planning practices is considered part of corporate accountability and can led to a positive relationship with the government, as well as a favorable public perception of companies (Triwacananingrum & Wijaya, 2022)

Effective governance at the country level helps to establish trust with all stakeholders and requires companies to provide more detailed information and transparency about their tax and CSR data (Cahan et al., 2016). Transparency is directly related to disclosure. In this regard, tax transparency has been defined as the state or outcome achieved by disclosing tax-related information, where tax disclosure is defined as the communication of tax-related information, from an issuer to one or more recipients, either on a mandatory or voluntary basis (Müller et al., 2020).

Public disclosure of information related to a firm's tax activities has become an increasingly significant aspect of a company's overall sustainability strategy and reporting (Adams et al., 2022). Sustainability reports are one of the ways in which companies disclose their CSR activities regarding environmental, social, and governance issues (ESG). As far as tax transparency in non-financial statements is concerned, the existing most influential CSR standards are by the Global Reporting Initiative (GRI), an international non-profit organization that issues the most widely adopted sustainability reporting standards.

In 2019, GRI updated its standards to include GRI 207: Tax 2019, which covers for reporting on tax practices. This standard has been in force since January 2021. Its purpose is to standardize the disclosure of information published in sustainability and CSR reports on the fiscal approach to governance and tax control, risk management and stakeholder engagement. It includes disclosures on tax strategy, governance, and risk management that meet different stakeholder reporting expectations. In addition to qualitative information on the approach to tax, tax governance, and risk management that meet different stakeholder reporting expectations, this standard requires public country-by-country reporting (CbCR), with a comprehensive list of tax-related items to be recorded for each country.

Much of debate about taxes and CSR focuses on the practice of corporate tax avoidance. The linkage between corporate tax avoidance and to tax disclosures in CSR reports is contentious. On one hand, some studies focusing on the relationship between CSR/tax disclosure and tax avoidance found that companies with aggressive tax strategies have higher levels of disclosure in an attempt to change stakeholders' perceptions (Lanis and Richardson, 2013; Santoso et al., 2019; Kao & Liao, 2021). Tax avoiders may try to improve their reputation by making socially responsible tax disclosures, whereas firms that engage in less tax avoidance are generally less motivated to disclose their tax information (Hardeck et al., 2020).

On the other hand, according to some studies, sustainability reporting has a negative effect on tax aggressiveness (Supriyati & Anggraini, 2021), so that firms who engage in tax avoidance schemes may be less likely to provide sustainability reports (Reiter, 2020; Ikoma, 2022). When CSR is of a higher quality, with more disclosures and transparency, it is less likely to be used as a smokescreen to conceal underlying tax avoidance behavior (Zeng, 2018). Watson (2015) argues that socially responsible firms tend to attract consumers and investors who share similar norms and values. As a result, these firms are more likely to disclose CSR information, to legitimize themselves before society, and additionally these firms are less likely to engage in tax-aggressive activities (Adams et al., 2022). Legitimacy theory suggests that aggressive tax planning is a violation of the social contract (low tax-paying firms are not paying their fair share). This leads to the prediction that tax-aggressive firms will disclose more information in an effort to retain/regain their legitimacy and/or to mitigate information asymmetries arising from the complexity of tax planning (Hardeck et al., 2020; Adams., 2022; Müller et al., 2020; Triwacananingrum & Wijaya, 2022).

As it is a recent standard, research on GRI-207 is still scarce, so the aim of this paper is to explore the new GRI standard on corporate tax reporting (GRI 207), analyzing what tax information companies should publicly disclose to their stakeholders in sustainability reports, as a mechanism for accountability and tax transparency. Thus, this paper aims to contribute to the debate on the link between corporate reporting and fiscal responsibility by providing an overview on the background and structure of the GRI 207 standard, which could serve as a basis for future research in the field of tax disclosure in sustainability reporting as well as contribute to the policy debate in the near future (Schnighter et al., 2021). Therefore, this paper could have political and practical implications, as transparency in the disclosure of tax-related issues in corporate reporting could mitigate the reputational tax risk of companies.

The remainder of the paper is structured as follows. In Section 1, we define tax transparency and its relationship with SR. Section 2 provides an overview of corporate tax disclosures regulatory initiatives. Section 3 illustrates the background and structure of the GRI 207 standard. Section 4 reviews the existing empirical literature on this topic. We conclude this chapter with the final remarks and some suggestions for future research.

CSR AND TAX TRANSPARENCY

Some major businesses are beginning to realize that taxation is an essential part of their social responsibilities. Corporate social responsibility (CSR) is a management concept that refers to the voluntary actions taken by companies to address social, environmental, and ethical issues in their operations and throughout their supply chain. CSR involves the integration of social and environmental concerns into business operations and decision-making processes, with the aim of creating sustainable value for all stakeholders, including employees, customers, shareholders, suppliers, and the wider society. Taxation, however, is the process by which governments collect revenue to fund public goods and services.

There is a close relationship between CRS and taxation, as taxation is one of the primary tools that governments use to promote social responsibility. By collecting taxes, governments can fund programs and services that benefit society, such as education, healthcare, infrastructure, and social welfare programs. In addition, taxation can be used to encourage social responsibility by incentivizing certain behaviors and discouraging others. For example, governments can provide tax breaks or incentives to businesses that adopt sustainable practices or invest in renewable energy sources. They can also impose taxes or penalties on businesses or individuals that engage in activities that are harmful to society, such as pollution or tax evasion.

Furthermore, social responsibility can also be reflected in the way that taxes are structured and collected. For instance, progressive taxation, where individuals with higher incomes pay a greater share of their income in taxes, can help promote greater equality and reduce social inequality. CSR implies recognition of the importance of disclosing information related to the various facets and areas of business activity, including taxation (Oats & Tuck, 2019). Moreover, "tax payments are often considered a fundamental [...] example of a company's citizenship behavior" (Dowling, 2014). Additionally, tax transparency and fairness can help to promote public trust and confidence in the tax system, which is an important aspect of social responsibility.

Information plays a crucial role in establishing an effective tax system, as taxes can only be enforced efficiently when authorities possess sufficient information on the nature and value of taxable transactions. In this regard, stakeholders and governments, are increasingly urging companies to disclose their tax-related information publicly (De la Cuesta-González & Pardo, 2019).

For the International Monetary Fund (IMF, 2022a), corporate transparency through the disclosure of reliable financial and non-financial information is crucial to restore investor and market confidence, especially in contexts of economic and financial crises. In particular, tax transparency is key to economic and financial stability by increasing confidence and strengthening fiscal credibility (IMF, 2022b).

Internationally, the focus on tax transparency reporting has resulted in Base Erosion and Profit Shifting (BEPS)[1] by large multinational firms (Stiglingh et al., 2017). BEPS refers to a set of practices that multinational companies use to exploit gaps and inconsistencies in tax rules and regulations in different countries, often using complex corporate structures and legal arrangements to reduce their tax bills. The BEPS project aims to ensure that all companies, regardless of their size, pay their fair share of taxes in the countries where they do business, and to prevent the erosion of tax bases and the shifting of profits to low-tax jurisdictions.

The Organization for Economic Cooperation and Development (OECD) developed a framework for addressing BEPS, which involves the implementation of several actions to enhance tax transparency between OECD and non-OECD member countries. According to Action 12 of the OECD (2013), taxpayers are required to disclose more specific information about their tax planning strategies. Furthermore, disclosing initiatives are recognized as an effective measure to enhance communication about tax risks to tax administrations and policymakers. Action 13 is another measure that focuses on tax transparency and aims to promote regulations regarding transfer pricing documentation (OECD, 2013). Although many of the BEPS Actions relate to country-by-country reporting (CbCR), disclosures concerning tax strategies (including tax planning) and transfer pricing policies apply at company level.

The public disclosure of disaggregated corporate tax information is a pertinent issue being pursued by various European Union's (EU) initiatives aimed at combating tax avoidance practices and preventing corruption in industries with high exposure, such as the extractive and financial services industries (De la Cuesta-González & Pardo, 2019). The revised EU's Accounting and Transparency Directive, Directive 2013/50/EU, introduces specific disclosure requirements for payments to governments by listed and large non-listed companies with activities in the extractive industry and the logging of primary forests. Regarding the financial services industry, the EU Capital Requirement Directive, Directive 2013/36/EU, requires European credit institutions to disclose tax-related data on a country-by-country basis in their annual financial statements, as well as other information necessary for contextualization. Financial institutions are required to report information on the interest income earned by their clients to the relevant tax authorities.

In addition, the EU has adopted the Common Reporting Standard (CRS), which is a global agreed standard for the automatic exchange of financial account information between tax authorities. It was developed by the OECD and approved by its Council on July 15, 2014. Under the CRS, financial institutions are required to identify and report information on the financial accounts held by non-resident individuals and entities to the relevant tax authorities. The aim of the CRS is to prevent tax evasion and improve the transparency of the global financial system.

Considering that "transparency is a crucial element in securing fairer taxation" (EC, 2015), the European Commission established a Tax Transparency Package in 2015 to promote the disclosure of information related to compliance with good tax governance standards. Furthermore, the Anti-Tax Avoidance Directive (ATAD) was introduced in 2016 and aims to combat aggressive tax planning and ensure a level playing field for businesses in the EU. The ATAD sets out a number of measures to tackle tax avoidance, such as rules on controlled foreign companies, interest limitation, and exit taxation. Financial institutions are required to comply with these rules and report any relevant information to the tax authorities.

Recently, the Directive (EU) 2021/2101 (commonly known as the public Country-By-Country Reporting (CbCR) Directive) amended the Directive 2013/34/EU regarding disclosure of income tax information by certain undertakings and branches. The Directive guarantees the publication of tax amounts and profits on a country-by-country basis (CBCR) for multinationals or stand-alone companies with consolidated revenues exceeding 750 million euros for the past two fiscal years in the EU where they have large or medium-sized entities. The CbCR Directive is part of the EU's wider efforts to combat tax avoidance and aggressive tax planning by multinational companies.

The tax transparency reporting initiatives referred to above focus either on country-by- country reporting (CbCR), or on additional reporting requirements for firms in specific sectors. Given the current global emphasis on tax transparency, the reputational risks associated with non-compliance, Stiglingh et al. (2017) contend that one effective means for firms to demonstrate their tax transparency is through disclosure in their corporate reports. This is the topic we will explore next.

CSR TAX DISCLOSURES IN CORPORATE REPORTING

Corporate reporting is crucial in the context of corporate tax responsibility, as national tax authorities can access public information and penalize companies for non-compliance regarding tax payments. Hence, from a CSR perspective, governments can promote responsible tax behavior by encouraging public disclosure of tax information (De la Cuesta-González & Pardo, 2019), as a means of promoting good tax conduct and combating tax avoidance practices.

CSR tax disclosures are becoming increasingly important for companies as stakeholders, including investors, customers, and the general public, are demanding greater transparency in corporate tax practices. By disclosing their tax-related information, companies can demonstrate their commitment to responsible tax behavior and accountability, and build trust with stakeholders.

The OECD Guidelines for Multinational Enterprises was the first instrument on CSR focusing extensively on tax matters and recognizing tax as being an important aspect of CSR (Albuquerque & Neves, 2021). The original version of the OECD Guidelines for Multinational Enterprises, issued in

1976, already included a two-paragraph section on taxation. It stated that enterprises should provide the necessary information to tax authorities and avoid abusive transfer pricing practices. The latest version of the OECD Guidelines (2011) contains a slightly more extensive chapter on taxation, advocating transparency in tax matters. It calls for cooperation with tax authorities, transparency in tax matters, and recognizes the significant financial, regulatory, and reputational risks that enterprises face in tax matters.

Incorporating taxes as a component of CSR reporting can facilitate the presentation of tax information in a more comprehensible manner. Companies that provide detailed information about their tax practices in their reports demonstrate transparent behavior to society (Scheiwiller & Symons, 2014). The sustainability report serves as a communication tool for sharing information about the different dimensions of CSR. It promotes transparency and legitimacy, which, in turn, increases trust in the organization (Zeng, 2018).

According to some scholars (Faúndez-Ugalde et al., 2022; Stiglingh et al., 2017) increasing transparency in sustainability reports could serve as an effective mechanism for mitigating the reputational tax risk. Therefore, companies that choose to incorporate tax transparency procedures into their decision-making processes through sustainability reporting can demonstrate their commitment to social well-being and enhance their reputation as an essential asset (Faúndez-Ugalde et al., 2022). Furthermore, the inclusion of information on tax issues in CSR or sustainability reports would contribute to reducing the information asymmetry gap (Gribnau & Jallai, 2019). Thus, CSR or sustainability reports appear as a vehicle that allows companies to improve the level and quality of transparency of non-financial information, including that related to tax matters (Hardeck & Kirn, 2016). According to Rudyanto and Pirzada (2021), voluntary sustainability reporting acts as an indication of an organization's sustainability which might be useful for companies to reduce shareholders' negative reaction to tax avoidance.

In the sustainability reporting field, the Global Reporting Initiative (GRI) intends to empower organizations transparency worldwide and encourage them to take responsibility for their impact through the release of standards/guidelines for their reports. Since 1997, the GRI has issued guidelines for the preparation of high-quality sustainability reports (Kundeliene & Leitoniene, 2015), making the elaboration of sustainability reports a standard practice in organizations around the world and being the most widely used reporting framework worldwide (KPMG, 2022).

In 2013, the GRI G4 was the last version launched before the United Nations 2030 Agenda adoption in 2015. GRI G4 intended to give the report a higher level of relevant and concentrated information. The GRI G4 Guidelines categorize taxation as a component of the Economic Performance Aspect and also address it under the Anti-Competitive Behaviour and Compliance Aspects under the Society sub-category.

GRI G4 includes two tax-related indicators within the economic dimension. The first is the direct economic value generated and distributed (G4-EC1), which requires companies to report, on a country-by-country basis, the payments made to governments. These include all taxes paid, such as corporate, income, and property taxes, as well as any related penalties. The GRI Implementation Manual emphasizes the importance of reporting taxes paid in each country for organizations that operate in multiple countries (GRI, 2013). The second is financial assistance received from governments (G4-EC4), which requires the reporting of the total monetary value of financial assistance received by the organization from governments. As part of the Compliance Aspect, even though it does not specifically address taxation, the G4-EN29 standard focuses on reporting significant fines and non-monetary sanctions for non-compliance.

Over the years, the GRI guidelines have evolved. These standards emerged in 2016 to replace the latest generation, G4, in response for the Sustainable development Goals (SDG) of the 2030 Agenda, namely the goal 12.6 that calls for corporate transparency mentioning the benefits of reporting sustainability actions and encouraging companies to integrate sustainability information into their reporting cycles. The GRI Standards have a modular, flexible, and inter-related set which starts with the universal standards and can be completed with a variety of other topics in economic, environmental, or social dimensions.

While G4 guidelines may not provide significant value for sustainability reporting related to tax compliance, the standard GRI 207: Tax (2019), which were introduced after the G4 Guidelines, delivers the first global standard for tax transparency. It directly addresses taxation as a CSR matter and provides more specific guidance on reporting tax-related information.

The GRI 207 (2019), oriented to multinational corporations, provides guidelines for organizations to disclose information related to their taxation policies, practices, and payments. The aim of this standard is to increase transparency and accountability in the area of taxation and to provide stakeholders with a clear understanding of an organization's tax-related activities.

Compliance with GRI is voluntary and each organization decides the material issues that will be covered in its sustainability report. If tax is a material topic for a company reporting 'in accordance' with the GRI standards, it will have to consider GRI 207: Tax 2019 and its disclosure requirement to report on tax aspects (detailed in the next point).

Nowadays, it is widely accepted that taxes are central to a country's macroeconomic stability: taxes are a vital instrument to fund defence, security, and health care. Therefore, tax collection plays a vital role in achieving the SDGs set out by the 2030 Agenda (Gribnau & Jallai, 2019; Schnighter et al., 2021). Indeed, according to the Platform for Collaboration on Tax (2018), taxation helps to finance the SDGs. The target 10.4, which is part of Goal 10 "Reduce inequality within and among countries," states that countries should "Adopt policies, especially fiscal, wage, and social protection policies, and progressively achieve greater equality." In response, many countries are now taking steps to institutionalize tax transparency.

Additionally, among the goals of the 2030 Agenda, the target 12.6 stands out, which encourages organizations to adopt sustainable practices and integrate sustainability information into their reporting cycle. Therefore, it is important to align the firm's strategy and corporate reporting with the SDGs, making stakeholders aware of the level of business commitment to the 2030 Agenda and reporting on progress in achieving the SDGs (García-Sánchez et al., 2022).

As part of the SDGs Compass, GRI and the United Nations Global Compact (UN Global Compact) provides a document "Business Reporting on the SDGs. Analysis of Goals and Targets" (2017) that links all SDGs and their specific targets to GRI indicators. The 2022 version of the document includes an updated list of disclosures per SDG which adds to the first edition. This document can be considered a guide for companies to communicate their contributions to Agenda 2030 to stakeholders in their sustainability reports, in a comparable and effective way.

In the case of tax disclosures, such a document points to a clear interconnection of GRI 2007, namely standards 207-1, 207-2, 207-3, 207-4, with some targets incorporated in SDG 1- No poverty, SDG10- Reduction of inequalities, and SDGs 17- Partnerships. Therefore, linking the SDGs to the GRI framework can be an added value (GRI, 2020). GRI 207 emphasized that achieving the SDGs is not possible without including tax as a sustainability topic, with an appropriate standard to report on.

Some authors argue that the disclosure of non-financial information in separated reports does not clarify how sustainability is meant to be integrated in the business model and the organizational strategy. Furthermore, stakeholders are increasingly demanding more rigorous, transparent and comprehensive sustainability information from organisations. For these reasons, the adoption of integrated reporting (IR) by the business community has increased (Eccles & Serafeim, 2017).

In 2013, the first IR framework was presented by the International Integrated Reporting Council (IIRC). It provides broad principles without prescribing any specific reporting requirement for the preparation of an integrated report, which is intended to inform stakeholders how organisations create value over time. For some scholars, IR is the future of business reporting since it provides more transparency about sustainability integration within organizations' activities and complements the gaps existing in separated reporting practices (Izzo et al., 2020).

Thinking about taxes in an integrated manner may result in managers considering various trade-offs. On the one hand, managers may want to minimize taxes in order to maximize firm value for shareholders, but they also need to consider the risks associated with tax audits and penalties. On the other hand, firms have a social responsibility to pay their taxes, even if it means sacrificing short-term profitability (Venter et al., 2017).

While "tax payments" is listed as one of the outcomes of integrated reporting, the IIRC Framework does not specifically refer to 'taxation' elsewhere (Stiglingh et al., 2017; Venter et al., 2017). The integrated reporting framework requires firms to consider how taxes relate to or affect the firm's ability to create or destroy value across the six forms of capital, without prescribing specific disclosure requirements (Venter et al., 2017).

Considering the above, it is worth noting that among CSR standards, only GRI 207 comprehensively and specifically addresses taxation as a CSR matter. In the following point, we will outline the structure of this standard.

TAX IN SUSTAINABILITY REPORTING: THE GRI 207 STANDARD

Since 2016, GRI has provided a comprehensive framework for sustainability reporting structured as follows: three universal standards, which apply to all organizations; sectoral standards, applicable to specific sectors; and topic-specific standards, focused on their material topics. It is worth noting that the GRI has since updated its standards, with the latest version being the GRI Standards 2021. However, the general structure and approach of the GRI Standards remains largely consistent across versions.

In the case of universal standards, they include the core reporting requirements that apply to all organizations, regardless of their size, sector, or location. The Universal Standards cover the organization's profile, governance, ethics, and management approach, as well as its impacts on human rights, labor practices, the environment, and society. The universal standards correspond to the following: GRI 101: Fundamentals, which includes essential information on how to use and make reference to the standards; GRI 102: General Contents, which concerns the disclosure of contextual information about the organization; and GRI 103: Management Focus, which is meant to provide information about the management of each thematic standard.

Sector-Specific Standards include guidance and requirements for reporting on sustainability issues that are specific to certain sectors, such as mining, oil and gas, and financial services. These standards are designed to provide more tailored guidance and requirements for organizations in these sectors.

The topic-Specific Standards include additional guidance and requirements for reporting on specific sustainability issues that are relevant to many organizations. They are organized in three series: GRI 200, economic themes; GRI 300, environmental themes; and GRI 400, social themes. The use of these standards to define which topics are material (materiality assessment) helps organizations to achieve sustainable development, by addressing issues related to management, anti-corruption policy, anti-competitive behaviors, water and effluents, security practices, and customer privacy, among others.

A company may describe its reporting as being 'in accordance' with the GRI standards, when it meets: (a) the requirements of the three mandatory universal standards and include management approach disclosures, as well as at least one topic-specific disclosure from each of the three categories: Economic, Environmental, and Social (core option) or, (b) in addition, all material topic-specific disclosures from the thirty-three different topic-specific standards (comprehensive option).

Regarding taxation, GRI Standards 2016 suggested that companies indicate their economic contribution, and they specifically mention payments to governments (disclosure 201-1), as well as any tax relief or subsidies received from governments (disclosure 201-4). In 2019, GRI updated its standards, including the tax topic index in GRI 207, which is derived from the economic issues standard series 200.

GRI 207 came into force for sustainability reporting from January 2021 (sustainability reports for the fiscal year 2020), and for the first-time tax disclosure requirements in sustainability reporting were defined, which would allow companies to be compared to each other. GRI 207 enables companies to report on tax payments on a country-by-country basis along with their approach to tax management. Thus, GRI 207 includes disclosures on the tax management approach (GRI 207- 1, 2 and 3), as well as topic-specific disclosures (GRI 207-4), as follows:

- *GRI 207-1 – Approach to tax*: Companies are required to provide general information on how they handle tax issues, including their governance and compliance approach, and how the tax, business, and sustainability strategies are linked.
- *GRI 207-2 – Tax governance, control and risk management*: Companies are required to report on their intra-company processes for managing tax risks and compliance, as well as addressing unethical and unlawful conduct within the company, and also the integration and reliability of these processes.
- *GRI 207-3 – Stakeholder engagement and management of concerns related to tax*: Companies are required to provide information about the cooperation with stakeholders such as the tax authorities, public policy advocacy, and corporate processes in order to consider the external stakeholders' concerns.
- *GRI 207-4 – Country- by-Country Reporting (CbCR):* In conjunction with Disclosure 102–45 of GRI 102: General Disclosures, companies are required to provide a list of tax jurisdictions where they are residents or have a place of management (Disclosure 207-4-a). Additionally, data reported should be based on the most recently audited or publicly available financial statements (Disclosure 207-4-b).

The reporting requirements of the GRI 207 are shown in Table 1.

EMPIRICAL STUDIES

The empirical literature linking the non-financial reporting disclosures of corporate tax and CSR has been emerging in recent years. Scheiwiller and Symons (2014) use an Anglo-American mining company as an example, as it reports the total tax contribution by country in its CSR reports as part of its economic dimension. This includes all the various taxes paid and collected, and explains how they are generated

Table 1. GRI 207 reporting requirements

Disclosure 207-1: Approach to tax A description of the approach to tax, including: i. whether the organization has a tax strategy and, if so, a link to this strategy if publicly available; ii. the governance body or executive-level position within the organization that formally reviews and approves the tax strategy, and the frequency of this review; iii. the approach to regulatory compliance; iv. how the approach to tax is linked to the business and sustainable development strategies of the organization.
Disclosure 207-2: Tax governance, control, and risk management **a)** A description of the tax governance and control framework, including: i. the governance body or executive-level position within the organization accountable for compliance with the tax strategy; ii. how the approach to tax is embedded within the organization; iii. the approach to tax risks, including how risks are identified, managed, and monitored; iv. how compliance with the tax governance and control framework is evaluated. **b)** A description of the mechanisms for reporting concerns about unethical or unlawful behavior and the organization's integrity in relation to tax. **c)** A description of the assurance process for disclosures on tax and, if applicable, a reference to the assurance report, statement, or opinion.
Disclosure 207-3: Stakeholder engagement and management of concerns related to tax A description of the approach to stakeholder engagement and management of stakeholder concerns related to tax, including: i. the approach to engagement with tax authorities; ii. the approach to public policy advocacy on tax; iii. the processes for collecting and considering the views and concerns of stakeholders, including external stakeholders.
Disclosure 207-4: Country-by-country reporting **a)** All tax jurisdictions where the entities included in the organization's audited consolidated financial statements, or in the financial information filed on public record, are resident for tax purposes. **b)** For each tax jurisdiction reported in Disclosure 207-4-a: i. Names of the resident entities; ii. Primary activities of the organization; iii. Number of employees, and the basis of calculation of this number; iv. Revenues from third-party sales; v. Revenues from intra-group transactions with other tax jurisdictions; vi. Profit/loss before tax; vii. Tangible assets other than cash and cash equivalents; viii. Corporate income tax paid on a cash basis; ix. Corporate income tax accrued on profit/loss; x. Reasons for the difference between corporate income tax accrued on profit/loss and the tax due if the statutory tax rate is applied to profit/loss before tax. **c)** The time period covered by the information reported in Disclosure 207-4.

Source: GRI 207- Tax 2019

across the life-cycle of a mining project. More recently, Stiglingh et al. (2017) found that 86% of the large firms in South Africa comply with more than 70% of the mandatory tax reporting requirements. Their results also show that 50% of the firms are transparent regarding their disclosure of tax strategy and risk management, tax figures and performance, their total tax contribution and the wider economic impact of their tax behavior. KPMG (2019) analyzes the state of corporate responsibility reporting by 4,900 companies across 49 countries. It includes a section on tax transparency, discussing the various tax-related disclosures made by companies in their corporate responsibility reports.

Muller et al. (2020) present a review of empirical evidence on the determinants, informativeness, and effects of tax-related disclosures. The evidence suggests an ambiguous relationship between tax aggressiveness and disclosure practices, reflecting a trade-off that firms face. Tax-aggressive firms may: (a) reduce the quality of, or even fully comply with, mandatory disclosures to keep sensitive information private; or (b) issue more disclosures that involve more qualitative and/or voluntary information to legitimize their tax arrangements or to reduce information asymmetry resulting from the complexity of tax planning. Regarding the informativeness of tax-related disclosures, the authors state that voluntary CbCRs, according to the new GRI Standard 207, may offer additional opportunities for future studies. Regarding the effects of tax transparency on firms and their stakeholders, on one hand, extant evidence shows that the disclosure is perceived as costly for firms; on the other, little is known about how stakeholders (e.g., investors, analysts and tax authorities) benefit or use information from tax-related disclosures.

Reiter (2020) analyzed tax disclosures in sustainability reports of Fortune 500 companies and concluded that only a few firms provide information about their tax policies. Instead, they focus on detailing the economic impacts of taxes paid and the social impacts of tax-supported programs. Neves and Albuquerque (2019) investigated the extent to which income tax disclosure was treated as a CSR issue among the most influential companies in Portugal's stock market index (PSI 20). They found that tax was not viewed as a significant concern to be addressed in the CSR reports of these companies. Only six out of the seventeen companies examined included a note on tax or a statement on tax payments in their CSR reports. Kao and Liao (2021) examine how a firm's tax disclosures in a CSR report are influenced by its tax avoidance behavior, using a sample of public U.K. firms. Their findings suggest that the positive association between tax avoidance and tax disclosures is attenuated when firms exhibit better CSR performance as well as stronger corporate governance.

Few studies highlight the importance of tax-related disclosures in sustainability reports. Faúndez-Ugalde et al. (2022) focuses on the contribution of sustainability reports to transparency in corporate decision-making and its evolution in Latin American countries. Specifically, it evaluates the sustainability reports, under the new disclosure GRI 207, of the top 30 Chilean companies listed on the stock market in 2020. The authors found that 37% of Chilean companies report sustainability reports under the GRI standard; 40% present integrated annual reports under the GRI standard, and 23% do not present sustainability reports. The study demonstrates opacity problems in their tax sustainability related to the GRI 207 standard. The results of the study indicate that the compliance of GRI 207-4 disclosures performs significantly better compared to GRI 207-1, 207-2, and 207-3 disclosures. The higher level of compliance in GRI 207-4 is due to the information provided by the financial statements, that require companies, in accordance with international accounting regulations, to consolidate their results of foreign investments. In contrast, the focus of disclosures GRI 207-1, 207-2, and 207-3 is on disclosing the tax strategy, which is viewed on a smaller scale. In contrast, in the case of GRI 207-1, 207-2, and 207-3 standards, the focus is on disclosing the tax strategy transversally, which is presented to a lesser extent. The authors conclude that taxpayers employ their CSR activities primarily to mitigate reputation risks, rather than to promote tax transparency.

Ikoma (2022) analyze the tax disclosures (based on GRI 207) in the sustainability reports published by companies from countries with high (Japan and Norway) and low (Germany and Spain) tax morale and examine its association with tax avoidance and national culture. The results show that the higher the level of tax avoidance, the lower the level of tax disclosure, and that national culture also has a direct impact on the level of tax disclosure. Adams et al. (2022) examine the association of voluntary tax disclosures in sustainability reports and annual reports captured by Corporate Register and various country-level institutional and societal characteristics. Within these reports, the authors used a content analysis based on a list of tax transparency keywords that also include GRI 207 tax reporting framework. The findings suggest that more tax aggressive firms generally provide less tax-related disclosure. Therefore, there is a negative correlation between aggressive international income shifting and the companies' decision to offer more detailed country-by-country reporting (CbCR) disclosures. These findings are consistent with the notion that firms with higher tax payments may want to signal their virtue by disclosing more tax information, or alternatively, aggressive tax strategists may prefer not to be forthcoming about tax activities that could be viewed as violations of the social contract or that may lead to costly regulatory or reputational consequences.

Supriyati and Anggraini (2021) examine the effect of sustainability reporting on tax aggressiveness, using a sample of public companies listed on the Indonesia Stock Exchange. Results prove that the disclosure of sustainability reporting has a significant negative effect on tax aggressiveness. Hardeck et al. (2020) examine influences on the decision to voluntarily include socially responsible tax disclosures in the CSR reports from 22 countries supplied to the GRI's database. Contrary to Adams et al. (2022), Hardeck et al. (2020) found no relation between tax aggressiveness and tax disclosure in CSR reports, except in the case of firms headquartered in liberal market economies (USA, Canada, and Australia). They found that these firms use socially responsible tax disclosures to offset reputational costs arising from corporate tax avoidance, whereas other firms do not appear to engage in this behavior. Those firms used these disclosures to greenwash their tax avoidance activities, which could be considered in line with legitimacy theory.

Santoso et al. (2019) examined the mediation role of assurance in the association between disclosure of social responsibility and tax aggressiveness in developing countries. They considered a sample of publicly traded companies in Thailand that issued stand-alone sustainability reports based on the GRI for the 2013-2018 period. The study found a positive correlation between CSR disclosure and tax aggressiveness. However, it was also discovered that the presence of assurance against CSR disclosure mitigated these results.

FUTURE RESEARCH DIRECTIONS

Future research should aim to further develop a deeper analysis of tax-related items in CSR/sustainability reports. It might consider a content analysis of sustainability reports published since 2021, under the new disclosure GRI 207 on tax sustainability, in order to evaluate the level of compliance of GRI disclosures, and to identify their determinants factors (which may include characteristics, such as size, activity, auditor, and country). Future research adopting qualitative methodologies like case studies could enrich the field of tax-related sustainability reports by obtaining detailed information on companies that do not disclose such information.

CONCLUSION

This research focuses on the topic of taxes from the perspective of CSR and transparency. It outlines the sustainability reporting guidelines that can be utilized in this context and provides insights into the sparse empirical research on this subject. Among the CSR standards, the GRI Standards 207 specifically and extensively address taxation as a CSR matter. They emphasize the importance of tax compliance behavior and place the topic of taxes at the forefront of CSR reporting.

The academic literature on tax as a CSR matter is emerging, with the authors accepting that tax is a concern of CSR. International research provides evidence that firms that are more socially responsible tend to be less likely to engage in tax-aggressive activities when they provide more disclosures and transparency. Socially responsible tax disclosures in CSR reports appear to be symbolic, and driven by reputational concerns. Companies are publishing more tax-related data to preserve or enhance their reputations, as investors and other stakeholders increasingly interpret the lack of transparency as a signal of irresponsible tax practices.

Research has mainly been focused on the issues of tax, tax avoidance, tax reporting, and CSR reporting separately. The current literature on tax reporting focuses mainly on the mandatory IFRS tax-related requirements disclosures in financial reports (Neves & Albuquerque, 2019; Albuquerque & Neves, 2021). It becomes evident that scholars are still only at the beginning of empirical research on tax transparency and tax-related disclosures in sustainability reports. Studies demonstrating the importance of tax reporting using the GRI framework are scarce (Faúndez-Ugalde et al. 2022; Ikoma, 2022; Adams et al., 2022), and they suggest that greater tax transparency can enhance reputational capital (Hardeck et al., 2020).

ACKNOWLEDGMENT

The authors acknowledge the financial support of FCT - Foundation for Science and Technology, I.P., within the scope of multi-annual funding UIDB/04043/2020

REFERENCES

AdamsJ.DemersE.KlassenK. (2022), Tax Aggressive Behavior and Voluntary Tax Disclosure: Evidence from Corporate Sustainability Reporting. SSRN. https://ssrn.com/abstract=4284813 doi:10.2139/ssrn.4284813

Albuquerque, F., & Neves, J. (2021). Tax Disclosures in Financial and CSR Reporting as a Deterrence for Evasion. Handbook of Research on Theory and Practice of Financial Crimes. IGO Global. doi:10.4018/978-1-7998-5567-5.ch021

Alm, J., & Torgler, B. (2011). Do Ethics Matter? Tax Compliance and Morality. *Journal of Business Ethics*, *101*(4), 635–651. doi:10.100710551-011-0761-9

Branco, M. C., & Delgado, C. J. M. (2014). *Informação sobre combate à corrupção nos relatórios de sustentabilidade das empresas portuguesas.* Reporting on the fight, 4th GECAMB - Conference on Environmental Management and Accounting, Leiria

Cahan, S., Villiers, C., Jeter, D. C., Naiker, V., & Staden, C. (2016). Are CSR disclosure value relevant? Cross-country evidence. *European Accounting Review*, *25*(3), 579–611. doi:10.1080/09638180.2015.1064009

De la Cuesta-González, M., & Pardo, E. (2019). Corporate tax disclosure on a CSR basis: A new reporting framework in the post-BEPS era. *Accounting, Auditing & Accountability Journal*, *32*(7), 2167–2192. doi:10.1108/AAAJ-12-2017-3282

Dowling, G. R. (2014). The curious case of corporate tax avoidance. Is it socially irresponsible? *Journal of Business Ethics*, *124*(1), 173–184. doi:10.100710551-013-1862-4

Eccles, R. G. & Serafeim, G. (2017). Corporate and integrated reporting: A functional perspective. *Corporate Stewardship: Achieving Sustainable Effectiveness.*

European Commission. (2015). A Fair and Efficient Corporate Tax System in the 22 European Union: 5 Key Areas for Action. *Communication from the Commission to the European Parliament and the Council*. EC.

Faúndez-Ugalde, A., Toledo-Zúñiga, P., & Castro-Rodríguez, P. (2022). Tax Sustainability: Tax Transparency in Latin America and the Chilean Case. *Sustainability (Basel)*, *14*(4), 2107. doi:10.3390u14042107

García-Sánchez, I. M., Aibar-Guzmán, B., Aibar-Guzmán, C., & Somohano-Rodríguez, F. M. (2022). The drivers of the integration of the sustainable development goals into the non-financial information system: Individual and joint analysis of their influence. *Sustainable Development (Bradford)*, *30*(4), 513–524. doi:10.1002d.2246

Global Reporting Initiative. (2013). *G4 Sustainability Reporting Guidelines: Reporting Principles and Standard Disclosures*. GRI.

Global Reporting Initiative (GRI). *GRI Standards*. GRI.

Global Reporting Initiative (GRI, 2020). Linking the SDGs and the GRI Standards GRI and the United Nations Global Compact (UN Global Compact) (2017). Business Reporting on the SDGs. Analysis of Goals and Targets. GRI and the United Nations Global Compact (UN Global Compact) (2022). Business Reporting on the SDGs. *Analysis of Goals and Targets, updated version*. UN.

Gribnau, H., & Jallai, A. G. (2019). Sustainable tax governance and transparency. In *Challenges in Managing Sustainable Business* (pp. 337–369). Palgrave Macmillan. doi:10.1007/978-3-319-93266-8_15

Hardeck, I., Inger, K. K., Moore, R. D., & Schneider, J. (2020). *Do the bad guys report? Examining whether CSR performance influences the use of socially responsible tax disclosures*. Working paper.

Hardeck, I., & Kirn, T. (2016). Taboo or technical issue? An empirical assessment of taxation in sustainability reports. *Journal of Cleaner Production*, *133*, 1337–1351. doi:10.1016/j.jclepro.2016.06.028

Ikoma, A. (2022). *Voluntary tax disclosure from the perspective of tax avoidance and national culture. Master Dissertation in Finance and Taxation. FEP, Oporto University International Monetary Fund (IMF) (2022a). Fiscal Transparency*. International Monetary Fund. https://www.state.gov/department-of-state-releases-2022-fiscal-transparency-report/

International Monetary Fund (IMF). (2022b). *How Does the IMF Encourage Greater Fiscal Transparency?* IMF. https://www.imf.org/en/About/Factsheets/Sheets/2016/07/27/15/46/Encouraging-Greater-Fiscal-Transparency

Izzo, M., Dello Strologo, A., & Graná, F. (2020). Learning from the Best : New Challenges and Trends in IR Reporters ' Disclosure and the Role of SDGs. *Sustainability (Basel)*, *12*(14), 5545. doi:10.3390u12145545

Kao, W., & Liao, C. (2021). Tax Avoidance and Tax Disclosures in Corporate Social Responsibility Reports in the United Kingdom. *Journal of International Accounting Research*, *20*(3), 59–80. doi:10.2308/JIAR-2020-036

KPMG (2019). *The KPMG Survey of Corporate Responsibility Reporting 2019*. KPPMG.

KPMG (2022). *Big shifts, small steps: The KPMG survey of sustainability reporting 2022*. KPMG.

Kundeliene, K., & Leitoniene, S. (2015). Business Information Transparency: Causes and Evaluation Possibilities. *Procedia: Social and Behavioral Sciences*, *213*, 340–344. doi:10.1016/j.sbspro.2015.11.548

Lanis, R., & Richardson, G. (2013). Corporate social responsibility and tax aggressiveness: A test of legitimacy theory. *Accounting, Auditing & Accountability Journal*, *26*(1), 75–100. doi:10.1108/09513571311285621

Müller, R., Spengel, C., & Vay, H. (2020). *On the determinants and effects of corporate tax transparency: Review of an emerging literature, ZEW Discussion Papers 20-063*. ZEW - Leibniz Centre for European Economic Research.

Neves, J., & Albuquerque, F. (2019). Income Tax of Portuguese Listed Companies in Financial and Corporate Social Responsibility Reports. *Revista Brasileira de Gestão de Negócios*, *21*(3), 464–485. doi:10.7819/rbgn.v21i3.4009

Oats, L., & Tuck, P. (2019). Corporate tax avoidance: Is tax transparency the solution? *Accounting and Business Research*, *49*(5), 565–583. doi:10.1080/00014788.2019.1611726

OECD. (Organisation for Economic Co-operation and Development). (2013). *BEPS action plan*. OECD. https://www.oecd.org/ctp/BEPSActionPlan.pdf

Organisation for Economic Co-operation and Development. (2011). *OECD Guidelines for Multinational Enterprises*. OECD. https://www.oecd.org/daf/inv/mne/48004323.pdf

Platform for Collaboration on Tax. (2018). *Taxation and the Sustainable Development Goals*. OECD. https://www.oecd.org/tax/platform-for-collaboration-on-tax-conference-statement-2018.pdf

Reiter, S. (2020). Tax disclosures in sustainability reports. *Journal of Accounting and Finance*, 20(87), 51-64

Rudyanto, A., & Pirzada, K. (2021). The role of sustainability reporting in shareholder perception of tax avoidance. *Social Responsibility Journal*, *17*(5), 669–685. doi:10.1108/SRJ-01-2020-0022

Santoso, E., Laturette, K., & Mastan, S. (2019). Corporate Social Responsibility Disclosure, Tax Aggressiveness and Sustainability Report Assurance: Evidence from Thailand. *International Journal of Economics and Management Engineering*, *13*(11), 1413–1418.

Scheiwiller, T., & Symons, S. (2014). Corporate responsibility and paying tax. *The OECD Observer. Organisation for Economic Co-Operation and Development*, *1*. http://www.oecdobserver.org/news/archivestory.php/aid/3132/Corporate_responsibility_and_paying_tax.html

Schnighter, A., Holle, F., & Kockrow, M. (2021). Tax and Transparency: Reporting in Accordance with the Global Reporting Initiative. *Intertax*, *49*(8 & 9), 702–7212.

Stiglingh, M., Venter, E. R., Penning, I., Smit, A., Schoeman, A., & Steyn, T. (2017). Tax transparency reporting by the top 50 JSE-listed firms. *South African Journal of Accounting Research*, *31*(2), 151–168. doi:10.1080/10291954.2016.1160196

Supriyati, S., & Anggraini, D. (2021). Sustainability Reporting and Tax Aggressiveness: Evidence from a Public Company in Indonesia. *Indonesian Journal of Sustainability Accounting and Management, 5*(1), 71–80. doi:10.28992/ijsam.v5i1.249

Triwacananingrum, W. & Wijaya, G. (2022). Sustainability reporting and tax aggressiveness before and during Covid-19: GCG moderating variable. *Jurnal Akuntansi dan Keuangan Indonesia, 19* (1), 96-119

Venter, E., Stiglingh, M., & Smit, A. (2017). Integrated Thinking and the Transparency of Tax Disclosures in the Corporate Reports of Firms. *Journal of International Financial Management & Accounting, 28*(3), 394–427. doi:10.1111/jifm.12064

Watson, L. (2015). Corporate social responsibility, tax avoidance, and earnings performance. *The Journal of the American Taxation Association, 37*(2), 1–21. doi:10.2308/atax-51022

Zeng, T. (2018). Relationship between corporate social responsibility and tax avoidance: International evidence. *Social Responsibility Journal, 15*(2), 244–257. doi:10.1108/SRJ-03-2018-0056

ADDITIONAL READING

Bird, R., & Davis-Nozemack, K. (2018). Tax Avoidance as a Sustainability Problem. *Journal of Business Ethics, 151*(4), 1009–1025. doi:10.100710551-016-3162-2

Hauque, S. & Di Russo, L. (2020). Responsabilidad social organizacional y tributacion. Reflexiones sobre algunos aspectos de la norma GRI 207. *XIV Jornadas de investigación de la FCE*, 20-21 octubre

Hongler, P., & Regli, F. & Berndt, T. (2021). *Tax Reporting and Sustainability.* FF-HSG Working Paper No. 2021-6. Institute of Public Finance, Fiscal Law and Law & Economics (IFF-HSG)

Laksmi, D., & Narsa, N. (2021). Corporate Social Responsibility, Capital Intensity, and Tax Aggresiveness: Evidence from Indonesia. *AKRUAL: Jurnal Akuntansi, 13*(2), 132–143. doi:10.26740/jaj.v13n2.p132-143

Netjes, W., & Freyer, D. (2022). Tax Transparency Is Here to Stay: An Analysis of the Public CbCR Directive. *Intertax, 50*(8 & 9), 612–618. doi:10.54648/TAXI2022060

KEY TERMS AND DEFINITIONS

Corporate Social Responsibility (CSR): This refers to voluntary measures taken by companies to address social, environmental, and ethical issues both in their operations and throughout their supply chain.

Global Reporting Initiative (GRI): An international standard-setting organization that offers a framework for sustainability reporting, by providing guidelines for identifying ESG issues and reporting on them.

GRI 207- Tax: This standard was created to address the crucial impact of tax contributions on sustainable development and to meet the growing demand for tax transparency from stakeholders. It establishes guidelines for disclosing tax payments on a country-by-country basis, as well as outlining tax strategy and governance.

Sustainability Reporting: is the process of communicating a company's environmental, social, and governance (ESG) performance to its various stakeholders. Sustainability reporting can take various forms, such as annual reports, stand-alone sustainability reports or integrated reports that combine financial and sustainability information.

Tax Aggressiveness: is a deliberate act by taxpayers who adopt aggressive or borderline positions to minimize their tax liabilities in breach of current tax regulations.

Tax Avoidance: is the legal practice of minimizing one's tax liability by using various tax planning strategies and taking advantage of the loopholes in tax legislation.

Tax Transparency: refers to the act of disclosing tax-related information, whether voluntary or mandatory.

ENDNOTE

1 It is a tax-related term used to describe the tax planning strategies employed by multinational companies to artificially shift their profits from high-tax jurisdictions to low or no-tax jurisdictions in order to minimize their overall tax liability.

Chapter 15
Tourist Taxes and Sustainability:
A Systematic Literature Review and Future Research

Eva Maria Miranda
Marketing and Tourism School, Polytechnic Institute of Cávado and Ave, Portugal

Soraia Gonçalves
https://orcid.org/0000-0001-7385-438X
Management School, Polytechnic Institute of Cávado and Ave, Portugal & UNIAG, Portugal

Laurentina Vareiro
https://orcid.org/0000-0001-8945-1593
Management School, Polytechnic Institute of Cávado and Ave, Portugal & UNIAG, Portugal

ABSTRACT

As the global tourism industry continues to grow, concerns about its impact on the environment and local communities have increased. In response to these concerns, many destinations have implemented tourist taxes as a policy tool to manage the negative impacts of tourism on destinations and promote sustainable tourism. This chapter presents a systematic literature review on the relationship between tourist taxes and sustainability, using the Bibliometrix R software and a content analysis. The aim of this review is to provide a comprehensive analysis of the existing research, identifying gaps in the literature and suggesting areas for future research. The results demonstrate two main categories, concerning demand and supply dimensions. The issue of sustainability is transversal to the articles analyzed, highlighting its evolution in a holistic way and encompassing the economic, social, and environmental pillars.

DOI: 10.4018/978-1-6684-8592-7.ch015

INTRODUCTION

The literature on tourist taxes and sustainability is multidisciplinary and diverse, including perspectives from economics, tourism studies, environmental sciences, and public policy. The key concepts and debates in this field include the definition and classification of tourist taxes (Cetin et al., 2017; Durán-Román et al., 2021a; Soares et al., 2022), the rationale and objectives of tourist taxes (Cárdenas-García et al., 2022), the impacts of tourist taxes on stakeholders and destinations (Adedoyin et al., 2023; Pulido-Fernández & Cárdenas-García, 2012; Durán-Román et al., 2020, 2021b; Nunkoo & Gursoy, 2017; Pulido-Fernández & López-Sánchez, 2016; Song et al., 2019), the distributional effects of tourist taxes (Adedoyin et al., 2023; Durán-Román et al., 2020, 2021a; Pulido-Fernández et al., 2022), and the policy design and implementation of tourist taxes (Nunkoo & Gursoy, 2017; Pulido-Fernández & Cárdenas-García, 2012). These studies have addressed these themes from different perspectives, namely through case studies, surveys, experiments, and modeling.

This chapter presents a systematic literature review on the relationship between tourist taxes and sustainability, using the *Bibliometrix* R software, which discloses a range of indicators and visualizations for bibliographic data, and also a content analysis. The aim of this review is to provide a comprehensive analysis of the existing research, identifying gaps in the literature and suggesting areas for future research.

The systematic literature review process followed the PRISMA guidelines and included a comprehensive search of the electronic database Web of Science for articles published until 2023. The search terms included combinations of keywords related to tourist taxes and sustainability. The selection criteria included the relevance and quality of the articles, based on their titles, abstracts, and keywords.

After the introduction, this chapter is structured in two sections. The first section describes the research methodology used, followed by the second section in which the study results of the bibliometric and content analysis are presented. The chapter ends with the main conclusions.

RESEARCH METHODOLOGY

To identify the themes and trends in a particular field of study provides an appropriate purpose for this study. The aim of this systematic literature review is to develop a research strategy to identify and analyze relevant literature to address the research question on tourist taxes evolution in the tourism literature and the implications to tourism sustainability. Scholars can identify potential research gaps and opportunities for future research by mapping and evaluating the body of literature through a systematic process (Tranfield et al., 2003). This research presents a systematic literature review methodology based on the Preferred Reporting Items for Systematic Reviews and Meta-Analyses (PRISMA) guidelines, followed by a bibliometric analysis using the *Bibliometrix* R software. The PRISMA approach is distinct from the traditional literature reviews once it is more transparent and replicable, involving several explicit steps, such as utilizing a standardized method to identify all potentially relevant publications (Moher et al., 2009). Researchers have employed different methods to conduct literature reviews, and, among these, the "popularity-based approach" has emerged as a significant perspective, which involves the use of bibliometric analysis. This method was developed to examine the titles and keywords used by authors in published articles across different research fields. By analyzing data such as authors' affiliations, popular words, keywords and their frequency of use, bibliometric analysis can provide unique insights that may not be captured or evaluated by other types of reviews (Park & Jeong, 2019).

In the first phase, based on earlier research, a range of keywords was selected. Accordingly, three keywords were identified, namely "tax* touris*", "touris* tax*" and "sustainab*" based on the topic and objectives of this study, to ensure a comprehensive coverage of relevant articles. The combination "tax* touris*" addresses the relationship between taxes and tourism, encompassing various forms of the terms like "taxes" "taxation", "tourists", and "tourism". The keyword "sustainab*" ensures that the review includes literature related to sustainable tourism and the alignment of tourist taxes with sustainable practices. By using asterisks (*), the search captures different word variations, making the review more complete and effective in gathering pertinent information.

Hence, the following process involved searching academic publications during April 2023 in the database Web of Science, using a combination of search terms. The Boolean term, including assessing numerous search techniques, such as truncation and proximity operators, was built and evaluated in the search process. The search was restricted to the *topic*, which includes *title*, *abstract*, *author keywords* and *keywords plus*.

The first criterion of exclusion was to eliminate all literature not referring to the document type article. The option of including only articles ensures rigorous research by relying on peer-reviewed and well-documented studies. The second exclusion criterion was language, considering only English and Spanish articles, since the authors do not have linguistic competence on Turkish, Croatian, and Russian, the other languages in which the articles on these topics were written. The third exclusion criterion was applied regarding the Web of Science Categories, considering only the following ones: Hospitality Leisure Sport Tourism, Economics, Management, Public Administration and Political Science. There were no restrictions about the publishing date in the current view. From this analysis, 65 articles were selected.

A rigorous bibliometric analysis on these 65 records was conducted using the software, and then a content analysis of key papers was carried out. To identify these key papers two inclusion criteria were used, first the five most cited, and cumulatively the five most cited in the last five years. The first criterion allowed to identify five papers published between 2003 and 2012, so there was a need to consider a second criterion in order to obtain more current works.

To describe the outline of this research, an adaptation of the PRISMA flow diagram was used as a means of presenting all the relevant information that emerged from the selection phases (Figure1).

RESULTS

Bibliometric Analysis

This subsection emphasizes the results generated via the *Bibliometrix* tool in identifying trends in the literature on tourist taxes and sustainability. First, it is provided a comprehensive view of publication trends and citation performance, followed by the most productive journals, countries and authors and their affiliations, finishing with the most influential articles to identify current tendencies in the study of the tourist taxes and sustainability.

Figure 1. The PRISMA flow diagram (adapted)

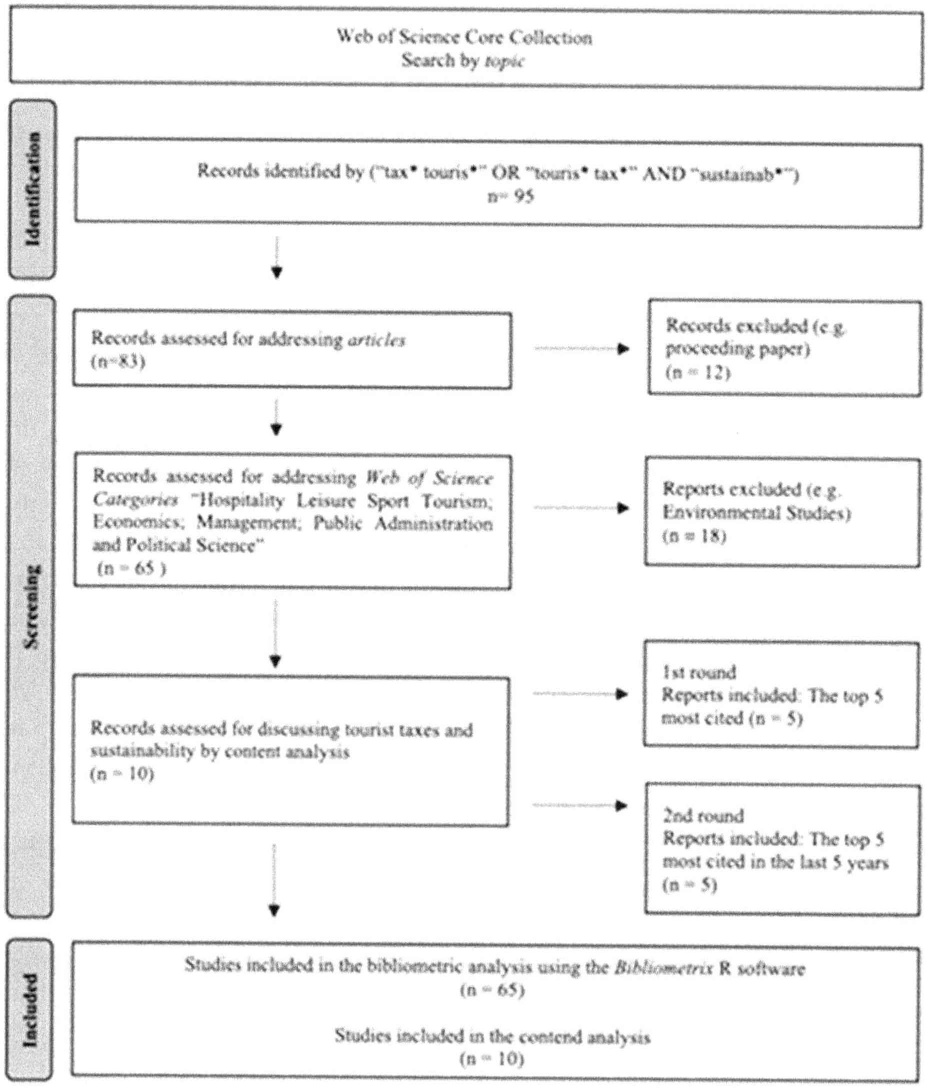

Annual Scientific Production and Citation

Figure 2 shows the number of publications and the average citations per year of the tourist taxes and sustainability within the period of study. According to the systematic review carried out, the first paper on this area was published in 1986; altogether, there were three articles until 2000. Fujii et al (1985) published an article in *National Tax Journal* that examines the incidence and exportability of an ad valorem hotel room occupancy tax for Hawaii vis-à-vis alternative tourist taxes. Bird (1992), in the journal *World Development,* argues that, in principle, there is a strong economic case in many countries regarding taxing tourism, and also that the nature of the industry and administrative difficulties set serious limits to what can be done in practice.

Bonham and Gangnes (1996) published an article in *Applied Economics* that studied the effect on hotel revenues of the Hawaii room tax using time series intervention analysis, finding no evidence of statistically significant tax impacts.

Publications in the tourist taxes and sustainability fields start to gain consistency only from the end of the first decade of the 21st century, with publication peaks in 2017 (when problems with overtourism arise), 2021 and 2022 (in the post-COVID 19 pandemic recovery period).

Figure 2. Annual scientific production and average citations per year (1985–2023)

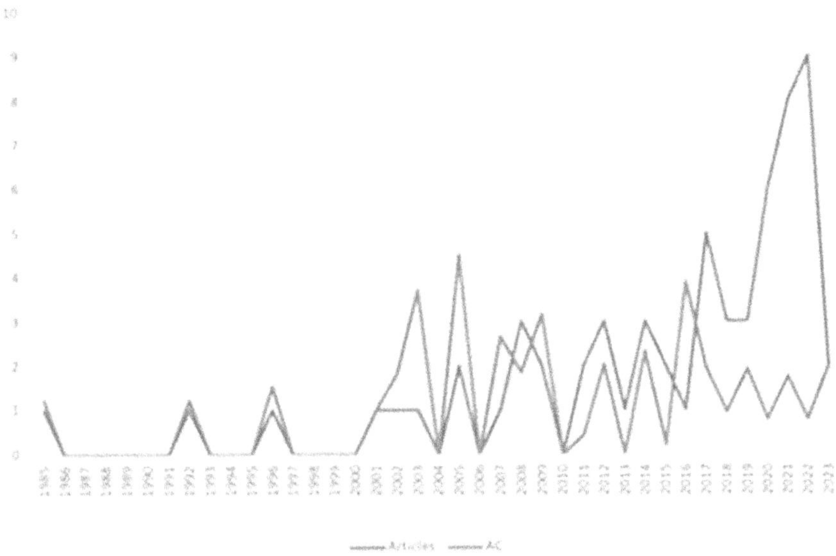

Most productive journals

In this study, 65 journal articles, published in 35 journals, were analyzed. Table 1 shows the top five most influential journals in the area of the tourist taxes and sustainability, ranked according to the number of publications. There are 31 articles published in the five journals, accounting for 47.7% among all publications. *Tourism Management* has published the largest number of articles, nine publications, followed by *Journal of Travel Research* (seven) and *Tourism Economics* (seven).

The number of citations for each journal is computed and shown in Table 1, along with the average citations. For example, the publications from *Tourism Management* gained 392 citations and ranked first in terms of citations, followed by *Journal of Travel Research* (60) and *Journal of Destination Marketing & Management* (57). Consequently, these journals also ranked as the top three journals with the highest average citations, with 43.56, 8.57, and 11.4, respectively (in terms of average citations, there was a switch between 2nd and 3rd places). The dynamic trend of the journals, according to the analysis of the publications per year of articles on tourism fees linked to sustainability, is presented in Figure 3.

Table 1. The top five most influential journals

Sources	TP	TC	AC	PY_S
Tourism Management	9 (14%)	392	43.56	2001
Journal of Travel Research	7 (11%)	60	8.57	2011
Tourism Economics	7 (11%)	46	6.57	2008
Journal of Destination Marketing & Management	5 (8%)	57	11,4	2014
International Journal of Tourism Research	3 (5%)	14	4,67	2017

Notes: TP represents the total publication/articles (percentage), TC the total citation, AC the ratio of total citation per total publication, PY_S publication year started in publishing articles of tourist taxes and sustainability.

Figure 3. Sources' production over time

Authors, affiliations, and countries

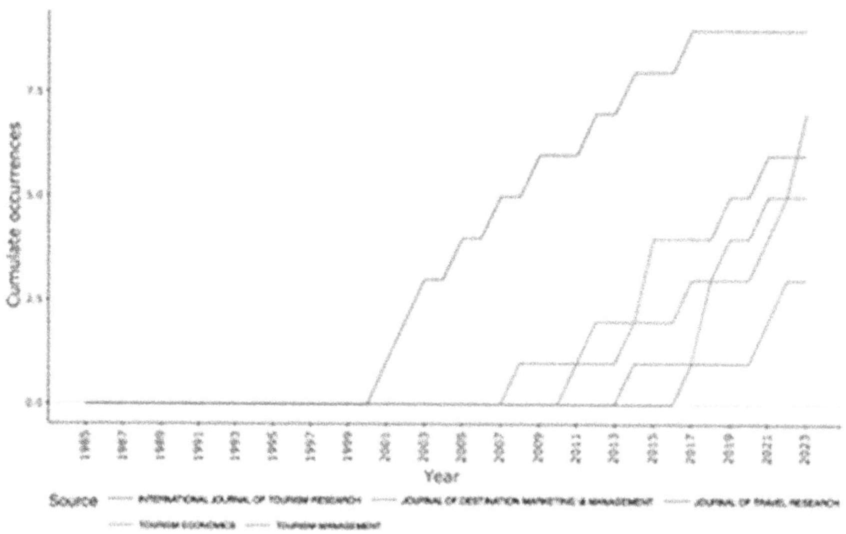

The 65 articles included in this analysis were authored by 136 researchers. As stated in Table 2, 14 papers have single authorship (from 13 authors), and the others are multi-authorship. 121 authors wrote only one article on the topic, and only eight authors wrote more than three articles (Tables 2 and 3).

In total, 86 affiliations were found from 19 countries. Most prolific countries and universities are listed in Tables 4 and 5. Most of the research work is concentrated within Spain, with a total of 18 publications. The USA emerged as a country with the second-highest number of publications (11), followed by the United Kington (seven) and China (six). Publications from Spain recorded the highest number of citations (330), followed by the United Kington (209) and the USA (143).

Table 2. Authorship

	Authorship	No. of documents
	Single-authored docs	14
	Authors	136
Documents written	**No. of authors**	**Proportion of authors**
1	121	0,88970588
2	7	0,05147059
3	5	0,03676471
4	2	0,01470588
5	1	0,00735294

Table 3. Prominent authors

Authors	PY_S	TP	Institutions
Pulido-Fernández, J.I.	2020	5	Universidad de Jaen
Durán-Román, J.L.	2020	4	Universidad de Jaen
Mak, J.	1985	4	University of Hawaii Manoa
Beritelli, P.	2014	3	University of St Gallen, Switzerland
Cárdenas-García, P.J.	2020	3	Universidad de Jaen
Laesser, C.	2014	3	University of Wollongong, Australia
Riera, A.	2003	3	Universidad des Illes Balears
Seetaram, N.	2014	3	Bournemouth University

Notes: PY_S represents the publication year the author started in publishing articles of tourist taxes and sustainability; TP represents the total publication/articles.

Table 4. The most productive countries

Country	TP	SCP	MCP	Country	TC	AC
Spain	18	16	2	Spain	330	18.33
USA	11	10	1	United Kington	209	29.86
United Kington	7	2	5	USA	143	13
China	6	4	2	China	72	12
Australia	3	1	2	Portugal	56	56
Italy	3	2	1	Mauritius	45	45
Switzerland	3	1	2	Australia	40	13.33
Czech Republic	2	2	0	Canada	39	39
Ukraine	2	1	1	Switzerland	34	11.33
Others	1	1	0	Thailand	31	31

Notes: TP represents the total publication/articles, SCP Single Country Publication, MCP Multiple Country Publication, TC the total citation, AC the ratio of total citation per total publication.

The United Kington has the highest number of publications in collaboration with other countries (five), followed by Australia and Switzerland, in which two of the three publications are in collaboration. Spain, with 18 publications, only has two in collaboration with other countries (with Belgium, Chile, China, and United Kington). Figure 4 also shows that the United Kington has collaborations with Denmark, Greece, Italy and the United Arab Emirates; Australia with Canada, Indonesia and Switzerland, and Switzerland also collaborates with Sweden.

Figure 4. A network analysis of countries co-authorship based on documents weights

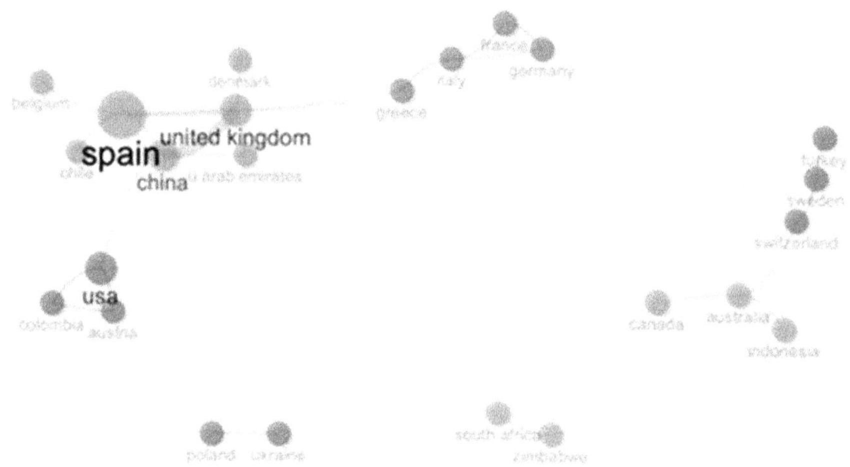

Table 5. The top ten most productive universities

Affiliation	TP
Universidad de Jaen	15
University of Hawaii Manoa	8
Universidad de Illes Balears	8
Universidad de Las Palmas Gran Canaria	8
Bournemouth University	7
University of Sassari	6
University of St Gallen	6
Istanbul University	5
University of Edinburgh	4
University of Macau	4

Most Influential Articles

Table 6 presents the top five highly cited articles involving tourist taxes and sustainability. Among the 65 articles, the article with the highest number of citations is "Economics of Tourism Taxation - Evidence from Mauritius" by Gooroochurn and Sinclair (2005), published in *Annals of Tourism Research*, with a total of 106 citations. Concurrently, this article by Gooroochurn and Sinclair also reported the highest average citation (5.58). The "Tourism and Environmental Taxes with Special Reference to the "Balearic Ecotax" by Palmer and Riera (2003) ranked as the second-highest cited article (78), with an average citation of 3.71, followed by "The Short-Term Price Effect of a Tourist Tax Through a Dynamic Demand Model: The Case of the Balearic Islands" (65) by Aguilo et al. (2005), both published in *Tourism Management.*

Table 6. The top five most cited articles

Paper	PY	Journal	TC	TC_Y
Gooroochurn & Sinclair	2005	Annals of Tourism Research	106	5.58
Palmer & Riera	2003	Tourism Management	78	3.71
Aguilo, Riera & Rossello	2005	Tourism Management	65	3.42
Do Valle et al.	2012	Tourism Management	56	4.67
Gago et al.	2009	Tourism Management	56	3.73

Notes: PY represents the publication year, TC the total citation, TC_Y the ratio of total citation per year.

Considering that the five most cited articles dated from the first decade of the 21st century, except for one article from 2012, it was considered appropriate to extend the analysis to the five most cited articles from the last five years (Table 7).

Table 7. The top five most cited articles, in the last 5 years

Paper	PY	Journal	TC	TC_Y
Durán-Román et al.	2021	Journal Destination Marketing & Management	14	4.67
Nepal & Nepal	2021	Tourism Geographies	14	4.67
Zhang & Yang	2019	Tourism Economics	14	2.80
Reinhold et al.	2018	Journal Destination Marketing & Management	13	2.17
Song et al.	2019	Journal Destination Marketing & Management	9	1.80

Notes: PY represents the publication year, TC the total citation, TC_Y the ratio of total citation per year.

The articles with the highest number of citations in the five last years were "Tourists' willingness to pay to improve sustainability and experience at destination" by Durán-Román et al. (2021a), published in the *Journal of Destination Marketing & Management*, and "Managing overtourism through economic taxation: policy lessons from five countries", by Nepal and Nepal (2021), published in *Tourism Geogra-*

phies, both with a total of 14 citations and 4.67 citation per year. Concurrently, the article "Prescribing for the tourism-induced Dutch disease: A DSGE analysis of subsidy policies" by Zhang and Yang (2019) also reported a total of 14 citations but only 2.8 citation per year. "The 2016 St. Gallen Consensus on Advances in Destination Management" by Reinhold et al. (2018) ranked as the fourth-highest cited article (13), with an average citation of 2.17, followed by "The Short-Term Price Effect of a Tourist Tax Through a Dynamic Demand Model: The Case of the Balearic Islands" (nine) by Song et al. (2019), both published in the *Journal of Destination Marketing & Management*.

Content Analysis

To answer the main objective of this study, a more in-depth analysis of the five most cited articles in the period of research was carried out as well as of the five most cited articles in the last five years (Table 8).

From the content analysis, in which three units of analysis were considered - study's objectives, research gaps and future research - it is possible to identify two main categories. These categories conferred some coherence to the analysis of the selected articles, concerning the demand and supply dimensions.

Regarding the demand dimension, it is highlighted the impact that tourism taxation has on consumer behavior, namely the tax compliance and willingness to pay taxes, as well as on the spending behavior in other tourism demand components.

It is also important to stress the importance of further studies linking transparency in the application of tourist taxes (earmarked) to the willingness to pay. This tax compliance is related to the individual characteristics of tourists and, therefore, the need to reinforce clustering studies.

Regarding the supply dimension, studies aim at identifying an optimal level of tourism-related taxation which should consider the form of taxation, general or specific, the local context and competitiveness of destination, and the tax burden on residents, tourists, and sectors of economy.

The issue of sustainability is transversal to the studies analyzed. In the first articles published, the environmental damage caused by tourism is highlighted; in later studies the focus on sustainability extends to all its pillars (economic, social and environmental), encompassing the research on the sustainability of destinations as a whole.

FUTURE RESEARCH DIRECTIONS AND CONCLUSION

This study carried out a systematic literature review on tourist taxes and sustainability, using *Bibliometrix* R software. The search process and the bibliometric analysis present the evolution of the number of publications on these topics over time, providing a useful tool for mapping the existing literature and identifying knowledge gaps as well opportunities for collaboration, regarding future research areas.

Considering publications on tourist taxes and sustainability, the results show that studies started to gain consistency from the end of the first decade of the 21st century, with publication peaks in 2017, 2021 and 2022. These studies were published in 35 journals, highlighting that five journals account for almost 50% among all publications. *Tourism Management* has published the largest number of articles, followed by *Journal of Travel Research,* and *Tourism Economics*.

The articles analyzed were authored by 136 researchers. Most of the papers are multi-authorship, and only eight authors wrote more than three articles. Regarding affiliations, it was found that most of the research work concentrates within Spain, followed by the USA, the United Kington and China.

Table 8. A summary of studies objectives, research gaps and future research areas

Paper	Study's objectives	Research gaps	Future research areas
Palmer & Riera, 2003	To confirm the existence of external environmental costs and the need to internalize them using appropriate economic instruments; to review a proposal to introduce a tourist tax as an environmental tax.	To comprehend the pressing issues in mature tourist destinations, like the Balearic Islands, it is essential to consider external costs of tourism and public service provision problems.	A new perspective is being introduced to the analysis of environmental damage caused by tourism and policies to mitigate it, where further investigation is needed.
Aguilo et al., 2005	To investigate the elasticity of tourism determinants in the Balearic Islands using a new method involving the transmission of information to tourists.	Demand models are important for interpreting regression coefficients, but tourism research has lacked well-defined variables and relationships, limiting progress to "preliminary investigations". More structured theories are needed to advance knowledge.	The general effects of implementing a tourist tax are examined, assuming hotels can pass on the cost to consumers, and suggesting further research on possible oligopolistic pressures and supply effects.
Gooroochurn & Sinclair, 2005	To discuss the economic concepts, objectives, principles, and effects of tourism taxation; to provide an overview of the literature and measure the associated effects.	Research on tourism taxes has mainly focused on their role as a source of revenue, burden distribution, and tax exportability, with most studies using a partial equilibrium framework. The study is innovative in using a Computable General Equilibrium (CGE) approach to analyze tourism taxation in Mauritius.	Design an optimal tourism tax and highlight the importance of scrutinizing proposed policies due to the different effects they have on tourists, residents, and sectors in the economy.
Gago et al., 2009	To discuss whether the introduction or increase of tourism taxation should be addressed through specific or general indirect taxation.	The use of a CGE model to assess the effects of two hypothetical taxes affecting tourism: a specific tax aimed at tourism (hotel room tax) and an indirect tax reform that affects tourism through changes in VAT rates.	Explore the difference between taxing tourism through a general tax such as VAT or through a specific tax such as hotel room tax. Study which different alternatives of taxation are most efficient and equitable.
Do Valle et al., 2012	To study tourist attitudes towards an accommodation tax earmarked for environmental protection in the Algarve.	By segmenting tourists according to their willingness to pay the earmarked accommodation tax, this analysis has important policy implications, such as defining where and how to implement this type of tourism tax.	Address the attitudes of tourists towards an earmarked accommodation tax in the context of a contingent valuation survey, with scenarios of improved environmental quality.
Reinhold et al., 2018	To examine the main insights in destination management regarding tourism taxation and regulation.	Tax regimes and fees vary according to tourism destination and differ in five main features – the object taxed, filing and reporting requirements, the use of collected tax revenues, actual benefits of tax revenues for tourism, and auditing frequency of taxed subjects.	Determine the antecedent concepts to design the optimal level of tourism taxation on specific destination settings; issues considering behavioral implications of tax compliance; further studies on taxes and competitiveness implications for destinations and on the capacity to invest tax revenues for long-term viability; address gaps concerning destination governance, destination branding and communication and also issues on sustainability
Song et al., 2019	To analyze whether consumers absorb the additional cost of travel by reducing their consumption concerning other demand components.	Study how tourist taxation influences the spending behavior of individual tourists, highlighting the microeconomic tourism demand approach.	Replicate the study to other travelers, namely business and outbound travelers, as well as travelers from other countries.

continues on following page

Table 8. Continued

Paper	Study's objectives	Research gaps	Future research areas
Zhang & Yang, 2019	To analyze the effects of external inbound tourism booms on the national economic account of a small open economy with two sectors: tourism and manufacturing.	Apply the dynamic stochastic general equilibrium (DSGE) model, quite used in main-stream macroeconomics, to tourism economics.	Extend, theoretical and empirical tourism issues, by applying the DSGE methodology, namely, to know the impacts of the tourism sector to the rest of economy.
Durán-Román et al., 2021b)	To determine tourists' willingness to pay taxes and public fees, allowing the financing of policies to improve sustainability to the destination as well as the tourism experience.	Provide a new means of approaching issues related to sustainability and the financing of tourism activity in tourism destinations; therefore, it is necessary to determine the suitability of fiscal instruments associated to tourism activity concerning the demand perspective.	Study tourists' willingness to pay by tourist cluster considering groups of individuals with similar characteristics, and their willingness to pay for specific taxes and/or public fees.
Nepal & Nepal, 2021	To analyze overtourism management policies focusing on economic taxation as instrument in tourism management in popular destinations	Identify suitable instruments or strategies to redress overtourism concerns.	Modelling CGE studies to analyze the economy impacts of a rise in tourist taxes and entrance fees; estimate the price elasticities of demand to tourism to account for the demand-side behavior related to taxes and fees; construct destination-specific indicators of overtourism concerning local contexts.

The top five highly cited articles involving tourist taxes and sustainability are authored by Gooroochurn and Sinclair (2005), Palmer and Riera (2003) Aguilo et al. (2005), Do Valle et al. (2012), and Gago et al. (2009. Taking into account that the five most cited articles dated from the first decade of the 21st century (except for one article from 2012), it was considered appropriate to extend the analysis to the five most cited articles from the last five years. These articles are authored by Durán-Román et al. (2021a), Nepal and Nepal (2021), Zhang and Yang (2019), Reinhold et al. (2018), and Song et al. (2019).

In the content analysis, three units of analysis were considered – the study's objectives, research gaps and future research – and two main categories were identified, concerning the demand and supply dimensions. The demand dimension highlights the impact that tourism taxation has on consumer behavior. The supply dimension aims at identifying an optimal level of tourism-related taxation. The issue of sustainability is transversal to the analysis undertaken, stressing the environmental damage caused by tourism in the first studies; in the subsequent studies, the focus on sustainability extends to economic, social and environmental pillars.

Overall, the authors expect that the results of this study highlight the importance of studying tourist taxation and sustainability from an interdisciplinary and international perspective and provide directions for future research and policy debates.

Relying on only one database, in this case the Web of Science, and the methodology of selecting search terms as well as inclusion/exclusion on criteria during the literature search are two limitations of this study, to the extent that it may introduce bias and potentially disregard other possible pertinent sources. Nevertheless, recognizing these limitations, this study could serve as a foundation for initiating future research studies.

REFERENCES

Adedoyin, F. F., Seetaram, N., Disegna, M., & Filis, G. (2023a). The Effect of Tourism Taxation on International Arrivals to a Small Tourism-Dependent Economy. *Journal of Travel Research*, *62*(1), 135–153. doi:10.1177/00472875211053658

Aguilo, E., Riera, A., & Rossello, J. (2005). The short-term price effect of a tourist tax through a dynamic demand model. The case of the Balearic Islands. *Tourism Management*, *26*(3), 359–365. doi:10.1016/j.tourman.2003.07.005

Bird, R. M. (1992). Taxing tourism in developing countries. *World Development*, *20*(8), 1145–1158. doi:10.1016/0305-750X(92)90006-H

Bonham, C. S., & Gangnes, B. (1996). Intervention analysis with cointegrated time series: The case of the Hawaii hotel room tax. *Applied Economics*, *28*(10), 1281–1293. doi:10.1080/000368496327831

Cárdenas-García, P. J., Pulido-Fernández, J. I., Durán-Román, J. L., & Carrillo-Hidalgo, I. (2022). Tourist taxation as a sustainability financing mechanism for mass tourism destinations. *International Journal of Tourism Research*, *24*(4), 577–592. doi:10.1002/jtr.2523

Durán-Román, J. L., Cárdenas-García, P. J., & Pulido-Fernández, J. I. (2021a). Tourist tax to improve sustainability and the experience in mass tourism destinations: The case of Andalusia (Spain). *Sustainability (Basel)*, *13*(1), 42. doi:10.3390u13010042

Durán-Román, J. L., Cárdenas-García, P. J., & Pulido-Fernández, J. I. (2021b). Tourists' willingness to pay to improve sustainability and experience at destination. *Journal of Destination Marketing & Management*, *19*, 100540. doi:10.1016/j.jdmm.2020.100540

Durán-Román, J. L., Pulido-Fernández, J. I., & Cárdenas-García, P. J. (2020). Delimitation of applicable government revenue in Spanish tourist destinations. *Investigaciones Regionales*, *2020*(47), 179–199. doi:10.38191/iirr-jorr.20.016

Fujii, E., Khaled, M., & Mak, J. (1985). The exportability of hotel occupancy and other tourist taxes. *National Tax Journal*, *38*(2), 169–177. doi:10.1086/NTJ41792007

Gooroochurn, N., & Sinclair, M. T. (2005). Economics of tourism taxation: Evidence from Mauritius. *Annals of Tourism Research*, *32*(2), 478–498. doi:10.1016/j.annals.2004.10.003

Moher, D., Liberati, A., Tetzlaff, J., & Altman, D. G. (2009). Preferred reporting items for systematic reviews and meta-analyses: The PRISMA statement. *BMJ (Clinical Research Ed.)*, *339*(7716), 332–336. doi:10.1136/bmj.b2535 PMID:19622551

Nepal, R., & Nepal, S. K. (2021). Managing overtourism through economic taxation: Policy lessons from five countries. *Tourism Geographies*, *23*(5–6), 1094–1115. doi:10.1080/14616688.2019.1669070

Nunkoo, R., & Gursoy, D. (2017). Political trust and residents' support for alternative and mass tourism: An improved structural model. *Tourism Geographies*, *19*(3), 318–339. doi:10.1080/14616688.2016.1196239

Palmer, T., & Riera, A. (2003). Tourism and environmental taxes. With special reference to the "Balearic ecotax.". *Tourism Management*, *24*(6), 665–674. doi:10.1016/S0261-5177(03)00046-3

Park, J., & Jeong, E. B. (2019). Service quality in tourism: A systematic literature review and keyword network analysis. *Sustainability (Basel)*, *11*(13), 3665. Advance online publication. doi:10.3390u11133665

Pulido-Fernández, J. I., & Cárdenas-García, P. J. (2012). *Política económica del turismo*. Pirámide.

Pulido-Fernández, J. I., Román, J. L. D., Cárdenas-García, P. J., & Carrillo-Hidalgo, I. (2022). Taxation as a stimulus for sustainability in mature tourist destinations. *Revista de Turismo y Patrimonia Cultural*, *20*(3), 681–697. doi:10.25145/j.pasos.2022.20.047

Reinhold, S., Laesser, C., & Beritelli, P. (2018). The 2016 St. Gallen Consensus on Advances in Destination Management. *Journal of Destination Marketing & Management*, *8*, 426–431. doi:10.1016/j.jdmm.2017.03.001

Soares, J. R. R., Remoaldo, P., Perinotto, A. R. C., Gabriel, L., Lezcano-Gonzalez, M. E., & Sanchez-Fernandez, M. D. (2022). Residents' Perceptions Regarding the Implementation of a Tourist Tax at a UNESCO World Heritage Site: A Cluster Analysis of Santiago de Compostela (Spain). *Land (Basel)*, *11*(2), 189. Advance online publication. doi:10.3390/land11020189

Song, H., Seetaram, N., & Ye, S. (2019). The effect of tourism taxation on tourists' budget allocation. *Journal of Destination Marketing & Management*, *11*, 32–39. doi:10.1016/j.jdmm.2018.12.001

Tranfield, D., Denyer, D., & Smart, P. (2003). Towards a Methodology for Developing Evidence-Informed Management Knowledge by Means of Systematic Review. *British Journal of Management*, *14*(3), 207–222. doi:10.1111/1467-8551.00375

Zhang, H., & Yang, Y. (2019). Prescribing for the tourism-induced Dutch disease: A DSGE analysis of subsidy policies. *Tourism Economics*, *25*(6), 942–963. doi:10.1177/1354816618813046

ADDITIONAL READING

Bernal, A., Montero, P., & Recio, L. (2022). Do tourist companies support a greater direct tax burden? The case of Spain. *Current Issues in Tourism*, *25*(4), 579–591. doi:10.1080/13683500.2021.1892605

Caponi, V. (2022). The economic and environmental effects of seasonality of tourism: A look at solid waste. *Ecological Economics*, *192*, 107262. doi:10.1016/j.ecolecon.2021.107262

Durán-Román, J. L., Pulido-Fernández, J. I., Rey-Carmona, F. J., & Núñez-Tabales, J. M. (2022). Willingness to Pay by Tourist Companies for Improving Sustainability and Competitiveness in a Mature Destination. *Leisure Sciences*, 1–22. doi:10.1080/01490400.2022.2123072

Mapingure, C., & du Plessis, E. (2022). Tourist Taxonomies of International Tourists Visiting Zimbabwe Using an Extended International Tourist Role Scale. *Journal of Hospitality & Tourism Research (Washington, D.C.)*, *46*(2), 344–369. doi:10.1177/1096348021996786

Mohan, P., & Strobl, E. (2023). Tourism and tax revenue: Evidence from stay-over tourists in the Eastern Caribbean. *Current Issues in Tourism*, 1–23. doi:10.1080/13683500.2023.2202307

Sharma, A., Perdue, R. R., & Nicolau, J. L. (2022). The Effect of Lodging Taxes on the Performance of US Hotels. *Journal of Travel Research*, *61*(1), 108–119. doi:10.1177/0047287520971044

Yang, C., Lin, Z., Huang, J., & Cheng, T. (2023). A Dynamic Tax Model Based on Destination Lifecycle for Sustainable Tourism Development. *Journal of Travel Research*, *62*(1), 217–232. doi:10.1177/00472875211057596

KEY TERMS AND DEFINITIONS

Cluster Analysis: Multivariate statistical method used to group cases or variables into clusters based on their common characteristics. This technique allows the creation of homogeneous groups in which the cases of a cluster are similar to each other and different from the cases of other clusters.

Content Analysis: Qualitative research method that can be used to analyze data from various discourses and communication types, regardless of their format. Through this approach, researchers aim to identify the underlying characteristics, structures, or patterns present within the fragments of messages being analyzed.

Overtourism: The impact of tourism on destinations, influencing excessively and/or negatively the alleged quality of citizens' life and/or the quality of visitors' experience.

Sustainability: Encompasses current and future generations by meeting the needs of current generations without compromising the needs of the future ones, ensuring a balance between economic development, environmental care and social well-being.

Systematic Literature Review: Research methodology that examines the data and results of other authors, using a comprehensive and rigorous method to address a specific research question.

Tax Compliance: Willingness of individuals and other taxable entities to act in accordance with tax laws and regulations.

Tourism Taxes: Charges levied on tourists used to address environmental impacts of tourism and to finance the provision of public goods and services whose demand is increased by tourists.

Willingness to Pay: Used to assess the value of non-marketable goods, based on the idea of rational decision-making and utility maximization.

Chapter 16
Ways of Using the Tribute for Environmental Protection

Rute Neto Cabrita e Gil Saraiva
University of Lisbon, Portugal

Victor Gustavo Rocha Nylander
University of Lisbon, Portugal

ABSTRACT

As a result of major climate problems, this chapter aims to analyze the ways of using taxation to achieve its objectives, especially to combat externalities. The methodology used is bibliographical. The first chapter explained the social function of the tribute in the socio-environmental state, which would have as its function the fulfillment of its socio-environmental function. The second chapter deals with the objectives of environmental taxation, which is with the generation of dividends, benefits, for society with the correction of externalities. And the last chapter shows the ways to use environmental taxation to reduce externalities, such as use of direct taxation on emissions, use taxes, indirect taxation, taxation at different rates, funds. Finally, an analysis is made of the best way to use environmental taxation.

INTRODUCTION

The tax has a primary objective (fiscal objective), which is to collect resources from the economy for the sustenance of the State itself. There is also the side effect (extra fiscal) to be instituted, which is to generate stimuli, behaviors. Because of the impossibility of having a tax with absolute fiscal neutrality, the first being the objective and the second as a side effect, but the reverse can also happen, the State create a tax to generate a certain stimulus without the due concern of being a great source of tax revenue.

When the State wants to change a certain behavior or achieve some goal, with constitutional legal basis, and the parallel effect is the collection of resources, the two can also be objectives of the tax, be revenue and encourage behavior change.

DOI: 10.4018/978-1-6684-8592-7.ch016

Within the current scenario, the State has the obligation to formulate public policies to protect society from possible risks. These risks are generated by the production process of the economy, in its consumption preferences, in its way of living and relating in national and international society.

The tribute needs to be constituted from this new scenario in which the State is obliged to formulate material solutions. For this it is necessary to understand the function of the tribute in this new State, concerned with all sectors of society (economy, social and environmental) for the current and future generation.

Due to the new needs of the State to achieve the legal objectives, it is necessary to pay attention to the sustainability of public accounts, in the relationship between revenue and expenditure, which is the way by which the State will do all its financial activity in meeting public needs.

To ensure sustainability, taxes are used to meet public needs, in addition to ensuring sustainability, reducing spending over time, and generating revenue to finance the financial activity of the State. One of these paths is environmental taxation, a tool by which the State uses taxes to seek to achieve this sustainability and meet public needs.

However, there are several forms of use, which will be the object of this article, the understanding about the forms of use of environmental taxation to be able to correct the negative externalities. With the use essentially of the bibliographic methodology.

This article is divided into three parts. The first explains the social function of taxes in the Socio-Environmental State model. The second part deals with the objectives of environmental taxation. And finally, in the last part, the ways of using environmental taxation.

SOCIAL FUNCTION OF TAXES IN THE SOCIO-ENVIRONMENTAL STATE

The social function of tribute is mixed with the function of each kind of State. It is perceived that, in the Liberal State, the idea of non-intervention of the State over the economy and society prevailed, idealized by the Enlightenment ideals, the tribute also could not intervene in private decision-making, because it is a tool of the State. In the Social State, the tax was used to reduce inequalities, regulate competition, benefit the poorest layers of society, among other objectives. Now, in the Socio-Environmental State, the social function of the tribute expands to another *area*, the environmental, so that the tax can also be used as an indirect form of environmental protection, to avoid economic, social, environmental risks and meet resources for several generations.

Otherwise, taxes will be used for the realization of fundamental rights, regardless of generation, either as public revenue for entry into the public coffers of the State and this through its financial activity applies these taxes in society through public services or by making goods available to society for its use and consequent materialization of rights. Also, by inducing behavior so that an economic or social agent has the incentive to act in a certain way, making him realize his rights and / or rights of others, in the most diverse perspectives.

It can be stated that the primary function of the tax is to be public revenue and must enter the public coffers, however, when faced with situations relevant to the rights and to the State itself in its full operation, the sustainability of the State or the financing of rights or both simultaneously, or, as in vogue in Europe, may come into question it may be facing relevant economic issues, which makes this “primary function” no longer really primary[1].

There are no impediments for the State to use its own tax burden to encourage changes in the behavior of economic agents for the realization of one or more rights (individual, social, collective, or diffuse), including, and the State may stop collecting to achieve this purpose, through the increase[2] or reduction of the tax burden or the use of tax benefits. It is important to note that by using tax benefits, society, through the figure of the State, transfers public resources to certain taxpayer(s), guided by its government policies and ideology applied when observing the concrete case.

There are no impediments for the State to analyze the social, economic, and environmental issues for the applicability of the form of taxation. It is necessary to analyze the factual situations to achieve its objectives, otherwise the power to tax, which holds the power to destroy, will result in self-destruction of the State and rights by preventing the development of innovations, commercial, environmental, and social relations, which are the basic points of collection for the sustainability of the State itself and rights[3]. If the various perspectives are not observed, the Tax State may become an Indebted State[4].

When the social function is approached from the perspective of the Socio-environmental State, the social function of the tribute is not changed, it is only expanded to more fields of analysis, beyond merely reaching the economic, social, and financial fields of the State.

Taxes also need to meet environmental demands, which have become the source of various anxieties in society, including high technological development, the exacerbated use of various natural and human resources (capital) for development. They have resulted in an increase in the temperature of the planet, melting of the polar ice caps, alteration of the environmental and political climate, in addition to resulting in more intense crises in smaller time intervals, problems that were previously non-existent and, as a social, economic, financial and environmental result, there is a society that is at constant risk that cannot carry out planning or programming of its activities, they are unable to make more democratic decisions, given the flexibility and high volatility that the future has become in the present society, jointly compromising the actions of the State itself, due to the concentration of financial, economic and cultural decisions being concentrated in a few economic agents and dependent on the positioning of international markets, specifically their *players* (players), either in their financial bias, or as a guarantor of rights, in the current generation as in the future. Tribute operates in the age of uncertainty.

The social function of the tribute is one of cooperation and not of domination over the activities of the State, to assist the State within this complex interconnected social system, almost in a single world system, to achieve development outside this constant social, economic, environmental, political, and cultural risk.

Development from the perspective of sustainability, which focuses on ensuring the improvement of the quality of life for longer and avoiding the loss of this in periods of crisis. Bring back legal certainty for taxpayers and a harmonious relationship between State, society, economy and environment, mischaracterizing the image of the "lion" (Tax Administration) the hunt for "food" (resources of society and economy) to survive without caring if there will be "food" (resources) the next day, because when there is this relationship the hunted goes after more resources to be able to survive away from the pursuit of the hunter. A striking scenario demonstrated in the global financial crisis (*subprime* crisis) of 2008 (which became a global economic crisis, and in later years became the crisis of the Social State), with several austerity policies and increased tax burden without proper analysis of the social situation of the country. To be able to bring the sense of justice in the performance of the State, whether social, fiscal, environmental or, in the most recent, in an environmental fiscal justice, to the taxpayer back and, more broadly, to the world society, the tribute needs to analyze the variables of its point of origin from the very complexity of the system, even without knowing for sure the exact results, but the possible ones, from past experiences and future probabilities.

To achieve this end, the State will need to use its public policies, in question of fiscal policies, to achieve the social function of the tax and consequently the materialization of the rights and sustainability of the Socio-Environmental State. It can be affirmed in the current scenario that the tax has **a socio-environmental function**, for every decision to be postulated should be analyzed in its environmental, social, and economic issues.

With all these perspectives, the socio-environmental function of the tax can be a tool to help reduce risks of society, State, economy, environment, and the world itself in which this tax is allocated to produce effects and effects produced on it.

OBJECTIVE(S) OF THE USE OF TAXATION FOR ENVIRONMENTAL PROTECTION

Addressing the objective(s) of environmental taxation or environmental taxation can be as simple as it is difficult, concomitantly. It will follow the objectives of traditional fiscal policies, which involve social and economic issues. It will also follow the issues of environmental policies, whose purpose is to avoid environmental risk, having principles of Tax and Environmental Law, without resulting in distortion in the premises of one or the other system, where solidarity, cooperation and integration will be fundamental pieces for this harmonious realization towards the sustainable development of the economic, financial, social, environmental, and public finance system.

In this *area*, taxes will be used, which are instruments of fiscal policies, to achieve environmental objectives, however, it cannot abandon social, economic, and public finance issues, needing to find a balance between the four, shaped in the holistic view of the model for the materialization of the socio-environmental tax function.

Society has several environmental, social, and economic problems, caused by itself. Environmental taxation will sometimes function as an economic counter cycle, that is, it will be used in the opposite direction of the world movement with strong intervention on social and economic decision-making, to correct the problems generated by the market, the State and society. Because of this characteristic, taxes will sometimes be *ad valorem* (countercyclical) to achieve a paradigmatic transition in the economic, social, environmental, and public finance context. At other times the environmental taxes will be procyclical, at which time the continuity of the behavior will be encouraged since it is already in accordance with environmental and tax issues, only when worked together, benefits from all *fields* without the loss of freedoms is that the issue of sustainable development can be addressed, as demonstrated by Ignacy Sachs, Jeffrey Sachs and Amartya Sen.

Environmental taxation will also act as an inducer/catalyst for behaviors already conforming to the molds of sustainable development. At this moment there will no longer be a strong intervention to curb, but to induce behavior in favor of these practices being reproduced by other economic agents or that conditions are created for their maintenance, without resulting in externality. If it happens, it will lose the bias of sustainability or simple non-intervention because the practice is already sustainable without state intervention or induction.

This conception has existed since the development of the notion of environmental externality, so that academic efforts have been around correcting the market failures generated by the external costs of production or consumption, creating a series of theories based on property rights and fiscal impositions, encouraging the use of economic instruments, whose purpose is to internalize social costs and alter the imposable base, ecologically nefarious (TUPIASSU, 2006, p. 91).

The objective of environmental taxation is the internalization of externalities as a form of solution with the compensation or restraint of these (GROOSMAN, 2000, p. 539), which cause damage to the environment and to society itself, by the resulting social problems, and the economy, by moving away from the ideal of perfect economy, in addition to inducing society, economy and State to adopt behaviors consistent with the model of sustainable development with this increase in marginal costs, in which environmental principles will be materialized within the tax bias. The environmental tribute will be the point of intercession.

With the advent of the concept of sustainable development, it was necessary to change the structure of the economy combined with environmental and socio-economic fiscal policies, eliminating the use without strategy of regulatory policies, which resulted in competitive market distortions, without compliance with the principles of environmental taxation.

Green or environmental fiscal policies emerge to optimize regulatory policies and may or may not increase revenues for the financing of public spending, although there is a tendency for revenues to be reduced, spending and public debt to fall. With environmental taxation dividends, the first to be identified were between preservation of environments and the generation of employment.

The concern with job creation has existed since the fiscal policies of the New Deal, focusing on the reduction of direct taxation (income and labor) and greater indirect taxation (TUPIASSU, 2006, pp. 92-94). This exchange of taxation has an environmental strategy, which cannot be ignored, the criteria have come to be to reduce and / or avoid an externality in the various species of environment, in addition to carrying out a clean tax reform, for the existence of the double dividend (FERREIRA, 2005, p. 272).

The double dividend of the environmental tax allows for environmental improvements, as well as reducing unemployment and increasing economic efficiency. Environmental taxes will not distort economic decisions but will guide them in a way that discourages economic losses with the proportion of economic burdens and economic gains, creating the possibility of reducing the tax burden of other taxes that are too high, so that the environmental tax, within the taxation transcends the merely environmental bias and enters the other areas, initially not worked (ROBAINA. ROSETA-PALMA, 2014, p. 5; UNION EUROPEAN, 1997).

The double dividend brought by the environmental tax would not cause damage to tax collection, as is reported by Lawrence Goulder (1994, p. 4), who gives three definitions regarding the double dividend:

a) **Weak,** in that the use of environmental tax revenues is to finance reductions in the marginal rates of another existing distorted tax, for cost savings compared to the case where taxes are returned to taxpayers in general.
b) **Intermediate**, in which a distorted tax is found and to carry out the neutral substitution involves a zero or negative gross cost.
c) **Strong**, in which you can remove any distorted or representative tax at zero or negative cost. With the revenues and the resulting possibilities, there is room for the materialization of environmental restoration and projects in need of financing, such as social benefits (SARAIVA, 2010, p. 687).

The exemption, based on the principle of the receiving protector, and increase, by the PPP, can be used as engines to reduce environmental degradation and improve environmental quality, in addition to using the administrative structure (already existing for the application of tax executions) to reduce administrative costs and supervision over command-and-control solution.

Another point is that revenues can relieve taxation on productive sectors and redistribute resources and generate economic stability, maximizing the generation of jobs through public spending, without losing the focus of changing the behavior of issuers, with a possible tendency to reduce revenue (SARAIVA, 2010, p. 687), as there may also be an increase in other sectors based on the principles of selectivity, progressivity and the polluter pays.

It is important to argue, although there is no exact proof of the double dividend, there are also possibilities for more dividends, such as: the fight against poverty; poor living conditions; in addition to the revenue obtained by environmental taxes will not have as its sole destination the replacement of the rec and it as obtained on the income of employees, as it will serve as a stimulus to affirmative action in favor of the environment and the reconstruction of environmental and economic public policies in general, meet the objectives of the Socio-environmental State, among them sustainable development (economic sustainability, social and environmental with maintenance and increase of freedoms (TUPIASSU, 2009. pp. 100-101).[5]

Other possible benefits of the environmental tax are: reduction of public spending; reduction of administrative costs; reduction of the number of taxpayers (optimizing tax collection and inspection, in addition, the taxpayers themselves may use tax avoidance as a way to avoid a higher tax burden, but this avoidance will be in order to use methods and technologies compatible with the healthy and sustainable environment); increase in the investment of clean development mechanisms (with the use of wind, solar, tidal, hydraulic, geothermal, nuclear, biogas and biofuel) or use of new PPE's (preferable products environmental), the environmental tax is in consensus that the environmental/ecological risk/damage should be reduced (TUPIASSU, 2009, pp. 102-104).

The dividends mentioned in the two paragraphs above are just some of the possible ones, there is no absolute certainty of the amount of dividends to be generated by each tax or tax system modeled from the environmental perspective. There is a probability that other indirect dividends will be generated, such as the improvement of the quality of the family environment, due to the economic, employment and environmental stability, possibility of stability of this environment as of the other species, are consequences of the dividends generated by taxation. Although there are limits on how to use the tax, the benefits will need to be recalculated to understand the real scope of each tax or the scope of the effects of the reformulated environmental tax system.

In summary, the tax in environmental perspective is at the center of the international discussion of environmental taxation and green tax reform with the adoption of Smart Tax Policies[6], and the environmental tax should be planned with analysis of the specific macro-good, the correlates in which it is being developed, its micro-goods and how the composition of these reflects on the macro-goods. How these influence the others (without forgetting the volatility and flexibility of the microgood), within the complexity of the system, to achieve the sustainability of the entire social, economic, environmental, and public finance system. This complexity will need to be analyzed in order not to violate the principle of tax neutrality.

The implementation of the environmental tax or an environmental tax reform is no guarantee that all these dividends will occur because it depends on several factors, such as those discussed above. Even with the implementation as addressed, there is no exact guarantee that the system will generate jobs, it is not known if there will be a positive impact on the environment and the materialization of other dividends, so it is necessary to build not only taxation, but other sectors, especially society to effect and legitimize the reform.

As is brought by Don Fullerton (1997), by various circumstances environmental taxes may, in fact, bring environmental improvements and reduce the overall tax burden of the system, however, in others may increase the burden of the tax system, so each reform will need to be evaluated on its own merits, in which the public revenue of the tax system may take a back seat, for all the harmony and functionality of the tax and economic system will be at risk, and thus the objective of reform may be lost.

The tax must be broad, with the aim of avoiding the generation of distortions in parallel markets, since capital will need to be relatively immobile, in an international perspective, otherwise the flexibility of the world financial system would allow the capital invested in one country could, to avoid the tax burden, move to another country.

Four are the reasons why capital can choose a particular market when taxes on labor are cut (OECD, 1999-2000): the degree of competitiveness in the markets for goods and factors of production (lower costs and higher employment); deadline for the introduction of green tax reform (so that there is a deadline for adaptation); existence of a "social consensus" for higher taxation related to the environment, whose purpose is to avoid pressure for wage compensation above that offered by the revenue recycling package; and international coordination of tax reforms would weaken compensation arguments based on informed concerns of sectoral competitiveness.

The purpose of the use of environmental taxes may be in the proper sense (prevention) or improper (recovery (SOARES, 2002. p. 15). The first acts on behaviors, creating incentives to change behaviors so that they are compatible with the environment and prevent future damage to the *natural continuum* (SOARES, 2002. p. 15). While the second focuses on goods or behaviors associated with external cost and inelastic demand, to allow the financing of the State without distorting the economy, signaling to economic agents the meaning of the technological evolution desired by society, in addition to enabling the constitution of ecological funds susceptible to financing the recovery of environmental balance (SOARES, 2002. p. 15). Based on the use of taxes, green or environmental tax policy may also follow these purposes depending on the instrument used, the object and the environment.

Cláudia Dias Soares (2002, p. 15) argues that one cannot abdicate the explicitness of the objective of the instrument, about the possibility of not having effectiveness and / or efficiency, in addition to deception of the taxpayer, because a tax that is placed as its own and aims at this objective can not be limited by the Laffer curve, by the fact that it can never fix prices to influence the conduct of the taxable person, failing to redirect the functioning of the economy, on the grounds of maintaining the level of revenues.

The question of exceeding the limit set by the Laffer curve goes beyond the mere question of falling revenue, it needs to ask the correct question, why did revenue fall? If it fell as a result of the tax avoidance of taxpayers for products and services that do not result in externalities, if it fell as a result of tax evasion, if it fell as a result of the outflow of capital from the country due to the effect of environmental taxation instituted or some effect that shook the international economy, such as a pandemic, or if it fell as a result of increased unemployment, it is necessary to understand the reasons why revenue has fallen in general and sectoral perspective, as well as the importance (proportion) of environmental taxation to this fall. If the answer is the first, then the change of behavior has materialized. If the answer is any other mentioned above, it should analyze the functioning of the country's economy and restructure the tax and/or the tax system to be able to create incentives for agents to change their process of choices.

Based on the principles of contributory capacity, non-confiscation and neutrality, the tax has a margin of incidence, the environmental tax will have a greater margin of incidence, approaching as close as possible to confiscation and as much as possible to the existential minimum, without entering into these margins, in which there is a reduction of freedoms and a reduction in the quality of life worthy of the taxpayer, may result in increased poverty, an objective that should not be set by any tax, to avoid encouraging the taxpayer to carry out aggressive tax planning (XAVIER, 2016, pp. 30-35)[7]. For this not to happen it will be necessary to create alternatives for the taxpayer, increase and reduction in other areas for the choice of the taxpayer, so as not to occur the loss of the purpose of the environmental tax when created or instrumentalized.

Among the ways to reduce and avoid tax evasion and aggressive tax planning, the principle of equivalence, applied in all tax species, with emphasis on the principle of benefit and cost, will be important for the effectiveness of environmental fiscal policy. You will need to carry out constant analyses between cost and benefit, in which the environmental tax will be better accepted. As a rule, when the benefits received are greater than the costs, as a result it will encourage the receipt of capital from foreign markets, which will enable the ease in the generation of employment and the tax burden originated from the environmental tax will be more acceptable in moral and economic perspective, in addition to being able to create balance in the trade balance and other sources for the financing of public revenues.

The purpose of the tax is in conjunction with that of the tax system itself and the state. It needs to meet the same biases of the socio-environmental function of the tax, analyze the collection, the extrafiscal effects it causes and the way the revenue is spent. Regarding the extrafiscal effects and public spending, in addition to the socioeconomic effect will also be analyzed the environmental impact of the spheres that the fiscal policy will affect social, economic, environmental, and public finances, for the short, medium, and long term.

As much as environmental issues are in vogue, the guarantee of other rights and the environment itself passes through the capacity of the State to collect and apply this resource, as it is formed in the Socio-Environmental State, built from the perspective of sustainable development that does not place the environment as a "sacred temple" and untouchable, nor does it create an overload of rights impossible to be fulfilled, It also aims to curb the economic speed, especially of the production system, so that it becomes possible to accompany it and never stagnate.

There is thus an environmental and fiscal solidarity[8], which should not only be analyzed from the present or national perspective, going beyond these limits. Solidarity must be beyond political borders, with international cooperation, overcoming current concerns to safeguard future generations who will inherit contemporary resources, instruments, and knowledge.

Solidarity became more impactful from the middle of the twentieth century, with the Universal Declaration of Human Rights (1948), in its article 1[9]. With it came the question of the dignity of the human person endowed with intellect, preventing treatment outside of respect for the person in his sphere of rights, from the perspective of fraternity (solidarity). In the year 1950, in the European Convention on Human Rights (2000), the Charter of Fundamental Rights of the European Union in its preamble[10] brings not only solidarity, but freedom, equality and the dignity of the human person as inviolable bases of the European Union.

Solidarity expresses the fundamental need for the existence of the human being in a society, elaborating a system of intersubjective and social connections traced in the state community, far beyond a mere juridical-normative relationship, as the main conditioning factor for the development of society and the rule of law itself, whose vanguard is the protection and promotion of fundamental rights, whether in the present generation or in the future generation, because this principle does not act in isolation in the normative system, but in conjunction with other principles and legal values, especially redistributive and corrective social justice, from the perspective of the dignity of the human person and environmental protection (SARLET. FENSTERSEIFER, 2014, pp. 69-70).

The environmental tax will be within the framework of solidarity, cooperation of the whole society and sectors to build the model of sustainable development with its objectives. Clearly, the objectives of the tribute are not limited to economic and environmental issues, they are in a plan in which joint action is required with the other policies, coordinated, and integrated, whose objective is to bring together man and the environment for their own survival.

There is no obstacle between tax and environmental principles to be used in harmony. The unifying principles are the principles of solidarity, existing in both Environmental Law and Tax Law, and the principle of integration. The tax principles used under the genesis of the Liberal State and the environmental principles under the genesis of the Socio-Environmental State cannot be used in a classical way, without adaptation, but must be used as in the model of the Socio-Environmental Tax State, with the intelligent and strategic use to achieve the objectives of the current State model, which is the use of tax tools to reduce social, economic, financial and environmental risk, that is, both will need to be instrumentalized in the new social reality, in its contemporary complexity, without forgetting the probable futures.

WAYS OF TAXATION USED TO ACHIEVE THE OBJECTIVE(S)

In all forms of taxation, you will be faced with two choices, the public and the private, State and Market, "iron hand" and "invisible hand". The first will check the possibilities of using this revenue before and after it is collected, if highly efficient, before being collected will be able to reduce the risk and should not bring any damages. At the same time, it has the private choice of no longer wanting to use capital for the high costs, which result in externalities, and new market failures. If it is not possible, the revenue will have to be collected for subsequent application of revenue in a public expenditure, in which the choices of the State gain greater prominence, because the problems may have already occurred or not, so the choice between wanting to do some preventive act or not is strictly state, with the possibility of falling into the tragedy of public choice or not, the way the entire fund system, from tax application to the generation of public spending, will work.

Environmental and fiscal principles will need to be observed when using the environmental tax in any way.

Direct Taxes on Emissions

Taxes on emissions create fiscal impositions from the direct relationship with the volume of emission or with respect to the dumping directly or indirectly affecting the air, water, soil, cities, the work environment, family, among others, that is, any form of emission of element that makes impossible or hinders the proper environmental balance, of the economy and society, resulting in externalities, is also known as output taxation (profit and distribution), which can be developed through taxes and contributions (RNCREQ, 1998, p. 14). This form of taxation is more efficient in relation to the producer, the one who develops the economic activity and directly is the cause of that inappropriate emission for the environment (MONTERO, 2011. p. 172; EUROPEAN UNION, 1997. p. 4).

An example of taxation on emissions is on carbon dioxide (CO2), which is usual to try to change consumption and increase revenue through taxation on gasoline, diesel, coal, and natural gas (GUZMAN,

2002, p. 49). Another way is through the taxation of noise caused by motor vehicles, which, when they exceed certain decibels (dB), suffer the incidence of a certain tax burden.

Taxation on emissions creates the possibility of inducing the development and implementation of new techniques and strategies, whose purpose is to exchange the consumption of environmentally degrading products for more sustainable ones, with the possibility of causing an increase in the prices of these polluting products in the market, increasing the competitiveness of less polluting and more environmentally friendly products (CARDOSO, 2015, p. 42).

It is important to raise that its application is only feasible when the reasons of the polluting sources and their measurements are identified and there is no doubt as to their origin to apply the correct taxation on the emission, otherwise it may tax the generating fact of those who do not emit any type of emission. It is not advisable to apply it when the high number of elements results in a high administrative cost, so that a prior analysis of both the amount to be taxed and the administrative cost is necessary to be able to collect this amount (CARDOSO, 2015, p. 42).

Taxation without the certainty of the elements or agents causing damage to the environment may result in the violation of the principle of neutrality (by unjustifiably intervening in the individual freedoms of the taxpayer that does not cause damage to the economy, possibility of increasing inequality), polluter pays principle (only those who are not contributing in a specific act may be encumbered), recipient protector (one who already has practice according to sustainable development can not be specifically burdened and be created presumption, so as to also hurt the principle of real taxation), prevention and precaution (will not be preventing any act, on the contrary, the State will give signs that it does not seek to avoid a damage, but only collect), equivalence (when it results in the loss of unjustified revenue, without bringing benefits, in addition to having the cost increased) and unduly reduce the freedoms of taxpayers, in addition to possibly causing the opposite effect of what was intended.

The application may be better when there is the issuance of the respective harmful element, added to the due direct supervision to ascertain the respective issuance, to avoid possible tax litigation, in which there will be a discussion of how much was issued, in addition to the State needing evidence to be able to demand the proper application. These questions are within the principle of real income taxation.[11]

The presumption or estimate may be of complicated application, however much it is possible and has a low cost, which can be calculated based on the income in a ratio of proportion and may encourage the taxpayer to acquire PPE and install them in their production process or change the consumption of harmful intermediate goods and services.

To avoid the high cost, the calculation can be carried out by sample or by proportion, the second based on the taxpayer's income statement and the equipment used, while the first is by the supervision of the public administration. Despite this possibility, there will remain a certain degree of uncertainty, the possibility of the taxpayer being able to reduce the tax burden through administrative defenses before the tax authority or in the judiciary. To avoid possible disputes, electronic, digital, and on-site surveillance should be used for the correct application of the tax burden to internalize the environmental cost and encourage a different conduct in its business procedure.

Usage Fees

This form of taxation is more specific than the previous one, because in the former there is a range of forms of taxation, especially the use of taxes and contributions. In this second, it will be in specific the use of the fees, which will focus on the use, which will be payments by a natural person (physical) or legal for the benefit of a specific and divisible service provided by the public power or of a public good made available directly proportional to the amount paid, having its efficiency in protecting the environment and reducing the use of natural resources, because of the cost recovery function, with the usage fees being a direct application of the polluter pays principle (RNCREQ, 1998, p. 24).

Taxes can be applied on products such as raw materials, on intermediate inputs of the production process and on products. Taxes on products have existed for many years, especially in the field of energy for integration between the balanced and healthy environment with respect to the energetic ones, for example, taxes on gasoline, diesel and heating oils and electricity (EU, 1997, pp. 4-5).

The taxes, in addition to focusing on products, can also be public services available or effectively provided to the taxpayer, for example, domestic waste collection service, tourist tax for the conservation of tourist spots, among others.

When a tax has as its generating event the act of polluting, based on the polluter pays principle, the requirement to pay this fee may result in a change in the intention to pollute, discouraging it (COURY-CHANEL, 2016. p. 5). Can analyze some aspects according to Coury-Chanel:

a) **Fact of polluting:** it will necessarily result in tax contribution of the taxable person of this tax relationship, which is the polluter, perceiving the direct relationship between the act of polluting in relation to the act to be performed by the public power as a response to the act of polluting situation or the risk of polluting (COURY-CHANEL, 2016. p. 11).
b) **Passive amount:** defined by the sum of the damages caused by the polluter to third parties or to society by its activity resulting in actual or potential pollution (COURY-CHANEL, 2016. p. 11).

When it is analyzed from the perspective of the polluter pays, when using the public goods available, it will not find “friction” with the tax legal system. As can be demonstrated in the previous paragraph, there are two characteristics of the tax levied on externalities, the fact that it pollutes and the amount to be calculated based on the damage caused.

There may be two obstacles: the first, if the rate is calculated in relation to the goods made available and the second when the provision of service by the public administration occurs. Both can only be legitimate if they are calculated according to the public service of restoration of the degraded environment or service performed for its conservation, precaution or prevention, or the public service to make available the respective public good, if there was a production of an essential or useful public good by the public power may exist the fee based on the costs of production.

Regarding the first case (service), the calculation of actual or potential damages caused to third parties can be applied, at which time the tax will be based on full calculation.

Regarding the second case (goods) of making a good available, the costs to maintain the environmental balance that have been drawn from the environment for delivery to the private initiative may be included. We need to consider the administrative costs, which should be as low as possible.

As a result of not being able to forget the synallagmatic character of the fees, in which there is a direct service provision for the amount paid or the public good made available by the government, this cost analysis is based on the principles of fiscal equivalence and neutrality, which will be extended beyond the issue of administrative cost as well as touch on the environmental benefit generated.

The tax may not be calculated or applied as an attempt to restore a *status quo before* the polluting effect of that activity and the public power seeks to revert to those who suffered directly or indirectly the damage. There are financial instruments applicable for applicability such as fines and fines imposed by environmental inspection bodies. The essential idea is that of retribution, the absence of this element makes it possible to claim illegitimacy of the fee (TUPIASSU-MERLIN, 2009, p. 274).

Although there is ample scope for the application of the rate, there are crucial points that hinder the application of the respective tax tool: a) confirmation of the real polluter, when there is a conglomerate of agents, to make complex, sometimes impossible to measure the volume of individualized and specified damage (SOARES, 2002, p. 75); b) clearly define the radius of impact of the damage, which will affect the relation to the cost of depollution or to avoid it, and may. (TUPIASSU-MERLIN, 2009, p. 276)

The principle of contributory capacity and the prohibition of confiscation may be two other principles limiting the internalization of costs by the polluting agent, although the principle of equivalence, neutrality, prevention, precaution, and PPP is not an obstacle. With the possibility of identification and determined calculations, because the first principle will inform the minimum and the maximum in which the respective tax can be used to internalize the cost of the damage, while the second principle will work with the first to totally curb the excessive taxation on the polluting agent in such a way as to be an impossible amount to be paid according to its capacity. In this scenario, there is an upward limit of how much the government can charge, which may make the total internalization ineffective, if it turns out to be too large, and possibly efficient in minor damages or in a preventive way.

Although the rate is limiting, it does not mean that it should be abandoned in the context of green fiscal policies, since in the medium and long term it has the capacity to have effects for the polluting agent to change the polluting habit. It is only shown that it is not the only method.

Indirect Taxation

Indirect taxes, on products and services, can be an alternative to direct taxation on emissions and fees, since the economic situation does not allow direct taxation or the cause of the environmental/ecological damage is not clear or cannot be taxed directly, consequently values the environment, and corrects externalities, in addition to reducing environmental/ecological damage (RNCREQ, 1998, p. 25)

Taxes on goods and services have as scope a greater extension by their differential characteristic, such as taxing raw materials, in which the radius of influence may be greater, due to being a good of common use of various activities and production processes, which may occur in a more punctual way on the consumption of intermediate good and on final consumption, when the actual taxpayer/consumer acquires that good or service (TUPIASSU-MERLIN, 2009, p. 281).

The taxes levied on the production chain end by being passed on to the final consumer, in the end the consumer will be influenced to a lesser or greater degree to consume goods and services resulting in fewer externalities, which will also be the cheapest, because it is less taxed. In the first two cases it will be to maintain competitiveness in the market and in the last case the consumer will choose to switch to products and services less harmful to the environment (TUPIASSU-MERLIN, 2009, p. 281).

Indirect taxation can be used when pollution is related to consumer demand, such as disposable products or vehicles with low fuel consumption, in which the tax will be linked to the product itself and its potential components or in fact harmful to the environment, when raw materials or intermediate goods are affected by taxation, taxes levied on *commodities* (RNCREQ, 1998, p. 25).

For example, the taxation on automobiles is a good example, because these can cause emission of contaminants in the atmosphere, noise, congestion and accidents, to solve can happen the taxation in three ways (GUZMAN, 2002, pp. 51-54): a) **on the sale**, when the motor vehicle powered by combustion potentially causing more ecological damage will have its taxation increased, while those who do not use combustion for their movement will receive lower tax burden, as with pure and hybrid electric vehicles; b) **annually** on the automobile, annually increases the ownership of the motor vehicle, which may be based on its generating fact the power of the engine, the number of displacements or weight of the automobile; c) recycling and destruction of the **automobile**, when the taxation will be based on the refundable deposit system, at the time of purchase of the vehicle the corresponding tax is paid and the amount collected enters a waste management financing fund Coming from motor vehicles, when depositing the vehicle for its destruction in one of the places authorized by the public administration, the owner will receive a certificate, which allows to receive a sum of money.

In indirect taxation, taxes and contributions gain prominence, because they can be levied indirectly on products and services, and may have the unilateral characteristic, as are taxes, or the synallagmatic character, as is the case of contributions.

Taxation at Different Rates

This model can work based on the structuring of indirect, direct, specific taxes on the sale of products or services or on added value to protect the environment. With respect to those goods or services, as in income, when they come from practices that cause environmental/ecological damage, they may have a higher load or even a lower load (RNCREQ, 1998, p. 25)

The application of different tax burdens makes it possible to increase costs in a uniform way. It will be increased for all products, services and income that contain specific characteristics stipulated for higher or lower taxation, not being advised for each producer, because it is not possible, because indirect taxation is being worked. Regarding taxation on the producer, there is taxation on emission, which is a direct taxation, so, based on the differentiated rates, there will be encumbrance only of the final product and not in its production chain (CARDOSO, 2015, p. 43).

In addition to the use of different tax rates, several tax benefits may be used, such as exemptions, subsidies, compensations, among others. For example, as demonstrated in the Portuguese tax reform of 2014, rates based on the essentiality of goods and services for the incidence of VAT, which partially or fully exempt the use of electric and hybrid vehicles.

The different rates may better touch the distribution of the costs internalization charges, because, according to the practice, consumption, income and the economic, social and environmental conjuncture there will be, in specific situations, increase to a greater or lesser degree, internalize more or less, so that the respective form enters the other ways as another variable to be analyzed within each form of use of taxes, be they direct, indirect, real and the fees.

Fund

When funds are approached, one is working directly with contributions, which will have characteristics like taxes, but with indirect synallagmatic provision, in which first the revenue is collected for later application of a specific public service, which can benefit an individual, a collectivity or be diffuse.

In the environmental bias, it will be used when there are no conditions for the provision of direct and specific service by the public power nor can the responsibility (SOARES, 2001, p. 181. SOARESb, 2001. pp. 64-67) be applied to the entity causing the environmental/ecological damage or even to build a reserve plan for the risk that lies ahead.

The taxes that will be used as revenue to compose the fund do not fail to use the same effects of direct taxation, such as the IRS and the IRC, in the system, to seek the improvement of the economic and environmental conjuncture, and it may be autophagic or be collected for later application.

In an "ideal" or "perfect" economy, the contribution and the fund should not exist given the proper functioning of the markets, but as this is not what happens, it is necessary to take them into account and they are a way of using the tax, the contribution, for the application of green fiscal public policies, now related to the collection and the subsequent public expenditure to be made to achieve the environmental purpose. Other revenues may be linked to the constitution of these funds, such as taxes and public loans and international loans to international organizations and/or other countries, or even the State itself uses its original revenue to constitute it, such as the activity carried out by Public Enterprises, or by intergovernmental transfers.

The application of the fund for the repair of any environmental/ecological damage may not be effective against the damage that has occurred, as the fund may be preventive or repressive. In the preventive and repressive possibility, the difficulty in measuring the radius of beneficiaries is very great, in addition to the proportion of the benefit, not being sure if the one who is closer to the core of the application of financial resources will benefit more than the one in a more peripheral region. The very delimitation of the periphery of the effects of the policy is difficult to be defined because it is not known at least the reaction of the microbens after the new scenario built with the financial resources of the fund, because they transcend the macro goods.

Another crucial point is the possibility that the financial resources are not sufficient and/or their being applied inefficiently in the repair or prevention of risk/damage. This second is a matter of tax management to be applied, in which the fund itself needs to have sufficient resources to bear the administrative costs and be added value within the macroeconomic, social, and environmental conjuncture. If the total costing of administrative and tax costs does not occur, it will be necessary to withdraw resources from another source of revenue, generating the possibility of imbalance in other environmental, social, and economic sectors.

On the Externality Generated by Heritage

This taxation will be distinct, because it will not reach the externality directly, but will use the taxpayer's conduct so that its patrimony is not an ecological and environmental aggressor. Higher and lower tax rates may be used to encourage conduct, in addition to tax benefits. If it is not possible to measure exactly the amount of issuance and/or damage caused to the taxpayer's assets, the increase in taxes on the patrimony may be applied.

The tax levied on the patrimony is that it will receive the benefit and not the taxpayer in analysis of its income, but on the environmental, economic, social effects that the patrimony generates from externalities, which differs from the direct taxation in which it touches on intermediate and final goods and services. As an example, there is the green property tax in some Brazilian municipalities, where there will be exemptions for those who use photovoltaic panels in their homes or buildings. Andin Portugal there is a similar policy and uses the IMI exemption for buildings that use photovoltaic panels. Another example, the reduction of IPVA and ISV rates on electric and hybrid vehicles. Both situations founded on the principle of the receiving protector. Regarding the PPP, the property tax in Brazil may be increased when in the sense of progressivity, the social function of the property is not fulfilled, being at the discretion of each Municipality, according to its master plan, to delimit what will be the social function of each property in each region of the Municipality.

Taxation may also be increased, if the property is an emitter of gases, through taxes, contributions or fees, but not be measured according to the ecological / environmental damage, as is presented in paragraph "a", although it can use the amount of emissions and damages to verify the applicability of greater or lesser tax burden, it is only a way to increase the cost, due to inappropriate behavior towards the objective of the Socio-Environmental State, using the PPP.

Improved Form(s) of Tax Instrumentalization for the Objective(s)

It is common to expect the existence of an ideal way to use the tax to correct market failures, or at least criteria of which would be the best. Despite this optimism, the answer depends on how the market failure, the externality, the economic, social, and environmental conjuncture behaves to be able to define an environmental fiscal policy that can internalize the environmental costs and materialize, change behaviors.

The forms of use are general, which can still be combined, such as the question of contributions and taxes. The first must be linked to future public expenditure and the second not (being provided, according to the legal system). In addition to constituting funds may have the effect of direct taxation, with double application, however, may have its effectiveness questioned, and can only be proven in the medium and long term, because, at first, it may not be a suicidal tax (will have collection), to then be (with the drop in collection), to the detriment of the change of behaviors of economic agents (producer and / or consumer), In addition to the use of different rates, progressively or not, according to the potential damage and risk to be caused, the tax burden may be changed.

Indirect taxation may, as a rule, be more effective with the taxpayer of lower income, because it is more subject to the effects of costs, but this form will not work if there are no other policies that facilitate this change of habit, otherwise it will result in increased poverty, it will be contrary to fiscal policy as environmental.

In conjunction with indirect taxation, it is possible to use fees, as Portugal did with the fees for plastic bags (increase of € 0.08 on the price of plastic bags), and divided between the State, the Fund for the Conservation of Nature and Biodiversity, the Portuguese Environment Agency and IGAMAOT (General Inspectorate of Agriculture, the Sea, the Environment and Spatial Planning). This fee will focus on ensuring resources for proper government management of the environment and the constitution of a fund. This form of use is still legitimate, because there is consideration (retribution) by the direct public administration related to the consumption of plastic bags.

Another possibility is the set of all forms, for example, a business society from its activity that causes externalities have to contribute with direct and indirect taxes, on equity, fees and contributions to the constitution of funds, without the occurrence of double taxation. However, here there are already limitations of possibilities of placing in a single tax, as is the case of direct and indirect taxation, by the optics of contributory capacities are different, in the first there is the personhood and in the second not, already to change much how the tax will act, there being no impediment to the accumulation of taxation on the patrimony, direct, contributions and fees. That is why it will be necessary to organize at least two groups: direct taxation and indirect taxation, which in each may have the accumulation of forms of taxation.

Another way of taxing is from the tripartite organization of the forms of taxing externalities: direct, indirect and patrimony (the latter also called real taxation, in allusion to real rights). With the possibility of using short-term revenues to finance funds or not, depending on the need, in addition to being strategically organized to increase marginal costs. In addition to giving the taxpayer the possibility of choosing the most appropriate solution for your case (SOARES, 2002, p. 67).

The possibility of explaining the issue of a single tax for all issues is out of the question, much less a single tax with standard rates for all, for failing to achieve the various characteristics of taxpayers, an issue that could violate in individual perspective of the ability to contribute, PPP, recipient protector, progressiveness, cooperation, solidarity, taxation on real income, as well as causing more distortions. Only a single tribute would be possible in the existence of homogeneity and an extremely simplified economic system, but it would lose one of its great assets which would be the heterogeneity to meet the diverse individual and social needs and desires. Although the single tax simplifies tax administration, it cannot be done at the expense of the principle of isonomy and in the light of the prohibition of social and environmental regression.

Claudia Dias Soares (2002, pp. 538-554) reports that in some situations you will not be able to use taxation to change behavior for environmental purposes or even if it is used it will have its effect softened or ineffective, such as: polluting emissions with persistent effect[12]; when the emission occurs from a compound of polluting elements[13]; emission of substances causing non-linear damage[14]; economies of scale and monopoly[15]; random variables and different assimilative capacity of the environment[16]; concentration of polluting sources in a given area[17]; lack of literality or discontinuity of the CME function and multiple optimal location situations[18]; concentration of pollutant emissions in a given period of time[19]. In summary, when the situations are not clear or are extremely risky to ensure well-being, you will need to use other methods, which have short-term effects.

In summary, the forms of taxation, at some point, end up repeating themselves and mixing themselves to be able to instrumentalize the fiscal tool. There is no better way to carry out taxation, what would be better is the intelligent use, interconnected, to be able to internalize and move the behaviors to other types of consumption and production, with care not to cause more market distortions to the point of making it more imperfect. It is still necessary to have the vision interconnected with other environmental economic policies and financial policies so that it can have its potential increased exponentially and not undermined, such as the use of the public budget and intergovernmental transfers.

CONCLUSION

This article aimed to analyze the possible ways of using environmental taxation to achieve its objectives within the model of Socio-Environmental State. The methodology used was bibliographical.

In the first chapter it was addressed about the function of taxes in the Socio-Environmental State. It was analyzed in this model that the tribute has, what could be called, socio-environmental function, since it has once constituted precise to be built concerned with economic, social, and environmental issues. This function can be an aid tool to combat the risks of society, including climate change.

The second chapter discusses the objective(s) of the use of taxation for environmental protection. There are several objectives: such as the internalization of externalities, generation of dividends, functioning as catalysts, sometimes as countercyclical, stimulators/discouragers of behaviors.

Of all these objectives, the most sought for a proper analysis is the amount of dividend generation since there may be a double dividend: environmental improvements and job creation. It is also possible to generate more dividends or even indirect dividends, depending on how the tax structure is built in that society.

The tax needs to be built in an 'intelligent' way so that it has the generation of dividends, functioning as prevention or repression of environmental damage, so that fiscal and environmental generational solidarity is maintained, whose objective is that there is no overload between generations.

The last chapter deals with ways to use taxation to achieve your goals:

a) the first form is direct taxation on emissions: by which there would be a higher tax burden for those who emit more polluting gases, the advantage of this form of taxation is that the burden is only on those that generate externalities but would have a high administrative cost of inspection and verification.
b) The second form is the use fee: in which the taxpayer pays for the raw material made available or for the service provided by the public administration to avoid certain damage or repair the damage caused by the taxpayer. The advantage of this taxation lies in the specificity of the polluter, but the existing disadvantage lies in the difficulty of measuring the passive amount to be calculated.
c) The third form is indirect taxation: unlike the previous two, this taxation does not directly affect the polluter, but the goods that are used for the pollution. The advantage is to be able to tax products that affect the pollution chain, being able to interrupt, but the disadvantage is in sometimes getting the proper identification of the good that enhances pollution.
d) The fourth way is the use of different rates: instead of creating a specific tax, the existing ones are used, in which the most polluting and largest polluting products will have higher rates to be paid when compared to other products. The advantage is that of better distribution of the cost's internalization charges, the disadvantage is that there is a limit to these differentiations.
e) The fifth form is funds: there is first the collection of the tax, in which part or all the tax revenue from one or several taxes are used for a fund to make the necessary investments. The advantage is that there is a concentration of revenue for later service by the State, but the disadvantage is that if the tax is not constructed intelligently, it may be merely tax revenue, without being able to discourage conduct.
f) The sixth and final form is the externality generated by wealth: in which there would be higher taxation on goods that are polluting or have pollution potential. The advantage lies in being able to identify the specific capital goods that result in some type of pollution, but the difficulty is still in being able to measure the level of pollution so that the tax burden is increased proportionally.

Finally, an analysis is made as to what would be the best way to use to correct market failures, improve the environment. The conclusion is that it is necessary to use the various techniques, as seen none is perfect and all present advantages of disadvantages.

REFERENCES

Cardoso, A. C. C. (2015). *O imposto como instrumento de desenvolvimento sustentável.* [Unpublished master dissertation]. University of Porto, Porto, Portugal.

Carvalho, C. (2018). *Teoria da decisão tributária*. Almedina.

Coury-Chanel, E. (2016). *Esboço de uma teoria fiscal da Ecotaxa. Tributação, meio ambiente e desenvolvimento. Forensics*. CESUPA.

Dourado, A. P. (2016). *Direito fiscal: lições*. Almedina.

European Union. (1997). Commission for European Community Affairs. *Environmental Taxes and Charges in the Single Market*. http://aei.pitt.edu/4785/1/4785.pdf

Ferreira, E. P. (2005). *Ensinar finanças públicas numa faculdade de direito: relatório sobre o programa, conteúdo e métodos de ensino da disciplina*. Almedina.

Fullerton, D. (1997). *Environmental Taxes and the Double-Dividend Hypothesis: Did You Really Expect Something for Nothing?* University of Texas. https://ase.tufts.edu/economics/papers/9706.pdf doi:10.3386/w6199

Goulder, L. H. (1994) Environmental taxation and the "double dividend:" a reader's guide. *NBER Working Paper Series*. Standford.

Groosman, B. (2000). *Pollution Tax – entry 2500. The 'Encyclopaedia of Law and Economics*. Edward Elgar.

Guzman, T. J. (2002) Fiscalidade ambiental: análisis y efectos distributivos. Granada.

Hyun, J. K. (2006). Tax Compliances in Korea and Japan: Why are they so different? *The journal of the korean economy*.

Montero, C. E. P. (2011). *Extrafiscalidade e meio ambiente: O tributo como instrumento de proteção ambiental. Reflexões sobre a tributação ambiental no Brasil e na Costa Rica*. [Unpublished doctoral dissertation]. State University of Rio de Janeiro, Rio de Janeiro, Brazil.

NABAIS. J. C. (1998). *O dever fundamental de pagar impostos*. Almedina.

OECD, Organization for Economic Co-operation, and Development. (1999-2000). *Greening tax mixes in OECD countries: a preliminary assessment*. OECD. https://www.oecd.org/env/tools-evaluation/2385866.pdf

OECD, Organization for Economic Cooperation and Development. (2005). *La reforme fiscale écologique axée sur la réduction de la pauvreté*. OECD. https://www.oecd-ilibrary.org/docserver/9789264008717-fr.pdf?expires=1571321286&id=id&accname=guest&checksum=1289F017793C27021D9BCF44DF485EBD

Palil, R. M., & Mustapha, A. F. (2011). The Evolution and Concept of Tax Compliance in Asia and Europe. *Australian Journal of Basic and Applied Sciences*.

RNCREQ. (1998). *Les instruments économiques et la protection de l'environnement*. RNCREQ. http://www.rncreq.org/pdf/instruments.pdf

Robaina, M., & Roseta-Palma, C. (2014). *Impostos ambientais e o duplo dividendo: experiências europeias*. (Working Papers in Economics.)

Saraiva, R. C. G. (2010). *A Herança de Quioto em Clima de Incerteza: análise jurídico-económica do mercado de emissões num quadro de desenvolvimento sustentado*. [Unpublished doctoral dissertation]. University of Lisbon, Lisbon, Portugal.

Sarlet, I. W., & Fernsterseifer, T. (2014). *Princípios do direito ambiental*. Saraiva.

Soares, C. D. (2001). *Instrumentos de promoção da qualidade ambiental: o imposto ecológico*. [Unpublished master dissertation]. University of Coimbra, Coimbra, Portugal.

Soares, C. D. (2002). *O imposto ambiental: direito fiscal do ambiente*. Almedina.

Soares, C. D. (2014). *Smart tax policies: environmental financial reform in times of economic crisis. Does austerity cure? Does austerity kill? SOARESb, C. D. (2001). A receita dos impostos ambientais*. Scientia Ivridica Separata.

Streeck, W. (2013). *Tempo comprado: A crise adiada do capitalismo democrático*. Conjunctura Actual Editora.

Tupiassu, L. V. (2006). *Tributação ambiental: a utilização de instrumentos econômicos e fiscais do direito ao meio ambiente saudável*. Renovar.

Tupiassu, L. V. (2009). *A reforma fiscal verde e duplo dividendo. Direito Tributário e Econômicos aplicado ao meio ambiente e à mineração*.

Tupiassu-Merlin, L. (2009). *Contribution de la fiscalité a sa realisation: Le cas brésilien*. [Unpublished doctoral dissertation]. University of Toulouse Capital-1, Toulouse, France.

Xavier, C. A. C. (2016). *Planeamento fiscal agressivo: delimitação conceptual de uma nova realidade*. [Unpublished master dissertation]. New University of Lisbon, Lisbon, Portugal.

KEY TERMS AND DEFINITIONS

Dividends: Benefits generated.

Environmental taxation: Technique of building taxation to meet environmental public needs.

Smart Tax Policies: Tax policies that do not only seek to collect revenue, but also reflect on other areas.

Social function: Collective or diffuse allocation given to a particular good or right.

Socio-environmental function: Collective or diffuse environmental destination given to a particular asset or right.

Socio-Environmental State: The form of State that seeks to meet public economic, social, and environmental needs.

Socio-environmental tax function: Collective or diffuse allocation of taxation from an environmental perspective.

ENDNOTES

1 This discussion is extremely difficult, due to the complexity that is the economic structure, to be able to delimit a practical situation, in an objective way, that the economic issue is superior to taxation, because both have their objectives as already mentioned in the course of this work, each

one will have a function and both are relevant in the feedback system between State, society and economy.

2 When the State raises taxation, it needs to be too careful, because it may generate incentives for its own desertion, from which, from the perspective of cost-benefit rationality, the taxpayer will require benefits greater than the amounts paid, otherwise he will probably choose to avoid the tax burden in a lawful (with the use of tax planning) or illicit (through fraud) simulation and dissimulation), with physical change of the taxpayer or his resources to more favored jurisdictions, such as "tax havens". (CARVALHO, 2018, p. 237).

3 As is demonstrated by Goldscheid, from the point of view of the sciences of finance, on the Marxist theorem, between the growing contradiction of the capitalist mode of production and the relations of production, the Fiscal State will reach its maximum capacity to extort from its citizens the resources necessary for the provision of the services required for the provision of the State. At some point they would cease to be sufficient and could result in an indebted state (STREECK, 2013, pp. 116-118).

4 When the Fiscal State becomes an Indebted State, it ceases to depend solely on its taxpayers and begins to depend on the confidence of the creditors of the markets, among them the one of great influence since the 70s of the twentieth century, the financial markets, which is at the transnational level, connected to the countries through contracts, that is, linked as investors and not as citizens, completely changing the relationship of the State, placing it as dependent on private initiative that is not totally under its jurisdiction, already enters the State under contracts and conditions (STREECK, 2013, pp. 129-130).

5 "La RFE peut directement contribuer à réduire la pauvreté en aidant à faire face aux problèmes d'environnement (tels que la contamination de l'eau et la pollution de l'air) don't sont victimes les pauvres. Elle peut également y contribuer indirectement, en générant ou en libérant des ressources permettant de financer les investissements 'infrastructures essentiels pour les démunis: distribution d'eau, assainissement *and so on*. Elle peut par ailleurs concourir à libérer des ressources budgétaires pour d'autres investissements utiles aux défavorisés dans des domaines tels que la santé et l'éducation. La RFE peut donc jouer un rôle important dans la réalisation des objectifs du millénaire pour le développement visant à éradiquer l'extrême pauvreté et la faim et à assurer un environnement durable." (OECD, 2005, p. 56).

6 "... a qual se baseia em cinco princípios essenciais: (i) a redução dos impostos sobre o trabalho e sobre o rendimento das empresas; (ii) o aumento da fiscalidade ambiental; (iii) o aumento da base tributação; (iv) a redução do efeito que a fiscalidade provoca no favorecimento do endividamento; (v) a redução da fraude e da evasão fiscal e aumento da eficácia da cobrança fiscal" (SOARES, 2014, p. 311).

7 Other policies can also be used to prevent tax evasion, such as tax compliance (HYUN, 2006. pp. 135-153; PALIL. MUSTAPHA, 2011. pp. 557-563).

8 As fiscal solidarity has already been addressed, only environmental solidarity will be addressed.

9 "Artigo 1. Todas os seres humanos nascem livres e iguais em dignidade e direitos. São dotados de razão e consciência e devem agir em relação uns aos outros com espírito de fraternidade."

10 "Consciente do seu património espiritual e moral, a União baseia-se nos valores indivisíveis e universais da dignidade do ser humano, da liberdade, da igualdade e da solidariedade; assenta nos princípios da democracia e do Estado de direito. Ao instituir a cidadania da União e ao criar um espaço de liberdade, de segurança e de justiça, coloca o ser humano no cerne da sua acção."

11 The principle is clearer within the legal system Portuguese. Found in Article 104(2) of the CRP, "“A tributação das empresas incide fundamentalmente sobre o seu rendimento real”.". There is a distinction between taxation on real income and normal (or average) income. In the first (real) there is a presumption that it would be real, it is admitted proof to the contrary, calculated based on the difference between income and expenses, calculated by direct methods such as accounting. While the second (normal or average) taxation leads to situations of intolerable inequity. (NABAIS, 1998, pp. 501-502; DOURADO, 2016, p. 196).

12 “... for emitida uma qualquer substância cujos efeitos sejam persistentes, isto é, cuja componente danosa para o equilíbrio ecológico se caracterize por uma grande longevidade, a tributação não será o instrumento mais adequado. Pois, caso sejam emitidos elementos poluentes não biodegradáveis e dotados de efeitos susceptíveis de perdurar no tempo, as lesões ambientais serão continuadas. Assim, a dimensão do dano apurado no momento em que as emissões em causa são geradas, a qual serve de base de cálculo do imposto, será inferior ao verdadeiro custo ambiental a que aquelas irão dar causa.” (SOARES, 2002, p. 539).

13 “A tributação desdobrada de cada uma delas individualmente, atendendo apenas ao seu efeito isolado, não permitirá atingir o nível óptimo de poluição, em virtude de, neste caso, em regra, diverso daquele que provocará uma emissão combinada dessa mesma quantidade com outras substâncias poluentes, sendo o impacto ecológico, normalmente, maior no último caso.” (SOARES, 2002, p. 540).

14 “...só ocorre um dano ambiental quando os indivíduos reconhecem que sofreram uma perda de bem-estar. E quando as emissões poluentes geram um impacto reduzido no equilíbrio ecológico, os sujeitos podem não se aperceber do efeito destas sobre o seu bem-estar (...). Quando existe um limite abaixo do que qual o dano ambiental causado pelas emissões poluentes não é significativo mas acima do qual graves consequências se produzem no equilíbrio ecológico, a utilização destes instrumentos não é desejável também em virtude de a flexibilidade que os mesmos conferem aos agentes económicos não permitir que se saiba com certeza, à partida, qual o montante exacto de poluição que será gerado, não se podendo, assim, garantir que a fronteira perigosa não será ultrapassada. Nestes casos, serão preferíveis elementos de imposição...” (SOARES, 2002, p. 542).

15 “No caso da empresa poluidora deter uma posição monopolística no mercado, o uso dos instrumentos fiscais pode resultar numa diminuição excessiva dos níveis de produção (...), vai provocar uma redução do bem-estar”. (SOARES, 2002, p. 543-544).

16 “Outro aspecto a ter em conta é o de que existe um conjunto de variáveis exógenas e/ou apenas dificilmente perceptíveis que tornam complexo ou impossível antecipar com um grau mínimo de certeza as consequências da emissão de determinado tipo de elementos contaminantes. E os impostos ambientais também apresentam uma fragilidade quando os poluentes emitidos não geram um impacto uniforme no espaço ou no tempo. (...) A eficiência dos impostos ambientais, nestes casos, depende da diferenciação da taxa que grava cada realidade em função das suas características. Pois, a taxa de imposto aplicada a um poluidor deve ser igual ao dano marginal causados pelas suas emissões. O facto pelas suas emissões. Estas dependem de uma combinação produtiva utilizada e do produto final obtido.” (SOARES, 2002, pp. 547-548).

17 “O problema da concentração de sujeitos poluidores numa determinada área (*hot spot*), como o inerente aumento do custo social marginal causado por cada nova unidade de poluição, não pode ser resolvido através de um imposto sobre emissões. Pois, a entrada e a saída do mercado são livres

e o uso deste instrumento não impede a localização de mais poluidores numa área já com graves problemas de desequilíbrio ecológico." (SOARES, 2002, p. 549).

18 "Quando a uma determinada intensidade fiscal pode corresponder mais do que um nível de emissões, devido à falta de linearidade ou à descontinuidade da função do custo de controlo das mesas, o imposto não será um instrumento eficiente, por ser incapaz de assegurar que o nível óptimo de controlo será atingido. Tal como acontece quando se verificam múltiplos óptimos locais. Porquanto, neste caso, o imposto pode permitir atingir níveis de poluição que correspondem apenas a um óptimo local, e não um óptimo global. Não se consegue, então, garantir que se alcança o menor dos custos líquidos mínimos." (SOARES, 2002, p. 552).

19 "O dano ambiental causado pelas emissões lançadas na atmosfera não depende só da quantidade destas, mas também da sua concentração no tempo. (...) tal como acontece na presença de variáveis aleatórias de variáveis aleatórias e de diferentes capacidades assimilativas do meio, um imposto uniforme e indesejável. Há que construir um sistema não linear, cujas taxas sejam diferenciadas de acordo com os custos a suportar pelo poluidor em função do nível de concentração das emissões lançadas no meio. Concentração esse que é determinada não só pelo comportamento de cada agente adoptada por todos os demais que fazem uso da mesma capacidade assimilativa. O que torna o desenho do tributo (sobre emissões, e não sobre produtos ou processos) extremamente custoso, pela enorme quantidade de informação que será necessário recolher." (SOARES, 2002, pp. 553-554).

Compilation of References

Abreu, L. C. (2011). *Fiscalidade e protecção do ambiente*. In G. Teixeira & A. F. Cunha (Eds.), *I Congresso de Direito Fiscal* (pp. 233-245). Vida Económica.

Acciarini, C., Borelli, F., Capo, F., Cappa, F., & Sarrocco, C. (2022). Can digitalization favour the emergence of innovative and sustainable business models? A qualitative exploration in the automotive sector. *Journal of Strategy and Management, 15*(3), 335–352. doi:10.1108/JSMA-02-2021-0033

AdamsJ.DemersE.KlassenK. (2022), Tax Aggressive Behavior and Voluntary Tax Disclosure: Evidence from Corporate Sustainability Reporting. SSRN. https://ssrn.com/abstract=4284813 doi:10.2139/ssrn.4284813

Adamson, Th., & Jordans, F. (2019). France to slap new 'ecotax' on plane tickets from 2020. *Houston Chronicle. Associated Press.* https://apnews.com/article/be0605b81b214502b2fce419869ea09e

Adebayo, T. S., Oladipupo, S. D., Adeshola, I., & Rjoub, H. (2022). Wavelet analysis of impact of renewable energy consumption and technological innovation on CO_2 emissions: Evidence from Portugal. *Environmental Science and Pollution Research International, 29*(16), 23887–23904. doi:10.100711356-021-17708-8 PMID:34817815

Adedoyin, F. F., Seetaram, N., Disegna, M., & Filis, G. (2023a). The Effect of Tourism Taxation on International Arrivals to a Small Tourism-Dependent Economy. *Journal of Travel Research, 62*(1), 135–153. doi:10.1177/00472875211053658

Adshead, F., Thorpe, A., & Rutter, J. (2006). Sustainable development and public health. *Public Health, 11*(12), 1102–1105. doi:10.1016/j.puhe.2006.10.003 PMID:17097119

Afum, R.; Agyabeng-Mensah, Y.; Baah, C.; Agyapong, G.; Armas, A. & Al Farooque, O. (2022). Prioritizing zero-waste performance and green differentiation advantage through the Prism of circular principles adoption: A mediated approach. *Journal of Cleaner Production, 361*(August 10), 132182. doi:10.1016/j.jclepro.2022.132182

Ağcakaya, S. & Kaya, I. (2022). Sürdürülebilir kalkinma ve yeşil ekonomi perspektifinden yeşil maliye politikalari uygulamalari, *Çukurova Üniversitesi Sosyal Bilimler Enstitüsü Dergisi, 2*, 512-525.

Agência para a Modernização Administrativa, IP (2023). *Obter informações sobre Organizações Não Governamentais de Ambiente (ONGA)*. ePortugal.https://eportugal.gov.pt/pt/servicos/informar-sobre-organizacoes-nao-governamentais-de-ambiente-onga-

Aguiar, N. (2016). O Problema da Tributação do Rendimento das Cooperativas – reflexão a partir do direito português. *Revista Cooperativismo e Economía Social, 28*, 163-190.

Aguilo, E., Riera, A., & Rossello, J. (2005). The short-term price effect of a tourist tax through a dynamic demand model. The case of the Balearic Islands. *Tourism Management, 26*(3), 359–365. doi:10.1016/j.tourman.2003.07.005

Akbaş, H. E., & Canikli, S. (2019). Determinants of voluntary greenhouse gas emission disclosure: An empirical investigation on Turkish firms. *Sustainability (Basel)*, *11*(1), 107. doi:10.3390u11010107

Akdag, S., & Yıldırım, H. (2020). Toward a sustainable mitigation approach of energy efficiency to greenhouse gas emissions in the European countries. *Heliyon*, *6*(3), e03396. doi:10.1016/j.heliyon.2020.e03396 PMID:32154402

Alborg, I. R. (2015). *Contabilidad de los derechos de emisión de carbono*. Instituto de Contabilidad y Auditoría de Cuentas.

Albuquerque, F., & Neves, J. (2021). Tax Disclosures in Financial and CSR Reporting as a Deterrence for Evasion. Handbook of Research on Theory and Practice of Financial Crimes. IGO Global. doi:10.4018/978-1-7998-5567-5.ch021

Alm, J., & Torgler, B. (2011). Do Ethics Matter? Tax Compliance and Morality. *Journal of Business Ethics*, *101*(4), 635–651. doi:10.100710551-011-0761-9

Alola, A. A., & Ozturk, I. (2021). Mirroring risk to investment within the EKC hypothesis in the United States. *Journal of Environmental Management*, *293*, 112890. doi:10.1016/j.jenvman.2021.112890 PMID:34082348

Alonso Madrigal, J. (2012). Los derechos de emisión de CO2 y la fiscalidad española. ICADE. *Revista de la Facultad de Derecho (Montevideo)*, *86*, 55–90.

Alonso-Martinez, D., De Marchi, V., & Di Maria, E. (2021). The sustainability performances of sustainable business models. *Journal of Cleaner Production*, *323*(129145), 129145. doi:10.1016/j.jclepro.2021.129145

Amorim, J. D. C. (2018). Os incentivos fiscais em matérias de fiscalidade verde. *RCEJ Rebules, 30*, 303–318. https://parc.ipp.pt/index.php/rebules/article/view/3165/1085

Amorim, J. C., & Azevedo, P. A. (2016). *Lições de Direito Fiscal*. Calendário de Letras.

Andrew, J., Kaidonis, M. A., & Andrew, B. (2010). Carbon tax: Challenging neoliberal solutions to climate change. *Critical Perspectives on Accounting*, *21*(7), 611–618. doi:10.1016/j.cpa.2010.03.009

Antón, Á. (2022). EU energy taxation framework revision in the context of the green deal: Towards a green, efficient, socially just, and inclusive economy transition. *VISUAL REVIEW. International Visual Culture Review / Revista Internacional de Cultura Visual*, *9,* 1–13. https://journals.eagora.org/revVISUAL/article/view/3670

Antón, Á. A. (2022). Review of the EU energy taxation framework in the context of the Green Deal. Towards a green, efficient, socially just and inclusive economic transition. *International Visual Culture Review*, *1*(2), 2–13. doi:10.37467/revvisual.v9.3670

Antonio-Paulo, U. A. (2022). *"4R" sustainability and quality teaching using Free Software*. University of Zaragoza. doi:10.37467/revtechno.v11.3794

APA - Agência Portuguesa do Ambiente. (2022a). *Relatório Anual de Resíduos Urbanos 2021*. APA. https://apambiente.pt/sites/default/files/_Residuos/Producao_Gest%C3%A3o_Residuos/Dados%20RU/RARU_2021.pdf [Accessed in February 2023].

APA - Agência Portuguesa do Ambiente. (2022b). *PERNU 2030: Plano Estratégico para os Resíduos Não Urbanos*. APA. https://participa.pt/contents/consultationdocument/PERNU%202030.pdf

APA. (2019). *European Emissions Trading (ETS)*. APA. https://apambiente.pt/index.php?ref=17&subref=295 (2019).

APA. (2020). *Green Taxation*. APA. https://apambiente.pt/index.php?ref=17&subref=1104 (2020).

Apleni, L., Mxunyelwa, S., & Vallabh, D. (2017). The implications of the proposed carbon tax in South Africa's tourism industry. *African Journal of Hospitality, Tourism and Leisure*, *6*(1), 1–12.

Arnold, D. (2016). Corporate Responsibility, Democracy, and Climate Change. *Midwest Studies in Philosophy*, XL.

Aspen Institute. (2022). *Green Agenda for the Western Balkans. The Road Toward Effective and Sustainable Implementation.* Aspen Institute. https://www.aspeninstitute.de/wp-content/uploads/Green-Agenda-for-the-Western-Balkans_2023.pdf

Avison, Z. (2022). *A green tax road map for the future*. Green Alliance. https://green-alliance.org.uk/wp-content/uploads/2022/11/A-green-tax-roadmap-for-the-future.pdf

Ayaz, H. (2017). Analysis of carbon emission accounting practices of leading carbon emitting European Union companies. *Athens Journal of Business & Economics*, *3*(4), 463–486. doi:10.30958/ajbe.3.4.5

Aydin, C., & Esen, Ö. (2018). Reducing CO_2 emissions in the EU member states: Do environmental taxes work? *Journal of Environmental Planning and Management*, *61*(13), 2396–2420. doi:10.1080/09640568.2017.1395731

Aziz, S., Rahman, M., Hussain, D., & Nguyen, D. K. (2021). Does corporate environmentalism affect corporate insolvency risk? The role of market power and competitive intensity. *Ecological Economics*, *189*(July), 107182. doi:10.1016/j.ecolecon.2021.107182

Baah, C., Afum, E., Agyabeng-Mensah, Y., & Agyeman, D. O. (2021). Stakeholder influence on adoption of circular economy principles: Measuring implications for satisfaction and green legitimacy. *Circular Economy and Sustainability*, *2*(1), 91–111. doi:10.100743615-021-00093-2

Badulescu, A., Badulescu, D., Saveanu, T., & Hatos, R. (2018). The relationship between firm size and age, and its social responsibility actions-Focus on a developing country (Romania). *Sustainability (Basel)*, *10*(3), 805. doi:10.3390u10030805

Baker, R., & Green, J. (2019). The role of financial incentives in promoting green businesses. *Journal of Sustainable Business*, *15*(3), 187–202.

Balsalobre-Lorente, D., Driha, O. M., Leitão, N. C., & Murshed, M. (2021). The carbon dioxide neutralizing effect of energy innovation on international tourism in EU-5 countries under the prism of the EKC hypothesis. *Journal of Environmental Management*, *298*, 113513. doi:10.1016/j.jenvman.2021.113513 PMID:34403918

Bancoldex. (2018). *First Green Bonds Report*. Bancoldex - Banco De Comercio Exterior De Colombia, Bogotá, Colombia. https://www.bancoldex.com/sites/default/files/11097_green_bonds_report_-_bancoldex_2018_copy.pdf

Baranzini, A., Goldemberg, J., & Speck, S. (2000). A future for carbon taxes. *Ecological Economics*, *32*(3), 395–412. doi:10.1016/S0921-8009(99)00122-6

Baranzini, A., van den Bergh, J. C. J. M., Carattini, S., Howarth, R. B., Padilla, E., & Roca, J. (2017). Carbon pricing in climate policy: Seven reasons, complementary instruments, and political economy considerations. *Wiley Interdisciplinary Reviews: Climate Change*, *8*(4), e462. doi:10.1002/wcc.462

Barcellos de Paula, L., & Marins, F. A. S. (2018). Algorithms applied in decision-making for sustainable transport. *Journal of Cleaner Production*, *176*, 1133–1143. doi:10.1016/j.jclepro.2017.11.216

Barragán Terán, D., & Muñoz Ávila, L. (2018). Forest governance in Colombia and Ecuador. *Journal of Environmental Law*, (09), 93–117. doi:10.5354/0719-4633.2018.49747

Bashir, M. F., Ma, B., Bashir, M. A., Radulescu, M., & Shahzad, U. (2022). Investigating the role of environmental taxes and regulations for renewable energy consumption: Evidence from developed economies. *Ekonomska Istrazivanja*, *35*(1), 1262–1284. doi:10.1080/1331677X.2021.1962383

Bashir, M. F., Ma, B., Bilal, Komal, B., & Bashir, M. A. (2021). Analysis of environmental taxes publications: A bibliometric and systematic literature review. *Environmental Science and Pollution Research International*, *28*(16), 20700–20716. doi:10.100711356-020-12123-x PMID:33405155

Bassey, E., Mulligan, E., & Ojo, A. (2022). A conceptual framework for digital tax administration - A systematic review. *Government Information Quarterly*, *39*(4), 101754. doi:10.1016/j.giq.2022.101754

BBC. (2006). Miliband Draws Up Green Tax Plan: Environment Secretary David Miliband Has Confirmed the Government Is Holding Discussions On Tackling Climate Change Using Green Taxes. http://news.bbc.co.uk/2/hi/uk_news/6095680.stm

Beauregard-Tellier, F. (2006). Ecological fiscal reform (EFR), *Library of Parliament Canada*. https://publications.gc.ca/site/eng/9.560939/publication.html

Bebbington, J., & Larrinaga-González, C. (2008). Carbon Trading: Accounting and Reporting Issues. *European Accounting Review*, *17*(4), 697–717. doi:10.1080/09638180802489162

Bednarčíková, D., & Repiská, R. (2021). Digital Transformation in the Context of the European Union and the Use of Digital Technologies as a Tool for Business Sustainability. *SHS Web of Conferences*, 115. 10.1051hsconf/202111501001

Bellon, M., Dabla-Norris, E., Khalid, S., & Lima, F. (2022). Digitalization to improve tax compliance: Evidence from VAT e-Invoicing in Peru. *Journal of Public Economics*, *210*, 104661. doi:10.1016/j.jpubeco.2022.104661 PMID:35711616

Bilbao Estrada, I. & Mateos Ansótegui, A. (2006a). El impuesto sobre sociedades y los derechos de emisión de gases de efecto invenadero. Retos y oportunidades. *Revista de contabilidad y tributación*, *285*, 83-128.

Bilbao Estrada, I. & Mateos Ansótegui, A. (2006b). La planificación fiscal en el Impuesto sobre Sociedades a través de los derechos de CO2. *Tribuna fiscal*, *194*, 16-31.

Bilbao Estrada, I. (2010a). Los impuestos mundiales sobre el CO2. *Revista Aranzadi de derecho ambiental*, *17*, 115-133.

Bilbao Estrada, I. (2010b). La tributación de los derechos de emisión en el IVA y en el IS. *Partida doble*, *226*, 84-97.

Bilbao Estrada, I., & Mateos Ansótegui, A. (2009). Los derechos de emisión y su contabilización: especial referencia a la provisión por derechos de emisión tras el nuevo Plan General Contable y la reforma del Impuesto sobre Sociedades, en La Fiscalidad de los Derechos de Emisión: Estado de Situación y Perspectivas de Futuro, Instituto de Estudios Fiscales, Madrid.

Bilbao Estrada, I., & Mateos Ansótegui, A. I. (2007). Planificación contable y derechos de emisión de gases de efecto invernadero. *Revista Universo Contábil*, *3*(3), 101–122.

Bilbao Estrada, I., & Rodríguez Márquez, J. (2010). IVA y derechos de emisión: Especial referencia a su utilización como instrumento de fraude. *Crónica Tributaria*, *136*, 37–84.

Bird, R. M. (1992). Taxing tourism in developing countries. *World Development*, *20*(8), 1145–1158. doi:10.1016/0305-750X(92)90006-H

Biyan, Ö., & Gök, M. (2014). Çevre politikaları kapsamında Avrupa Birliği ve Türkiye'de çevre vergilerinin uygulanışı: Karşılaştırmalı bir analiz. *Hitit Üniversitesi Sosyal Bilimler Enstitüsü Dergisi*, *2*, 281–310.

Black, C. M. (2013). Accounting for carbon emission allowances in the European Union: In search of consistency. *Accounting in Europe*, *10*(2), 223–239. doi:10.1080/17449480.2013.834730

Blau, A. & Janssen, C. (2020). *Regional Baseline Assessment. Region III: Albania, Bosnia and Herzegovina, Israel, Montenegro and Turkey*. SwitchMed. https://switchmed.eu/wp-content/uploads/2020/12/GCB_Regional-Baseline-Assessment-03.pdf

Boas, S., & Monteiro, S. (2012). O relato sobre licenças de emissão de CO2: o caso das empresas portuguesas do PNALE II, *XV Encontro AECA - Novos caminhos para a Europa: O papel das empresas e dos governos*, Ofir, 20-21 de Setembro.

Böhringer, C., Garcia-Muros, X., & González-Eguino, M. (2019). *Greener and fairer: a progressive environmental tax reform for Spain.* https://www.econstor.eu/handle/10419/191538

Bonham, C. S., & Gangnes, B. (1996). Intervention analysis with cointegrated time series: The case of the Hawaii hotel room tax. *Applied Economics, 28*(10), 1281–1293. doi:10.1080/000368496327831

Bontems, P., & Bourgeon, J. M. (2005). Optimal environmental taxation and enforcement policy. *European Economic Review, 49*(2), 409–435. doi:10.1016/S0014-2921(03)00036-9

Borges, F. G. L., & Merlin, L. V. D. C. T. (2018). The (IN) viability of green tax reform in Brazil. *Paths of Law, 15*(33), 347–375. doi:10.18623/rvd.v15i33.1243

Borghei, Z., Leung, P., & Guthrie, J. (2018). Voluntary greenhouse gas emission disclosure impacts on accounting-based performance: Australian evidence. *Australasian Journal of Environmental Management, 25*(3), 321–338. doi:10.1080/14486563.2018.1466204

Borrego, A. (2016). Diploma of green taxation - Finally an environmental taxation integrated in Portugal? *Journal of Social and Environmental Management, 19.* https://www.researchgate.net/publication/312128897%0ADiploma

Borrego, A. (2016). Diploma of green taxation - Finally an environmental taxation integrated in Portugal? *Revista de Gestão Social e Ambiental, 10*(3), 51–69. https://go.gale.com/ps/i.do?id=GALE%7CA554040581&sid=googleScholar&v=2.1&it=r&linkaccess=abs&issn=1981982X&p=IFME&sw=w

Borrego, A. (2016). Diploma of green taxation - Finally an environmental taxation integrated in Portugal? *Revista de Gestão Social e Ambiental, 10*(3), 51–68. doi:10.24857/rgsa.v10i3.1163

Bosello, F., & Carraro, C. (2001). Recycling energy taxes: Impacts on a disaggregated labour market. *Energy Economics, 23*(5), 569–594. doi:10.1016/S0140-9883(00)00083-9

Bovenberg, A. Lans, Mooij, R. A. (1994). Environmental Levies and Distortionary Taxation. *American Economic Review, 84*(4), 1085–1089. http://search.ebscohost.com/login.aspx?direct=true&db=bth&AN=9410250177&site=eds-live

Bovenberg, A. L. (1998). Environmental taxes and the double dividend. *Empirica, 25*(1), 15–35. doi:10.1023/A:1006836720884

Bovenberg, A. L., & Mooij, R. A. (1994). Environmental Levies and Distortionary Taxation. *The American Economic Review, 84*, 1085–1089.

Bovenberg, A. L., & van der Ploeg, F. (1994). Environmental policy, public finance and the labour market in a second-best world. *Journal of Public Economics, 55*(3), 349–390. doi:10.1016/0047-2727(93)01398-T

Bovenberg, A. L., & Van der Ploeg, F. (1998). Consequences of environmental tax reform for unemployment and welfare. *Environmental and Resource Economics, 12*(2), 137–150. doi:10.1023/A:1016040327622

Branco, M. C., & Delgado, C. J. M. (2014). *Informação sobre combate à corrupção nos relatórios de sustentabilidade das empresas portuguesas.* Reporting on the fight, 4th GECAMB - Conference on Environmental Management and Accounting, Leiria

Briguglio, M. (2021). Taxing household waste: Intended and unintended consequences. *Journal of Cleaner Production, 304*, 127034. doi:10.1016/j.jclepro.2021.127034

Brown, M., Smith, K., & Johnson, L. (2021). Government support for green businesses: A comparative analysis. *Environmental Policy Review, 27*(2), 45–60.

Brunetti, F., Matt, D., Bonfanti, A., De Longhi, A., Pedrini, G., & Orzes, G. (2020). Digital transformation challenges: Strategies emerging from a multi-stakeholder approach. *The TQM Journal*, *32*(4), 697–724. doi:10.1108/TQM-12-2019-0309

Bruninx, K., & Ovaere, M. (2022). COVID-19, Green Deal and recovery plan permanently change emissions and prices in EU ETS Phase IV. *Nature Communications*, *13*(1), 1165. doi:10.103841467-022-28398-2 PMID:35246534

Buck, C., Clarke, J., de Oliveira, R. T., Desouza, K. C., & Maroufkhani, P. (2023). Digital transformation in asset-intensive organisations: The light and the dark side. *Journal of Innovation & Knowledge*, *8*(100335), 100335. Advance online publication. doi:10.1016/j.jik.2023.100335

Burrows, P. (1979). Pigovian taxes, polluter subsidies, regulation, and the size of a polluting industry. *The Canadian Journal of Economics. Revue Canadienne d'Economique*, *12*(3), 494–501. doi:10.2307/134738

Cahan, S., Villiers, C., Jeter, D. C., Naiker, V., & Staden, C. (2016). Are CSR disclosure value relevant? Cross-country evidence. *European Accounting Review*, *25*(3), 579–611. doi:10.1080/09638180.2015.1064009

Cai, H., Wang, X., Adriaens, P., & Xu, M. (2019). Environmental benefits of taxi ride sharing in Beijing. *Energy*, *174*, 503–508. doi:10.1016/j.energy.2019.02.166

Calise, F., Cappiello, F. L., Cartenì, A., Dentice d'Accadia, M., & Vicidomini, M. (2019). A novel paradigm for a sustainable mobility based on electric vehicles, photovoltaic panels and electric energy storage systems: Case studies for Naples and Salerno (Italy). *Renewable & Sustainable Energy Reviews*, *111*, 97–114. doi:10.1016/j.rser.2019.05.022

Camilleri, M. A. (2022). The rationale for ISO 14001 certification: A systematic review and a cost–benefit analysis. *Corporate Social Responsibility and Environmental Management*, *29*(4), 1067–1083. doi:10.1002/csr.2254

Cao, K., Xu, B., He, Y., & Xu, Q. (2020a). Optimal carbon reduction level and ordering quantity under financial constraints. *International Transactions in Operational Research*, *27*(5), 2270–2293. doi:10.1111/itor.12606

Cao, K., Xu, B., & Wang, J. (2020b). Optimal trade-in and warranty period strategies for new and remanufactured products under carbon tax policy. *International Journal of Production Research*, *58*(1), 180–199. doi:10.1080/00207543.2019.1617448

Carattini, S.; Baranzini, A. & Lalive, R. (2018). Is Taxing Waste a Waste of Time? Evidence from a Supreme Court Decision. *Ecological Economics*, *148*, 131.151. doi:10.1016/j.ecolecon.2018.02.001

Cárdenas-García, P. J., Pulido-Fernández, J. I., Durán-Román, J. L., & Carrillo-Hidalgo, I. (2022). Tourist taxation as a sustainability financing mechanism for mass tourism destinations. *International Journal of Tourism Research*, *24*(4), 577–592. doi:10.1002/jtr.2523

Cardoso, A. C. (2015). *O imposto ambiental como instrumento de desenvolvimento sustentável*. [Master's thesis, FDUP]. https://repositorio-aberto.up.pt/bitstream/10216/81181/2/37025.pdf

Cardoso, A. C. C. (2015). *O imposto como instrumento de desenvolvimento sustentável*. [Unpublished master dissertation]. University of Porto, Porto, Portugal.

Carvalho, C. (2018). *Teoria da decisão tributária*. Almedina.

Castellani, V., & Sala, S. (2012). Ecological Footprint and Life Cycle Analysis in the sustainability assessment of tourism activities. *Ecological Indicators*, *16*, 135–147. doi:10.1016/j.ecolind.2011.08.002

Castiglione, C., Infante, D., Minervini, M. T., & Smirnova, J. (2014). Environmental taxation in Europe: What does it depend on? *Cogent Economics & Finance*, *2*(1), 967362. doi:10.1080/23322039.2014.967362

Castilho, A. F. A. N. (2016). A extrafiscalidade no Brasil. *Revista Jurídica Luso Brasileira, 2*(4), 135-159. https://www.cidp.pt/revistas/rjlb/2016/4/2016_04_0135_0159.pdf

Čavoški, A. (2022). The European Green Deal and technological determinism. *Environmental Law Review, 24*(3), 201–213. doi:10.1177/14614529221104558

Cazorla Prieto, L. M., & Cazorla González-Serrano, L. (2009). *El tratamiento en el ordenamiento jurídico español del incremento de los precios de la energía eléctrica como consecuencia de la internalización de los derechos de emisión de CO2. Tratado de regulación del sector eléctrico* (Vol. 1). Aranzadi, Navarra.

Chacón, E. (2022). Mypes and sustainability in times of pandemic in Arequipa, Peru A structural equation model. *Visual Review.* doi:10.37467/revvisual.v9.3624

Chang, N., & Han, C. (2020). Cost-push impact of taxing carbon in China: A price transmission Perspective. *Journal of Cleaner Production, 248*, 1–9. doi:10.1016/j.jclepro.2019.119194

Chaturvedi, A., Saluja, M. S., Banerjee, A., & Arora, R. (2014). Environmental fiscal reforms. *IIMB Management Review, 26*(3), 193–205. doi:10.1016/j.iimb.2014.06.004

Chen, H., & Wang, Y. (2018). The impact of tax incentives on green businesses: Evidence from a panel data analysis. *Journal of Environmental Economics, 42*(1), 78–92.

Chen, W., & Hu, Z.-H. (2018). Using evolutionary game theory to study governments and manufacturers' behavioral strategies under various carbon taxes and subsidies. *Journal of Cleaner Production, 201*, 123–141. doi:10.1016/j.jclepro.2018.08.007

Chien, F., Ananzeh, M., Mirza, F., Bakar, A., Vu, H. M., & Ngo, T. Q. (2021). The effects of green growth, environmental-related tax, and eco-innovation towards carbon neutrality target in the US economy. *Journal of Environmental Management, 299*(113633), 113633. Advance online publication. doi:10.1016/j.jenvman.2021.113633 PMID:34492439

Chiroleu-Assouline, M., & Fodha, M. (2009). Double dividend and distribution of welfare: Advanced results and empirical considerations. *Economie Internationale, 120*(4), 91–108.

Chiu, Y., & Zhang, W. (2023). Moderating Effect of Financial Development on the Relationship between Renewable Energy and Carbon Emissions. *Energies, 16*(3), 1467. https://www.mdpi.com/1996-1073/16/3/1467. doi:10.3390/en16031467

Choudhary, A., Sarkar, S., Settur, S., & Tiwari, M. K. (2015). A carbon market sensitive optimization model for integrated forward–reverse logistics. *International Journal of Production Economics, 164*, 433–444. doi:10.1016/j.ijpe.2014.08.015

Chu, L. K., & Le, N. T. M. (2022). Environmental quality and the role of economic policy uncertainty, economic complexity, renewable energy, and energy intensity: The case of G7 countries. *Environmental Science and Pollution Research International, 29*(2), 2866–2882. doi:10.100711356-021-15666-9 PMID:34382169

Church, B., Jiang, W., Kuang, X., & Vitalis, A. (2019). A Dollar for a Tree or a Tree for a Dollar? The Behavioral Effects of Measurement Basis on Managers' CSR Investment Decision. *The Accounting Review, 94*(5), 117–137. doi:10.2308/accr-52332

CIEC. Special Excise Tax Code . (2023). CIEC. https://www.pgdlisboa.pt/leis/lei_mostra_articulado.php?nid=1598&tabela=leis&so_miolo=

Cihat Onat, N. (2022). How to compare sustainability impacts of alternative fuel Vehicles? *Transportation Research Part D, Transport and Environment, 102*, 103129. doi:10.1016/j.trd.2021.103129

CIRC. Corporate Income Tax Code. (2023). CIRC. https://info.portaldasfinancas.gov.pt/pt/informacao_fiscal/codigos_tributarios/Cod_download/Documents/CIRC.pdf

CIRS. Personal Income Tax Code. (2023). CIRS. https://info.portaldasfinancas.gov.pt/pt/informacao_fiscal/codigos_tributarios/Cod_download/Documents/CIRS.pdf

CISV. Motor Vehicle Tax Code. (2023). CISV. https://www.pgdlisboa.pt/leis/lei_mostra_articulado.php?nid=976&tabela=leis

CIUC. Single Circulation Tax Code. (2023). CIUC. https://www.pgdlisboa.pt/leis/lei_mostra_articulado.php?nid=976&tabela=leis

CIVA. Value Added Tax Code. (2023). CIVA. https://info.portaldasfinancas.gov.pt/pt/informacao_fiscal/codigos_tributarios/Cod_download/Documents/CIVA.pdf

Clinch, J. P., Dunne, L., & Dresner, S. (2006). Environmental and wider implications of political impediments to environmental tax reform. *Energy Policy*, *34*(8), 960–970. doi:10.1016/j.enpol.2004.08.048

Coman, D. M., Ionescu, C. A., Duică, A., Coman, M. D., Uzlau, M. C., Stanescu, S. G., & State, V. (2022). Digitization of Accounting: The Premise of the Paradigm Shift of Role of the Professional Accountant. *Applied Sciences (Basel, Switzerland)*, *12*(7), 1–25. doi:10.3390/app12073359

Corson, C., MacDonald, K. I., & Neimark, B. (2013). Grasping "green": Markets, environmental governance and the materialization of natural capital. *Human Geographies*, *6*(1), 1–15. doi:10.1177/194277861300600101

Costa, T., Fernández-Llimós, S., & Lopes, M. (2012). *Bibliometrics and the evaluation of scientific production: indicators and tools,* 19–37 NIH.

Costa, T., Lopes, S., Fernández-Llimós, F., Amante, M., & Lopes, P. (2012). Bibliometrics and the Evaluation of Scientific Production: Indicators and tools. *ACTAS – National Congress of Bibliotecários. Archivists and Documentalists*, *11*, 1–7.

Costello, K. W. (2019). Essay on climate apocalypse and a carbon tax. *The Electricity Journal*, *32*(10), 1–6. doi:10.1016/j.tej.2019.106669

Cottrell, J., Kai, S., Matthias, R., & Alexander, M. (2016). Environmental tax reform in developing, emerging and transition economies, The German Development Institute, 1-126.

Coury-Chanel, E. (2016). *Esboço de uma teoria fiscal da Ecotaxa. Tributação, meio ambiente e desenvolvimento. Forensics*. CESUPA.

Craizer, R., Frantzis, I., & Bacolas, N. C. (2020). *Economic instruments to improve waste management in Greece – Including a pre-feasibility study on a deposit refund system, Final Report*. Brussels: European Commission.

Cropley, A. J. (2022). *Qualitative research methods: A practice-oriented introduction*. Editura Intaglio., doi:10.13140/RG.2.1.3095.6888/1

CRP. Constitution of the Portuguese Republic. (2005). CRP. https://www.pgdlisboa.pt/leis/lei_mostra_articulado.php?tabela=leis&nid=565&pagina=1&ficha=1

D'Atri, A. M. (2017). Keucheyan, Razmig. (2016). Nature is a battlefield. Finance, ecological crisis and new green wars. *Yearbook of the Faculty of Human Sciences*, *13*(13), 140. doi:10.19137/an1310

Datta, T. K., Nath, P., & Choudhury, K. (2020). A hybrid carbon policy inventory model with emission sourcebased green investments. *OPSEARCH*, *57*(1), 202–220. doi:10.100712597-019-00430-y

David, L., Nwulu, N., Aigbavboa, C., & Adepoju, O. (2022). Integrating fourth industrial revolution (4IR) technologies into the water, energy & food nexus for sustainable security: A bibliometric analysis. *Journal of Cleaner Production, 363*, 132522. doi:10.1016/j.jclepro.2022.132522

De la Cuesta-González, M., & Pardo, E. (2019). Corporate tax disclosure on a CSR basis: A new reporting framework in the post-BEPS era. *Accounting, Auditing & Accountability Journal, 32*(7), 2167–2192. doi:10.1108/AAAJ-12-2017-3282

Decree-Law n.º 102-D/2020, 10th December, General waste management regime. https://www.pgdlisboa.pt/leis/lei_mostra_articulado.php?nid=3398&tabela=leis&so_miolo=S

Decree-Law n.o 78/2021, 24th September, Transposes the Directive (EU) 2019/904 on the reduction of the impact of certain plastic products on the environment. https://files.dre.pt/1s/2021/09/18700/0000300027.pdf

Delgado, F. J., Freire-gonzález, J., & Presno, M. J. (2022). Environmental taxation in the European Union : Are there common trends? *Economic Analysis and Policy, 73*, 670–682. doi:10.1016/j.eap.2021.12.019

Depren, Ö., Kartal, M. T., Ayhan, F., & Kılıç Depren, S. (2023). Heterogeneous impact of environmental taxes on environmental quality: Tax domain based evidence from the nordic countries by nonparametric quantile approaches. *Journal of Environmental Management, 329*, 117031. doi:10.1016/j.jenvman.2022.117031 PMID:36528942

Deyganto, K. O. (2022). The effect of tax incentives practices on the sustainability of micro, small and medium enterprises in Ethiopia during the outbreak of corona virus pandemic. *Journal of Innovation and Entrepreneurship, 11*(1), 8. doi:10.118613731-022-00194-8 PMID:35106268

Di Foggia, G., & Beccarello, M. (2023). Designing circular economy-compliant municipal solid waste management charging schemes. *Utilities Policy, 81*, 101506. doi:10.1016/j.jup.2023.101506

Dias, A., Montero, P., & Quirós, J. (2016). O relato financeiro das licenças de emissão atribuídas gratuitamente no período 2008-2012. *XVII Encuentro AECA*, Bragança, 22-23 setembro.

Dias, J. E. O. F. (2007). *Direito Constitucional e Administrativo do Ambiente*. Almedina.

Dieguez, T., Ferreira, L. P., Silva, F. J. G., & Tjahjono, B. (2020). Open innovation and sustainable development through industry-academia collaboration: A case study of automotive sector. *Procedia Manufacturing, 51*, 1773–1778. doi:10.1016/j.promfg.2020.10.246

Dijkgraaf, E., & Gradus, R. (2016). An EU recycling target: What does the Dutch evidence tell us? *Environmental and Resource Economics, 68*(3), 501–526. doi:10.100710640-016-0027-1

Diller, M., Asen, M., & Späth, T. (2020). The effects of personality traits on digital transformation: Evidence from German tax consulting. *International Journal of Accounting Information Systems, 37*(100455), 100455. Advance online publication. doi:10.1016/j.accinf.2020.100455

Dinç, G. (2008). *Avrupa insan hakları sözleşmesi'ne göre çevre ve insan*. Türkiye Barolar Birliği Yayınları.

Doğan, B., Chu, L. K., Ghosh, S., Diep Truong, H. H., & Balsalobre-Lorente, D. (2022). How environmental taxes and carbon emissions are related in the G7 economies? *Renewable Energy, 187*, 645–656. doi:10.1016/j.renene.2022.01.077

Dourado, A. P. (2016). *Direito fiscal: lições*. Almedina.

Dowling, G. R. (2014). The curious case of corporate tax avoidance. Is it socially irresponsible? *Journal of Business Ethics, 124*(1), 173–184. doi:10.100710551-013-1862-4

Dufva, T., & Dufva, M. (2019). Grasping the future of the digital society. *Futures, 107*, 17–28. doi:10.1016/j.futures.2018.11.001

Durán-Román, J. L., Cárdenas-García, P. J., & Pulido-Fernández, J. I. (2021a). Tourist tax to improve sustainability and the experience in mass tourism destinations: The case of Andalusia (Spain). *Sustainability (Basel), 13*(1), 42. doi:10.3390u13010042

Durán-Román, J. L., Cárdenas-García, P. J., & Pulido-Fernández, J. I. (2021b). Tourists' willingness to pay to improve sustainability and experience at destination. *Journal of Destination Marketing & Management, 19*, 100540. doi:10.1016/j.jdmm.2020.100540

Durán-Román, J. L., Pulido-Fernández, J. I., & Cárdenas-García, P. J. (2020). Delimitation of applicable government revenue in Spanish tourist destinations. *Investigaciones Regionales, 2020*(47), 179–199. doi:10.38191/iirr-jorr.20.016

Dussaux, D. (2020), "The joint effects of energy prices and carbon taxes on environmental and economic performance: Evidence from the French manufacturing sector", *OECD Environment Working Papers*, No. 154, OECD Publishing, Paris, https://doi.org/ doi:10.1787/19970900

Dvarskas, A. (2017). Dynamically linking economic models to ecological condition for coastal zone management: Application to sustainable tourism planning. *Journal of Environmental Management, 188*, 163–172. doi:10.1016/j.jenvman.2016.12.014 PMID:27978443

Dwivedi, Y., Hughes, L., Kar, A., Baabdullah, A., Grover, P., Abbas, R., Andreini, D., Abumoghli, I., Barlette, Y., Bunker, D., Kruse, L., Constantiou, I., Davison, R., De', R., Dubey, R., Taylor, H., Gupta, B., He, W., Kodama, M., & Wade, M. (2022). Climate change and COP26: Are digital technologies and information management part of the problem or the solution? An editorial reflection and call to action. *International Journal of Information Management, 63*, 102456. doi:10.1016/j.ijinfomgt.2021.102456

Dwyer, L., Forsyth, P., & Spurr, R. (2012). Wither Australian tourism? Implications of the carbon tax. *Journal of Hospitality and Tourism Management, 19*(1), 15–30. doi:10.1017/jht.2012.18

Dwyer, L., Forsyth, P., Spurr, R., & Hoque, S. (2013). Economic Impacts of a Carbon Tax on the Australian Tourism Industry. *Journal of Travel Research, 52*(2), 143–155. doi:10.1177/0047287512461568

Eccles, R. G. & Serafeim, G. (2017). Corporate and integrated reporting: A functional perspective. *Corporate Stewardship: Achieving Sustainable Effectiveness.*

ECLAC. (2020). *MSMEs and COVID-19.* ECLAC. https://www.cepal.org/es/euromipyme/mipymes-covid-19

Edelmann, N., & Virkar, S. (2023). The Impact of Sustainability on Co-Creation of Digital Public Services. *Administrative Sciences, 13*(2), 43. doi:10.3390/admsci13020043

EEA. (2010). *EEA Briefing Note for ETR Workshop, Dublin.* EEA.

EEA. (2013). *EEA Briefing Note for ETR Workshop, Lisbon.* EEA.

EEA. (2019). *Atmospheric greenhouse gas concentrations.* EEA. https://www.eea.europa.eu/data-and-maps/indicators/atmospheric-greenhouse-gas-concentrations-6/assessment-1

EEA. (2022). *The role of (environmental) taxation in supporting sustainability transitions.* EEA. https://www.eea.europa.eu/publications/the-role-of-environmental-taxation

Eisenack, K., Edenhofer, O., & Kalkuhl, M. (2012). Resource rents: The effects of energy taxes and quantity instruments for climate protection. *Energy Policy, 48*, 159–166. doi:10.1016/j.enpol.2012.05.001

Ekeman, E. (1998). *Avrupa Birliği ve Türkiye'nin çevre politikalarının karşılaştırmalı incelemesi.* İktisadi Kalkınma Vakfı Yayını.

Ekins, P., & Zenghelis, · Dimitri. (2021). *The costs and benefits of environmental sustainability*. Springer. doi:10.1007/s11625-021-00910-5

Ekins, P., & Barker, D. D. T. (2001). Carbon taxes and carbon emissions trading. *Journal of Economic Surveys*, *15*(3), 325–376. http://www.scopus.com/scopus/openurl/link.url?ctx_ver=Z39.88-2004&ctx_enc=info:ofi/enc:UTF-8&svc_val_fmt=info:ofi/fmt:kev:mtx:sch_svc&svc.citedby=yes&rft_id=info:eid/2-s2.0-0034862828&rfr_id=http://search.ebscohost.com&rfr_dat=partnerID:NnvIuKwx&rfr_dat=m. doi:10.1111/1467-6419.00142

Elison, L., & Snyder, F. (2001). *The Montana State Constitution: A Reference Guide*. Greenwood Publishing Group.

Ellerman, D., Marcantonini, C. & Zaklan, A. (2015). The European Union Emissions Trading System: Ten Years and Counting. *Review of Environmental Economics and Policy*. *10*(1), 89–107.

Elliott, R., & Timulak, L. (2021). Why a generic approach to descriptive-interpretive qualitative research? In R. Elliott & L. Timulak (Eds.), *Essentials of descriptive-interpretive qualitative research: A generic approach* (pp. 3–14). American Psychological Association. doi:10.1037/0000224-001

Elsevier. (2022). *Pathways to Net Zero: Global South Research in the Transition to Clean Energy*. Elsevier. https://elsevierfoundation.org/report-pathways-net-zero-global-south-research-transition-clean-energy/

Engberg-Pedersen, P., & Zwart, R. (2018). *The 2030 Agenda and Development Co-operation Results*. OECD. https://www.oecd-ilibrary.org/content/paper/2f391534-en

EPA - U.S. Environmental Protection Agency. (2023). *Overview of Greenhouse Gases*. EPA. https://www.epa.gov/ghgemissions/overview-greenhouse-gases

EPI. (2022). *Environmental Performance Index*. Yale Center for Environmental Law & Policy. https://epi.yale.edu/

Ercan, E. (2015). Türkiye'de çevre vergileri mi çevre vergisi mi? *TBB Dergisi*, *119*, 209–226.

Esen, Ö., & Dündar, M. (2021). Do energy taxes reduce the carbon footprint? Evidence from Turkey. *Journal of Emerging Economies and Policy*, *6*(2), 179–186. https://dergipark.org.tr/joeep

Esen, Ö., Yıldırım, D. Ç., & Yıldırım, S. (2021). Pollute less or tax more? Asymmetries in the EU environmental taxes – Ecological balance nexus. *Environmental Impact Assessment Review*, *91*, 106662. Advance online publication. doi:10.1016/j.eiar.2021.106662

European Commission. (2010). *EUROPE 2020 Strategy for smart, sustainable and inclusive growth* (Vol. 0278). ReVision.

European Commission. (2015). A Fair and Efficient Corporate Tax System in the 22 European Union: 5 Key Areas for Action. *Communication from the Commission to the European Parliament and the Council*. EC.

European Commission. (2020). *Stepping up Europe's 2030 climate ambition Investing in a climate-neutral future for the benefit of our people*. EC. https://eur-lex.europa.eu/legal-content/EN/TXT/PDF/?uri=CELEX:52020DC0562

European Commission. (2021). *Special Eurobarometer 513: Climate change*. EC. https://europa.eu/eurobarometer/surveys/detail/2273

European Commission. (2022). *Energy Support package for the Western Balkans*. EC. https://www.wbif.eu/storage/app/media/Library/12.%20Economic%20and%20Investment%20Plan/WBIF%20Energy%20Support%20Package%20%20for%20the%20Western%20Balkans%202022.pdf

European Commission. (2022). Towards a green, digital and resilient economy: our European Growth Model (COM/2022/83 final). *Official Journal of the European Union*, 1–16. https://eur-lex.europa.eu/legal-content/EN/TXT/?uri=CELEX:52022DC0083

European Court of Auditors. (2021). *Special report 12/2021: The polluter pays principle: Inconsistent application across EU environmental policies and actions*. ECA. https://www.eca.europa.eu/en/publications?did=58811

European Environment Agency. (2005). *Market-based instruments for environmental policy in Europe*. EEA. https://www.eea.europa.eu/publications/technical_report_2005_8

European Environment Agency. (2022). *Decarbonising road transport: the role of vehicles, fuels and transport demand*. EEA. doi:10.2800/68902

European Union. (1997). Commission for European Community Affairs. *Environmental Taxes and Charges in the Single Market*. http://aei.pitt.edu/4785/1/4785.pdf

European Union. (2011). *Regulation (EU) n.º 691/2011 of the European Parliament and of the Council of 6 July 2011 on European environmental economic accounts*. EU. https://eur-lex.europa.eu/legal-content/EN/TXT/?uri=celex:32011R0691

European Union. (2012). *Consolidated version of the treaty on the functioning of the European Union*. EU. https://eur-lex.europa.eu/LexUriServ/LexUriServ.do?uri=CELEX:12012E/TXT:en:PDF

European Union. (2022). *Implementing Green Agenda for the Western Balkans*. EU. https://neighbourhood-enlargement.ec.europa.eu/system/files/2022-11/factsheet%20green%20agenda%20nov2022.pdf

Eurostat. (2010). *European System of Accounts – ESA 2010*. Eurostat.https://ec.europa.eu/eurostat/web/products-manuals-and-guidelines/-/KS-02-13-269

FA - Fundo Ambiental. (2019). *Relatório Final - Economia Circular em Freguesias (Programa JUNTAr+)*. FA. https://www.fundoambiental.pt/ficheiros/relatorio-final-juntar-mais-pdf.aspx .

FA - Fundo Ambiental. (2021). *Relatório Final - Economia Circular em Freguesias (Programa JUNTAr+)*. FA. https://www.fundoambiental.pt/ficheiros/relatorio-final-aviso-71822021-juntar--pdf.aspx

FA - Fundo Ambiental. (2023a). *RecolhaBio - Apoio à implementação de projetos de recolha seletiva de biorresíduos*. FA. https://www.fundoambiental.pt/avisos-anteriores/apoios-2022/residuos-e-economia-circular/recolhabio-apoio-a-implementacao-de-projetos-de-recolha-seletiva-de-biorresiduos.aspx [Accessed in March 2023].

FA - Fundo Ambiental. (2023b). *O que é o programa Recolhabio?* FA. https://www.fundoambiental.pt/recolhabio-apoio-a-implementacao-de-projetos-de-recolha-seletiva-de-biorresiduos/ficheiros/faqs-recolhabio-pdf.aspx [Accessed in March 2023].

Fabra, N., & Reguant, M. (2014). Pass-Through of Emissions Costs in Electricity Markets. *The American Economic Review*, *104*(9), 9, 2872–2899. doi:10.1257/aer.104.9.2872

Falcão, T. (2015). Is Carbon Tax Compatible with the WTO Treaties? Intertax, 43(10), 571-799. doi:10.54648/TAXI2015056

Falcão, T. (2019). Taxing Carbon Emissions from International Shipping. Intertax, 47(10), 832-844. doi:10.54648/TAXI2019085

Faúndez-Ugalde, A., Toledo-Zúñiga, P., & Castro-Rodríguez, P. (2022). Tax Sustainability: Tax Transparency in Latin America and the Chilean Case. *Sustainability (Basel)*, *14*(4), 2107. doi:10.3390u14042107

Fazey, I., Schäpke, N., Caniglia, G., Patterson, J., Hultman, J., van Mierlo, B., Säwe, F., Wiek, A., Wittmayer, J., Aldunce, P., Al Waer, H., Battacharya, N., Bradbury, H., Carmen, E., Colvin, J., Cvitanovic, C., D'Souza, M., Gopel, M., Goldstein, B., & Wyborn, C. (2018). Ten essentials for action-oriented and second order energy transitions, transformations and climate change research. *Energy Research & Social Science*, *40*, 54–70. doi:10.1016/j.erss.2017.11.026

Fede Organicos. (2008). *Green and sustainable business in Colombia*. FO. http://www.fedeorganicos.com/wp-content/uploads/2016/07/los-negocios-verdes-y-sostenibles-en-colombia-minambiente.pdf

Feng, H., Wang, F., Song, G., & Liu, L. (2022). Digital Transformation on Enterprise Green Innovation: Effect and Transmission Mechanism. *International Journal of Environmental Research and Public Health*, *19*(17), 10614. Advance online publication. doi:10.3390/ijerph191710614 PMID:36078329

Fernandes, F. V., & Moreira, T. M. (2022). *The evolution of green taxation in Portugal*. International Law Office. https://www.internationaltaxreview.com/

Fernandes, M. T. (2009). *O mercado de direitos de emissão de CO2: um estudo experimental*. [PhD Thesis in Economic Sciences, Universidade do Minho].

Fernandes, F. V. (2022). *Direito Fiscal do Terceiro Setor*. AAFDL Editora.

Fernández Cuesta, C., & Fronti de García, L. (2010). El cambio climático y su control económico en la empresa. *Partida Doble*, *226*, 54–69.

Fernández Cuesta, C., Larrinaga González, C., & Moneva Abadía, J. M. (2006). Derechos de emisión de gases efecto invernadero: Registro, valoración e información. *Partida Doble*, *182*, 88–97.

Fernando, A., Albuquerque, F., Almeida, M., & Delgado, S. (2010). Tratamento contabilístico das licenças de emissão de gases com efeito de estufa: o caso português, *XIV Encuentro AECA*. Coimbra.

Feroz, A. K., Zo, H., & Chiravuri, A. (2021). Digital transformation and environmental sustainability: A review and research agenda. *Sustainability*, *13*(3), 1–20. doi:10.3390u13031530 PMID:34676112

Ferreira, E. P. (2005). *Ensinar finanças públicas numa faculdade de direito: relatório sobre o programa, conteúdo e métodos de ensino da disciplina*. Almedina.

Florindo, A., Lemos, K., Monteiro, S., & Ribeiro, V. (2020). Corporate social responsibility (CSR) reporting and climate change: evidence of carbon emissions disclosures by Portuguese companies, in Book Conceptual and Theoretical Approaches to Corporate Social Responsibility, Entrepreneurial Orientation and Financial Performance. IGI Global.

Franco, R. C. (2015). *Coord.. Diagnóstico das ONG em Portugal*. Fundação Calouste Gulbenkian.

Freire-González, J.; Martinez-Sanchez, V. & Puig-Ventosa, I. (2022). Tools for a circular economy: Assessing waste taxation in a CGE multi-pollutant framework. *Waste Management*, 139 (February 15), 50-59. doi:10.1016/j.wasman.2021.12.016

Freire-González, J., & Ho, M. S. (2018). Environmental fiscal reform and the double dividend: Evidence from a dynamic general equilibrium model. *Sustainability (Basel)*, *10*(2), 501. doi:10.3390u10020501

Fridstrøm, L., & Østli, V. (2017). The vehicle purchase tax as a climate policy instrument. *Transportation Research Part A, Policy and Practice*, *96*, 168–189. doi:10.1016/j.tra.2016.12.011

Fuinhas, J. A., Koengkan, M., Leitão, N. C., Nwani, C., Uzuner, G., Dehdar, F., Relva, S., & Peyerl, D. (2021). Effect of Battery Electric Vehicles on Greenhouse Gas Emissions in 29 European Union Countries. *Sustainability*, *13*(24), 13611. doi:10.3390/su132413611

Fujii, E., Khaled, M., & Mak, J. (1985). The exportability of hotel occupancy and other tourist taxes. *National Tax Journal*, *38*(2), 169–177. doi:10.1086/NTJ41792007

Fullerton, D., & Wolverton, A. (2003, August), The Two-Part Instrument in a Second-Best World. *NCEE Working Paper Series*. (Working Paper # 03-04).

Fullerton, D. (1997). *Environmental Taxes and the Double-Dividend Hypothesis: Did You Really Expect Something for Nothing?* University of Texas. https://ase.tufts.edu/economics/papers/9706.pdf doi:10.3386/w6199

Fullerton, D., & Metcalf, G. (1998). Environmental taxes and the doule-dividend hypothesis: Did you really expect something for nothing? *Chicago-Kent Law Review*, *73*(189), 221–256.

Fusco, G. (2016). *Impostos especiais sobre o consumo em Portugal: A extrafiscalidade e a saúde*. [ISEG, Master's thesis]. http://hdl.handle.net/10400.26/14263

Gale, W., Brown, S., & Saltiel, F. (2013). Carbon taxes as part of the fiscal solution. Brookings Institution.

Gallego-Alvarez, I., Martínez-Ferrero, J., & Cuadrado-Ballesteros, B. (2016). Accounting Treatment for Carbon Emission Rights. *Systems*, *4*(1), 12. doi:10.3390ystems4010012

García, A. (2022). Private investment in green businesses: Trends and drivers. *Journal of Sustainable Finance*, *18*(4), 223–240.

García-Sánchez, I. M., Aibar-Guzmán, B., Aibar-Guzmán, C., & Somohano-Rodríguez, F. M. (2022). The drivers of the integration of the sustainable development goals into the non-financial information system: Individual and joint analysis of their influence. *Sustainable Development (Bradford)*, *30*(4), 513–524. doi:10.1002d.2246

Garcia-Torea, N., Giordano-Spring, S., Larrinaga, C., & Rivière-Giordano, G. (2022). Accounting for Carbon Emission Allowances: An Empirical Analysis in the EU ETS Phase 3. *Social and Environmental Accountability Journal*, *42*(1–2), 93–115. doi:10.1080/0969160X.2021.2012496

George, G., & Schillebeeckx, S. J. D. (2022). Digital transformation, sustainability, and purpose in the multinational enterprise. *Journal of World Business*, *57*(101326), 101326. Advance online publication. doi:10.1016/j.jwb.2022.101326

Ghazouani, A., Xia, W., Jebli, M., & Shahzad, U. (2020). Exploring the Role of Carbon Taxation Policies on CO2 Emissions: Contextual Evidence from Tax Implementation and Non-Implementation European Countries. *Sustainability (Basel)*, *12*(20), 8680. doi:10.3390u12208680

Giménez, J. V., & Marco, A. Z. (2020). A Dynamic Spatial Panel of Subnational GHG Emissions: Environmental Effectiveness of Emissions Taxes in Spanish Regions. *Sustainability (Basel)*, *12*(7), 2872. doi:10.3390u12072872

Giner Inchausti, B. (2007). La contabilidad de los derechos de emisión: Una perspectiva internacional. *Revista Española de Financiación y Contabilidad*, *36*(133), 175–193. doi:10.1080/02102412.2007.10779619

Giorgios, X., & Dimitrios, X. (2009). *Accounting treatment of emission credits*. International Hellenic University.

Global Reporting Initiative (GRI). *GRI Standards*. GRI.

Global Reporting Initiative (GRI, 2020). Linking the SDGs and the GRI Standards GRI and the United Nations Global Compact (UN Global Compact) (2017). Business Reporting on the SDGs. Analysis of Goals and Targets. GRI and the United Nations Global Compact (UN Global Compact) (2022). Business Reporting on the SDGs. *Analysis of Goals and Targets, updated version*. UN.

Global Reporting Initiative. (2013). *G4 Sustainability Reporting Guidelines: Reporting Principles and Standard Disclosures*. GRI.

Gonçalves, A., & Góis, C. (2012). Reconhecimento contabilístico das licenças de emissão de CO2: o caso português, *XV Encontro AECA - Novos caminhos para a Europa: O papel das empresas e dos governos*. Ofir.

Gonçalves, T. C., Dias, J., & Barros, V. (2022). Sustainability Performance and the Cost of Capital. *International Journal of Financial Studies*, *10*(3), 1–32. doi:10.3390/ijfs10030063

Gonzalez-Gonzalez, J., & Zamora Ramírez, C. (2016). Voluntary carbon disclosure by Spanish companies: An empirical analysis. *International Journal of Climate Change Strategies and Management*, *8*(1), 57–79. doi:10.1108/IJCCSM-09-2014-0114

Gonzalo Angulo, J.A. & Larriba Díaz-Azorita, A. (2007). Contabilización de los derechos de emisión de gases de efecto invernadero. *Cuadernos de sostenibilidad y patrimonio natural*, 12, 54-66.

Gooroochurn, N., & Sinclair, M. T. (2005). Economics of tourism taxation: Evidence from Mauritius. *Annals of Tourism Research*, *32*(2), 478–498. doi:10.1016/j.annals.2004.10.003

Gore, T., Urios, J. & Karamperi, M. (2022, February). Green and fair taxation in the EU - Green EU tax systems tend to also be more progressive. *Institute for European Environmental Policy.*

Goulder, L. H. (1994) Environmental taxation and the "double dividend:" a reader's guide. *NBER Working Paper Series.* Standford.

Goulder, L. H. (1995). Environmental taxation and the double dividend: A reader's guide. *International Tax and Public Finance*, *2*(2), 157–183. doi:10.1007/BF00877495

Goulder, L. H. (2013). Climate change policy's interactions with the tax system. *Energy Economics*, *40*, S3–S11. doi:10.1016/j.eneco.2013.09.017

Gouveia, E., & Palma, C. (2018). *Impacts of Environmental Taxation on the Portuguese Economy*. NiH.

Gray, W. B., & Metcalf, G. (2017). Carbon tax competitiveness concerns: Assessing a best practices carbon credit. *National Tax Journal*, *70*(2), 447–468. doi:10.17310/ntj.2017.2.08

Gribnau, H., & Jallai, A. G. (2019). Sustainable tax governance and transparency. In *Challenges in Managing Sustainable Business* (pp. 337–369). Palgrave Macmillan. doi:10.1007/978-3-319-93266-8_15

Groosman, B. (2000). *Pollution Tax – entry 2500. The 'Encyclopaedia of Law and Economics*. Edward Elgar.

GulinD.HladikaM.ValentaI. (2019). Digitalization and the Challenges for the Accounting Profession. *ENTRENOVA 2019 Proceeding, September*, 502–511. doi:10.2139/ssrn.3492237

Guo, B., Wang, Y., Feng, Y., Liang, C., Tang, L., Yao, X., & Hu, F. (2022). The effects of environmental tax reform on urban air pollution: A quasi-natural experiment based on the Environmental Protection Tax Law. *Frontiers in Public Health*, *10*, 967524. https://www.frontiersin.org/articles/10.3389/fpubh.2022.967524/full. doi:10.3389/fpubh.2022.967524 PMID:36033767

Gürlük, S. (2010). Sürdürülebilir kalkınma gelişmekte olan ülkelerde uygulanabilir mi? *Eskişehir Osmangazi Üniversitesi İktisadi ve İdari Bilimler Fakültesi Dergisi*, *5*, 85–99.

Gutman, V., & López, A. (2017). Green production and eco-innovation. *Eco-innovation and green production: a review of policies in Latin America and the Caribbean.* Santiago: ECLAC. https://repositorio.cepal.org/bitstream/handle/11362/41748/S1700072_es.pdf?sequence=1&isAllowed=y

Guyader, H., Friman, M., & Olsson, L. E. (2021). Shared Mobility: Evolving Practices for Sustainability. *Sustainability (Basel)*, *13*(21), 12148. doi:10.3390u132112148

Guzman, T. J. (2002) Fiscalidade ambiental: análisis y efectos distributivos. Granada.

Haites, E. (2018). Carbon taxes and greenhouse gas emissions trading systems: What have we learned? *Climate Policy*, *18*(8), 955–966. doi:10.1080/14693062.2018.1492897

Hájek, M., Zimmermannová, J., Helman, K., & Rozenský, L. (2019). Analysis of carbon tax efficiency in energy industries of selected EU countries. *Energy Policy*, *134*, 110955. doi:10.1016/j.enpol.2019.110955

Han, H., & Lee, T. (2022). Varieties of Green Stimulus Policies: Comparative Analysis of the Green Growth and Green New Deal Policies in South Korea. *Journal of Environment & Development*, *107049652211393*. doi:10.1177/10704965221139303

Hao, X., Wang, H., Lin, Z., & Ouyang, M. (2020). Seasonal effects on electric vehicle energy consumption and driving range: A case study on personal, taxi, and ridesharing vehicles. *Journal of Cleaner Production*, *249*(119403), 119403. doi:10.1016/j.jclepro.2019.119403

Hapsoro, D., & Falih, Z. (2020). The Effect of Firm Size, Profitability, and Liquidity on The Firm Value Moderated by Carbon Emission Disclosure. *Journal of Accounting and Investment*, *1*(2), 240–257. doi:10.18196/jai.2102147

Hardeck, I., Inger, K. K., Moore, R. D., & Schneider, J. (2020). *Do the bad guys report? Examining whether CSR performance influences the use of socially responsible tax disclosures*. Working paper.

Hardeck, I., & Kirn, T. (2016). Taboo or technical issue? An empirical assessment of taxation in sustainability reports. *Journal of Cleaner Production*, *133*, 1337–1351. doi:10.1016/j.jclepro.2016.06.028

Harris, S., Martin, M., & Diener, D. (2021). Circularity for circularity's sake? Scoping review of assessment methods for environmental performance in the circular economy. *Sustainable Production and Consumption*, *26*, 172–186. doi:10.1016/j.spc.2020.09.018

Harzendorf, F., Wulf, C., Haase, M., Baumann, M., Ersoy, H., & Zapp, P. (2022). Domestic value added as an indicator for sustainability assessment: A case study on alternative drivetrains in the passenger car sector. *Clean Technologies and Environmental Policy*, *24*(10), 3145–3169. doi:10.100710098-022-02402-1

Hassan, M., Oueslati, W., & Rousselière, D. (2020). Environmental taxes, reforms and economic growth: An empirical analysis of panel data. *Economic Systems*, *44*(3), 100806. https://www.sciencedirect.com/science/article/abs/pii/S093936251830400X?via%3Dihub. doi:10.1016/j.ecosys.2020.100806

Hassler, J., Krusell, P., & Olovsson, C. (2021). Suboptimal climate policy. *Journal of the European Economic Association*, *19*(6), 2895–2928. doi:10.1093/jeea/jvab048

Hassouna, F. M. A., & Assad, M. (2020). Towards a sustainable public transportation: Replacing the conventional taxis by a hybrid taxi fleet in the west bank, Palestine. *International Journal of Environmental Research and Public Health*, *17*(23), 1–11. doi:10.3390/ijerph17238940 PMID:33271931

Hawken, P. (2017). *Drawdown: The Most Comprehensive Plan Ever Proposed to Roll Back Global Warming*. Penguin USA.

Hawkins, R. (2000). The use of economic instruments and green taxes to complement an environmental regulatory regime. *Water, Air, and Soil Pollution*, *123*(1–4), 379–394. doi:10.1023/A:1005294917875

He, P., Ning, J., Yu, Z., Xiong, H., Shen, H., & Jin, H. (2019). Can Environmental Tax Policy Really Help to Reduce Pollutant Emissions? An Empirical Study of a Panel ARDL Model Based on OECD Countries and China. *Sustainability*, *11*(16), 4384. doi:10.3390/su11164384

He, Y., & Liu, R. (2023). The impact of the level of green finance development on corporate debt financing capacity. *Finance Research Letters*, *52*(November 2022), 1–8. doi:10.1016/j.frl.2022.103552

Hermawan, A., Aisyah, I., Gunardi, A., & Putri, W. (2018). Going Green: Determinants of Carbon Emission Disclosure in Manufacturing Companies in Indonesia. *International Journal of Energy Economics and Policy*, *8*(1), 55–61.

Hjøllund, L., & Svendsen, G. T. (2001). Why green taxation? *Energy & Environment*, *12*(1), 29–38. doi:10.1260/0958305011500571

Hörisch, J. (2013). Combating climate change through organisational innovation: An empirical analysis of internal emission trading schemes. *Corporate Governance (Bradford)*, *13*(5), 569–582. doi:10.1108/CG-06-2013-0077

Hou, Y., Jia, M., Tian, X., Wei, F., & Wei, K. (2015). Optimal decisions of countries with carbon tax and carbon tariff. *Journal of Industrial Engineering and Management*, *8*(3), 981–1001. doi:10.3926/jiem.1506

Hovelaque, V., & Bironneau, L. (2015). The carbon-constrained EOQ model with carbon emission dependent demand. *International Journal of Production Economics*, *164*, 285–291. doi:10.1016/j.ijpe.2014.11.022

Huang, L., & Xiao, E. (2021). Peer effects in public support for Pigouvian taxation. *Journal of Economic Behavior & Organization*, *187*, 192–204. https://www.sciencedirect.com/science/article/abs/pii/S0167268121001608?via%3Dihub. doi:10.1016/j.jebo.2021.04.019

Hu, X., Yang, Z., Sun, J., & Zhang, Y. (2020). Carbon tax or cap-and-trade: Which is more viable for Chinese remanufacturing industry? *Journal of Cleaner Production*, *243*, 1–16. doi:10.1016/j.jclepro.2019.118606

Hwang, J., Fontan, I. A., Oguz, A., & Vardanyan, V. (2023). *How can the Western Balkans boost green finance?* https://blogs.worldbank.org/europeandcentralasia/how-can-western-balkans-boost-green-finance

Hyun, J. K. (2006). Tax Compliances in Korea and Japan: Why are they so different? *The journal of the korean economy*.

Ikoma, A. (2022). *Voluntary tax disclosure from the perspective of tax avoidance and national culture. Master Dissertation in Finance and Taxation. FEP, Oporto University International Monetary Fund (IMF) (2022a). Fiscal Transparency*. International Monetary Fund. https://www.state.gov/department-of-state-releases-2022-fiscal-transparency-report/

IMF. (2022). *IMF Staff Country Reports 2022*. IMF. https://www.elibrary.imf.org/view/journals/002/2022/363/article-A001-en.xml

IMF/OECD (2021, April), Tax Policy and Climate Change: IMF/OECD Report for the G20 Finance Ministers and Central Bank Governors. *Communication from the Commission to the European Parliament and the Council - Business Taxation for the 21st Century*. OECD.

INE (Portuguese National Statistics Institute). (2022). *Environmentally related taxes reached 5 billion euro – 2021*. INE. https://www.ine.pt/xportal/xmain?xpid=INE&xpgid=ine_destaques&DESTAQUESdest_boui=414630553&DESTAQUESmodo=2&xlang=en

Instituto Nacional de Estatística. (2021, October). *Impostos e taxas com relevância ambiental 2020*. INE. www.ine.pt

Instituto Nacional de Estatística. (2022, October). *Impostos e taxas com relevância ambiental 2021*. INE. www.ine.pt

Inter-American Development Bank. (2019). *Nationally determined contributions in Colombia*. Inter-American Development Bank.

International Monetary Fund (IMF). (2022b). *How Does the IMF Encourage Greater Fiscal Transparency?* IMF. https://www.imf.org/en/About/Factsheets/Sheets/2016/07/27/15/46/Encouraging-Greater-Fiscal-Transparency

IPCC - Intergovernmental Panel on Climate Change. (2019). *Global Warming of 1.5°C*. IPCC. https://www.ipcc.ch/site/assets/uploads/sites/2/2022/06/SR15_Full_Report_HR.pdf

IPCC. (2005). *Carbon Dioxide Capture and Storage*. IPCC. https://www.ipcc.ch/site/assets/uploads/2018/03/srccs_wholereport-1.pdf (08/06/2020).

Islam, M., & Rahman, N., SarzulAlom Sodesh, M., & Saha, M. (2022). Green Taxation Impact On Industrial Sectors Which Reduces Carbon Emission In A Country. *Journal of Business and Management*, *24*(12), 30–46. doi:10.9790/487X-2412063046

Izzo, M., Dello Strologo, A., & Graná, F. (2020). Learning from the Best : New Challenges and Trends in IR Reporters ' Disclosure and the Role of SDGs. *Sustainability (Basel)*, *12*(14), 5545. doi:10.3390u12145545

Jackson, T. (2010). The employment and productivity effects of environmental taxation: Additional dividends or added distractions? *Journal of Environmental Planning and Management*, *43*(3), 389–406. doi:10.1080/09640560050010419

Jaramillo, C., & Patiño, A. (2020). Industrial engineering and the international green market. *Journal of Engineering Sciences of Athenea*, *1*(2), 37–40. https://athenea.autanabooks.com/index.php/revista/article/view/11/22

Jeffrey, C., & Perkins, J. D. (2015). The association between energy taxation, participation in an emissions trading system, and the intensity of carbon dioxide emissions in the European Union. *The International Journal of Accounting*, 1–21.

Jiang, H.-D., Hao, W.-T., Xu, Q.-Y., & Liang, Q.-M. (2020). Socio-economic and environmental impacts of the iron ore resource tax reform in China: A CGE-based analysis. *Resources Policy*, *68*, 101775. https://www.sciencedirect.com/science/article/abs/pii/S0301420719307251?via%3Dihub. doi:10.1016/j.resourpol.2020.101775

Jin, M., Shi, X., Emrouznejad, A., & Yang, F. (2018). Determining the optimal carbon tax rate based on data envelopment Analysis. *Journal of Cleaner Production*, *172*, 900–908. doi:10.1016/j.jclepro.2017.10.127

Ji, Q., Wang, C., & Fan, Y. (2022). Environmental and welfare effects of vehicle purchase tax: Evidence from China. *Energy Economics*, *115*, 106377. doi:10.1016/j.eneco.2022.106377

Jones, P. (2019). Financial instruments for supporting green businesses: A systematic review. *Environmental Finance Review*, *25*(1), 56–72.

Jr., R. A. (2020). Direito Internacional do Ambiente: a Responsabilidade no Antropoceno. *In Revista Jurídica Portucalense*. (No. 28). doi:10.34625/issn.2183-2705(28)2020.ic-05

Jurado Martín, J., Martín Zamora, M. & Cabello Léon, C. (2007b). *Tratamiento contable de los derechos de emisión: la aplicación de la resolución de 8 de febrero de 2006 del ICAC en el sector eléctrico español. VI Reunión en Investigación Social y Medioambiental*. Navarra.

Jurado Martín, J. A. (2016). *La información contable y los derechos de emisión: presupuesto de emisiones y la utilidad de la información revelada en las cuentas anuales*. Universidad de Huelva.

Jurado Martín, J. A., & Martín Zamora, M. P. (2006a). El Mercado de Derechos de Emisiones y su Incidencia en el Resultado Empresarial, Decisiones Basadas en el Conocimiento y el Papel Social de la Empresa. *XX Congreso Nacional y XVI Congreso Hispano Francés de AEDEM*, Palma de Mallorca (España.

Jurado Martín, J. A., & Martín Zamora, M. P. (2006b). *El Proceso Presupuestario y el Protocolo de Kioto: el Presupuesto de Emisiones de CO2. XVI Jornadas Luso-Espanholas de Gestao Científica, "Inovação e Empreendedorismo para o Séc*. XXI.

Jurado Martín, J. A., Martín Zamora, M. P., & Cabello León, C. (2007a). *Los Derechos de Emisión y su Repercusión en la Cuenta de Resultados en las Empresas del Sector Eléctrico Español. XI Jornada de Trabajo Sobre Contabilidad de Costes y de Gestión. X Jornada de Trabajo Sobre Contabilidad de Costes y Gestión*. ASEPUC.

Kahia, M., Jarraya, B., Kahouli, B., & Omri, A. (2023). Do Environmental Innovation and Green Energy Matter for Environmental Sustainability? Evidence from Saudi Arabia (1990–2018). *Energies*, *16*(3), 1376. doi:10.3390/en16031376

Kalita, U., & Barua, N. A. (2019). Determining a Carbon Tax Rate for India in the Context of Global Climate Change. *International Journal of Recent Technology and Engineering*, *8*(3), 8185–8191. doi:10.35940/ijrte.C6655.098319

Kang, S.-C., & Lee, H. (2019). Economic appraisal of implementing electric vehicle taxis in Seoul. *Research in Transportation Economics*, *73*, 45–52. doi:10.1016/j.retrec.2018.11.007

Kao, W., & Liao, C. (2021). Tax Avoidance and Tax Disclosures in Corporate Social Responsibility Reports in the United Kingdom. *Journal of International Accounting Research*, *20*(3), 59–80. doi:10.2308/JIAR-2020-036

Kaplan, A. (1991). Çevre sorunları ve ekonomi. *Ankara Üniversitesi Siyasal Bilgiler Fakültesi Dergisi*, *1*, 279–287.

Karatzoglou, B., & Karatzoglou, O. (2011). Carbon accounting in Greek companies participating in the European Union's Emissions Trading Scheme: Current practice and projected financial implications. In R. Burritt, S. Schaltegger, M. Bennett, T. Pohjola, & M. Csutora (Eds.), *Environmental management accounting and supply chain management. Eco-Efficiency in industry and science, 27*. Springer. doi:10.1007/978-94-007-1390-1_10

Kazancoglu, Y., Ozbiltekin-Pala, M., & Ozkan-Ozen, Y. D. (2021). Prediction and evaluation of greenhouse gas emissions for sustainable road transport within Europe. *Sustainable Cities and Society*, *70*, 102924. doi:10.1016/j.scs.2021.102924

Kim, S., & Park, J. (2021). The effects of tax incentives on green business performance: Evidence from a quasi-experimental design. *Journal of Sustainable Development*, *37*(4), 123–137.

Kirikkaleli, D., Awosusi, A. A., Adebayo, T. S., & Otrakçı, C. (2023). Enhancing environmental quality in Portugal: Can CO2 intensity of GDP and renewable energy consumption be the solution? *Environmental Science and Pollution Research International*, *30*(18), 53796–53806. doi:10.100711356-023-26191-2 PMID:36867330

Klok, J., Larsen, A., Hansen, K., & Dahl, A. (2006). Ecological Tax Reform in Denmark: History and social acceptability. *Energy Policy*, *34*(8), 905–916. doi:10.1016/j.enpol.2004.08.044

Knez, S., Štrbac, S., & Podbregar, I. (2022). Climate change in the Western Balkans and EU Green Deal: Status, mitigation and challenges. *Energy, Sustainability and Society*, *12*(1), 1. https://energsustainsoc.biomedcentral.com/articles/10.1186/s13705-021-00328-y. doi:10.118613705-021-00328-y

Kogels, H. (2019). *The Energy Transition and Fiscal Policy* (Vol. 1).

Koks, E., Carrera, L., Jonkeren, O., Aerts, J., Husby, T., Thissen, M., Standardi, G., & Mysiak, J. (2016). Regional disaster impact analysis: Comparing input–output and computable general equilibrium models. *Natural Hazards and Earth System Sciences*, *16*(8), 1911–1924. doi:10.5194/nhess-16-1911-2016

Kong, D., & Zhu, L. (2022). Governments' Fiscal Squeeze and Firms' Pollution Emissions. *Environmental and Resource Economics*, *81*(4), 833–866. doi:10.100710640-022-00656-3

Kortum, S., & Weisbach, D. (2017, June 01). Weisbach & The design of border adjustments for carbon prices. *National Tax Journal*, *70*(2), 421–446. doi:10.17310/ntj.2017.2.07

Koskela, E., Schöb, R., & Sinn, D. H. W. (1998). Pollution, Factor Taxation and Unemployment. *International Tax and Public Finance*, *5*(3), 379–396. doi:10.1023/A:1008642512728

KPMG (2019). *The KPMG Survey of Corporate Responsibility Reporting 2019*. KPPMG.

KPMG (2022). *Big shifts, small steps: The KPMG survey of sustainability reporting 2022*. KPMG.

Kundeliene, K., & Leitoniene, S. (2015). Business Information Transparency: Causes and Evaluation Possibilities. *Procedia: Social and Behavioral Sciences*, *213*, 340–344. doi:10.1016/j.sbspro.2015.11.548

Kuo, T. C., Hong, I.-H., & Lin, S. C. (2016). Of carbon taxes work? Analysis of government policies and enterprise strategies in equilibrium. *Journal of Cleaner Production, 139*, 337–346. doi:10.1016/j.jclepro.2016.07.164

Kuralbayeva, K. (2019). Environmental Taxation, Employment and Public Spending in Developing Countries. *Environmental and Resource Economics, 72*(4), 877–912. doi:10.100710640-018-0230-3

Kuşat, N. (2013). Yeşil sürdürülebilirlik için yeşil ekonomi: Avantaj ve dezavantajları- Türkiye incelemesi. *Journal of Yasar University, 29*, 4896–4916.

Labandeira, X., Labeaga, J. M., & López-Otero, X. (2019). New Green Tax Reforms: Ex-Ante Assessments for Spain. *Sustainability (Basel), 11*(20), 5640. doi:10.3390u11205640

Lai, Y. B. (2018). The Feasibility of the Double-Dividend Hypothesis in a Democratic Economy. *The Scandinavian Journal of Economics, 120*(1), 211–241. doi:10.1111joe.12213

Lanis, R., & Richardson, G. (2013). Corporate social responsibility and tax aggressiveness: A test of legitimacy theory. *Accounting, Auditing & Accountability Journal, 26*(1), 75–100. doi:10.1108/09513571311285621

Law n.º 82-D / 2014 , 31st December, Green Tax Law. https://www.pgdlisboa.pt/leis/lei_mostra_articulado.php?nid=2253&tabela=leis&so_miolo=

Lee, C. (2017). Tax policy for promoting green businesses: Lessons from international experience. *Green Economy Journal, 12*(2), 89–105.

Lee, J.-Y. (2019). Investing in carbon emissions reduction in the EOQ model. *The Journal of the Operational Research Society*, 1–12.

Le, H. P., & Ozturk, I. (2020). The impacts of globalization, financial development, government expenditures, and institutional quality on CO_2 emissions in the presence of environmental Kuznets curve. *Environmental Science and Pollution Research International, 27*(18), 22680–22697. doi:10.100711356-020-08812-2 PMID:32323231

Leitão, N. C. (2021). Testing the Role of Trade on Carbon Dioxide Emissions in Portugal. *Economies, 9*(1), 22. doi:10.3390/economies9010022

Leitão, N. C., Koengkan, M., & Fuinhas, J. A. (2022). The Role of Intra-Industry Trade, Foreign Direct Investment, and Renewable Energy on Portuguese Carbon Dioxide Emissions. *Sustainability, 14*(22), 15131. doi:10.3390/su142215131

Leitão, N. C., Dos Santos Parente, C. C., Balsalobre-Lorente, D., & Cantos Cantos, J. M. (2023). Revisiting the effects of energy, population, foreign direct investment, and economic growth in Visegrad countries under the EKC scheme. *Environmental Science and Pollution Research International, 30*(6), 15102–15114. doi:10.100711356-022-23188-1 PMID:36168012

Liang, Q.-M., Wang, T., & Xue, M.-M. (2016). Addressing the competitiveness effects of taxing carbon in China: Domestic tax cuts versus border tax adjustments. *Journal of Cleaner Production, 112*, 1568–1581. doi:10.1016/j.jclepro.2015.02.092

Liao, H., Tang, M., Luo, L., Li, C., Chiclana, F., & Zeng, X.-J. (2018). A Bibliometric Analysis and Visualization of Medical Big Data Research. *Sustainability (Basel), 10*(1), 1–18. doi:10.3390u10010166

Li, G., & Masui, T. (2019). Assessing the impacts of China's environmental tax using a dynamic computable general equilibrium model. *Journal of Cleaner Production, 208*, 316–324. doi:10.1016/j.jclepro.2018.10.016

Lin, B., & Li, X. (2011). The effect of carbón tax on per capita CO2 emissions. *Energy Policy, 39*(9), 5137–5146. doi:10.1016/j.enpol.2011.05.050

Lin, B., & Xu, D. M. (2019). Exploring the green total factor productivity of China's metallurgical industry under carbon tax: A perspective on factor substitution. *Journal of Cleaner Production, 233*, 1322–1333. doi:10.1016/j.jclepro.2019.06.137

Li, S., Gao, L., Hanb, C., Guptac, B., Alhalabid, W., & Almakdie, S. (2023). Exploring the effect of digital transformation on Firms' innovation performance. *Journal of Innovation & Knowledge& Knowledge*, *8*(100317), 100317. doi:10.1016/j.jik.2023.100317

Litan, H., Rong, K., Wu, Y., Xie, D., Zhang, H., & Zhao, D. (2023). Profit versus Sustainability in Bikeshare. *Sustainable Cities and Society*, *93*(March), 104512. doi:10.1016/j.scs.2023.104512

Liu, A. A. (2013). Tax evasion and optimal environmental taxes. *Journal of Environmental Economics and Management*, *66*(3), 656–670. doi:10.1016/j.jeem.2013.06.004

Liu, B., & Ge, J. (2023). The optimal choice of environmental tax revenue usage: Incentives for cleaner production or end-of-pipe treatment? *Journal of Environmental Management, 329*, 117106. doi:10.1016/j.jenvman.2022.117106 PMID:36566734

Liu, Q. (2022). Evaluating the effectiveness of financial and tax incentives for green businesses: A meta-analysis. *Journal of Sustainable Business*, *18*(1), 35–50.

Liu, X., Wang, C., Niu, D., Suk, S., & Bao, C. (2015). An analysis of company choice preference to carbon tax policy in China. *Journal of Cleaner Production, 103*, 393–400. doi:10.1016/j.jclepro.2014.12.084

Li, Y., Mao, J., Chen, S., & Yang, D. (2022). Tax-reducing incentive and corporate green performance: What we learn from China. *Renewable Energy*, *199*(April), 791–802. doi:10.1016/j.renene.2022.07.128

Li, Z., & Zhou, Z. (2022). Research on the Impact Mechanism of Fiscal and Tax Policies on Carbon Emissions under the Dual Carbon Goals. *Frontiers in Business. Economics and Management*, *6*(2), 32–37. doi:10.54097/fbem.v6i2.2637

Loganathan, N., Shahbaz, M., & Taha, R. (2014). The link between green taxation and economic growth on CO2 emissions: Fresh evidence from Malaysia. *Renewable & Sustainable Energy Reviews*, *38*, 1083–1091. doi:10.1016/j.rser.2014.07.057

López Gordo, M. G., & López Gordo, J. F. (2007). Normativa sobre el registro contable de los derechos de emisión. [Revista Científica y Técnica de Ecología y Medio Ambiente]. *Ecosistemas (Madrid)*, *16*(1), 90–96.

Lovell, H., Aguiar, T., Bebbington, J., & Larrinaga-Gonzalez, C. (2010). Accounting for Carbon. *Certified Accountants Educational Trust for the Association of Chartered Certified Accountants. Research Reports (Montgomery)*, 122.

Low, M., Seah, C., Cham, T., & Teoh, S. (2022). Digitalization adoption for digital economy: An examination of Malaysian small medium-sized enterprises through the technology–organization–environment framework. *Business Process Management Journal*, *28*(1), 1473–1494. doi:10.1108/BPMJ-06-2022-0282

Luckin, D. (2000). Environmental taxation and red-green politics. *Capital and Class*, *24*(3), 161–189. doi:10.1177/030981680007200108

Luís, S., Roseta-Palma, C., Matos, M., Lima, M. L., & Sousa, C. (2020). Psychosocial and economic impacts of a charge in lightweight plastic carrier bags in Portugal: Keep calm and carry on? *Resources, Conservation and Recycling, 161*, 104962. doi:10.1016/j.resconrec.2020.104962

Luppi, B., Parisi, F., & Rajagopalan, S. (2012). The rise and fall of the polluter-pays principle in developing countries. *International Review of Law and Economics*, *32*(1), 135–144. doi:10.1016/j.irle.2011.10.002

Lutfi, A., Alkelani, S., Al-Khasawneh, M., Alshira'h, A., Alshirah, M., Almaiah, M., Alrawad, M., Alsyouf, A., Saad, M., & Ibrahim, N. (2022). Influence of Digital Accounting System Usage on SMEs Performance: The Moderating Effect of COVID-19. *Sustainability (Basel)*, *14*(22), 15048. doi:10.3390u142215048

Lutz, C., & Bern, M. (2010). Environmental tax reform in the European Union: Import on CO2 emissions and the economy. *Z. Energiewirstseh*, *34*(1), 1–10. doi:10.100712398-010-0009-x

Mack, N., Woodsong, C., Macqueen, K. M., Guest, G., & Namey, E. (2005). *Qualitative Research Methods: A Data Collector's Field Guide*. Family Health International.

Madrigal, J. A. (2012). Los derechos de emisión de CO2 y la fiscalidad española. icade. *Revista de la Facultad de Derecho (Montevideo)*, (86), 55–90.

Magon, R. B., Thomé, A. M. T., Ferrer, A. L. C., & Scavarda, L. F. (2018). Sustainability and performance in operations management research. *Journal of Cleaner Production*, *190*, 104–117. doi:10.1016/j.jclepro.2018.04.140

Ma, N., Li, H., Wang, Y., Zhang, J., Li, Z., & Arif, A. (2020). The short-term roles of sectors during a carbon tax on Chinese economy based on complex network: An in-process analysis. *Journal of Cleaner Production*, *251*, 1–14. doi:10.1016/j.jclepro.2019.119560

Manjunatha, N. (2019). Descriptive Research. *Journal of Emerging Technologies and Innovative Research*, *6*(6), 863–867. https://www.jetir.org/papers/JETIR1908597.pdf

MAOTE. Ministry of the Environment, Spatial Planning and Energy. (2015). *Reforma fiscalidade verde*. MAOTE. https://www.crescimentoverde.gov.pt/wp-content/uploads/2014/10/ReformaFiscalidadeVerde_GreenTaxReform_emagazine.pdf

Mardones, C. (2022). Pigouvian taxes to internalize environmental damages from Chilean mining-A computable general equilibrium analysis. *Journal of Cleaner Production*, *362*, 132359. doi:10.1016/j.jclepro.2022.132359

Mardones, C., & Baeza, N. (2018). Economic and environmental effects of a CO2 tax in Latin American countries. *Energy Policy*, *114*, 262–273. doi:10.1016/j.enpol.2017.12.001

Markandya, A. (2002). Environmental implacations of non-environmental policies. Handbook of Environmental Economics, 3, Stockholm.

Marques, R. C., Simões, P., & Pinto, F. S. (2018). Tariff regulation in the waste sector: An unavoidable future. *Waste Management (New York, N.Y.)*, *78*, 292–300. doi:10.1016/j.wasman.2018.05.028 PMID:32559914

Martín Zamora, M. P., & Jurado Martín, J. A. (2008). *La Contabilización de los Impuestos Ecológicos. X Seminario Hispano-Luso de Economía Empresarial*. Servicio de Publicaciones de la Universidad de Huelva.

Martín Zamora, M., & Jurado Martín, J. A. (2005). Nuevos Procedimientos Contables: La Contabilidad del Mercado de Emisiones de CO2. *Técnica Contable*, *680*, 5–36.

Martínez, M. (2016). Derechos de emisión y tributos en la unión Europea. *Jurídicas Cuc*, *12*(1), 139–180.

Martinez, M., & López, R. (2020). Carbon pricing and emissions trading: Implications for green businesses. *Environmental Economics Review*, *28*(3), 152–168.

Martínez-Rubio, K., Delgado-Cruz, A., & Vargas-Martínez, E. E. (2021). Adoption of green technologies and their influence on environmental responsibility practices. Perceptions of hotel workers. *Management Studies*, *37*(161), 532–541. doi:10.18046/j.estger.2021.161.4071

Martins, C. Q. (2021). *ESNL – Entidades do Setor Não Lucrativo. Abordagem Contabilística e Fiscal*. APECA.

Mateos Ansótegui, A. (2006). La contabilidad de los derechos de emisión. *New Challeges of Accounting Ethics, Independece and Critical Thought*. XI Accounting Congress, Coimbra.

Matsumura, E., Prakash, R., & Vera-Muñoz, S. (2011). *Voluntary Disclosures and the Firm-Value Effects of Carbon Emissions*. University of Notre Dame.

Maxim, M. R. (2020). Environmental fiscal reform and the possibility of triple dividend in European and non-European countries: Evidence from a meta-regression analysis. *Environmental Economics and Policy Studies*, *22*(4), 633–656. doi:10.100710018-020-00273-8

Maxim, M. R., Zander, K. K., & Patuelli, R. (2019). Green tax reform and employment double dividend in European and non-European countries: A meta-regression assessment. *International Journal of Energy Economics and Policy*, *9*(4), 342–355. doi:10.32479/ijeep.7776

Maxwell, J. A. (1992). Understanding and validity in qualitative research. *Harvard Educational Review*, *62*(3), 279–300. doi:10.17763/haer.62.3.8323320856251826

McGuinn, J., Oulès, L., Bradley, H., & McNeill, A. (2017). *Effective multilevel environmental governance for a better implementation of EU environment legislation: Study*. Europa. https://cor.europa.eu/en/engage/studies/Documents/Environmental-governance.pdf

McGuire, W. (2014). The effect of ISO 14001 on environmental regulatory compliance in China. *Ecological Economics*, *105*(September), 254–264. doi:10.1016/j.ecolecon.2014.06.007

Meira, D. (2013). A Lei de Bases da Economia Social Portuguesa: do projeto ao texto final. *CIRIEC-España, Revista jurídica de economía social y cooperativa* (number 24/2013).

Meireles, M., Robaina, M., & Magueta, D. (2021). The Effectiveness of Environmental Taxes in Reducing CO2 Emissions in Passenger Vehicles: The Case of Mediterranean Countries. *International Journal of Environmental Research and Public Health*, *18*(10), 5442. doi:10.3390/ijerph18105442 PMID:34069643

Meng, X., Yao, Z., Nie, J., Zhao, Y., & Li, Z. (2018). Low-carbon product selection with carbon tax and competition: Effects of the power structure. *International Journal of Production Economics*, *200*, 224–230. doi:10.1016/j.ijpe.2018.03.029

Mesquita, M. (2014). *A tributação autónoma no CIRC: A sua (in)coerência* [Master's thesis, UCP]. https://repositorio.ucp.pt/handle/10400.14/16231

Metcalf, G., & Weisbach, D. (2009). *The Design of a Carbon Tax*. Reg-Markets Center Working Paper, 1-63.

Metcalf, E. G. (2003). Environmental levies and distortionary taxation: Pigou, taxation and pollution. *Journal of Public Economics*, *87*(2), 313–322. doi:10.1016/S0047-2727(01)00116-5

Metcalf, G. E. (2014). Using the tax system to address competition issues with a carbon tax. *National Tax Journal*, *67*(4), 779–806. doi:10.17310/ntj.2014.4.02

Metcalf, G. E. (2021). Carbon taxes in theory and practice. *Annual Review of Resource Economics*, *13*(1), 245–265. doi:10.1146/annurev-resource-102519-113630

Meyer, C. A. (1997). Public-Nonprofit Partnerships and North-South Green Finance. *Journal of Environment & Development*, *6*(2), 123–146. https://www.jstor.org/stable/44319261. doi:10.1177/107049659700600203

Miceikiene, A., Rauluskeviciene, J., Streimikine, D., & Ciuleviciene, V. (2018). Assessment of the effect of environmental taxes on environmental protection [Journal of Economics]. *Ekonomicky Casopis*, *66*, 286–308.

Milanés, P., Pérez-Calderón, E., & Dias, A. (2020). Transparency of Financial Reporting on Greenhouse Gas Emission Allowances: The Influence of Regulation. *International Journal of Environmental Research and Public Health, 17*(3), 893. doi:10.3390/ijerph17030893 PMID:32023942

Miller, R., & Blair, P. (2009). *Input-output analysis: Foundations and extensions* (2nd ed.). Cambridge University Press. doi:10.1017/CBO9780511626982

Ministry of Environment and Sustainable Development. (2021). *Report on green taxes in Colombia*. MESD.

Ministry of Environment. (2017). *Portfolio of goods and services of Green Business*. MoE. http://www.minambiente.gov.co/index.php/negocios-verdes-y-sostenibles/negocios-verdes/portafolio-de-bienes-y-servicios-de-negocios-verdes#portafolio-de-negocios

Ministry of the Environment. (2010). *Colombian green businesses*. MoE. http://www.minambiente.gov.co/index.php/negocios-verdes-y-sostenibles/negocios-verdes

Mishra, U., Wu, J.-Z., & Sarkar, B. (2020). A sustainable production-inventory model for a controllable carbon emissions rate under shortages. *Journal of Cleaner Production, 256*, 1–19. doi:10.1016/j.jclepro.2020.120268

Mitic, P., Fedajev, A., & Kojic, M. (2023). *Exploring the Economy -Environment Interactions in the Western Balkans: A Panel Data Analysis*. http://dx.doi.org/ doi:10.28934/ea.23.56.1.pp43-56

Moher, D., Liberati, A., Tetzlaff, J., & Altman, D. G. (2009). Preferred reporting items for systematic reviews and meta-analyses: The PRISMA statement. *BMJ (Clinical Research Ed.), 339*(7716), 332–336. doi:10.1136/bmj.b2535 PMID:19622551

Moioli, S., & Pellegrini, L. A. (2020). Fixed and Capture Level Reduction operating modes for carbon dioxide removal in a Natural Gas Combined Cycle power plant. *Journal of Cleaner Production, 254*, 1–12. doi:10.1016/j.jclepro.2020.120016

Monast, J., Anda, J., & Profeta, T. (2009). *U.S. Carbon Market Design: regulating emission allowances as financial instruments*. Working paper, Nicholas Institute for Environmental Policy Solutions, Duke University.

Monast, J., Murray, B., & Wiener, J. (2017). On morals, markets and climate change: Exploring Pope Francis' challenge. *Law and Contemporary Problems, 80*, 135–162.

Mondejar, M., Avtar, R., Diaz, H., Dubey, R., Esteban, J., Gómez-Morales, A., Hallam, B., Mbungu, N., Okolo, C., Prasad, K., She, Q., & Garcia-Segura, S. (2021). Digitalization to achieve sustainable development goals: Steps towards a Smart Green Planet. *The Science of the Total Environment, 794*(10), 148539. doi:10.1016/j.scitotenv.2021.148539 PMID:34323742

Mongeon, P., & Paul-Hus, A. (2016). The journal coverage of Web of Science and Scopus: A comparative analysis. *Scientometrics, 106*(1), 213–228. doi:10.100711192-015-1765-5

Montero, C. E. P. (2011). *Extrafiscalidade e meio ambiente: O tributo como instrumento de proteção ambiental. Reflexões sobre a tributação ambiental no Brasil e na Costa Rica*. [Unpublished doctoral dissertation]. State University of Rio de Janeiro, Rio de Janeiro, Brazil.

Moreira, T. M., & Fernandes, F. V. (2022, April). The evolution of green taxation in Portugal. *International Tax Review*. https://www.internationaltaxreview.com/article/2a6abcbtm 7tszx6frm680/the-evolution-of-green-taxation-in-portugal

Moreira, T., & Fernandes, F. (2022). *Evolution of green taxation in Portugal*. International Law Office. https://www.vda.pt/xms/files/05_Publicacoes/2022/Artigos/Evolution_of_green_taxation_in_Portugal.pdf

Mosquera Valderrama, I., & Balharová, M. (2021). *Tax Incentives in Developing Countries: A Case Study—Singapore and Philippines*. Springer. doi:10.1007/978-3-030-64857-2_7

Mota, P. M. F. (2016). *Avaliação do desempenho das políticas de Ambiente em Portugal*. UNL. https://run.unl.pt/bitstream/10362/18561/1/Mota_2016.pdf

Mota, D., & Palma, C. (2015). *Green tax reforms: an application to the Portuguese economy*. ISCTE.

Mottershead, D., Watkins, E., Gore, T., Andersen, M. S., Pedersen, A. B., Whiteoak, K., Bruyn, S., Vergeer, R., Chowdhury, T., Hummelen, S. V., & Chewpreecha, U. (2021). *Green taxation and other economic instruments – Internalizing environmental costs to make the polluter pay*. European Commission.

Moutinho, V., Varum, C., & Madaleno, M. (2017). How economic growth affects emissions? An investigation of the environmental Kuznets curve in Portuguese and Spanish economic activity sectors. *Energy Policy*, *106*, 326–344. doi:10.1016/j.enpol.2017.03.069

Mpofu, F. Y. (2022). Taxation of the Digital Economy and Direct Digital Service Taxes: Opportunities, Challenges, and Implications for African Countries. *Economies*, *10*(9), 219. Advance online publication. doi:10.3390/economies10090219

Müller, R., Spengel, C., & Vay, H. (2020). *On the determinants and effects of corporate tax transparency: Review of an emerging literature, ZEW Discussion Papers 20-063*. ZEW - Leibniz Centre for European Economic Research.

Murat Telatar, O., & Birinci, N. (2022). The effects of environmental tax on Ecological Footprint and Carbon dioxide emissions: A nonlinear cointegration analysis on Turkey. *Environmental Science and Pollution Research International*, *29*(29), 44335–44347. doi:10.100711356-022-18740-y PMID:35129752

N. De Sadeleer. (2022). Harmonisation of law, internal market and environment: the challenge of implementing the Green Deal. https://doi.org/ doi:10.20318/cdt.2022.6681

Nabais, J. C. (2005). *Por um Estado Fiscal Suportável Estudos de Direito Fiscal*. Almedina.

Nabais, J. C. (2021). Direito fiscal e tutela do ambiente em Portugal. *Fiscalidade: Revista de Direito e Gestão Fiscal*, (21). Business & Economics School. https://www.isg.pt/wp-content/uploads/2021/02/21_1_CASALTA.23-11-2004.pdf

NABAIS. J. C. (1998). *O dever fundamental de pagar impostos*. Almedina.

Nabais, J. C. (2021). *Direito Fiscal* (11th ed.). Almedina.

Nassaji, H. (2015). Qualitative and descriptive research: Data type versus data analysis. *Language Teaching Research*, *19*(2), 129–132. doi:10.1177/1362168815572747

National Planning Department. (2020). *Green growth policy*. NPD. https://colaboracion.dnp.gov.co/CDT/Prensa/Presentaci%C3%B3n%20Pol%C3%ADtica%20de%20Crecimiento%20Verde.pdf

Ndhaief, N., Nidhal, R., Hajji, A., & Bistorin, O. (2019). Environmental issue in an integrated production and maintenance control of unreliable manufacturing/remanufacturing systems. *International Journal of Production Research*, 1–19.

Nemati, M., Zheng, Y., & Hu, W. (2019, May). The ISO 14001 Standard and Firms' Environmental Performance. *Journal of Agricultural and Resource Economics*, *44*(2), 422–438. https://www.jstor.org/stable/26797565

Nepal, R., & Nepal, S. K. (2021). Managing overtourism through economic taxation: Policy lessons from five countries. *Tourism Geographies*, *23*(5–6), 1094–1115. doi:10.1080/14616688.2019.1669070

Neves, J., & Albuquerque, F. (2019). Income Tax of Portuguese Listed Companies in Financial and Corporate Social Responsibility Reports. *Revista Brasileira de Gestão de Negócios*, *21*(3), 464–485. doi:10.7819/rbgn.v21i3.4009

Nordhaus, W. D. (2010). Carbon taxes to move toward fiscal sustainability. *Economists' Voices*, 1-5.

Nordhaus, W. D. (1992). An Optimal Transition Path for Controlling Greenhouse Gases. *Science*, *258*(5086), 1315–1319. doi:10.1126cience.258.5086.1315 PMID:17778354

Novais, A. (2022). *Os impostos ambientais enquanto instrumento de proteção ambiental.* [Master's thesis, ISCAL]. http://hdl.handle.net/10400.21/15310

Novas, N., Garcia Salvador, R. M., Portillo, F., Robalo, I., Alcayde, A., Fernández-Ros, M., & Gázquez, J. A. (2022). Global Perspectives on and Research Challenges for Electric Vehicles. *Vehicles*, *4*(4), 1246–1276. doi:10.3390/vehicles4040066

Nunes, C. (2009). Fiscalidade vs. Licenças de emissão de CO2 no combate à mitigação. *da Sphera*, julho, 1-40.

Nunkoo, R., & Gursoy, D. (2017). Political trust and residents' support for alternative and mass tourism: An improved structural model. *Tourism Geographies*, *19*(3), 318–339. doi:10.1080/14616688.2016.1196239

Oats, L., & Tuck, P. (2019). Corporate tax avoidance: Is tax transparency the solution? *Accounting and Business Research*, *49*(5), 565–583. doi:10.1080/00014788.2019.1611726

Ochôa, J. (2014). Breve reflexão sobre a incidência subjetiva do imposto único de circulação. *Cadernos de Justiça Tributária, 6*(Outubro-Dezembro), 3-15. https://www.tfra.pt/wp-content/uploads/2020/08/Breve-reflexao-sobre-a-incidencia-sujectiva-do-Imposto-Unico-de-Circulacao-1.pdf

OECD (2006). Subsidy Reform and Sustainable, . doi:10.1787/9789264019379-en

OECD, Organization for Economic Cooperation and Development. (2005). *La reforme fiscale écologique axée sur la réduction de la pauvreté.* OECD. https://www.oecd-ilibrary.org/docserver/9789264008717-fr.pdf?expires=1571321286&id=id&accname=guest&checksum=1289F017793C27021D9BCF44DF485EBD

OECD, Organization for Economic Co-operation, and Development. (1999-2000). *Greening tax mixes in OECD countries: a preliminary assessment.* OECD. https://www.oecd.org/env/tools-evaluation/2385866.pdf

OECD. (2001). *Environmentally related taxes: issues and strategies Policy Brief. OECD.* www.oecd.org/pcd/2674642.pdf

OECD. (2009). *Environmentally related taxes in the OECD countries.* OECD. https://www.oecd.org/dataoecd/27/23/36966499.pdf

OECD. (2010). Forum on Tax Administration: *Guidance Standard Audit File. In Guidance for the Standard Audit File – Tax Version 2.0.* OECD. https://www.oecd.org/tax/administration/45045602.pdf

OECD. (2015). *Environmental Taxation A Guide for Policy Makers.* OECD. https://www.oecd.org/env/tools-evaluation/48164926.pdf

OECD. (2018). *Financing climate objectives in cities and regions to deliver sustainable and inclusive growth.* OECD.

OECD. (2020). *Tax Challenges Arising from Digitalisation – Economic Impact Assessment: Inclusive Framework on BEPS.* OECD/G20 Base Erosion and Profit Shifting Project. OECD Publishing, Paris. https://doi.org/https://doi.org/10.1787/0e3cc2d4-en

OECD. (Organisation for Economic Co-operation and Development). (2013). *BEPS action plan.* OECD. https://www.oecd.org/ctp/BEPSActionPlan.pdf

Oliveira, M. O. B. (2004). Os impostos ambientais – alguns aspectos da sua temática. *Fiscalidade: Revista de Direito e Gestão Fiscal* (no 19/20). Business & Economics School. https://www.isg.pt/wp-content/uploads/2021/02/19_20_3_Impostos-ambientais-Revista-Fiscalidade.pdf

Oliveira, V. (2016). *O impacte das tributações autónomas nas microentidades*. [Master's thesis, UA]. http://hdl.handle.net/10773/16772

Oliveira, D., & Mendes, I. (2017). Green taxation on competitiveness: The effect of the ISP tax on the Portuguese retail sector of road fuels. *International Journal of Energy Economics and Policy*, *7*(1), 178–184.

Olja, M. I., & Mirjana, G. (2012). Environmental Taxation as a Tool for Sustainable Development Policy-State Comparison of Serbia and Application of Ecological Taxation Reform in European Union. *Economic Analysis*, *45*(1), 32–44. https://www.semanticscholar.org/paper/Environmental-Taxation-as-a-Tool-for-Sustainable-of-Ivanovi%C4%87-Golu%C5%A1in/0924238c7e174f17307587f592002f3076e8ff21

Opland, L. E., Pappas, I. O., Engesmo, J., & Jaccheri, L. (2022). Employee-driven digital innovation: A systematic review and a research agenda. *Journal of Business Research*, *143*, 255–271. doi:10.1016/j.jbusres.2022.01.038

Order n.º 467 / 2010 , Definition of the acquisition cost or revaluation value of passenger or mixed-use vehicles . https://www.pgdlisboa.pt/leis/lei_mostra_articulado.php?artigo_id=2253A0024&nid=2253&tabela=leis&pagina=1&ficha=1&nversao=

Organisation for Economic Co-operation and Development. (2011). *OECD Guidelines for Multinational Enterprises*. OECD. https://www.oecd.org/daf/inv/mne/48004323.pdf

Organization for Economic Cooperation and Development. (2011). *Towards Green Growth A summary for policymakers*. Council at OECD ministerial level. https://www.oecd.org/greengrowth/49709364.pdf

Organization for Economic Cooperation and Development. (2013). *Putting green growth at the heart of development*. OECD Publications.

Orhan, G. (2013). Türkiye'de çevre politikaları: Değişen söylemler değişmeyen öncelikler. *Memleket Siyaset Yönetim*, *19*, 1–24.

Oueslati, W. (2015). Growth and welfare effects of environmental tax reform and public spending policy. *Economic Modelling*, *45*, 1–13. doi:10.1016/j.econmod.2014.10.040

Oueslati, W., Zipperer, V., Rousselière, D., & Dimitropoulos, A. (2017). Energy taxes, reforms and income inequality: An empirical cross-country analysis. *Inter Economics*, *150*, 80–95. doi:10.1016/j.inteco.2017.01.002

Özçağ, M., & Hotunluoğlu, H. (2015). Kalkınma anlayışında yeni bir boyut: Yeşil ekonomi. *CBÜ Sosyal Bilimler Dergisi*, *13*, 304–324.

Özdemir, B. (2009). Küresel kirlenme sürdürülebilir ekonomik büyüme ve çevre vergileri. *Maliye Dergisi*, *156*, 1–36.

Öz, E., & Kutbay, H. (2016). Ekolojik vergileme: Seçilmiş bazı dünya ülkeleri ile türkiye verilerinin karşılaştırılması. *Eskişehir Osmangazi Üniversitesi İktisadi ve İdari Bilimler Fakültesi Dergisi, Vol*, *11*, 247–271.

Özkanca Andıç, N., & Erasa Akça, İ. (2019). Küresel kamusal bir mal olan çevrenin korunmasına yönelik vergileme yeşil vergiler ve Türkiye uygulaması. *Journal of Social and Humanıtıes Sciences Research*, *6*, 3447–3453.

Palil, R. M., & Mustapha, A. F. (2011). The Evolution and Concept of Tax Compliance in Asia and Europe. *Australian Journal of Basic and Applied Sciences*.

Palma, C. (2012). As tributações autónomas vistas pelo Tribunal Constitucional. Comentário ao acórdão do Tribunal Constitucional n.o 310/2012, de 20 de junho de 2012. *Revista de Finanças Públicas e Direito Fiscal*, *5*(2), 241–255.

Palma, C. C. (2015). *Introdução ao Imposto Sobre o Valor Acrescentado* (6th ed.). Almedina.

Palmer, T., & Riera, A. (2003). Tourism and environmental taxes. With special reference to the "Balearic ecotax.". *Tourism Management, 24*(6), 665–674. doi:10.1016/S0261-5177(03)00046-3

Panarello, D., & Gatto, A. (2023). Decarbonising Europe – EU citizens' perception of renewable energy transition amidst the European Green Deal. *Energy Policy, 172*, 113272. doi:10.1016/j.enpol.2022.113272

Panzone, L., Ulph, A., Areal, F., & Grippo, V. (2021). A ridge regression approach to estimate the relationship between landfill taxation and waste collection and disposal in England. *Waste Management (New York, N.Y.), 129*, 95–110. doi:10.1016/j.wasman.2021.04.054 PMID:34051583

Park, J., & Jeong, E. B. (2019). Service quality in tourism: A systematic literature review and keyword network analysis. *Sustainability (Basel), 11*(13), 3665. Advance online publication. doi:10.3390u11133665

Park, S. J., Cachon, G. Q., Lai, G., & Seshadri, S. (2015). Supply Chain Design and Carbon Penalty: Monopoly vs. Monopolistic Competition. *Production and Operations Management, 24*(9), 1494–1508.

Parry, I., Morris, A., & Williams, R. (2015). Implementing a US carbon tax: Challenges and debates. Routledge. doi:10.4324/9781315747682

Pástor, Ľ., Stambaugh, R. F., & Taylor, L. A. (2021). Sustainable investing in equilibrium. *Journal of Financial Economics, 142*(2), 550–571. doi:10.1016/j.jfineco.2020.12.011

Patel, S. K., & Jhalani, P. (2023). Formulation of variables of environmental taxation: A bibliometric analysis of Scopus database (2001–2022). *Environment, Development and Sustainability*. https://link.springer.com/article/10.1007/s10668-023-03027-0

Pearce, D. (1991). THE ROLE OF CARBON TAXES IN ADJUSTING TO GLOBAL WARMING. *Economic Journal (London), 101*(407), 938–948. http://search.ebscohost.com/login.aspx?direct=true&db=bth&AN=4531951&site=eds-live. doi:10.2307/2233865

Pearce, D. W., Atkinson, G. D., & Dubourg, W. R. (1994). The economics of sustainable development. *Annual Review of Energy and the Environment, 19*(1), 454–474. doi:10.1146/annurev.eg.19.110194.002325

Peng, J. T., Wang, Y., Zhang, X., He, Y., Taketani, M., Shi, R., & Zhu, X. D. (2019). Economic and welfare influences of an energy excise tax in Jiangsu province of China: A computable general equilibrium approach. *Journal of Cleaner Production, 211*, 1403–1411. doi:10.1016/j.jclepro.2018.11.267

Peng, P., Zhu, L., & Fan, Y. (2017). Performance evaluation of climate policies in China: A study based on an integrated assessment model. *Journal of Cleaner Production, 164*, 1068–1080. doi:10.1016/j.jclepro.2017.07.016

Pereira, M. G. (2021). *Environmental Law and Practice in Portugal: Overview*. Practical Law. https://uk.practicallaw.thomsonreuters.com/3-500-2889?navId=AE5DEF8415A6B6C9866CE77

Pereira, A. M., & Pereira, R. M. (2014). Environmental Fiscal Reform and Fiscal Consolidation: The Quest for the Third Dividend in Portugal. *Public Finance Review, 42*(2), 222–253. doi:10.1177/1091142113485803

Pereira, A. M., & Pereira, R. M. (2023). Energy Taxation Reform with an Environmental Focus in Portugal. *Energies, 16*(3), 1232. doi:10.3390/en16031232

Pereira, A. M., Pereira, R. M., & Rodrigues, P. G. (2016). A new carbon tax in Portugal: A missed opportunity to achieve the triple dividend? *Energy Policy, 93*, 110–118. doi:10.1016/j.enpol.2016.03.002

Pereira, M. F. (2014). *Fiscalidade* (5th ed.). Almedina.

Pérez, A., & Soto, R. (2017a). Real Decreto 602/2016: Aplicaciones prácticas de las modificaciones contables (I). *Revista de Contabilidad*, (53), 14–57.

Pérez, A., & Soto, R. (2017b). Real Decreto 602/2016: Aplicaciones prácticas de las modificaciones contables (II y última parte). *Revista de Contabilidad*, (54), 8–37.

Perino, G. (2018). New EU ETS Phase 4 rules temporarily puncture waterbed. *Nature Climate Change*, *8*(4), 262–264. doi:10.103841558-018-0120-2

Pertuz, V., Francisco Miranda, L., & Sánchez Buitrago, J. O. (2021). Towards a conceptual understanding of green entrepreneurship. *Revista Venezolana de Gerencia*, *26*(94), 745–761. doi:10.52080/rvgv26n94.23

Pfister, N., & Mathys, N. A. (2022). Waste taxes at work: Evidence from the canton of Vaud in Switzerland. *Ecological Economics*, *193*, 107314. doi:10.1016/j.ecolecon.2021.107314

Pham, Q., & Vu, K. (2022). Digitalization in small and medium enterprise: A parsimonious model of digitalization of accounting information for sustainable innovation ecosystem value generation. *Asia Pacific Journal of Innovation and Entrepreneurship*, *16*(1), 2–37. doi:10.1108/APJIE-02-2022-0013

Pigou, A. C. (1920). The economics of welfare. Macmillan.

Pigou, A. C. (1920). *The Economics of Welfare* (1st ed.). Macmillan London.

Pillay, S., & Buys, P. W. (2013). Carbon tax pricing and the social cost of carbon: The case in the South African motor vehicle manufacturing industry. *Journal of Applied Business Research*, *29*(6), 1751–1762. doi:10.19030/jabr.v29i6.8212

Pillay, S., & Buys, P. W. (2015). Investigating the impact of carbon tax on socially responsible corporate governance: The case of south African motor vehicle manufacturers. *Corporate Ownership and Control*, *12*(2), 128–134. doi:10.22495/cocv12i2p10

Pinto-Bazurco, J. F. (2022). How to enforce the polluter-pays principle. *International Institute for Sustainable Development Earth Negotiations Bulletin*, 1–8. IISD. https://www.iisd.org/system/files/2022-02/still-one-earth-polluter-pays-principle.pdf

Pires, J. M. F. (2016). *Lições de Impostos sobre o Património e do Selo* (3rd ed.). Almedina.

Platform for Collaboration on Tax. (2018). *Taxation and the Sustainable Development Goals*. OECD. https://www.oecd.org/tax/platform-for-collaboration-on-tax-conference-statement-2018.pdf

Pogge, T. W. (2001). Eradicating Systemic Poverty: Brief for a global resources' dividend. *Journal of Human Development*, *2*(1), 59–77. doi:10.1080/14649880120050246

Portuguese Corporate Income Tax Code Rabadán Martín, I., Jurado Martín, J. A., & Martín Zamora, M. P. (2017). Análisis de la utilidad de la información contable relativa a los derechos de emisión: Propuesta de un indicador. *Revista International Business and Economics Review*.

Pulido-Fernández, J. I., & Cárdenas-García, P. J. (2012). *Política económica del turismo*. Pirámide.

Pulido-Fernández, J. I., Román, J. L. D., Cárdenas-García, P. J., & Carrillo-Hidalgo, I. (2022). Taxation as a stimulus for sustainability in mature tourist destinations. *Revista de Turismo y Patrimonia Cultural*, *20*(3), 681–697. doi:10.25145/j.pasos.2022.20.047

Purcinelli, L. M., Abreu, R., & Vasconcelos, A. L. (2020). Sustainable and Smart System: Rethinking Accounting and Taxation in Portugal. In D. Crowther & S. Seifi (Eds.), Governance and Sustainability: International Perspectives. Springer International Publishing. doi:10.1007/978-981-15-6370-6_4

Qi, Y., Zhang, J., & Chen, J. (2023). Tax incentives, environmental regulation and firms' emission reduction strategies: Evidence from China. *Journal of Environmental Economics and Management, 117*(November 2021). doi:10.1016/j.jeem.2022.102750

Qiu, J., Huang, K., & Hawkins, J. (2022). The taxi sharing practices: Matching, routing and pricing methods. *Multimodal Transportation, 1*(1), 100003. doi:10.1016/j.multra.2022.100003

Rafique, M. Z., Fareed, Z., Ferraz, D., Ikram, M., & Huang, S. (2022). Exploring the heterogenous impacts of environmental taxes on environmental footprints: An empirical assessment from developed economies. *Energy, 238*, 121753. doi:10.1016/j.energy.2021.121753

Rahman, M. M. (2022). The effect of taxation on sustainable development goals: Evidence from emerging countries. *Heliyon, 8*(9), e10512. doi:10.1016/j.heliyon.2022.e10512 PMID:36105472

Ramos Díaz, J. (2016). Green Economy and Employment: The labour potential of the "Ecological Transition" in Spain. *Notebooks of Labour Relations, 34*(2). Advance online publication. doi:10.5209/CRLA.53436

Rapuluchukwu, E. U., Belmondo, T. V., & Ibukun, B. (2017). Incentives and firm productivity: Exploring multidimensional fiscal incentives in a developing country. *World Development Perspectives, 5*, 56–59. doi:10.1016/j.wdp.2017.03.001

Real Decreto 1514/2007, de 16 de noviembre, por el que se aprueba el Plan General de Contabilidad (PGC).

Real Decreto 602/2016, de 2 de diciembre, por el que se modifican el Plan General de Contabilidad (PGC)

Rebelo, A. R. B. (2016). *Taxa sobre sacos de plástico leves e proibição da disponibilização gratuita de sacos de caixa: perceções e impacto nos hábitos de consumo dos portugueses*. [Master's thesis, ISCTE]. https://www.iscte-iul.pt/tese/13754

Reinhold, S., Laesser, C., & Beritelli, P. (2018). The 2016 St. Gallen Consensus on Advances in Destination Management. *Journal of Destination Marketing & Management, 8*, 426–431. doi:10.1016/j.jdmm.2017.03.001

Reis, C., & Lopes, A. (2011). Fatores económicos e contabilísticos que influenciam a qualidade da informação ambiental: um estudo com empresas admitidas à cotação na Euronext Lisbon que tenham títulos de Emissão de gases com efeito de estufa. *XVI Encuentro AECA, Granada.*

Reiter, S. (2020). Tax disclosures in sustainability reports. *Journal of Accounting and Finance*, 20(87), 51-64

Resolución de 28 de mayo de 2013, del Instituto de Contabilidad y Auditoría de Cuentas (ICAC), por la que se dictan normas de registro, valoración e información a incluir en la memoria del inmovilizado intangible.

Resolución de 8 de febrero de 2006 del Instituto de Contabilidad y Auditoría de Cuentas (ICAC), por la que se aprueban normas para el registro, valoración e información de los derechos de emisión de gases de efecto invernadero.

Resor, J. P. (1997). Debt For Nature Swaps: A Decade of Experience and New Directions for the Future. *Unasylva, 48*(188), 15–22.

Reyhan, A. S. (2014). Çevre ekonomisinde çevre vergileri uygulamaları. *Hitit Üniversitesi Sosyal Bilimler Enstitüsü Dergisi, 7*, 110–120.

Ribeiro, F. M., & Santos, S. F. (2013). *A Fiscalidade e as Organizações de Economia Social*. Vida Económica.

RNCREQ. (1998). *Les instruments économiques et la protection de l'environnement.* RNCREQ. http://www.rncreq.org/pdf/instruments.pdf

Robaina, M., & Roseta-Palma, C. (2014). *Impostos ambientais e o duplo dividendo: experiências europeias.* (Working Papers in Economics.)

Rodrigues, P., David, F., Soares, E., Monteiro, E., Melo, N., Gregório, J., & Rodrigues, R. (2022). Economic Aspects of a Collection Solution of Biowaste in an Association of Municipalities in Portugal. In: IWWG–International Waste Working Group (Eds). *Proceedings of the 9th International Symposium on Energy from Biomass and Waste*. Padova (Italy): Cisa Publisher.

Rodríguez, M., Robaina, M., & Teotónio, C. (2019). Sectoral effects of a Green Tax Reform in Portugal. *Renewable & Sustainable Energy Reviews*, *104*, 408–418. doi:10.1016/j.rser.2019.01.016

Rua-Usme, D., Ospina, A., Ramírez, C., Ortiz, P., & Patiño-Murillo, J. A. (2019). Scenario for the development of the line of green financial products in the Colombian economy. *CINTEX Magazine*, *24*(2), 64–90. doi:10.33131/24222208.352

Rudyanto, A., & Pirzada, K. (2021). The role of sustainability reporting in shareholder perception of tax avoidance. *Social Responsibility Journal*, *17*(5), 669–685. doi:10.1108/SRJ-01-2020-0022

Ruggiero, V. (2022). Sustainability and Financial Crime. *International Criminology*, *2*(2), 143–151. doi:10.100743576-022-00052-3

Ruiz-Pérez, F., Lleo, A., & Ormazabal, M. (2021). Employee sustainable behaviors and their relationship with Corporate Sustainability: A Delphi study. *Journal of Cleaner Production*, *329*, 129742. Advance online publication. doi:10.1016/j.jclepro.2021.129742

Runst, P., & Höhle, D. (2022). The German eco tax and its impact on CO_2 emissions. *Energy Policy*, *160*, 112655. doi:10.1016/j.enpol.2021.112655

Rupeika-Apoga, R., Bule, L., & Petrovska, K. (2022). Digital Transformation of Small and Medium Enterprises: Aspects of Public Support. *Journal of Risk and Financial Management*, *15*(2), 45. doi:10.3390/jrfm15020045

Saber Más Ser Más. (2021). *What are green financing lines?* ASOBANCARIA. https://www.sabermassermas.com/que-son-lineas-de-financiamiento-verde/

Saelim, S. (2019). Carbon tax incidence on household demand: Effects on welfare, income inequality and poverty incidence in Thailand. *Journal of Cleaner Production*, *234*, 521–533. doi:10.1016/j.jclepro.2019.06.218

Saka, C., Oshika, T., & Jimichi, M. (2020). Does tax avoidance diminish firms' sustainability? *Journal Global Policy and Governance*, *8*(2), 95–114. doi.org/10.14666/2194-7759-8-2-005

Saldivia-Gonzatti, L. I., Jannes, G., & Barreal, J. (2022). Factors influencing the rate of sorted solid waste collection: An empirical analysis towards local management in Catalonia (NE Spain). *Cities (London, England)*, *131*, 104038. doi:10.1016/j.cities.2022.104038

Sanches, J. L. Saldanha (2007). Manual de Direito Fiscal. (3.º ed.). Almedina.

Sandelowski, M. (2000). Whatever Happened to Qualitative Description? *Research in Nursing & Health*, *23*(4), 334–340. doi:10.1002/1098-240X(200008)23:4<334::AID-NUR9>3.0.CO;2-G PMID:10940958

Sandelowski, M. (2010). What's in a name? Qualitative description revisited. *Research in Nursing & Health*, *33*(1), 77–84. doi:10.1002/nur.20362 PMID:20014004

Sandmo, A. (2011). Atmospheric externalities and environmental taxation. *Energy Economics*, *33*(SUPPL. 1), S4–S12. doi:10.1016/j.eneco.2011.07.021

Sanfey, P., Milatovic, J., & Kresic, A. (2016). How the Western Balkans Can Catch Up. *EBRD Working Paper No.185*. https://papers.ssrn.com/sol3/papers.cfm?abstract_id=3119685

Sanoran, K. (2023). Corporate sustainability and sustainable growth: The role of industry sensitivity. *Finance Research Letters, 53*(103596), 103596. doi:10.1016/j.frl.2022.103596

Santos, M. L. (2021). *O setor dos resíduos em Portugal: o impacto da reforma da fiscalidade verde e da regulação sobre os consumidores finais* [Master's thesis, FEP]. https://hdl.handle.net/10216/136556

Santos, M. L. (2021). *O setor dos resíduos em Portugal: o impacto da Reforma da Fiscalidade Verde e da regulação sobre os consumidores finais.* Bit Stream. https://repositorio-aberto.up.pt/bitstream/10216/136556/2/502062.pdf

Santos, A. (2013). O comércio de licenças de emissão em Portugal. *International Business and Economics Review: Revista Internacional de Gestão e Comunicação, 4*, 62–77.

Santoso, E., Laturette, K., & Mastan, S. (2019). Corporate Social Responsibility Disclosure, Tax Aggressiveness and Sustainability Report Assurance: Evidence from Thailand. *International Journal of Economics and Management Engineering, 13*(11), 1413–1418.

Saraiva, R. C. G. (2010). *A Herança de Quioto em Clima de Incerteza: análise jurídico-económica do mercado de emissões num quadro de desenvolvimento sustentado.* [Unpublished doctoral dissertation]. University of Lisbon, Lisbon, Portugal.

Sarigül, S. S., & Topcu, B. A. (2021). The Impact of Environmental Taxes on Carbon Dioxide Emissions in Turkey. *International Journal of Business and Economic Studies, 3*(1), 43–54. 10.08.2021

Sarlet, I. W., & Fernsterseifer, T. (2014). *Princípios do direito ambiental.* Saraiva.

Sarmento, J. M. (2022). Preface. In R. Marques & S. Reis (Eds.), *Código Fiscal do Investimento Anotado e Comentado* (pp. 7–10). Almedina.

Scheiwiller, T., & Symons, S. (2014). Corporate responsibility and paying tax. *The OECD Observer. Organisation for Economic Co-Operation and Development, 1.* http://www.oecdobserver.org/news/archivestory.php/aid/3132/Corporate_responsibility_and_paying_tax.html

Schnighter, A., Holle, F., & Kockrow, M. (2021). Tax and Transparency: Reporting in Accordance with the Global Reporting Initiative. *Intertax, 49*(8 & 9), 702–7212.

Schöb, R. (2003, August). *The Double Dividend Hypothesis of Environmental taxes: A Survey.* (CESIFO Working Paper - NCEE Working Paper Series).

Segrelles, J. A. (2009). A reflection on the unsustainability of tourism activities in rural and natural areas. The cases of ecotourism and deep ecology*. *Human Geographies, 2*(1), 1–21. doi:10.1177/194277860900200109

Serdaroğlu, S. (2003). *Avrupa bütünleşmesine katkıda bulunan bir faktör olarak çevre politikası, dünden bugüne AB.* Boyut Yayınları.

Shafiei, E., Davidsdottir, B., Fazeli, R., Leaver, J., Stefansson, H., & Asgeirsson, E. I. (2018). Macroeconomic effects of fiscal incentives to promote electric vehicles in Iceland: Implications for government and consumer costs. *Energy Policy, 114*(May 2017), 431–443. doi:10.1016/j.enpol.2017.12.034

Shafi, M., Ramos-Meza, C. S., Jain, V., Salman, A., Kamal, M., Shabbir, M. S., & Rehman, M. (2023). The dynamic relationship between green tax incentives and environmental protection. *Environmental Science and Pollution Research International, 30*(12), 32184–32192. doi:10.100711356-023-25482-y PMID:36723845

Shahzad, U. (2020). Environmental taxes, energy consumption, and environmental quality: Theoretical survey with policy implications. *Environmental Science and Pollution Research, 27*(20), 24848–24862. doi:10.1007/s11356-020-08349-4

Sharif, A., Kartal, M. T., Bekun, F. V., Pata, U. K., Foon, C. L., & Kılıç Depren, S. (2023). Role of green technology, environmental taxes, and green energy towards sustainable environment: Insights from sovereign Nordic countries by CS-ARDL approach. *Gondwana Research, 117*, 194–206. doi:10.1016/j.gr.2023.01.009

Shayanmehr, S., Radmehr, R., Ali, E. B., Ofori, E. K., Adebayo, T. S., & Gyamfi, B. A. (2023). *How do environmental tax and renewable energy contribute to ecological sustainability? New evidence from top renewable energy countries.* Taylor & Francis. doi:10.1080/13504509.2023.2186961

Shi, H., Qiao, Y., Shao, X., & Wang, P. (2019). The effect of pollutant charges on economic and environmental performances: Evidence from Shandong Province in China. *Journal of Cleaner Production, 232*, 250–256. doi:10.1016/j.jclepro.2019.05.272

Shi, Q., Ren, H., Cai, W., & Gao, J. (2019). How to set the proper level of carbon tax in the context of Chinese construction sector? A CGE analysis. *Journal of Cleaner Production, 240*, 1–12. doi:10.1016/j.jclepro.2019.117955

Shmelev, S. E., & Speck, S. U. (2018). Green fiscal reform in Sweden: Econometric assessment of the carbon and energy taxation scheme. *Renewable & Sustainable Energy Reviews, 90*, 969–981. doi:10.1016/j.rser.2018.03.032

Shuang, Y., Diabat, A., & Liao, Y. (2019). A stochastic reverse logistics production routing model with emissions control policy selection. *International Journal of Production Economics, 213*, 201–216. doi:10.1016/j.ijpe.2019.03.006

Silva, P., & Chaves, C. (2015). *The Implementation of Environmental Tax Reform in Portugal: How Does it Compare with Other European Experiences?* University of Porto.

Silva, S. T., & Soares, C. D. (2015). *Regime Fiscal das Entidades da Economia Social e Civil.* Vida Económica.

Skjærseth, J. B. (2021). Towards a European Green Deal: The evolution of EU climate and energy policy mixes. *International Environmental Agreement: Politics, Law and Economics, 21*(1), 25–41. doi:10.100710784-021-09529-4

Smith, D. (2018). Government policies for supporting green businesses: An international perspective. *Journal of Sustainable Development, 35*(2), 87–102.

Smith, J., & Memon, A. (2017). CiteScore: A cite for sore eyes, or a valuable, transparent metric? *Scientometrics, 111*(1), 553–556. doi:10.100711192-017-2250-0

Soares, C. D. (2001). *Instrumentos de promoção da qualidade ambiental: o imposto ecológico.* [Unpublished master dissertation]. University of Coimbra, Coimbra, Portugal.

Soares, C. D. (2002). *O imposto ambiental: direito fiscal do ambiente.* Almedina.

Soares, C. D. (2014). *Smart tax policies: environmental financial reform in times of economic crisis. Does austerity cure? Does austerity kill? SOARESb, C. D. (2001). A receita dos impostos ambientais.* Scientia Ivridica Separata.

Soares, J. R. R., Remoaldo, P., Perinotto, A. R. C., Gabriel, L., Lezcano-Gonzalez, M. E., & Sanchez-Fernandez, M. D. (2022). Residents' Perceptions Regarding the Implementation of a Tourist Tax at a UNESCO World Heritage Site: A Cluster Analysis of Santiago de Compostela (Spain). *Land (Basel), 11*(2), 189. Advance online publication. doi:10.3390/land11020189

Sollund, S. (2007). *Environment taxes. United Nations Group of Experts on Domestic Resource Mobilisation*, Rome.

Song, H., Seetaram, N., & Ye, S. (2019). The effect of tourism taxation on tourists' budget allocation. *Journal of Destination Marketing & Management, 11*, 32–39. doi:10.1016/j.jdmm.2018.12.001

Sousa, A. S. N., & Silva, A. F. J. (2017). As tributações autónomas no sistema fiscal Português. *XVI Congresso Internacional de Contabilidade e Auditoria.* OCC. https://www.occ.pt/pt/dtrab/trabalhos/listacc/trabalhoscc.php

Sousa, S. (2021). *Environmental Taxation in Portugal: A Contribution to Sustainability*. Springer. doi:10.1007/978-3-030-63149-9_23

Spanish Corporate Income Tax Law, Steenkamp, N., Rahman, A., & Kashyap, V. (2011). Recognition, measurement and disclosure of carbon emission allowances under the EU ETS: an exploratory study.*10th Australasian CSEAR Conference, Session 15: Climate change*, University of Tasmania.

Spasić, V. (2022). Which Western Balkan countries intend to introduce carbon tax? *Balkan Green Engineering News*. https://balkangreenenergynews.com/which-western-balkan-countries-intend-to-introduce-carbon-tax/

Speck, S. (2007). Overview of environmental tax reforms in EU member states, COMERTFinal Report to the European Commission, DG Research and DG Taxation and Customs Union. EU.

Speck, S. (2022). *The role of environmental taxation in supporting sustainability transitions*. EEA. https://www.eea.europa.eu/publications/the-role-of-environmental-taxation

Stiglingh, M., Venter, E. R., Penning, I., Smit, A., Schoeman, A., & Steyn, T. (2017). Tax transparency reporting by the top 50 JSE-listed firms. *South African Journal of Accounting Research*, *31*(2), 151–168. doi:10.1080/10291954.2016.1160196

Streeck, W. (2013). *Tempo comprado: A crise adiada do capitalismo democrático*. Conjunctura Actual Editora.

Sun, Y., Mao, X., Liu, G., Yin, X., & Zhao, Y. (2020). Modelling the effects of energy taxes on ecological footprint transfers in China's foreign trade. *Ecological Modelling*, *431*, 109200. doi:10.1016/j.ecolmodel.2020.109200

Supriyati, S., & Anggraini, D. (2021). Sustainability Reporting and Tax Aggressiveness: Evidence from a Public Company in Indonesia. *Indonesian Journal of Sustainability Accounting and Management*, *5*(1), 71–80. doi:10.28992/ijsam.v5i1.249

Taborda, D., & Martins, A. (2023). As isenções em IRC das atividades exercidas pelas entidades do setor não lucrativo. *Revista Jurídica Portucalense, n.º 33*. https://revistas.rcaap.pt/juridica/article/view/28483

Tang, M., & Wang, Y. (2022). Tax incentives and corporate social responsibility: The role of cash savings from accelerated depreciation policy. *Economic Modelling*, *116*, 106040. doi:10.1016/j.econmod.2022.106040

Tang, Y., Cockerill, T. T., Pimm, A. J., & Yuan, X. (2021). Reducing the life cycle environmental impact of electric vehicles through emissions-responsive charging. *iScience*, *24*(12), 103499. doi:10.1016/j.isci.2021.103499 PMID:34927031

Tax Benefits Statute. (2023) EBF. https://info.portaldasfinancas.gov.pt/pt/informacao_fiscal/codigos_tributarios/Cod_download/Documents/EBF.pdf

Teles, N. S. (2015). *A Fiscalidade Verde: Do Princípio da Equivalência à Neutralidade Fiscal*. Bit Stream. https://repositorio-aberto.up.pt/bitstream/10216/83754/2/37256.pdf .

Terkla, D. (1984). The efficiency value of effluent tax revenues. *Journal of Environmental Economics and Management*, *11*(2), 107–123. doi:10.1016/0095-0696(84)90010-X

Toprak, D. (2006). Sürdürülebilir kalkınma çerçevesinde çevre politikaları ve mali araçlar. *Süleyman Demirel Üniversitesi Sosyal Bilimler Enstitüsü Dergisi*, *4*, 46–69.

Tozer, R. (2007). Fiscalidad de los derechos de emisión de CO2. *Cuadernos de Sostenibilidad y Patrimonio Natural*, 12.

Tracana, D., & Dias, C. C. (2020). Tributação do Setor Cooperativista em Portugal: análise dos principais aspetos e questões em aberto. *Revista Eletrónica de Fiscalidade da Associação Fiscal Portuguesa, II:2*. https://www.afp.pt/revista/edicoes/708-revista-electronica-de-fiscalidade-da-afp-ano-ii-n-2-2020?start=4

Tranfield, D., Denyer, D., & Smart, P. (2003). Towards a Methodology for Developing Evidence-Informed Management Knowledge by Means of Systematic Review. *British Journal of Management, 14*(3), 207–222. doi:10.1111/1467-8551.00375

Trenerry, B., Chng, S., Wang, Y., Suhaila, Z. S., Lim, S. S., Lu, H. Y., & Oh, P. H. (2021). Preparing Workplaces for Digital Transformation: An Integrative Review and Framework of Multi-Level Factors. *Frontiers in Psychology, 12*(March), 1–24. doi:10.3389/fpsyg.2021.620766 PMID:33833714

Trinks, A., Mulder, M., & Scholtens, B. (2022). External carbon costs and internal carbon pricing. *Renewable & Sustainable Energy Reviews, 168*(October), 112780. doi:10.1016/j.rser.2022.112780

Triwacananingrum, W. & Wijaya, G. (2022). Sustainability reporting and tax aggressiveness before and during Covid-19: GCG moderating variable. *Jurnal Akuntansi dan Keuangan Indonesia, 19* (1), 96-119

Truong, T. (2022). The impact of digital transformation on environmental sustainability. *Advances in Multimedia, 2022*, 1–12. doi:10.1155/2022/6324325

Tupiassu, L. V. (2009). *A reforma fiscal verde e duplo dividendo. Direito Tributário e Econômicos aplicado ao meio ambiente e à mineração.*

Tupiassu, L. V. (2006). *Tributação ambiental: a utilização de instrumentos econômicos e fiscais do direito ao meio ambiente saudável*. Renovar.

Tupiassu-Merlin, L. (2009). *Contribution de la fiscalité a sa realisation: Le cas brésilien.* [Unpublished doctoral dissertation]. University of Toulouse Capital-1, Toulouse, France.

Tyutyuryukov, V., & Guseva, N. (2021). From remote work to digital nomads: Tax issues and tax opportunities of digital lifestyle. *IFAC-PapersOnLine, 54*(13), 188–193. doi:10.1016/j.ifacol.2021.10.443

Ulucak, R., & Kassouri, Y. (2020). An assessment of the environmental sustainability corridor: Investigating the non-linear effects of environmental taxation on CO2 emissions. *Sustainable Development (Bradford), 28*(4), 1010–1018. doi:10.1002d.2057

Ulusoy, A., & Akdemir, T. (2013). Yerel yönetimlerin finansman sorunlarının çözümünde motorlu taşıtlar vergisi önerisi. *Sosyoekonomi, 19*, 87–116.

UNCTAD BioTrade Initiative. (2007). Biotrade principles and criteria. United Nations Conference on Trade and Development. *BioTrade Initiative* (pp. 6-14). Geneva, Switzerland: UNCTAD. http://www.asocam.org/sites/default/files/publicaciones/files/2807acffecd17b29a5d3ee02ca4395a5.pdf

UNDP. (2016). *Human Development Report 2016 Risk-Proofing the Western Balkans: Empowering People to Prevent Disasters*. UNDP. https://www.undp.org/albania/publications/human-development-report-2016-risk-proofing-western-balkans-empowering-people-prevent-disasters

UNEP. (2010). *Green economy: driving a green economy through public finance and fiscal policy reform.* UNEP. https://www.greengrowthknowledge.org/sites/default/files/downloads/resource//Driving_a_GE_through_public_finance_and_fiscal_policy_reform_UNEP.pdf

UNFCCC. THE PARIS AGREEMENT, L.9/ Rev.1 § (2015). https://doi.org/FCCC/CP/2015/L.9/Rev.1

United Nations Treaty Collection. (1997). *Kyoto Protocol to the United Nations Framework Convention on Climate Change.* Kyoto Protocol to the United Nations Framework Convention on Climate Change. https://treaties.un.org/Pages/ViewDetails.aspx?src=IND&mtdsg_no=XXVII-7-a&chapter=27&clang=_en

United Nations. (2015). *The Paris Agreement.* UNFCCC. https://unfccc.int/process-and-meetings/the-paris-agreement

United Nations. (2023). *Transforming our world: the 2030 Agenda for Sustainable Development.* UN. https://sdgs.un.org/2030agenda

Urrutia-Mosquera, J., & Fábrega, J. (2021). Impact of fiscal incentives in the consumption of low emission vehicles. *Case Studies on Transport Policy*, *9*(3), 1151–1159. doi:10.1016/j.cstp.2021.06.003

van Eck, G., de Jong, G., Wesseling, B., & van Meerkerk, J. (2019). Simulating the impact of tax incentives using a type choice model for lease cars. *Case Studies on Transport Policy*, *7*(4), 814–822. doi:10.1016/j.cstp.2019.07.014

van Eck, N. J., & Waltman, L. (2010). Software survey: VOSviewer, a computer program for bibliometric mapping. *Scientometrics*, *84*(2), 523–538. doi:10.100711192-009-0146-3 PMID:20585380

van Heerden, J., Blignaut, J., Bohlmann, H., Cartwright, A., Diederichs, N., & Mander, M. (2016). The economic and environmental effects of a carbon tax in South Africa: A dynamic CGE modelling approach. *Suid-Afrikaanse Tydskrif vir Ekonomiese en Bestuurswetenskappe*, *19*(5), 714–732. doi:10.4102ajems.v19i5.1586

van Vuuren, D. P., Zimm, C., Busch, S., Kriegler, E., Leininger, J., Messner, D., Nakicenovic, N., Rockstrom, J., Riahi, K., Sperling, F., Bosetti, V., Cornell, S., Gaffney, O., Lucas, P. L., Popp, A., Ruhe, C., von Schiller, A., Schmidt, J. O., & Soergel, B. (2022). Defining a sustainable development target space for 2030 and 2050. *One Earth*, *5*(2), 142–156. doi:10.1016/j.oneear.2022.01.003

Vanii, S., & Uçar, A. (2021). Çevre ve yeşil vergiler üzerine Türkiye özelinde değerlendirmeler, *International Journal of Social* [IJSPFR]. *Political and Financial Researches*, *1*, 68–81.

Vasques, S. (2011). *Manual de Direito Fiscal.* Almedina.

Vence, X., & López Pérez, S. J. (2021). Taxation for a Circular Economy: New Instruments, Reforms, and Architectural Changes in the Fiscal System. *Sustainability (Basel)*, *13*(8), 4581. doi:10.3390u13084581

Venter, E., Stiglingh, M., & Smit, A. (2017). Integrated Thinking and the Transparency of Tax Disclosures in the Corporate Reports of Firms. *Journal of International Financial Management & Accounting*, *28*(3), 394–427. doi:10.1111/jifm.12064

Vítor, J. (2014). *Os Desafios Ambientais do Sistema Fiscal.* Faculdade de Direito da Universidade Católica. https://repositorio.ucp.pt/bitstream/10400.14/16439/1/Os%20 desafios%20ambientais%20do%20sistema%20fiscal.pdf

Vona, F. (2021). *Managing the distributional effects of environmental and climate policies: The narrow path for a triple dividend.* OECD. https://www.oecd-ilibrary.org/environment/managing-the-distributional-effects-of-environmental-and-climate-policies_361126bd-en

Wagner, P. M., Torney, D., & Ylä-Anttila, T. (2021). Governing a multilevel and cross-sectoral climate policy implementation network. *Environmental Policy and Governance*, *31*(5), 417–431. doi:10.1002/eet.1942

Wang, C., Wang, W., & Huang, R. (2017). Supply chain enterprise operations and government carbon tax decisions considering carbon emissions. *Journal of Cleaner Production*, *152*, 271–280. doi:10.1016/j.jclepro.2017.03.051

Wang, L., Ma, P., Song, Y., & Zhang, M. (2023). How does environmental tax affect enterprises' total factor productivity? Evidence from the reform of environmental fee-to-tax in China. *Journal of Cleaner Production*, *413*, 137441. doi:10.1016/j.jclepro.2023.137441

Wang, S., Li, X., Li, Z., & Ye, Y. (2023, January 1–20). The effects of government support on enterprises' digital transformation: Evidence from China. *MDE. Managerial and Decision Economics*, *44*(5), 2520–2539. doi:10.1002/mde.3831

Warwick, P., & Ng, C. (2012). The 'Cost' of climate change: How carbon emissions allowances are accounted for amongst European Union companies. *Australian Accounting Review*, *60*(22), 54–67. doi:10.1111/j.1835-2561.2011.00158.x

Watson, L. (2015). Corporate social responsibility, tax avoidance, and earnings performance. *The Journal of the American Taxation Association*, *37*(2), 1–21. doi:10.2308/atax-51022

WB. (2022). Western Balkans Regular Economic Report, No.22, Fall 2022. *Western Balkans Regular Economic Report*, (22). doi:10.1596/38189

Weitzman, M. L. (2015). Internalizing the Climate Externality: Can a Uniform Price Commitment Help? *Economics of Energy & Environmental Policy*, *4*(2), 37–50. doi:10.5547/2160-5890.4.2.mwei

Williams, R. C., III. (2016). *Environmental taxation.* NBER Working Paper Series 22303. https://www.nber.org/papers/w22303/

Williams, I. D., & Blyth, M. (2023). Autogeddon or autoheaven: Environmental and social effects of the automotive industry from launch to present. *The Science of the Total Environment*, *159987*, 159987. doi:10.1016/j.scitotenv.2022.159987 PMID:36372167

Williams, R. C. III, Gordon, H., Burtraw, D., Carbone, J. C., & Morgenstern, R. D. (2014). The initial incidence of a carbon tax across U.S. States. *National Tax Journal*, *67*(4), 807–830. doi:10.17310/ntj.2014.4.03

Williams, R. C. III, Gordon, H., Burtraw, D., Carbone, J. C., & Morgenstern, R. D. (2015). The initial incidence of a carbon tax across income groups. *National Tax Journal*, *68*(1), 195–214. doi:10.17310/ntj.2015.1.09

Wolde-Rufael, Y., & Mulat-Weldemeskel, E. (2021). Do environmental taxes and environmental stringency policies reduce CO_2 emissions? Evidence from 7 emerging economies. *Environmental Science and Pollution Research International*, *28*(18), 22392–22408. doi:10.100711356-020-11475-8 PMID:33417133

Wolde-Rufael, Y., & Mulat-Weldemeskel, E. (2022). The moderating role of environmental tax and renewable energy in CO_2 emissions in Latin America and Caribbean countries: Evidence from method of moments quantile regression. *Environmental Challenges*, *6*, 100412. doi:10.1016/j.envc.2021.100412

World Bank. (2017) *Green bonds*. World Bank. https://www.bancomundial.org/es/results/2017/12/01/green-bonds

World Bank. (2018). *Municipal Solid Waste Management: A Roadmap for Reform for Policy Makers*. World Bank Group.

Wu, W., Hu, Y., & Wu, Q. (2023). Subsidies and tax incentives - Does it make a difference on TFP? Evidences from China's photovoltaic and wind listed companies. *Renewable Energy*, *208*(June 2022), 645–656. doi:10.1016/j.renene.2023.03.080

Xavier, C. A. C. (2016). *Planeamento fiscal agressivo: delimitação conceptual de uma nova realidade.* [Unpublished master dissertation]. New University of Lisbon, Lisbon, Portugal.

Xu, C., Wang, C., & Huang, D. R. (2020). Impacts of horizontal integration on social welfare under the interaction of carbon tax and green subsidies. *International Journal of Production Economics*, *222*, 1–13. doi:10.1016/j.ijpe.2019.09.027

Xu, L., Yilmaz, H. Ü., Wang, Z., Poganietz, W. R., & Jochem, P. (2020). Greenhouse gas emissions of electric vehicles in Europe considering different charging strategies. *Transportation Research Part D, Transport and Environment*, *87*, 102534. doi:10.1016/j.trd.2020.102534

Xu, X., Xu, X., & He, P. (2016). Joint production and pricing decisions for multiple products with cap-and-trade and carbon tax regulations. *Journal of Cleaner Production*, *112*, 4093–4106. doi:10.1016/j.jclepro.2015.08.081

Yamen, A., Coskun, A., & Mersni, H. (2022). Digitalization and tax evasion: the moderation effect of corruption. *Economic Research-Ekonomska Istrazivanja*, 1–24. . doi:10.1080/1331677X.2022.2142634

Yang, H., & Chen, W. (2018). Retailer-driven carbon emission abatement with consumer environmental awareness and carbon tax: Revenue-sharing versus Cost-sharing. *Omega*, *78*, 179–191. doi:10.1016/j.omega.2017.06.012

Yang, H., Luo, J., & Wang, H. (2017). The role of revenue sharing and first-mover advantage in emission abatement with carbon tax and consumer environmental awareness. *International Journal of Production Economics*, *193*, 691–702. doi:10.1016/j.ijpe.2017.08.032

Yang, L., Zhou, Y., Meng, B., Li, H., Zhan, J., Xiong, H., Zhao, H., Cong, W., Wang, X., Zhang, W., Lakshmanan, P., Deng, Y., Shi, X., Chen, X., & Zhang, F. (2022). Reconciling productivity, profitability and sustainability of small-holder sugarcane farms : A combined life cycle and data envelopment analysis. *Agricultural Systems*, *199*, 103392. Advance online publication. doi:10.1016/j.agsy.2022.103392

Yan, S. (2018). The economic and environmental impacts of tax incentives for battery electric vehicles in Europe. *Energy Policy*, *123*, 53–63. doi:10.1016/j.enpol.2018.08.032

Yan, S., & Eskeland, G. S. (2018). Greening the vehicle fleet: Norway's CO2-Differentiated registration tax. *Journal of Environmental Economics and Management*, *91*, 247–262. doi:10.1016/j.jeem.2018.08.018

Yavilioğlu, C. (2002). Kalkınmanın anlambilimsel tarihi ve kavramsal kökenleri, *C. Ü. İktisadi ve İdari Bilimler Dergisi*, *3*, 59-77.

Yegen, B. (2019). Toplum sağlığına vergisel bir müdahale: Obezite vergisi. *Fiscaoeconomia*, *2*, 146–158.

Yin, H., & Peter, J. (2009, November). Schmeidler.(2008). Why do standardized ISO 14001 environmental management systems lead to heterogeneous environmental outcomes? *Business Strategy and the Environment*, *18*(7), 469–486. https://onlinelibrary.wiley.com/doi/10.1002/bse.629. doi:10.1002/bse.629

Yu, B., Wang, J., Lu, X., & Yang, H. (2020). Collaboration in a low-carbon supply chain with reference emission and cost learning effects: Cost sharing versus revenue sharing strategies. *Journal of Cleaner Production*, *250*, 1–26. doi:10.1016/j.jclepro.2019.119460

Yüksel, C. (2013). *Kamu ekonomisi ve çevre sorunları*. Ekin Kitabevi Yayınları.

Yu, M., Cruz, J. M., & Li, D. M. (2019). The sustainable supply chain network competition with environmental tax policies. *International Journal of Production Economics*, *217*, 218–231. doi:10.1016/j.ijpe.2018.08.005

Yunus, S., Elijido-Ten, E., & Abhayawansa, S. (2016). Determinants of carbon management strategy adoption. *Managerial Auditing Journal*, *31*(2), 156–179. doi:10.1108/MAJ-09-2014-1087

Yu, V. F., Maglasang, R., & Tsao, Y.-C. (2018). Shelf space allocation problem under carbon tax and emission trading Policies. *Journal of Cleaner Production*, *196*, 438–451. doi:10.1016/j.jclepro.2018.05.165

Zakeri, A., Dehghanian, F., Fahimnia, B., & Sarkis, J. (2015). Carbon pricing versus emissions trading: A supply chain planning perspective. *International Journal of Production Economics*, *164*, 197–205. doi:10.1016/j.ijpe.2014.11.012

Zannoni, D. (2018). Fiscal Aid to Promote Clean Transport. *EC Tax Review*.

Zeng, T. (2018). Relationship between corporate social responsibility and tax avoidance: International evidence. *Social Responsibility Journal*, *15*(2), 244–257. doi:10.1108/SRJ-03-2018-0056

Zhang, D., & Xie, Y. (2022). Customer environmental concerns and profit margin: Evidence from manufacturing firms. *Journal of Economics and Business*, *120*(November 2021), 106057. doi:10.1016/j.jeconbus.2022.106057

Zhang, D. (2021). How environmental performance affects firms' access to credit: Evidence from EU countries. *Journal of Cleaner Production*, *315*, 128294. doi:10.1016/j.jclepro.2021.128294

Zhang, H., & Yang, Y. (2019). Prescribing for the tourism-induced Dutch disease: A DSGE analysis of subsidy policies. *Tourism Economics*, *25*(6), 942–963. doi:10.1177/1354816618813046

Zhang, J., & Zhang, D. Y. (2020a). Chinese tourism economic change under carbon tax scenarios. *Current Issues in Tourism*, *23*(7), 836–851. doi:10.1080/13683500.2018.1551339

Zhang, J., & Zhang, D. Y. (2020b). Examining the economic and environmental effects of emissions policies in China: A Bayesian DSGE model. *Journal of Cleaner Production*, *266*, 1–9. doi:10.1016/j.jclepro.2020.122026

Zhang, K., Wang, Q., Liang, Q.-M., & Chen, H. (2016). A bibliometric analysis of research on carbon tax from 1989 to 2014. *Renewable & Sustainable Energy Reviews*, *58*, 297–310. doi:10.1016/j.rser.2015.12.089

Zhang, Z., & Baranzini, A. (2004). What do we know about carbon taxes? An inquiry into their impacts on competitiveness and distribution of income. *Energy Policy*, *32*(4), 507–518. doi:10.1016/S0301-4215(03)00152-6

Zhao, G., & Baker, J. (2022). Effects on environmental impacts of introducing electric vehicle batteries as storage - A case study of the United Kingdom. *Energy Strategy Reviews*, *40*, 100819. doi:10.1016/j.esr.2022.100819

Zhao, L., Yang, C., Su, B., & Zeng, S. (2020). Research on a single policy or policy mix in carbon emissions Reduction. *Journal of Cleaner Production*, *267*, 1–10. doi:10.1016/j.jclepro.2020.122030

Zhou, G., & Zhang, D. Y. (2017). Integration and consolidation in air freight shipment planning: An economic and environmental perspective. *Journal of Cleaner Production*, *166*, 1381–1394. doi:10.1016/j.jclepro.2017.07.145

Zhou, S., Shi, M., Li, N., & Yuan, Y. (2011). Impacts of Carbon Tax Policy on CO2 Mitigation and Economic Growth in China. *Advances in Climate Change Research*, *2*(3), 124–133. doi:10.3724/SP.J.1248.2011.00124

Zhou, S., Zhou, P., & Ji, H. (2022). Can digital transformation alleviate corporate tax stickiness: The mediation effect of tax avoidance. *Technological Forecasting and Social Change*, *184*(122028), 122028. doi:10.1016/j.techfore.2022.122028

Zhou, Y., Hu, F., & Zhou, Z. (2018). Pricing decisions and social welfare in a supply chain with multiple competing retailers and carbon tax policy. *Journal of Cleaner Production*, *190*, 752–777. doi:10.1016/j.jclepro.2018.04.162

Zhu, J., & Liu, W. (2020). A tale of two databases: The use of Web of Science and Scopus in academic papers. *Scientometrics*, *123*(1), 321–335. doi:10.100711192-020-03387-8

Zhu, L., Duan, H.-B., & Fan, Y. (2015). CO2 mitigation potential of CCS in China and an evaluation based on an integrated assessment model. *Journal of Cleaner Production*, *103*, 934–947. doi:10.1016/j.jclepro.2014.08.079

Zhu, N., Bu, Y., Jin, M., & Mbroh, N. (2020a). Green financial behavior and green development strategy of Chinese power companies in the context of carbon tax. *Journal of Cleaner Production*, *245*, 1–7. doi:10.1016/j.jclepro.2019.118908

Zhu, N., Qian, L., Jiang, D., & Mbroh, N. (2020b). A simulation study of China's imposing carbon tax against American carbon tariffs. *Journal of Cleaner Production*, *243*, 1–12. doi:10.1016/j.jclepro.2019.118467

Zuo, H., Du, H., Broadstock, D., Guo, J., Gong, Y., & Mao, G. (2016). China's future energy mix and emissions reduction potential: A scenario analysis incorporating technological learning curves. *Journal of Cleaner Production*, *112*, 1475–1485. doi:10.1016/j.jclepro.2015.08.012

Zupic, I., & Čater, T. (2015). Bibliometric Methods in Management and Organization. *Organizational Research Methods*, *18*(3), 429–472. doi:10.1177/1094428114562629

About the Contributors

Ana Dinis is a Researcher in Taxation in the Research Centre on Accounting and Taxation (IPCA, Portugal). PhD in Business Management from the Faculty of Economics of the University of Coimbra (2020). Master in Accounting and Corporate Taxation (2013) and Postgraduate in Accounting and Corporate Taxation (2009) from the Coimbra Business School (ISCAC). Postgraduate in Taxation (2010) by APEU / FEUC - Association for University Extension of the University of Coimbra. Professor at the Polytechnic Institute of Cávado and Ave (IPCA), in the Accounting and Taxation Department of the School of Management, Barcelos, in the area of Taxation since 2014. Chartered Accountant, since 2003. Author and co-author in scientific papers, in national and international scientific journals, in the tax area.

Fátima David is an Accounting and Finance Professor in the School of Technology and Management - Polytechnic of Guarda, Portugal, since 1991. PhD in Accounting and Taxation from the University of Salamanca, Spain (2007). PhD registred on Nova University Lisbon, Portugal (2008). Master and Degree in Management from University Beira Interior, Portugal (1997). She is a effective member of the Ethics Committee of the Polytechnic of Guarda, since 2018. She develops activities with Social Responsibility Research Network (SRRN) and Research Centre on Accounting and Taxation (IPCA-CICF); her research has been published in several journals, and she participates in several conferences and meetings.

Liliana Pereira completed in April 2015 a PhD in "Basic Principles and Categories of Internal and International Taxation. The Tax procedures" by the University of Santiago de Compostela. Degree in Taxation in July 2005 by the Polytechnic Institute of Cávado e do Ave and Bachelor Professor at Instituto Politécnico do Cávado e do Ave in the area(s) of Social Sciences with an emphasis on Economics and Management with an emphasis on Organization and Management companies. In his Science Vitae curriculum, the most frequent terms in the context of scientific, technological and artistic-cultural production are: taxation; tax law; taxes; tax education; State aid; VAT; income taxation.

Sara Dias has a PhD in Law (University of Santiago de Compostela). Master in Tax Law (University of Minho). Law Degree (Law Faculty of Porto University). Lawyer (partner) at Yolanda Busse, Oehen Mendes & Associados, S.P., R.L.. Teacher at the Polytechnic Institute of Cávado and Ave (tax law, insolvency law and corporate law). Trainer in several advanced training courses in the area of Tax Law, organized by UNIFOJ - Centre for Social Studies of the University of Coimbra. Participation, as teacher, in other post-graduations and master courses. Member of the Portuguese Tax Association (AFP) and the International Fiscal Association (IFA). External reviewer of Portuguese Law Lagazines

(Revista Jurídica Portucalense and RED - Revista Electrónica de Direito). Participation, as a speaker, in several conferences/seminars and publication of several articles in magazines and books in the areas of Tax Law and Insolvency Law. Researcher in Research Centre for Justice and Governance (JusGov) .

***

Beatriz Aibar-Guzmán is an Associate Professor of Accounting and Finance at the University of Santiago de Compostela (Spain). Her research lines are related to corporate social responsibility, corporate transparency, and management accounting. She has published papers in top journals ranked with a high JCR impact factor She has presented more than 90 papers at conferences and other academic forums, some of which have been awarded, and contributed with chapters to books

Tania Cristina Azevedo is an Accountant, Master in Accounting and PhD in Territorial Planning and Social Development, with an Abroad Internship at the University of Salamanca (USAL)/Spain. Professor at the State University of Feira de Santana, Accounting Sciences course and Professional Master's Program in Territorial Planning. Adjunct Professor IV at the Federal University of Bahia in the undergraduate course of Accounting Sciences. Partner and manager of the Silveira Accounting Organization. Urban socio-environmental sustainability researcher, with emphasis on economic and tax instruments of urban policy; green taxes and environmental taxation with a view to tax extrafiscality. And urban policies integrated with the Sustainable Development Goals of the 2030/UN Agenda.

Edmira Cakrani is an Assistant Professor, PhD at the Canadian Institute of Technology, Faculty of Economy, Albania. She holds a PhD degree in Economics and an MBA from the University of Tirana, as well as a Bachelor degree in Business Administration from the University "Ismail Qemali" of Vlora. Edmira has around 20 years of full-time or part-time teaching experience at various universities in Albania. Her research interests include Financial Economics, Applied Macroeconomics, International Economics, Development Economics. She is the author of many research articles and has participated in many scientific conferences. She is also a member of various Editorial Committees, as well as a reviewer in various scientific journals.

Cecília Carmo has a PhD in Accounting from the University of Aveiro and the University of Minho. MSc in Accounting and Finance from the Aberta University. Degree in Business Management from the Beira Interior University. Adjunct Professor at the University of Aveiro - Higher Institute of Accounting and Administration. Lecturer in Financial Accounting, Financial and Nonfinancial Reporting at Undergraduate, Masters, and Doctoral levels at the University of Aveiro. Integrated member of the Competitiveness, Innovation and Sustainability group of the research unit on Governance, Competitiveness and Public Policy (GOVCOPP).

André Carvalho is a First Stage Researcher in Taxation at the Center for Research in Accounting and Taxation (IPCA, Portugal). Graduated in Public Management and Postgraduate student in Accounting and Taxation of Companies (2022/2023) at the Coimbra Business School (ISCAC).

Enriko Ceko holds a Post Doctorate Diploma (2022-La Laguna University. Tenerife. Spain), Ph.D. diploma in Management (2005 – Faculty of Economy, Tirana University, Albania), graduated bachelor and master the Agronomy (1989–Agriculture University, Tirana), Economy (1996-Agriculture University, Tirana), Law (2002–Tirana University), MBA (2000-Tirana University & Nebraska Lincoln University), with about 33 years of working experience and about the same experience lecturing, specialized on about 45 training courses (Albania, USA, UK, Israel, Germany, Austria), Senator of World Business Angels and Investments Forum, Lecturer at WBAF Business School, and at LMC-London Metropolitan College, N.R.Macedonia, author of around 40 scientific articles, participating in more than 80 international scientific conferences, author of "Total Quality Management", "Quality Management Tools", "Planning, Organizing & Controlling Quality Tools", "Managing Agribusiness Firms", "Tourism and Hospitality Management", "Academic writing", "Business and Administrative communication", "Small theatre, big theatre", "Where Albanian economy is going", "100 years of economic, trade and business relations between UK and Albania", "100 years of diplomatic relations between UK and Albania" books, his work focuses on lecturing (mainly economics & managerial disciplines), public & private organizations consultancy, ISO Standards included, heading several NGO's, as well as participating in civic society activities, etc.

Sérgio Cruz has a PhD in Accounting Assistant Teacher.

Maria Goreti Dâmaso is a professor at the Higher School of Management and Technology of the Polytechnic Institute of Santarém, Portugal. She has a PhD in Business Management (specialization in Finance) from the Faculty of Economics of the University of Coimbra. His research fields are accounting, finance, and taxation.

Paulo Dias has a PhD in Accounting and MBA in Management, by ISCTE-IUL, Post-Graduate in Finance, by CEMAF-ISCTE and Graduate in Management, by UML. He is coordinator of Courses of Taxation, Business Taxation, Company Tax Management, Principles and Procedures of Tax Litigation and Taxation of Non-Residents. He is Executive Director of the Executive MBA and Director of Post-Graduation in Tax Management at INDEG-ISCTE.

Isabel-María García-Sánchez is a Full Professor of Accounting and Finance at the University of Salamanca. Her main research line is focused on sustainability and aimed at determining the reasons and factors that lead companies to develop different strategies and their economic impacts, as well as on the design of aggregate indicators that measure the business commitment to the different dimensions of sustainability both at the country and the sector level. Using Scopus, she has 205 documents and 8,878 citations (H-Index: 50). The recognition of her research career by the academy shows a growing trajectory, position 6,114 in the Ranking of the World Scientists: World's Top 2% Scientists in 202. She has published papers in top journals ranked with a high JCR impact factor, such as British Management Journal, Corporate Governance: An International Review, Journal of Banking & Finance, Business & Society, Long Range Planning, Technological Forecasting and Social Change, among others.

Soraia Gonçalves holds a Ph.D. in Political Science. She is a member of the WP3 of the Regional University Network - European University (RUN-EU) - Future and Advanced Skills Academy (FASA). She is a Pro-president for Academic Management and Quality at the Polytechnic Institute of Cávado and Ave. Her current areas of interest include public finance, civic participation, and local democracy.

Nuno Carlos Leitão is a professor at the Higher School of Management and Technology of Santarem. He is PhD with an Aggregation (habilitation) in Economics and is a researcher at CEFAGE, Evora University and CEsA, Lisbon University. His research fields are international economics, energy economics, resources sustainability and tourism.

Kátia Lemos has a PhD in Accounting Professor at the Department of Accounting and Taxation Dean of the Master in Accounting and Finance at School of Management Institute Polytechnic of Cávado and Ave

Jose Martin is a Professor at the University of Huelva since 1999 in the area of Financial Economics and Accounting, specializing in Management and Cost Accounting, as well as Financial Accounting. PhD in Business Sciences since 2016, much of the research has focused on the accounting characteristics and costs of intangible assets, such as the case of forests and greenhouse gas emission rights. In this sense, it is worth mentioning the presentation in numerous congresses of studies related to the accounting implication of the emission rights market, as a flexibility mechanism of the Kyoto Protocol. In recent years, he has continued to work on sustainability projects, as well as on digital accounting.

María-Pilar Martín-Zamora holds a degree and is a doctor (PhD) in Business Administration from the University of Sevilla (Spain). At present, she is a full professor at the University of Huelva, in the Department of Accounting and Finance. Since 1985, when she started her career in higher education, she has taught different subjects in the field of Financial Accounting. Her basic research is oriented towards Tax Accounting and Business Management. She is the author of some books and articles published in journals and she has presented papers in national and international conferences.

Eva Miranda has a Master in Education Administration and Planning, Professor of the Department of Tourism and Marketing, School of Hospitality and Tourism, Polytechnic Institute of Cávado and Ave. Research interests include ethics, social responsibility and sustainability.

Victor Nylander is a PhD student in Law & Economics at the Faculty of Law of the University of Lisbon. Master in Law and Economic Legal Sciences – Specialization in Tax Law. Master in Economics from the OMMA Business School of Madrid. Master in Economics & Markets from Universidade Presbiteriana Mackenzie. Bachelor of Law from the University Center of the State of Pará. Corporate tax lawyer.

Gisela Oliveira has an MSc in Accounting from the Higher Institute of Accounting and Administration (ISCA-UA), Universidade de Aveiro. Degree in Economics from Universidade de Aveiro.

Nagihan Özkanca Andıç holds a PhD in Public Finance from University of Pamukkale. Currently she is a Asst. Professor at the Pamukkale University. Her research interests include public economics, welfare economics, green taxation and regulations. She possesses 12 years of experience as an consultant to multinational companies. She published more than 20 papers and book chapters.

João Sérgio Ribeiro has a PhD in Public Law (Universidade Nova de Lisboa), Master of Laws (The London School of Economics and Political Science), Bachelor of Laws (Portuguese Catholic University). He has also attended the Advanced Course in Tax Treaties at the International Tax Center Leiden (Leiden University). Member of the Scientific Council of the Portuguese Tax Association. Member of the directive board of the Master's Degree in Tax Law (University of Minho). Member of the International Fiscal Association (IFA). Member of the Academic Council of the Revista Direito Tributário Atual do Instituto Brasileiro de Direito Tributário (IBDT) and member of the Editorial Board of the Revista Direito Tributário Internacional Atual do Instituto Brasileiro de Direito Tributário (IBDT) .

Ana Paula Rocha has a PhD in Law by Universidade Católica Portuguesa and a Master in Tax Law by the same University. Lecturer at ISCAP in the scientific area of Tax Law and Member of CEOS. PP (Center for Organisational and Social Studies of the Polytechnic of Porto). Tax Arbitrator at CAAD - Portuguese Center for Administrative Arbitration. In the past, Ana Paula Rocha worked for several years as a tax consultant in a Big Four and as a technical assistant to the Office of Vice-Presidents of the Supreme Administrative Court. She has published several articles in the scientific area of Tax Law.

Pedro Miguel Santos Melo Rodrigues holds a PhD in Chemistry from the Faculty of Sciences of the University of Porto. He is a Professor at the Polytechnic Institute of Guarda since 1993 and responsible for LABMIA since 2010. Since 1993, it has been carrying out analytical work in the environmental area, namely in quality control through physical-chemical and microbiological analysis of water, soil, leaf and air, among others. He has extensive laboratory experience in several analytical techniques, such as, for example, spectroscopy (atomic absorption, UV-vis, FTIR), chromatography (gas and HPLC), potentiometry, as well as several microbiological techniques.

Franz Rodríguez Franky has a Master's Degree in Environmental Management and Auditing from the Universidad Internacional Iberoamericana. Specialist in Administration and Management in Quality Systems from the Universidad Santo Tomas and Agricultural Engineer from the Universidad de Cundinamarca. Research Professor at the Fundación Universitaria Compensar. Leader of research groups, tutor and jury of degree projects, peer evaluator of research work, academic production in papers, research articles and trajectory in research processes.

Iara Cardoso dos Santos is a Master in Accounting from ISCTE since 2019 and Certified Accountant since 2021. With experience in accounting, taxation and tax consulting, she currently works at PwC, one of the world's leading professional services companies. Its research area was Green Taxation, an increasingly relevant topic in the current context of environmental concerns. The author is a scholar passionate about the field of Accounting and Taxation and through this article hopes to contribute to the understanding and improvement of the debate on Green Taxation in the current context.

Rute Saraiva has a PhD in Law and Economic Legal Sciences from the Faculty of Law of the University of Lisbon. Master in Law and International Legal Sciences from the Faculty of Law of the University of Lisbon. Degree in Law from the Faculty of Law of the University of Lisbon. Assistant Professor at the Faculty of Law of the University of Lisbon.

Sara Serra has a Ph.D. in Accounting. Associate Professor at the Polytechnic Institute of Cávado and Ave, in the Auditing area. Member of the Research Center on Accounting and Taxation. Author of several articles in the area of financial auditing research.

Maria Elisabete Santos Soares is Professor in Surveying/Geomatics Engineering in the School of Technology and Management - Polytechnic of Guarda, Portugal, since 1995. She holds a PhD Degree in Civil Engineering from University of Minho, Portugal. In her PhD thesis entitled "A sustainable urban expansion index", she developed an urban model to assess the sustainability of urban sprawl. She develops activities and research in geospatial sciences and has skills in Geographic Information Systems, Remote Sensing, multicriteria analysis and urban sustainability.

Laurentina Vareiro, Professor of Economics at the School of Management, Polytechnic Institute of Cávado and Ave (IPCA), and research center: Applied Management Research Unit (UNIAG), Portugal. She has a PhD in Economics (2008) at the University of Minho, Braga. Her research interests include regional, rural and urban economics, tourism economics, rural and cultural tourism.

Index

www.ingramcontent.com/pod-product-compliance
Lightning Source LLC
LaVergne TN
LVHW060400300926
849162LV00006B/1311